Mazda 323 Owners Workshop Manual

Mark Coombs

Models covered
Mazda 323 Hatchback, Saloon and Estate front-wheel-drive models, including special/limited editions
1071 cc, 1296 cc, 1323 cc, 1490 cc, 1498 cc and 1597 cc petrol engines

Also covers major features of Estate models to May 1991
Does not cover DOHC Turbo 4x4 or revised model range introduced in October 1989

ABCDE
FGHIJ
KLMNO

Haynes Publishing Group
Sparkford Nr Yeovil
Somerset BA22 7JJ England

Haynes Publications, Inc
861 Lawrence Drive
Newbury Park
California 91320 USA

Acknowledgements

Thanks are due to Champion Spark Plug, who supplied the illustrations showing spark plug conditions, to Holt Lloyd Limited who supplied the illustrations showing bodywork repair, and to Duckhams Oils, who provided lubrication data. Certain other illustrations are the copyright of Mazda Cars (UK) Ltd and are used with their permission. Thanks are also due to Sykes-Pickavant Limited, who provided some of the workshop tools, and to all those people at Sparkford who helped in the production of this manual.

© Haynes Publishing Group 1992

A book in the **Haynes Owners Workshop Manual Series**

Printed by J. H. Haynes & Co. Ltd., Sparkford, Nr Yeovil, Somerset BA22 7JJ, England

All rights reserved. No part of this book may be reproduced or transmitted in any form or by any means, electronic or mechanical, including photocopying, recording or by any information storage or retrieval system, without permission in writing from the copyright holder.

ISBN 1 85010 608 8

British Library Cataloguing in Publication Data
A catalogue record for this book is available from the British Library.

We take great pride in the accuracy of information given in this manual, but vehicle manufacturers make alterations and design changes during the production run of a particular vehicle of which they do not inform us. No liability can be accepted by the authors or publishers for loss, damage or injury caused by any errors in, or omissions from, the information given.

Restoring and Preserving our Motoring Heritage

Few people can have had the luck to realise their dreams to quite the same extent and in such a remarkable fashion as John Haynes, Founder and Chairman of the Haynes Publishing Group.

Since 1965 his unique approach to workshop manual publishing has proved so successful that millions of Haynes Manuals are now sold every year throughout the world, covering literally thousands of different makes and models of cars, vans and motorcycles.

A continuing passion for cars and motoring led to the founding in 1985 of a Charitable Trust dedicated to the restoration and preservation of our motoring heritage. To inaugurate the new Museum, John Haynes donated virtually his entire private collection of 52 cars.

Now with an unrivalled international collection of over 210 veteran, vintage and classic cars and motorcycles, the Haynes Motor Museum in Somerset is well on the way to becoming one of the most interesting Motor Museums in the world.

A 70 seat video cinema, a cafe and an extensive motoring bookshop, together with a specially constructed one kilometre motor circuit, make a visit to the Haynes Motor Museum a truly unforgettable experience.

Every vehicle in the museum is preserved in as near as possible mint condition and each car is run every six months on the motor circuit.

Enjoy the picnic area set amongst the rolling Somerset hills. Peer through the William Morris workshop windows at cars being restored, and browse through the extensive displays of fascinating motoring memorabilia.

From the 1903 Oldsmobile through such classics as an MG Midget to the mighty 'E' Type Jaguar, Lamborghini, Ferrari Berlinetta Boxer, and Graham Hill's Lola Cosworth, there is something for everyone, young and old alike, at this Somerset Museum.

Haynes Motor Museum

Situated mid-way between London and Penzance, the Haynes Motor Museum is located just off the A303 at Sparkford, Somerset (home of the Haynes Manual) and is open to the public 7 days a week all year round, except Christmas Day and Boxing Day.

Contents

	Page
Preliminary sections	
Acknowledgements	2
About this manual	6
Introduction to the Mazda 323	6
General dimensions and weights	7
Jacking, towing and wheel changing	8
Buying spare parts and vehicle identification numbers	11
Safety first!	12
General repair procedures	14
Tools and working facilities	15
Booster battery (jump) starting	19
Conversion factors	20
Fault diagnosis	21
MOT test checks	28
Chapter 1 Routine maintenance and servicing	32
Lubricants, fluids and capacities	35
Maintenance schedule	36
Maintenance procedures	47
Chapter 2 Engine	66
Part A: E series engine – in-car engine repair procedures	72
Part B: B series engine – in-car engine repair procedures	85
Part C: Engine removal and general engine overhaul procedures	100
Chapter 3 Cooling, heating and ventilation systems	112
Chapter 4 Fuel, exhaust and emission control systems	127
Part A: Carburettor engines	130
Part B: Fuel-injected engines	150
Part C: Emission control systems	159
Chapter 5 Ignition system	160
Part A: Contact breaker ignition system	161
Part B: Electronic igition system	164
Chapter 6 Clutch	171
Chapter 7 Manual gearbox and Automatic transmission	179
Part A: Manual gearbox	180
Part B: Automatic transmission	186
Chapter 8 Driveshafts	191
Chapter 9 Braking system	198
Chapter 10 Suspension and steering	219
Chapter 11 Bodywork and fittings	244
Chapter 12 Electrical system	268
Wiring diagrams	291
Index	337

Spark plug condition and bodywork repair colour pages between pages 32 and 33

Mazda 323 1300 Hatchback

Mazda 323 1500 GT Hatchback

Mazda 323 1500 GLX Estate

Mazda 323 1.3 LX Hatchback

About this manual

Its aim

The aim of this manual is to help you get the best value from your vehicle. It can do so in several ways. It can help you decide what work must be done (even should you choose to get it done by a garage), provide information on routine maintenance and servicing, and give a logical course of action and diagnosis when random faults occur. However, it is hoped that you will use the Manual by tackling the work yourself. On simpler jobs it may even be quicker than booking the car into a garage and going there twice, to leave and collect it. Perhaps most important, a lot of money can be saved by avoiding the costs a garage must charge to cover its labour and overheads.

The Manual has drawings and descriptions to show the function of the various components so that their layout can be understood. Then the tasks are described and photographed in a clear step-by-step sequence.

Its arrangement

The Manual is divided into Chapters, each covering a logical sub-division of the vehicle. The Chapters are each divided into Sections, numbered with single figures, eg 5; and the Sections into paragraphs (or sub-Sections), with decimal numbers following on from the Section they are in, eg 5.1, 5.2, 5.3 etc.

It is freely illustrated, especially in those parts where there is a detailed sequence of operations to be carried out. There are two forms of illustration: figures and photographs. The figures are numbered in sequence with decimal numbers, according to their position in the Chapter – eg Fig. 6.4 is the fourth drawing/illustration in Chapter 6. Photographs carry the same number (either individually or in related groups) as the Section or sub-Section to which they relate.

There is an alphabetical index at the back of the manual as well as a contents list at the front. Each Chapter is also preceded by its own individual contents list.

References to the 'left' or 'right' of the vehicle are in the sense of a person in the driver's seat facing forward.

Unless otherwise stated, nuts and bolts are removed by turning anti-clockwise, and tightened by turning clockwise.

Vehicle manufacturers continually make changes to specifications and recommendations, and these, when notified, are incorporated into our Manuals at the earliest opportunity.

We take great pride in the accuracy of information given in this Manual, but vehicle manufacturers make alterations and design changes during the production run of a particular vehicle of which they do not inform us. No liability can be accepted by the authors or publishers for loss, damage or injury caused by any errors in, or omissions from, the information given.

Project vehicles

The main project vehicles used in the preparation of this manual, and appearing in many of the photographic sequences were a 1988 Mazda 323 Hatchback and a pre-September 1985 323 Hatchback. Additional work was carried out and photographed on a 1500 GLX Estate, and 1.6i Hatchback.

Introduction to the Mazda 323

Introduced into the UK in the spring of 1981 the Mazda 323 quickly established itself as a leading contender in the small to medium car market and soon became the most popular Japanese car in its class in Europe.

The cars featured front-wheel drive from a transverse engine/transmission arrangement with independent front and rear suspension, dual circuit brakes and four- or five-speed manual or three-speed automatic transmission, depending on the model.

The model range included three, or five-door Hatchback or four-door Saloon versions with a choice of 1100, 1300 or 1500 cc engines and a top of the range 1500 GT twin carburettor sports model.

The model range remained largely unchanged until September 1985 when it underwent its first major update. The new models were visually sleeker with a redesigned interior. The original E series engine was retained but a new carburettor which was equipped with an automatic choke was fitted. The suspension was also slightly modified and ventilated front disc brakes were fitted.

In November 1985 a new top of the range sports model, the 1600 cc 1.6i model was introduced into the UK. This three-door hatchback was fitted with a new B type engine and was equipped with a Mazda Electronic Gasoline Injection (EGI) fuel injection system. The new model was comprehensively equipped and was fitted with disc brakes all round.

In May 1986 a 1500 cc Estate model was added to the range.

In July 1987 a new Mazda 323 range was introduced into the UK to supersede the earlier models. The new models had a slightly redesigned body shape and were available in 1300 and 1500 cc carburettor versions along with the fuel injected 1600 cc 1.6i model. All models were fitted with the B type engine, the 1300 and 1500 cc versions being very similar to the earlier 1600 cc engine except that Hydraulic Lash Adjusters (HLA) were fitted to automatically adjust the valve clearances hence alleviating the need to periodically adjust them at the required service interval.

All Mazda 323 models were designed with the emphasis on economical motoring, with a high standard of handling, performance and comfort. The car is quite conventional in design and the DIY home mechanic should find most work straightforward.

General dimensions and weights

Dimensions
Overall length:
  Hatchback models:
    Pre-September 1985 .... 3955 mm
    September 1985 onward .... 3900 mm
  Saloon models:
    Pre-September 1985 .... 4155 mm
    September 1985 onward .... 4195 mm
  Estate models .... 4225 mm
Overall width:
  Pre-September 1985 .... 1630 mm
  September 1985 onward .... 1645 mm
Overall height:
  Hatchback and Saloon models:
    Pre-September 1985 .... 1375 mm
    September 1985 onward .... 1390 mm
  Estate models .... 1430 mm
Wheelbase:
  Pre-September 1985 .... 2365 mm
  September 1985 onward .... 2400 mm
Front track .... 1390 mm
Rear track:
  Pre-September 1985 .... 1395 mm
  September 1985 onward .... 1415 mm

Weights
Kerb weight:
  1100 cc 3-door Hatchback models:
    Pre-September 1985 .... 820 kg
    September 1985 onward .... 865 kg
  1300 cc 3-door Hatchback models:
    Pre-September 1985 .... 825 kg
    September 1985 to July 1987 .... 865 kg
    July 1987 onward .... 875 kg
  1300 cc 5-door Hatchback models:
    Pre-September 1985 .... 840 kg
    September 1985 to July 1987 .... 880 kg
    July 1987 onward .... 895 kg
  1300 cc 4-door Saloon models .... 845 kg
  1500 cc 3-door Hatchback models .... 865 kg
  1500 cc 5-door (manual) Hatchback models .... 945 kg
  1500 cc 5-door (auto) Hatchback models:
    September 1985 to July 1987 .... 930 kg
    July 1987 onward .... 970 kg
  1500 cc 4-door (manual) Saloon models:
    Pre-September 1985 .... 885 kg
    September 1985 to July 1987 .... 925 kg
    July 1987 onward .... 955 kg
  1500 cc 4-door (auto) Saloon models .... 980 kg
  1500 cc Estate models .... 950 kg
  1600 cc 3-door Hatchback models:
    September 1985 to July 1987 .... 975 kg
    July 1987 onward .... 960 kg
Maximum braked towing weight:
  1100 cc models:
    Pre-September 1985 .... 800 kg
    September 1985 onward .... 700 kg
  1300 cc models:
    Pre-September 1985 .... 900 kg
    September 1985 to July 1987 .... 800 kg
    July 1987 onward .... 1350 kg
  1500 cc Saloon and Hatchback models:
    Pre-July 1987 .... 1000 kg
    July 1987 onward .... 1450 kg
  1500 cc Estate models .... 1000 kg
  1600 cc models:
    September 1985 to July 1987 .... 1100 kg
    July 1987 onward .... 1550 kg
Maximum unbraked towing weight (all models) .... 400 kg

Jacking, towing and wheel changing

To change a wheel, apply the handbrake and chock the wheel diagonally opposite the one to be changed. On automatic transmission models, place the selector lever in the P position. Make sure that the car is located on firm level ground. Remove the spare wheel, jack handle, jack and wheelbrace from the luggage compartment and then slightly loosen the wheel nuts with the brace provided (where applicable remove the trim first). Locate the jack head in the jacking point nearest to the wheel to be changed and raise the jack using the jack handle. When the wheel is clear of the ground remove the nuts and lift off the wheel. Fit the spare wheel and moderately tighten the bolts. Lower the car and

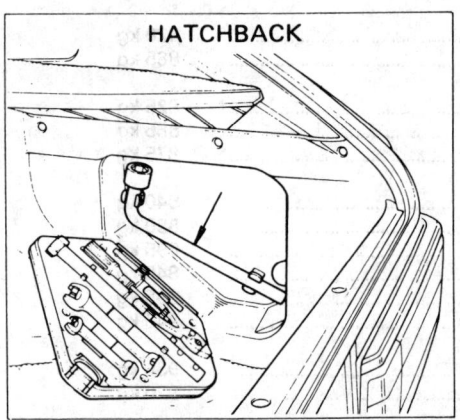

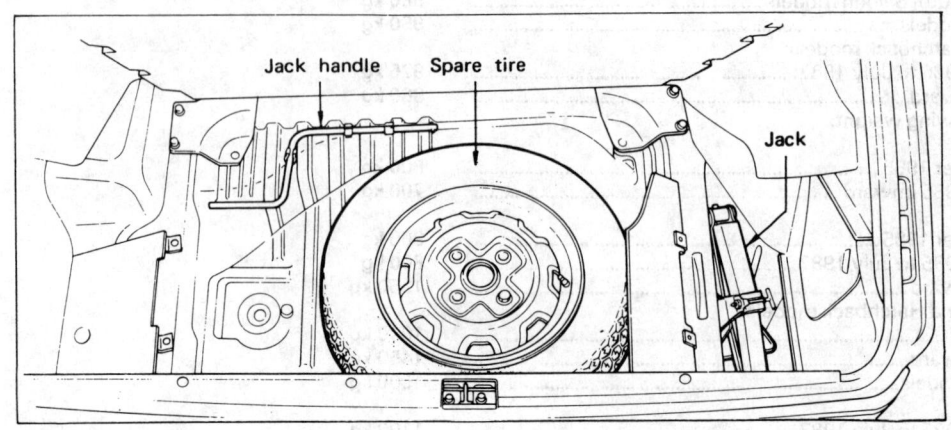

Jack, tool kit and spare wheel locations

Jacking, towing and wheel changing

Front towing hook

Rear towing hook

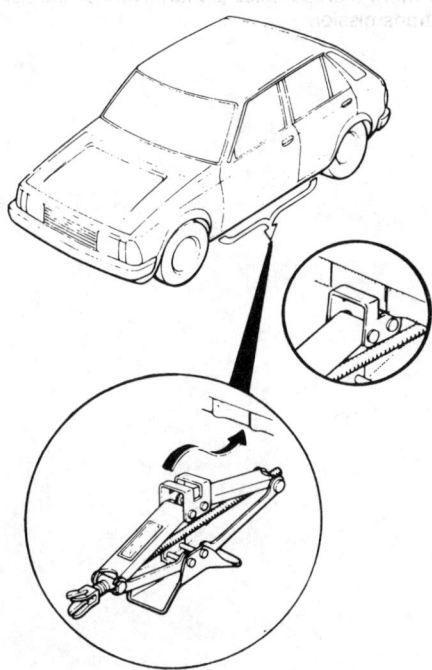
Front and rear side jacking points

Raising the front of the car using the front crossmember jacking point

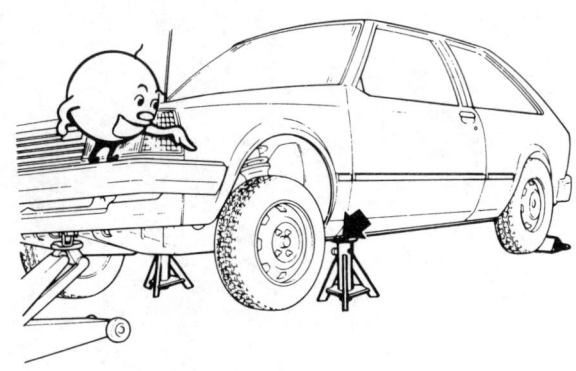
Front axle stand positions

then tighten the nuts fully. Refit the trim where applicable. With the spare wheel in position, remove the chock and stow the jack and tools in the luggage compartment.

When jacking up the car to carry out repair or maintenance tasks position the jack as follows.

If the front of the car is to be raised, position the jack head under the crossmember front mounting bolts. When the car is raised, supplement the jack with axle stands positioned under both the front jacking points on the front edge of the sills.

To raise the rear of the car, position the jack head centrally under the rear crossmember. When the car is raised, supplement the jack with axle stands positioned along side the jack under the crossmember or under the rear jacking points on the sills. *Never work under, around or near a raised car unless it is supported in at least two places with axle stands or suitable sturdy blocks.*

The car may be towed for breakdown recovery purposes only using the towing hook(s) positioned at the front of the vehicle (photo). On early models there is a single front towing hook which is situated on the front crossmember mounting point whereas on later models there is a

Jacking, towing and wheel changing

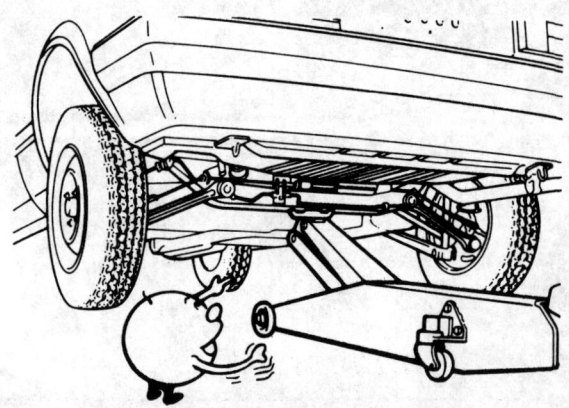

Raising the rear of the car using the rear crossmember jacking point

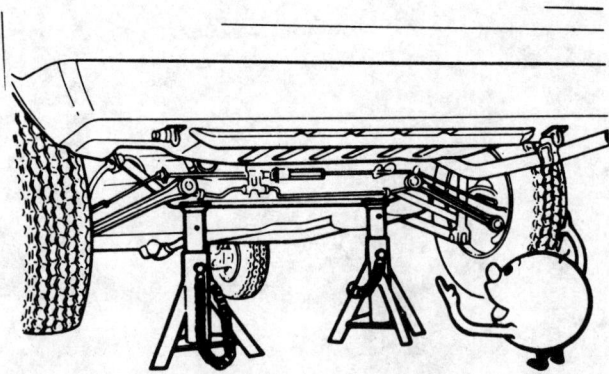

Rear axle stands positions

towing hook on both the right- and left-hand side of the front of the vehicle (photo). The rear towing hooks should be used only in an emergency situation, e.g. to pull the car out from a ditch, and not to tow the car any distance. All towing hooks are intended for towing loads only and must not be used for lifting the car either directly or indirectly. As a rule the vehicle must only be towed with its front wheels off the ground. If excessive vehicle damage or other conditions prevent the vehicle being towed with the front wheels off the ground, the front wheels should be supported on wheel dollies. If the car is to be towed with all four wheels on the ground it may only be towed in a forward direction, noting that it can not be towed at a speed exceeding 35 mph (56 km/h) or for more than 50 miles (80 km) without the risk of serious damage to the transmission.

Buying spare parts and vehicle identification numbers

Buying spare parts

Spare parts are available from many sources; for example, Mazda garages, other garages and accessory shops, and motor factors. Our advice regarding spare part sources is as follows.

Officially appointed Mazda garages – This is the best source for parts which are peculiar to your car, and are not generally available (eg complete cylinder heads, internal gearbox components, badges, interior trim etc). It is also the only place at which you should buy parts if the vehicle is still under warranty. To be sure of obtaining the correct parts, it will be necessary to give the storeman your car's Vehicle Identification Number (VIN), and if possible, take the old parts along for positive identification. Many parts are available under a factory exchange scheme – any parts returned should always be clean. It obviously makes good sense to go straight to the specialists on your car for this type of part, as they are best equipped to supply you.

Other garages and accessory shops – These are often very good places to buy materials and components needed for the maintenance of your car (eg oil filters, spark plugs, bulbs, drivebelts, oils and greases, touch-up paint, filler paste, etc). They also sell general accessories, usually have convenient opening hours, charge lower prices and can often be found not far from home.

Motor factors – Good factors will stock all the more important components which wear out comparatively quickly (eg exhaust systems, brake pads, seals and hydraulic parts, clutch components, bearing shells, pistons, valves etc). Motor factors will often provide new or reconditioned components on a part exchange basis – this can save a considerable amount of money.

Vehicle identification numbers

Modifications are a continuing and unpublicised process in vehicle manufacture, quite apart from major model changes. Spare parts manuals and lists are compiled upon a numerical basis, the individual vehicle identification numbers being essential to correct identification of the component concerned.

When ordering spare parts, always give as much information as possible. Quote the car model, year of manufacture, body and engine numbers as appropriate.

The *vehicle identification plate* is located on one or two plates attached to the front right-hand side of the engine compartment bulkhead.

The *chassis number* is located in the same position.

The *body number and paint code numbers* are located on the vehicle identification plate.

The *engine number* is stamped on the upper rear left-hand edge of the cylinder block just below the distributor.

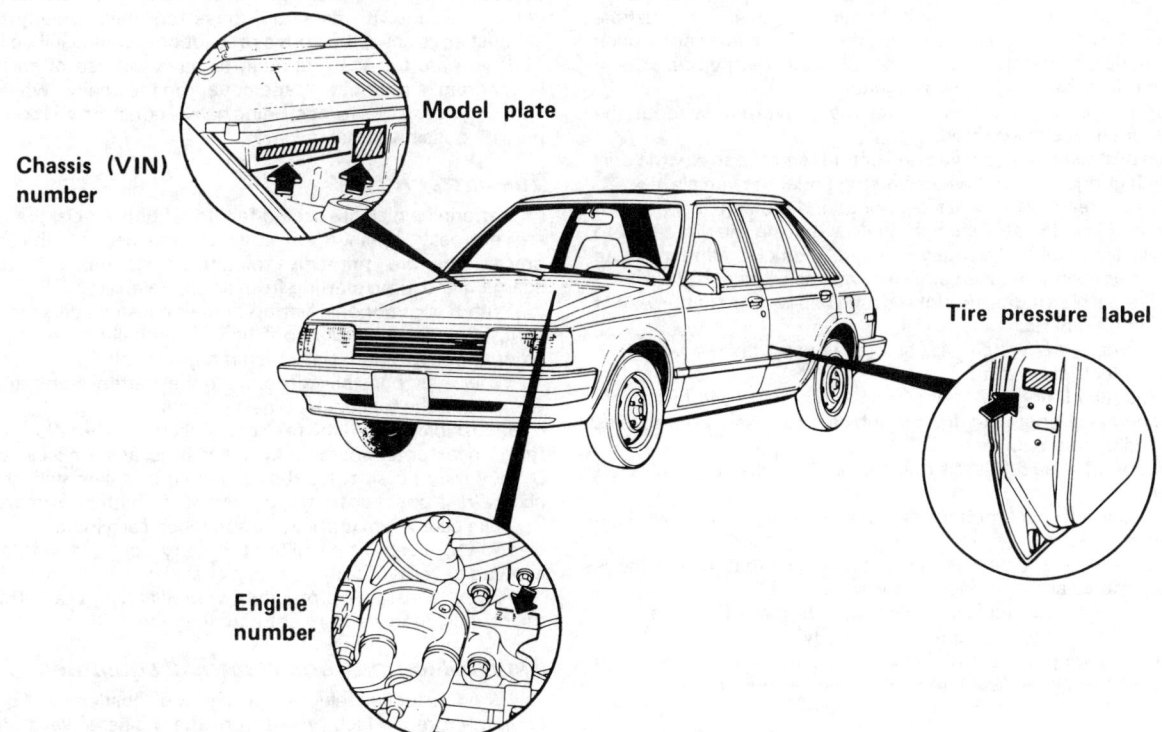

Vehicle identification number and data locations

Safety first!

However enthusiastic you may be about getting on with the job in hand, do take the time to ensure that your safety is not put at risk. A moment's lack of attention can result in an accident, as can failure to observe certain elementary precautions. There will always be new ways of having accidents, and the following points do not pretend to be a comprehensive list of all dangers; they are intended rather to make you aware of the risks and to encourage a safety-conscious approach to all work you carry out on your vehicle.

Essential DOs and DON'Ts

DON'T rely on a single jack when working underneath the vehicle. Always use reliable additional means of support, such as axle stands, securely placed under a structural part of the vehicle that you know will not give way.
DON'T attempt to loosen or tighten high-torque nuts (eg wheel hub nuts) while the vehicle is on a jack; it may be pulled off.
DON'T start the engine without first ascertaining that the transmission is in neutral (or 'Park' where applicable) and the handbrake applied.
DON'T suddenly remove the filler cap from a hot cooling system – cover it with a cloth and release the pressure gradually first, or you may get scalded by escaping coolant.
DON'T attempt to drain oil, automatic transmission fluid, or coolant until you are sure it has cooled sufficiently to avoid scalding you.
DON'T grasp any part of the engine, exhaust or catalytic converter without first ascertaining that it is sufficiently cool to avoid burning you.
DON'T allow brake fluid or antifreeze to contact vehicle paintwork.
DON'T syphon toxic liquids such as fuel, brake fluid or antifreeze by mouth, or allow them to remain on your skin.
DON'T inhale dust – it may be injurious to health (see *Asbestos* below).
DON'T allow any spilt oil or grease to remain on the floor – wipe it up straight away, before someone slips on it.
DON'T use ill-fitting spanners or other tools which may slip and cause injury.
DON'T attempt to lift a heavy component which may be beyond your capability – get assistance.
DON'T rush to finish a job, or take unverified short cuts.
DON'T allow children or animals in or around an unattended vehicle.
DON'T park vehicles with catalytic converters over combustible materials such as dry grass, oily rags, etc if the engine has recently been run. As catalytic converters reach extremely high temperatures, any such materials in close proximity may ignite.
DON'T run vehicles equipped with catalytic converters without the exhaust system heat shields fitted.
DO wear eye protection when using power tools such as an electric drill, sander, bench grinder etc, and when working under the vehicle.
DO use a barrier cream on your hands prior to undertaking dirty jobs – it will protect your skin from infection as well as making the dirt easier to remove afterwards; but make sure your hands aren't left slippery. Note that long term contact with used engine oil can be a health hazard.
DO keep loose clothing (cuffs, tie etc) and long hair well out of the way of moving mechanical parts.
DO remove rings, wristwatch etc, before working on the vehicle – especially the electrical system.
DO ensure that any lifting tackle or jacking equipment used has a safe working load rating adequate for the job, and is used precisely as recommended by the manufacturer.
DO keep your work area tidy – it is only too easy to fall over articles left lying around.
DO get someone to check periodically that all is well when working alone on the vehicle.
DO carry out work in a logical sequence and check that everything is correctly assembled and tightened afterwards.
DO remember that your vehicle's safety affects that of yourself and others. If in doubt on any point, get specialist advice.
IF, in spite of following these precautions, you are unfortunate enough to injure yourself, seek medical attention as soon as possible.

Asbestos

Certain friction, insulating, sealing, and other products – such as brake linings, brake bands, clutch linings, gaskets, etc – contain asbestos. *Extreme care must be taken to avoid inhalation of dust from such products since it is hazardous to health.* If in doubt, assume that they *do* contain asbestos.

Fire

Remember at all times that petrol is highly flammable. Never smoke, or have any kind of naked flame around, when working on the vehicle. But the risk does not end there – a spark caused by an electrical short-circuit, by two metal surfaces contacting each other, by careless use of tools, or even by static electricity built up in your body under certain conditions, can ignite petrol vapour, which in a confined space is highly explosive.

Whenever possible disconnect the battery earth terminal before working on any part of the fuel or electrical system, and never risk spilling fuel on to a hot engine or exhaust.

It is recommended that a fire extinguisher of a type suitable for fuel and electrical fires is kept handy in the garage or workplace at all times. Never try to extinguish a fuel or electrical fire with water.

Note: *Any reference to a 'torch' appearing in this manual should always be taken to mean a hand-held battery-operated electric lamp or flashlight. It does NOT mean a welding/gas torch or blowlamp.*

Fumes

Certain fumes are highly toxic and can quickly cause unconsciousness and even death if inhaled to any extent, especially if inhalation takes place through a lighted cigarette or pipe. Petrol vapour comes into this category, as do the vapours from certain solvents such as trichloroethylene. Any draining or pouring of such volatile fluids should be done in a well ventilated area.

When using cleaning fluids and solvents, read the instructions carefully. Never use materials from unmarked containers – they may give off poisonous vapours.

Never run the engine of a motor vehicle in an enclosed space such as a garage. Exhaust fumes contain carbon monoxide which is extremely poisonous; if you need to run the engine, always do so in the open air or at least have the rear of the vehicle outside the workplace. Although vehicles fitted with catalytic converters have greatly reduced toxic exhaust emissions, the above precautions should still be observed.

If you are fortunate enough to have the use of an inspection pit, never drain or pour petrol, and never run the engine, while the vehicle is standing over it; the fumes, being heavier than air, will concentrate in the pit with possibly lethal results.

The battery

Batteries which are sealed for life require special precautions which are normally outlined on a label attached to the battery. Such precautions are primarily related to situations involving battery charging and jump starting from another vehicle.

With a conventional battery, never cause a spark, or allow a naked light, in close proximity to it. It will normally be giving off a certain amount of hydrogen gas, which is highly explosive.

Whenever possible disconnect the battery earth terminal before working on the fuel or electrical systems.

If possible, loosen the filler plugs or cover when charging the battery from an external source. Do not charge at an excessive rate or the battery may burst. Special care should be taken with the use of high charge-rate boost chargers to prevent the battery from overheating.

Take care when topping up and when carrying the battery. The acid electrolyte, even when diluted, is very corrosive and should not be allowed to contact clothing, eyes or skin.

Always wear eye protection when cleaning the battery to prevent the caustic deposits from entering your eyes.

Mains electricity and electrical equipment

When using an electric power tool, inspection light, diagnostic equipment etc, which works from the mains, always ensure that the appliance is correctly connected to its plug and that, where necessary, it is properly earthed. Do not use such appliances in damp conditions and, again, beware of creating a spark or applying excessive heat in the vicinity of fuel or fuel vapour. Also ensure that the appliances meet the relevant national safety standards.

Ignition HT voltage

A severe electric shock can result from touching certain parts of the ignition system, such as the HT leads, when the engine is running or being cranked, particularly if components are damp or the insulation is defective. Where an electronic ignition system is fitted, the HT voltage is much higher and could prove fatal, especially to wearers of cardiac pacemakers.

Jacking and vehicle support

The jack provided with the vehicle is designed primarily for emergency wheel changing, and its use for servicing and overhaul work on the vehicle is best avoided. Instead, a more substantial workshop jack (trolley jack or similar) should be used. Whichever type is employed, it is essential that additional safety support is provided by means of axle stands designed for this purpose. Never use makeshift means such as wooden blocks or piles of house bricks, as these can easily topple or, in the case of bricks, disintegrate under the weight of the vehicle. Further information on the correct positioning of the jack and axle stands is provided in the *Jacking, towing and wheel changing* Section.

If removal of the wheels is not required, the use of drive-on ramps is recommended. Caution should be exercised to ensure that they are correctly aligned with the wheels, and that the vehicle is not driven too far along them so that it promptly falls off the other ends or tips the ramps.

General repair procedures

Whenever servicing, repair or overhaul work is carried out on the car or its components, it is necessary to observe the following procedures and instructions. This will assist in carrying out the operation efficiently and to a professional standard of workmanship.

Joint mating faces and gaskets

When separating components at their mating faces, never insert screwdrivers or similar implements into the joint between the faces in order to prise them apart. This can cause severe damage which results in oil leaks, coolant leaks, etc upon reassembly. Separation is usually achieved by tapping along the joint with a soft-faced hammer in order to break the seal. However, note that this method may not be suitable where dowels are used for component location.

Where a gasket is used between the mating faces of two components, ensure that it is renewed on reassembly and fit it dry unless otherwise stated in the repair procedure. Make sure that the mating faces are clean and dry with all traces of old gasket removed. When cleaning a joint face, use a tool which is not likely to score or damage the face, and remove any burrs or nicks with an oilstone or fine file.

Make sure that tapped holes are cleaned with a pipe cleaner and keep them free of jointing compound, if this is being used, unless specifically instructed otherwise.

Ensure that all orifices, channels or pipes are clear and blow through them, preferably using compressed air.

Oil seals

Oil seals can be removed by levering them out with a wide flat-bladed screwdriver or similar implement. Alternatively, a number of self-tapping screws may be screwed into the seal and these used as a purchase for pliers or some similar device in order to pull the seal free.

Whenever an oil seal is removed from its working location, either individually or as part of an assembly, it should be renewed.

The very fine sealing lip of the seal is easily damaged and will not seal if the surface it contacts is not completely clean and free from scratches, nicks or grooves. If the original sealing surface of the component cannot be restored, and the manufacturer has not made provision for slight relocation of the seal relative to the sealing surface, the component should be renewed.

Protect the lips of the seal from any surface which may damage them in the course of fitting. Use tape or a conical sleeve where possible. Lubricate the seal lips with oil before fitting and, on dual-lipped seals, fill the space between the lips with grease.

Unless otherwise stated, oil seals must be fitted with their sealing lips toward the lubricant to be sealed.

Use a tubular drift or block of wood of the appropriate size to install the seal and, if the seal housing is shouldered, drive the seal down to the shoulder. If the seal housing is unshouldered, the seal should be fitted with its face flush with the housing top face (unless otherwise instructed).

Screw threads and fastenings

Seized nuts, bolts and screws are quite a common occurrence where corrosion has set in, and the use of penetrating oil or releasing fluid will often overcome this problem if the offending item is soaked for a while before attempting to release it. The use of an impact driver may also provide a means of releasing such stubborn fastening devices when used in conjunction with the appropriate screwdriver bit or socket. If none of these methods works, it may be necessary to resort to the careful application of heat, or the use of a hacksaw or nut splitter device.

Studs are usually removed by locking two nuts together on the threaded part and then using a spanner on the lower nut to unscrew the stud. Studs or bolts which have broken off below the surface of the component in which they are mounted can sometimes be removed using a proprietary stud extractor. Always ensure that a blind tapped hole is completely free from oil, grease, water or other fluid before installing the bolt or stud. Failure to do this could cause the housing to crack due to the hydraulic action of the bolt or stud as it is screwed in.

When tightening a castellated nut to accept a split pin, tighten the nut to the specified torque, where applicable, and then tighten further to the next split pin hole. Never slacken the nut to align the split pin hole unless stated in the repair procedure.

When checking or retightening a nut or bolt to a specified torque setting, slacken the nut or bolt by a quarter of a turn, and then retighten to the specified setting. However, this should not be attempted where angular tightening has been used.

For some screw fastenings, notably cylinder head bolts or nuts, torque wrench settings are no longer specified for the latter stages of tightening, 'angle tightening' being called up instead. Typically, a fairly low torque wrench setting will be applied to the bolts/nuts in the correct sequence, followed by one or more stages of tightening through specified angles.

Locknuts, locktabs and washers

Any fastening which will rotate against a component or housing in the course of tightening should always have a washer between it and the relevant component or housing.

Spring or split washers should always be renewed when they are used to lock a critical component such as a big-end bearing retaining bolt or nut. Locktabs which are folded over to retain a nut or bolt should always be renewed.

Self-locking nuts can be reused in non-critical areas, providing resistance can be felt when the locking portion passes over the bolt or stud thread. However, it should be noted that self-locking stiffnuts tend to lose their effectiveness after long periods of use, and in such cases should be renewed as a matter of course.

Split pins must always be replaced with new ones of the correct size for the hole.

When thread-locking compound is found on the threads of a fastener which is to be re-used, it should be cleaned off with a wire brush and solvent, and fresh compound applied on reassembly.

Special tools

Some repair procedures in this manual entail the use of special tools such as a press, two or three-legged pullers, spring compressors etc. Wherever possible, suitable readily available alternatives to the manufacturer's special tools are described, and are shown in use. In some instances, where no alternative is possible, it has been necessary to resort to the use of a manufacturer's tool and this has been done for reasons of safety as well as the efficient completion of the repair operation. Unless you are highly skilled and have a thorough understanding of the procedures described, never attempt to bypass the use of any special tool when the procedure described specifies its use. Not only is there a very great risk of personal injury, but expensive damage could be caused to the components involved.

Environmental considerations

When disposing of used engine oil, brake fluid, antifreeze etc, give due consideration to any detrimental environmental effects. Do not, for instance, pour any of the above liquids down drains into the general sewage system or onto the ground to soak away. Many local council refuse tips provide a facility for waste oil disposal as do some garages. If none of these facilities are available, consult your local Environmental Health Department for further advice.

With the universal tightening-up of legislation regarding the emission of environmentally harmful substances from motor vehicles, most current vehicles have tamperproof devices fitted to the main adjustment points of the fuel system. These devices are primarily designed to prevent unqualified persons from adjusting the fuel/air mixture with the chance of a consequent increase in toxic emissions. If such devices are encountered during servicing or overhaul, they should, wherever possible, be renewed or refitted in accordance with the vehicle manufacturer's requirements or current legislation.

Tools and working facilities

Introduction

A selection of good tools is a fundamental requirement for anyone contemplating the maintenance and repair of a motor vehicle. For the owner who does not possess any, their purchase will prove a considerable expense, offsetting some of the savings made by doing-it-yourself. However, provided that the tools purchased meet the relevant national safety standards and are of good quality, they will last for many years and prove an extremely worthwhile investment.

To help the average owner to decide which tools are needed to carry out the various tasks detailed in this Manual, we have compiled three lists of tools under the following headings: *Maintenance and minor repair*, *Repair and overhaul*, and *Special*. Newcomers to practical mechanics should start off with the *Maintenance and minor repair* tool kit and confine themselves to the simpler jobs around the vehicle. Then, as confidence and experience grow, more difficult tasks can be undertaken, with extra tools being purchased as, and when, they are needed. In this way, a *Maintenance and minor repair* tool kit can be built up into a *Repair and overhaul* tool kit over a considerable period of time without any major cash outlays. The experienced do-it-yourselfer will have a tool kit good enough for most repair and overhaul procedures and will add tools from the *Special* category when it is felt that the expense is justified by the amount of use to which these tools will be put.

Maintenance and minor repair tool kit

The tools given in this list should be considered as a minimum requirement if routine maintenance, servicing and minor repair operations are to be undertaken. We recommend the purchase of combination spanners (ring one end, open-ended the other); although more expensive than open-ended ones, they do give the advantages of both types of spanner.

Combination spanners:
 Metric – 8, 9, 10, 11, 12, 13, 14, 15, 17 & 19 mm
Adjustable spanner – 35 mm jaw (approx)
Spark plug spanner (with rubber insert)
Spark plug gap adjustment tool
Set of feeler gauges
Brake bleed nipple spanner
Screwdrivers:
 Flat blade – approx 100 mm long x 6 mm dia
 Cross blade – approx 100 mm long x 6 mm dia
Combination pliers
Hacksaw (junior)
Tyre pump
Tyre pressure gauge
Grease gun (where applicable)
Oil can
Oil filter removal tool
Fine emery cloth
Wire brush (small)
Funnel (medium size)

Repair and overhaul tool kit

These tools are virtually essential for anyone undertaking any major repairs to a motor vehicle, and are additional to those given in the *Maintenance and minor repair* list. Included in this list is a comprehensive set of sockets. Although these are expensive, they will be found invaluable as they are so versatile – particularly if various

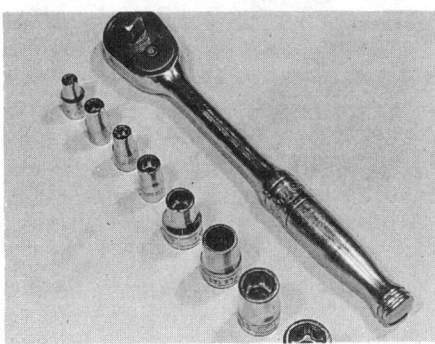

Sockets and reversible ratchet drive

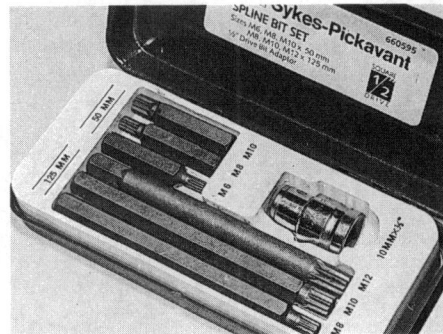

Spline bit set

Spline key set

Tools and working facilities

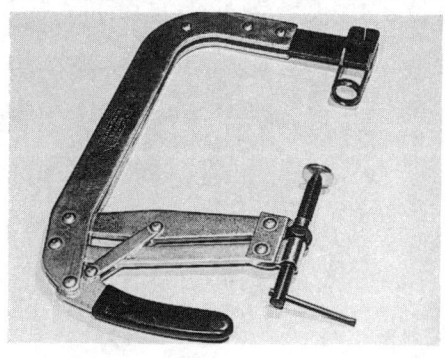
Valve spring compressor

Piston ring compressor

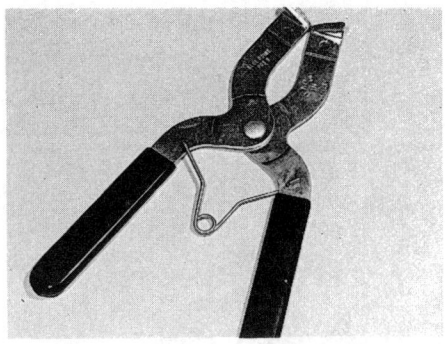
Piston ring removal/installation tool

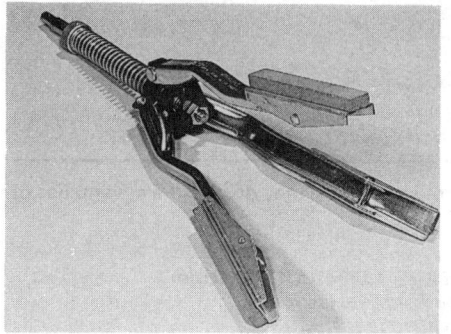
Cylinder bore hone

Three-legged hub and bearing puller

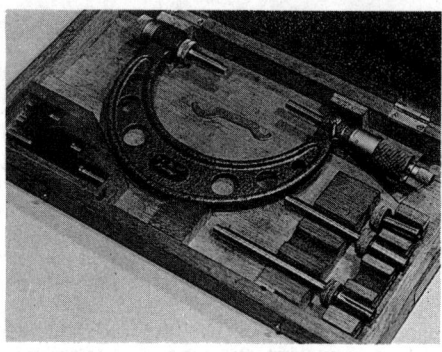
Micrometer set

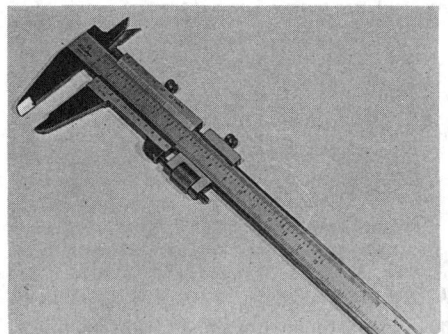
Vernier calipers

Dial test indicator and magnetic stand

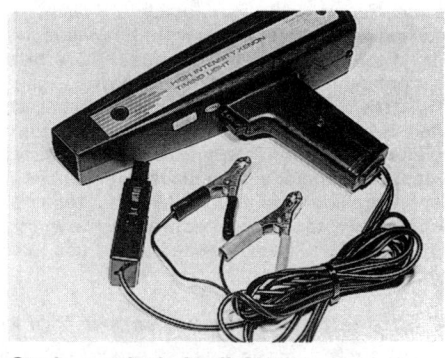
Stroboscopic timing light

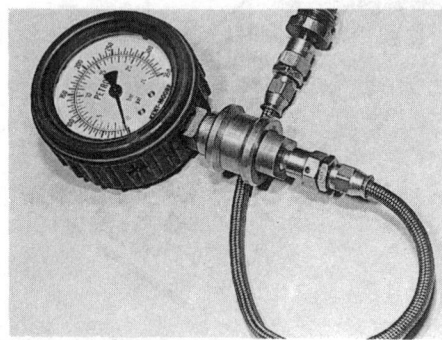
Compression gauge

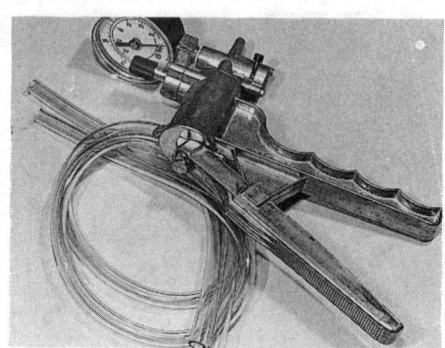
Vacuum pump and gauge

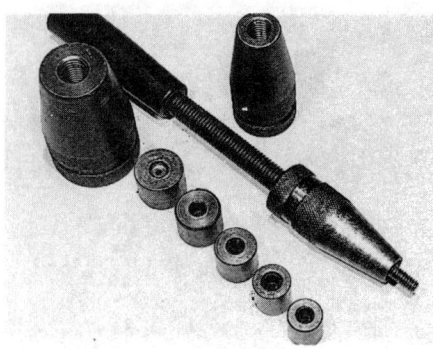
Clutch plate alignment set

Tools and working facilities

Bush and bearing removal/installation set

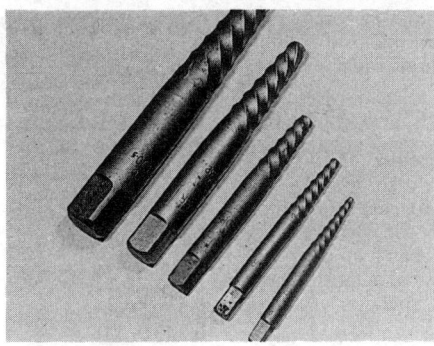
Stud extractor set

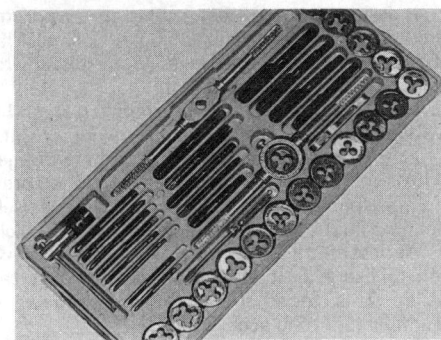
Tap and die set

drives are included in the set. We recommend the ½ in square-drive type, as this can be used with most proprietary torque wrenches. If you cannot afford a socket set, even bought piecemeal, then inexpensive tubular box spanners are a useful alternative.

The tools in this list will occasionally need to be supplemented by tools from the *Special* list.

Sockets (or box spanners) to cover range in previous list
Reversible ratchet drive (for use with sockets) (photo).
Extension piece, 250 mm (for use with sockets)
Universal joint (for use with sockets)
Torque wrench (for use with sockets)
Self-locking grips
Ball pein hammer
Soft-faced mallet (plastic/aluminium or rubber)
Screwdrivers:
    Flat blade – long & sturdy, short (chubby), and narrow (electricians) types
    Cross blade – long & sturdy, and short (chubby) types
Pliers:
    Long-nosed
    Side cutters (electricians)
    Circlip (internal and external)
Cold chisel – 25 mm
Scriber
Scraper
Centre punch
Pin punch
Hacksaw
Brake hose clamp
Brake/clutch bleeding kit
Selection of twist drills
Steel rule/straight-edge
Allen keys (inc. splined/Torx type) (photos).
Selection of files
Wire brush
Axle-stands
Jack (strong trolley or hydraulic type)
Light with extension lead

Special tools

The tools in this list are those which are not used regularly, are expensive to buy, or which need to be used in accordance with their manufacturers' instructions. Unless relatively difficult mechanical jobs are undertaken frequently, it will not be economic to buy many of these tools. Where this is the case, you could consider clubbing together with friends (or joining a motorists' club) to make a joint purchase, or borrowing the tools against a deposit from a local garage or tool hire specialist. It is worth noting that many of the larger DIY superstores now carry a large range of special tools for hire at modest rates.

The following list contains only those tools and instruments freely available to the public, and not those special tools produced by the vehicle manufacturer specifically for its dealer network. You will find occasional references to these manufacturer's special tools in the text of this Manual. Generally, an alternative method of doing the job without the vehicle manufacturers' special tool is given. However, sometimes there is no alternative to using them. Where this is the case and the relevant tool cannot be bought or borrowed, you will have to entrust the work to a franchised garage.

Valve spring compressor (photo).
Valve grinding tool
Piston ring compressor (photo).
Piston ring removal/installation tool (photo).
Cylinder bore hone (photo).
Balljoint separator
Coil spring compressors (where applicable)
Two/three-legged hub and bearing puller (photo).
Impact screwdriver
Micrometer and/or vernier calipers (photos).
Dial test indicator (photo).
Stroboscopic timing light (photo).
Dwell angle meter/tachometer
Universal electrical multi-meter
Cylinder compression gauge (photo).
Hand-operated vacuum pump and gauge (photo).
Clutch plate alignment set (photo).
Bush and bearing removal/installation set (photo).
Stud extractors (photo).
Tap and die set (photo).
Lifting tackle
Trolley jack

Buying tools

For practically all tools, a tool factor is the best source since he will have a very comprehensive range compared with the average garage or accessory shop. Having said that, accessory shops often offer excellent quality tools at discount prices, so it pays to shop around.

Remember, you don't have to buy the most expensive items on the shelf but it is always advisable to steer clear of the very cheap tools. There are plenty of good tools around at reasonable prices, but always aim to purchase items which meet the relevant national safety standards. If in doubt, ask the proprietor or manager of the shop for advice before making a purchase.

Care and maintenance of tools

Having purchased a reasonable tool kit, it is necessary to keep the tools in a clean and serviceable condition. After use, always wipe off any dirt, grease and metal particles using a clean, dry cloth, before putting the tools away. Never leave them lying around after they have been used. A simple tool rack on the garage or workshop wall for items such as screwdrivers and pliers is a good idea. Store all normal spanners and sockets in a metal box. Any measuring instruments, gauges, meters, etc, must be carefully stored where they cannot be damaged or become rusty.

Take a little care when tools are used. Hammer heads inevitably become marked and screwdrivers lose the keen edge on their blades from time to time. A little timely attention with emery cloth or a file will soon restore items like this to a good serviceable finish.

Working facilities

Not to be forgotten when discussing tools is the workshop itself. If anything more than routine maintenance is to be carried out, some form of suitable working area becomes essential.

It is appreciated that many an owner mechanic is forced by circumstances to remove an engine or similar item without the benefit of a garage or workshop. Having done this, any repairs should always be done under the cover of a roof.

Wherever possible, any dismantling should be done on a clean, flat workbench or table at a suitable working height.

Any workbench needs a vice; one with a jaw opening of 100 mm (4 in) is suitable for most jobs. As mentioned previously, some clean dry storage space is also required for tools, as well as for any lubricants, cleaning fluids, touch-up paints and so on, which become necessary.

Another item which may be required, and which has a much more general usage, is an electric drill with a chuck capacity of at least 8 mm ($\frac{5}{16}$ in). This, together with a good range of twist drills, is virtually essential for fitting accessories.

Last, but not least, always keep a supply of old newspapers and clean, lint-free rags available, and try to keep any working area as clean as possible.

Spanner jaw gap and bolt size comparison table

Jaw gap – in (mm)	Spanner size	Bolt size
0.197 (5.00)	5 mm	M 2.5
0.216 (5.50)	5.5 mm	M 3
0.218 (5.53)	$\frac{7}{32}$ in AF	
0.236 (6.00)	6 mm	M 3.5
0.250 (6.35)	$\frac{1}{4}$ in AF	
0.275 (7.00)	7 mm	M 4
0.281 (7.14)	$\frac{9}{32}$ in AF	
0.312 (7.92)	$\frac{5}{16}$ in AF	
0.315 (8.00)	8 mm	M 5
0.343 (8.71)	$\frac{11}{32}$ in AF	
0.375 (9.52)	$\frac{3}{8}$ in AF	
0.394 (10.00)	10 mm	M 6
0.406 (10.32)	$\frac{13}{32}$ in AF	
0.433 (11.00)	11 mm	M 7
0.437 (11.09)	$\frac{7}{16}$ in AF	$\frac{1}{4}$ in SAE
0.468 (11.88)	$\frac{15}{32}$ in AF	
0.500 (12.70)	$\frac{1}{2}$ in AF	$\frac{5}{16}$ in SAE
0.512 (13.00)	13 mm	M 8
0.562 (14.27)	$\frac{9}{16}$ in AF	$\frac{3}{8}$ in SAE
0.593 (15.06)	$\frac{19}{32}$ in AF	
0.625 (15.87)	$\frac{5}{8}$ in AF	$\frac{7}{16}$ in SAE
0.669 (17.00)	17 mm	M 10
0.687 (17.44)	$\frac{11}{16}$ in AF	
0.709 (19.00)	19 mm	M 12
0.750 (19.05)	$\frac{3}{4}$ in AF	$\frac{1}{2}$ in SAE
0.781 (19.83)	$\frac{25}{32}$ in AF	
0.812 (20.62)	$\frac{13}{16}$ in AF	
0.866 (22.00)	22 mm	M 14
0.875 (22.25)	$\frac{7}{8}$ in AF	$\frac{9}{16}$ in SAE
0.937 (23.79)	$\frac{15}{16}$ in AF	$\frac{5}{8}$ in SAE
0.945 (24.00)	24 mm	M 16
0.968 (24.58)	$\frac{31}{32}$ in AF	
1.000 (25.40)	1 in AF	$\frac{11}{16}$ in SAE
1.062 (26.97)	1 $\frac{1}{16}$ in AF	$\frac{3}{4}$ in SAE
1.063 (27.00)	27 mm	M 18
1.125 (28.57)	1 $\frac{1}{8}$ in AF	
1.182 (30.00)	30 mm	M 20
1.187 (30.14)	1 $\frac{3}{16}$ in AF	
1.250 (31.75)	1 $\frac{1}{4}$ in AF	$\frac{7}{8}$ in SAE
1.260 (32.00)	32 mm	M 22
1.312 (33.32)	1 $\frac{5}{16}$ in AF	
1.375 (34.92)	1 $\frac{3}{8}$ in AF	
1.418 (36.00)	36 mm	M 24
1.437 (36.49)	1 $\frac{7}{16}$ in AF	1 in SAE
1.500 (38.10)	1 $\frac{1}{2}$ in AF	
1.615 (41.00)	41 mm	M 27

Booster battery (jump) starting

When jump starting a car using a booster battery, observe the following precautions.

(a) *Before connecting the booster battery, make sure that the ignition is switched off.*
(b) *Ensure that all electrical equipment (lights, heater, wipers etc) is switched off.*
(c) *Make sure that the booster battery is the same voltage as the discharged one in the vehicle.*
(d) *If the battery is being jump started from the battery in another vehicle, the two vehicles MUST NOT TOUCH each other.*
(e) *Make sure that the transmission is in Neutral (manual gearbox) or Park (automatic transmission).*

Connect one jump lead between the positive (+) terminals of the two batteries. Connect the other jump lead first to the negative (−) terminal of the booster battery, and then to a good earthing point on the vehicle to be started, such as a bolt or bracket on the engine block, at least 45 cm (18 in) from the battery if possible. Make sure that the jump leads will not come into contact with the fan, drivebelts or other moving parts of the engine.

Start the engine using the booster battery, then with the engine running at idle speed, disconnect the jump leads in the reverse order of connection.

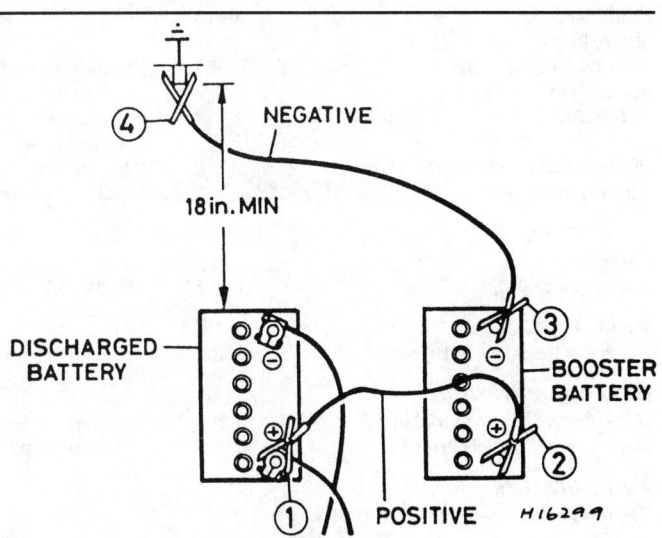

Jump start lead connections for negative earth vehicles – connect the leads in the order shown

Conversion factors

Length (distance)
Inches (in)	X	25.4	= Millimetres (mm)	X	0.0394	= Inches (in)
Feet (ft)	X	0.305	= Metres (m)	X	3.281	= Feet (ft)
Miles	X	1.609	= Kilometres (km)	X	0.621	= Miles

Volume (capacity)
Cubic inches (cu in; in^3)	X	16.387	= Cubic centimetres (cc; cm^3)	X	0.061	= Cubic inches (cu in; in^3)
Imperial pints (Imp pt)	X	0.568	= Litres (l)	X	1.76	= Imperial pints (Imp pt)
Imperial quarts (Imp qt)	X	1.137	= Litres (l)	X	0.88	= Imperial quarts (Imp qt)
Imperial quarts (Imp qt)	X	1.201	= US quarts (US qt)	X	0.833	= Imperial quarts (Imp qt)
US quarts (US qt)	X	0.946	= Litres (l)	X	1.057	= US quarts (US qt)
Imperial gallons (Imp gal)	X	4.546	= Litres (l)	X	0.22	= Imperial gallons (Imp gal)
Imperial gallons (Imp gal)	X	1.201	= US gallons (US gal)	X	0.833	= Imperial gallons (Imp gal)
US gallons (US gal)	X	3.785	= Litres (l)	X	0.264	= US gallons (US gal)

Mass (weight)
Ounces (oz)	X	28.35	= Grams (g)	X	0.035	= Ounces (oz)
Pounds (lb)	X	0.454	= Kilograms (kg)	X	2.205	= Pounds (lb)

Force
Ounces-force (ozf; oz)	X	0.278	= Newtons (N)	X	3.6	= Ounces-force (ozf; oz)
Pounds-force (lbf; lb)	X	4.448	= Newtons (N)	X	0.225	= Pounds-force (lbf; lb)
Newtons (N)	X	0.1	= Kilograms-force (kgf; kg)	X	9.81	= Newtons (N)

Pressure
Pounds-force per square inch (psi; lbf/in^2; lb/in^2)	X	0.070	= Kilograms-force per square centimetre (kgf/cm^2; kg/cm^2)	X	14.223	= Pounds-force per square inch (psi; lbf/in^2; lb/in^2)
Pounds-force per square inch (psi; lbf/in^2; lb/in^2)	X	0.068	= Atmospheres (atm)	X	14.696	= Pounds-force per square inch (psi; lbf/in^2; lb/in^2)
Pounds-force per square inch (psi; lbf/in^2; lb/in^2)	X	0.069	= Bars	X	14.5	= Pounds-force per square inch (psi; lbf/in^2; lb/in^2)
Pounds-force per square inch (psi; lbf/in^2; lb/in^2)	X	6.895	= Kilopascals (kPa)	X	0.145	= Pounds-force per square inch (psi; lbf/in^2; lb/in^2)
Kilopascals (kPa)	X	0.01	= Kilograms-force per square centimetre (kgf/cm^2; kg/cm^2)	X	98.1	= Kilopascals (kPa)
Millibar (mbar)	X	100	= Pascals (Pa)	X	0.01	= Millibar (mbar)
Millibar (mbar)	X	0.0145	= Pounds-force per square inch (psi; lbf/in^2; lb/in^2)	X	68.947	= Millibar (mbar)
Millibar (mbar)	X	0.75	= Millimetres of mercury (mmHg)	X	1.333	= Millibar (mbar)
Millibar (mbar)	X	0.401	= Inches of water (inH$_2$O)	X	2.491	= Millibar (mbar)
Millimetres of mercury (mmHg)	X	0.535	= Inches of water (inH$_2$O)	X	1.868	= Millimetres of mercury (mmHg)
Inches of water (inH$_2$O)	X	0.036	= Pounds-force per square inch (psi; lbf/in^2; lb/in^2)	X	27.68	= Inches of water (inH$_2$O)

Torque (moment of force)
Pounds-force inches (lbf in; lb in)	X	1.152	= Kilograms-force centimetre (kgf cm; kg cm)	X	0.868	= Pounds-force inches (lbf in; lb in)
Pounds-force inches (lbf in; lb in)	X	0.113	= Newton metres (Nm)	X	8.85	= Pounds-force inches (lbf in; lb in)
Pounds-force inches (lbf in; lb in)	X	0.083	= Pounds-force feet (lbf ft; lb ft)	X	12	= Pounds-force inches (lbf in; lb in)
Pounds-force feet (lbf ft; lb ft)	X	0.138	= Kilograms-force metres (kgf m; kg m)	X	7.233	= Pounds-force feet (lbf ft; lb ft)
Pounds-force feet (lbf ft; lb ft)	X	1.356	= Newton metres (Nm)	X	0.738	= Pounds-force feet (lbf ft; lb ft)
Newton metres (Nm)	X	0.102	= Kilograms-force metres (kgf m; kg m)	X	9.804	= Newton metres (Nm)

Power
Horsepower (hp)	X	745.7	= Watts (W)	X	0.0013	= Horsepower (hp)

Velocity (speed)
Miles per hour (miles/hr; mph)	X	1.609	= Kilometres per hour (km/hr; kph)	X	0.621	= Miles per hour (miles/hr; mph)

Fuel consumption
Miles per gallon, Imperial (mpg)	X	0.354	= Kilometres per litre (km/l)	X	2.825	= Miles per gallon, Imperial (mpg)
Miles per gallon, US (mpg)	X	0.425	= Kilometres per litre (km/l)	X	2.352	= Miles per gallon, US (mpg)

Temperature
Degrees Fahrenheit = (°C x 1.8) + 32 Degrees Celsius (Degrees Centigrade; °C) = (°F - 32) x 0.56

* It is common practice to convert from miles per gallon (mpg) to litres/100 kilometres (l/100km), where mpg (Imperial) x l/100 km = 282 and mpg (US) x l/100 km = 235

Fault diagnosis

Contents

Engine 1
Engine fails to rotate when attempting to start
Engine rotates but will not start
Engine difficult to start when cold
Engine difficult to start when hot
Starter motor noisy or excessively rough in engagement
Engine starts but stops immediately
Engine idles erratically
Engine misfires at idle speed
Engine misfires throughout the driving speed range
Engine hesitates on acceleration
Engine stalls
Engine lacks power
Engine backfires
Oil pressure warning light illuminated with engine running
Engine runs-on after switching off
Engine noises

Cooling system 2
Overheating
Overcooling
External coolant leakage
Internal coolant leakage
Corrosion

Fuel and exhaust system 3
Excessive fuel consumption
Fuel leakage and/or fuel odour
Excessive noise or fumes from exhaust system

Clutch 4
Pedal travels to floor – no pressure or very little resistance
Clutch fails to disengage (unable to select gears)
Clutch slips (engine speed increases with no increase in vehicle speed)
Judder as clutch is engaged
Noise when depressing or releasing clutch pedal

Manual gearbox 5
Noisy in neutral with engine running
Noisy in one particular gear
Difficulty engaging gears
Jumps out of gear
Vibration
Lubricant leaks

Automatic transmission 6
Fluid leakage
Transmission fluid brown or has burned smell
General gear selection problems
Transmission will not downshift (kickdown) with accelerator fully depressed
Engine will not start in any gear, or starts in gears other than Park or Neutral
Transmission slips, shifts roughly, is noisy or has no drive in forward or reverse gears

Driveshafts 7
Clicking or knocking noise on turns (at slow speed on full lock)
Vibration when accelerating or decelerating

Braking system 8
Vehicle pulls to one side under braking
Noise (grinding or high-pitched squeal) when brakes applied
Excessive brake pedal travel
Brake pedal feels spongy when depressed
Excessive brake pedal effort required to stop vehicle
Judder felt through brake pedal or steering wheel when braking
Brakes binding
Rear wheels locking under normal braking

Suspension and steering systems 9
Vehicle pulls to one side
Wheel wobble and vibration
Excessive pitching and/or rolling around corners or during braking
Wandering or general instability
Excessively stiff steering
Excessive play in steering
Lack of power assistance (where equipped)
Tyre wear excessive

Electrical system 10
Battery will not hold a charge for more than a few days
Ignition warning light remains illuminated with engine running
Ignition warning light fails to come on
Lights inoperative
Instrument readings inaccurate or erratic
Horn inoperative or unsatisfactory in operation
Windscreen/tailgate wipers inoperative or unsatisfactory in operation
Windscreen/tailgate washers inoperative or unsatisfactory in operation
Electric windows inoperative or unsatisfactory in operation
Central locking system inoperative or unsatisfactory in operation

Introduction

The vehicle owner who does his or her own maintenance according to the recommended service schedules should not have to use this Section of the Manual very often. Modern component reliability is such that, provided those items subject to wear or deterioration are inspected or renewed at the specified intervals, sudden failure is comparatively rare. Faults do not usually just happen as a result of sudden failure, but develop over a period of time. Major mechanical failures in particular are usually preceded by characteristic symptoms over hundreds or even thousands of miles. Those components which do occasionally fail without warning are often small and easily carried in the vehicle.

With any fault finding, the first step is to decide where to begin investigations. Sometimes this is obvious, but on other occasions a little detective work will be necessary. The owner who makes half a dozen haphazard adjustments or replacements may be successful in curing a fault (or its symptoms), but will be none the wiser if the fault recurs and ultimately may have spent more time and money than was necessary. A calm and logical approach will be found to be more satisfactory in the long run. Always take into account any warning signs or abnormalities that may have been noticed in the period preceding the fault – power loss, high or low gauge readings, unusual smells, etc – and remember that failure of components such as fuses or spark plugs may only be pointers to some underlying fault.

The pages which follow provide an easy reference guide to the more common problems which may occur during the operation of the vehicle. These problems and their possible causes are grouped under headings denoting various components or systems, such as Engine, Cooling system, etc. The Chapter and/or Section which deals with the problem is also shown in brackets. Whatever the fault, certain basic principles apply. These are as follows:

Verify the fault. This is simply a matter of being sure that you know what the symptoms are before starting work. This is particularly important if you are investigating a fault for someone else who may not have described it very accurately.

Don't overlook the obvious. For example, if the vehicle won't start, is there petrol in the tank? (Don't take anyone else's word on this particular point, and don't trust the fuel gauge either!) If an electrical fault is indicated, look for loose or broken wires before digging out the test gear.

Cure the disease, not the symptom. Substituting a flat battery with a fully charged one will get you off the hard shoulder, but if the underlying cause is not attended to, the new battery will go the same way. Similarly, changing oil-fouled spark plugs for a new set will get you moving again, but remember that the reason for the fouling (if it wasn't simply an incorrect grade of plug) will have to be established and corrected.

Don't take anything for granted. Particularly, don't forget that a 'new' component may itself be defective (especially if it's been rattling around in the boot for months), and don't leave components out of a fault diagnosis sequence just because they are new or recently fitted. When you do finally diagnose a difficult fault, you'll probably realise that all the evidence was there from the start.

1 Engine

Engine fails to rotate when attempting to start

- Battery terminal connections loose or corroded (Chapter 12).
- Battery discharged or faulty (Chapter 12).
- Broken, loose or disconnected wiring in the starting circuit (Chapter 12).
- Defective starter solenoid or switch (Chapter 12).
- Defective starter motor (Chapter 12).
- Starter pinion or flywheel ring gear teeth loose or broken (Chapter 12).
- Engine earth strap broken or disconnected (Chapter 12).
- Automatic transmission not in Park/Neutral position or starter inhibitor switch faulty (Chapter 7).

Engine rotates but will not start

- Fuel tank empty.
- Battery discharged (engine rotates slowly) (Chapter 12).
- Battery terminal connections loose or corroded (Chapter 12).
- Ignition components damp or damaged (Chapters 1 and 5).
- Broken, loose or disconnected wiring in the ignition circuit (Chapters 1 and 5).
- Worn, faulty or incorrectly gapped spark plugs (Chapter 1).
- Choke mechanism sticking, incorrectly adjusted, or faulty – carburettor models (Chapter 4).
- Dirty or incorrectly gapped contact breaker points – pre 1982 models (Chapter 5).
- Faulty condenser – pre 1982 models (Chapter 5).
- Major mechanical failure (eg camshaft drive) (Chapter 2).

Engine difficult to start when cold

- Battery discharged (Chapter 12).
- Battery terminal connections loose or corroded (Chapter 12).
- Worn, faulty or incorrectly gapped spark plugs (Chapter 1).
- Choke mechanism sticking, incorrectly adjusted, or faulty – carburettor models (Chapter 4).
- Other ignition system fault (Chapters 1 and 5).
- Low cylinder compressions (Chapter 2).

Engine difficult to start when hot

- Air filter element dirty or clogged (Chapter 1).
- Choke mechanism sticking, incorrectly adjusted, or faulty – carburettor models (Chapter 4).
- Carburettor float chamber flooding (Chapter 4).
- Low cylinder compressions (Chapter 2).

Starter motor noisy or excessively rough in engagement

- Starter pinion or flywheel ring gear teeth loose or broken (Chapter 12).
- Starter motor mounting bolts loose or missing (Chapter 12).
- Starter motor internal components worn or damaged (Chapter 12).

Engine starts but stops immediately

- Insufficient fuel reaching fuel system (Chapter 4).
- Loose or faulty electrical connections in the ignition circuit (Chapters 1 and 5).
- Vacuum leak at the carburettor/throttle housing or inlet manifold (Chapter 4).
- Blocked carburettor jet(s) or internal passages (Chapter 4).
- Fuel injection system faulty (Chapter 4).

Engine idles erratically

- Incorrectly adjusted idle speed and/or mixture settings (Chapter 1).
- Air filter element clogged (Chapter 1).
- Vacuum leak at the carburettor, inlet manifold or associated hoses – carburettor models (Chapter 4).
- Vacuum leak at the throttle housing, surge tank, inlet manifold or associated models – fuel injected models (Chapter 4).
- Worn, faulty or incorrectly gapped spark plugs (Chapter 1).
- Incorrectly adjusted valve clearances – B6 and E series engines (Chapter 1).
- Faulty HLA adjuster(s) – B3 and B5 engines (Chapter 2).
- Uneven or low cylinder compressions (Chapter 2).
- Camshaft lobes worn (Chapter 2).
- Timing belt/chain incorrectly tensioned (Chapter 2).

Engine misfires at idle speed

- Worn, faulty or incorrectly gapped spark plugs (Chapter 1).
- Faulty spark plug HT leads (Chapter 1).
- Incorrectly adjusted idle mixture settings (Chapter 1).
- Incorrect ignition timing (Chapter 5).
- Vacuum leak at the carburettor, inlet manifold or associated hoses – carburettor models (Chapter 4).
- Vacuum leak at the throttle housing, surge tank, inlet manifold or associated hoses – fuel injected models (Chapter 4).
- Distributor cap cracked or tracking internally (Chapter 1).
- Incorrectly adjusted valve clearances – B6 and E series engines (Chapter 1).
- Faulty HLA adjuster(s) – B3 and B5 engines (Chapter 2).
- Uneven or low cylinder compressions (Chapter 2).
- Disconnected, leaking or perished crankcase ventilation hoses (Chapters 1 and 4).

Engine misfires throughout the driving speed range

- Blocked carburettor jet(s) or internal passages (Chapter 4).
- Fuel injection system faulty (Chapter 4).
- Carburettor worn or incorrectly adjusted (Chapters 1 and 4).
- Fuel filter choked (Chapter 1).
- Fuel pump faulty or delivery pressure low (Chapter 4).
- Fuel tank vent blocked or fuel pipes restricted (Chapter 4).
- Vacuum leak at the carburettor, inlet manifold or associated hoses – carburettor models (Chapter 4).
- Vacuum leak at the throttle housing, surge tank, inlet manifold and associated hoses – fuel injected models (Chapter 4).
- Worn, faulty or incorrectly gapped spark plugs (Chapter 1).
- Faulty spark plug HT leads (Chapter 1).
- Dirty or incorrectly gapped contact breaker points – pre 1982 models (Chapter 5).
- Faulty condenser – pre 1982 models (Chapter 5).
- Distributor cap cracked or tracking internally (Chapter 1).
- Faulty ignition coil (Chapter 5).
- Uneven or low cylinder compressions (Chapter 2).

Engine hesitates on acceleration

- Worn, faulty or incorrectly gapped spark plugs (Chapter 1).
- Carburettor accelerator pump faulty (Chapter 4).
- Blocked carburettor jets or internal passages (Chapter 4).
- Fuel injection system fault (Chapter 4).
- Vacuum leak at the carburettor, inlet manifold or associated hoses (Chapter 4).
- Vacuum leak at the throttle housing, surge tank, inlet manifold or associated hoses (Chapter 4).
- Carburettor worn or incorrectly adjusted (Chapters 1 and 4).

Fault diagnosis

Engine stalls
- Incorrectly adjusted idle speed and/or mixture settings (Chapter 1).
- Blocked carburettor jet(s) or internal passages (Chapter 4).
- Fuel injection system fault (Chapter 4).
- Vacuum leak at the carburettor, inlet manifold or associated hoses (Chapter 4).
- Vacuum leak at the throttle housing, surge tank, inlet manifold or associated hoses (Chapter 4).
- Fuel filter choked (Chapter 1).
- Fuel pump faulty or delivery pressure low (Chapter 4).
- Fuel tank vent blocked or fuel pipes restricted (Chapter 4).
- Throttle positioner/idle-up system (where fitted) incorrectly adjusted (Chapters 1 and 4).

Engine lacks power
- Incorrect ignition timing (Chapter 1).
- Carburettor worn or incorrectly adjusted (Chapter 1).
- Timing belt/chain incorrectly fitted or tensioned (Chapter 2).
- Fuel filter choked (Chapter 1).
- Fuel pump faulty or delivery pressure low (Chapter 4).
- Uneven or low cylinder compressions (Chapter 2).
- Worn, faulty or incorrectly gapped spark plugs (Chapter 1).
- Vacuum leak at the carburettor, inlet manifold or associated hoses – carburettor models (Chapter 4).
- Vacuum leak at the throttle housing, surge tank, inlet manifold or associated hoses – fuel injected models (Chapter 4).
- Brakes binding (Chapters 1 and 9).
- Clutch slipping (Chapter 6).
- Automatic transmission fluid level incorrect (Chapter 1).

Engine backfires
- Ignition timing incorrect (Chapter 1).
- Timing belt/chain incorrectly fitted or tensioned (Chapter 2).
- Carburettor worn or incorrectly adjusted (Chapter 1).
- Vacuum leak at the carburettor, inlet manifold or associated hoses – carburettor models (Chapter 4).
- Vacuum leak at the throttle housing, surge tank, inlet manifold or associated hoses – fuel injected models (Chapter 4).

Oil pressure warning light illuminated with engine running
- Low oil level or incorrect grade (Chapter 1).
- Faulty oil pressure transmitter (sender) unit (Chapter 2).
- Worn engine bearings and/or oil pump (Chapter 2).
- High engine operating temperature (Chapter 3).
- Oil pressure relief valve defective (Chapter 2).
- Oil pick-up strainer clogged (Chapter 2).

Engine runs-on after switching off
- Idle speed excessively high (Chapter 1).
- Faulty fuel cut-off solenoid – carburettor models (Chapter 4).
- Excessive carbon build-up in engine (Chapter 2).
- High engine operating temperature (Chapter 3).

Engine noises

Pre-ignition (pinking) or knocking during acceleration or under load
- Ignition timing incorrect (Chapter 1).
- Incorrect grade of fuel (Chapter 4).
- Vacuum leak at the carburettor, inlet manifold or associated hoses – carburettor models (Chapter 4).
- Vacuum leak at the throttle housing, surge tank, inlet manifold or associated hoses – fuel injected models (Chapter 4).
- Excessive carbon build-up in engine (Chapter 2).
- Worn or damaged distributor or other ignition system component (Chapter 5).
- Carburettor worn or incorrectly adjusted (Chapter 1).

Whistling or wheezing noises
- Leaking inlet manifold or carburettor gasket – carburettor models (Chapter 4).
- Leaking inlet manifold or throttle housing/surge tank gasket – fuel injected models (Chapter 4).
- Leaking exhaust manifold gasket or pipe to manifold joint (Chapter 1).
- Leaking vacuum hose (Chapters 4, 5 and 9).
- Blowing cylinder head gasket (Chapter 2).

Tapping or rattling noises
- Incorrect valve clearances – B6 and E series engines (Chapter 1).
- Faulty HLA adjuster(s) – B3 and B5 engines (Chapter 2).
- Worn valve gear or camshaft (Chapter 2).
- Worn timing chain or tensioner – E series engines (Chapter 2).
- Ancillary component fault (water pump, alternator etc) (Chapters 3 and 12).

Knocking or thumping noises
- Worn big-end bearings (regular heavy knocking, perhaps less under load) (Chapter 2).
- Worn main bearings (rumbling and knocking, perhaps worsening under load) (Chapter 2).
- Piston slap (most noticeable when cold) (Chapter 2).
- Ancillary component fault (alternator, water pump etc) (Chapters 3 and 12).

2 Cooling system

Overheating
- Insufficient coolant in system (Chapter 3).
- Thermostat faulty (Chapter 3).
- Radiator core blocked or grille restricted (Chapter 3).
- Electric cooling fan or thermoswitch faulty (Chapter 3).
- Pressure cap faulty (Chapter 3).
- Water pump drivebelt worn, or incorrectly adjusted (Chapter 1).
- Ignition timing incorrect (Chapter 1).
- Inaccurate temperature gauge sender unit (Chapter 3).
- Air lock in cooling system (Chapter 1).

Overcooling
- Thermostat faulty (Chapter 3).
- Inaccurate temperature gauge sender unit (Chapter 3).

External coolant leakage
- Deteriorated or damaged hoses or hose clips (Chapter 1).
- Radiator core or heater matrix leaking (Chapter 3).
- Pressure cap faulty (Chapter 3).
- Water pump seal leaking (Chapter 3).
- Boiling due to overheating (Chapter 3).
- Core plug leaking (Chapter 2).

Internal coolant leakage
- Leaking cylinder head gasket (Chapter 2).
- Cracked cylinder head or cylinder bore (Chapter 2).

Corrosion
- Infrequent draining and flushing (Chapter 1).
- Incorrect antifreeze mixture or inappropriate type (Chapter 1).

3 Fuel and exhaust system

Excessive fuel consumption
- Air filter element dirty or clogged (Chapter 1).
- Carburettor worn or incorrectly adjusted (Chapter 4).
- Choke mechanism incorrectly adjusted or choke sticking – carburettor models (Chapter 4).
- Ignition timing incorrect (Chapter 1).
- Tyres underinflated (Chapter 1).

Fuel leakage and/or fuel odour
- Damaged or corroded fuel tank, pipes or connections (Chapter 1).
- Carburettor float chamber flooding (Chapter 4).

Fault diagnosis

Excessive noise or fumes from exhaust system
- Leaking exhaust system or manifold joints (Chapter 1).
- Leaking, corroded or damaged silencers or pipe (Chapter 1).
- Broken mountings causing body or suspension contact (Chapter 1).

4 Clutch

Pedal travels to floor – no pressure or very little resistance
- Broken clutch cable – cable operated clutch (Chapter 6).
- Low hydraulic fluid level – hydraulically operated clutch (Chapter 6).
- Incorrect clutch cable adjustment (Chapter 6).
- Air in the hydraulic system (Chapter 6).
- Broken clutch release bearing or fork (Chapter 6).
- Broken diaphragm spring in clutch pressure plate (Chapter 6).

Clutch fails to disengage (unable to select gears)
- Incorrect clutch cable adjustment (Chapter 6).
- Incorrect clutch pedal adjustment (Chapter 6).
- Clutch disc sticking on gearbox input shaft splines (Chapter 6).
- Clutch disc sticking to flywheel or pressure plate (Chapter 6).
- Faulty pressure plate assembly (Chapter 6).
- Gearbox input shaft seized in crankshaft spigot bearing (Chapter 2).
- Clutch release mechanism worn or incorrectly assembled (Chapter 6).

Clutch slips (engine speed increases with no increase in vehicle speed)
- Incorrect clutch cable adjustment (Chapter 6).
- Incorrect clutch pedal adjustment (Chapter 6).
- Hydraulic fluid level too high (Chapter 6).
- Clutch disc linings excessively worn (Chapter 6).
- Clutch disc linings contaminated with oil or grease (Chapter 6).
- Faulty pressure plate or weak diaphragm spring (Chapter 6).

Judder as clutch is engaged
- Clutch disc linings contaminated with oil or grease (Chapter 6).
- Clutch disc linings excessively worn (Chapter 6).
- Clutch cable sticking or frayed (Chapter 6).
- Clutch master or operating cylinder piston sticking (Chapter 6).
- Faulty or distorted pressure plate or diaphragm spring (Chapter 6).
- Worn or loose engine or gearbox mountings (Chapter 2).
- Clutch disc hub or gearbox input shaft splines worn (Chapter 6).

Noise when depressing or releasing clutch pedal
- Worn clutch release bearing (Chapter 6).
- Worn or dry clutch pedal bushes (Chapter 6).
- Faulty pressure plate assembly (Chapter 6).
- Pressure plate diaphragm spring broken (Chapter 6).
- Broken clutch disc cushioning springs (Chapter 6).

5 Manual gearbox

Noisy in neutral with engine running
- Input shaft bearings worn (noise apparent with clutch pedal released but not when depressed) (Chapter 7).*
- Clutch release bearing worn (noise apparent with clutch pedal depressed, possibly less when released) (Chapter 6).

Noisy in one particular gear
- Worn, damaged or chipped gear teeth (Chapter 7).*

Difficulty engaging gears
- Clutch fault (Chapter 6).
- Worn or damaged gear linkage (Chapter 7).
- Incorrectly adjusted gear linkage (Chapter 7).
- Worn synchroniser units (Chapter 7).*

Jumps out of gear
- Worn or damaged gear linkage (Chapter 7).
- Incorrectly adjusted gear linkage (Chapter 7).
- Worn synchroniser units (Chapter 7).*
- Worn selector forks (Chapter 7).*

Vibration
- Lack of oil (Chapter 1).
- Worn bearings (Chapter 7).*

Lubricant leaks
- Leaking differential output oil seal (Chapter 7).
- Leaking housing joint (Chapter 7).*
- Leaking input shaft oil seal (Chapter 7).*

* Although the corrective action necessary to remedy the symptoms described is beyond the scope of the home mechanic, the above information should be helpful in isolating the cause of the condition so that the owner can communicate clearly with a professional mechanic.

6 Automatic transmission

Note: *Due to the complexity of the automatic transmission, it is difficult for the home mechanic to properly diagnose and service this unit. For problems other than the following, the vehicle should be taken to a dealer service department or automatic transmission specialist.*

Fluid leakage
- Automatic transmission fluid is usually deep red in colour. Fluid leaks should not be confused with engine oil which can easily be blown onto the transmission by air flow.
- To determine the source of a leak, first remove all built-up dirt and grime from the transmission housing and surrounding areas using a degreasing agent or by steam cleaning. Drive the vehicle at low speed so air flow will not blow the leak far from its source. Raise and support the vehicle and determine where the leak is coming from. The following are common areas of leakage.

    (a) Oil pan (Chapters 1 and 7).
    (b) Dipstick tube (Chapters 1 and 7).
    (c) Transmission to oil cooler fluid pipes/unions (Chapter 7).

Transmission fluid brown or has burned smell
- Transmission fluid level low or fluid in need of renewal (Chapter 1).

General gear selection problems
- Chapter 7, Part B, deals with checking and adjusting the selector linkage on automatic transmissions. The following are common problems which may be caused by a poorly adjusted linkage.

    (a) Engine starting in gears other than Park or Neutral.
    (b) Indicator on gear selector lever pointing to a gear other than the one actually being used.
    (c) Vehicle moves when in Park or Neutral.
    (d) Poor gear shift quality or erratic gear changes.

- Refer to Chapter 7, Part B for the selector linkage adjustment procedure.

Transmission will not downshift (kickdown) with accelerator pedal fully depressed
- Low transmission fluid level (Chapter 1).
- Incorrect selector mechanism adjustment (Chapter 7, Part B).
- Incorrect kickdown switch adjustment (Chapter 7, Part B).

Fault diagnosis

Engine will not start in any gear, or starts in gears other than Park or Neutral
- Incorrect starter/inhibitor switch adjustment (Chapter 7, Part B).
- Incorrect selector mechanism adjustment (Chapter 7, Part B).

Transmission slips, shifts roughly, is noisy or has no drive in forward or reverse gears
- There are many probable causes for the above problems, but the home mechanic should be concerned with only one possibility – fluid level. Before taking the vehicle to a dealer or transmission specialist, check the fluid level and condition of the fluid as described in Chapter 1. Correct the fluid level as necessary or change the fluid and filter if needed. If the problem persists, professional help will be necessary.

7 Driveshafts

Clicking or knocking noise on turns (at slow speed on full lock)
- Lack of constant velocity joint lubricant (Chapter 8).
- Worn outer constant velocity joint (Chapter 8).

Vibration when accelerating or decelerating
- Worn inner constant velocity joint (Chapter 8).
- Bent or distorted driveshaft (Chapter 8).

8 Braking system

Note: *Before assuming that a brake problem exists, make sure that the tyres are in good condition and correctly inflated, the front wheel alignment is correct and the vehicle is not loaded with weight in an unequal manner.*

Vehicle pulls to one side under braking
- Worn, defective, damaged or contaminated front or rear brake pads/shoes on one side (Chapter 1).
- Seized or partially seized front or rear brake caliper/wheel cylinder piston (Chapter 9).
- A mixture of brake pad/shoe lining materials fitted between sides (Chapter 1).
- Brake caliper mounting bolts loose (Chapter 9).
- Rear brake backplate mounting bolts loose (Chapter 9).
- Worn or damaged steering or suspension components (Chapter 10).
- Faulty dual proportioning valve (Chapter 9).

Noise (grinding or high-pitched squeal) when brakes applied
- Brake pad or shoe friction lining material worn down to metal backing (Chapter 1).
- Excessive corrosion of brake disc or drum. (May be apparent after the vehicle has been standing for some time (Chapter 1).
- Foreign object (stone chipping etc) trapped between brake disc and splash shield (Chapter 1).

Excessive brake pedal travel
- Inoperative rear brake self-adjust mechanism (Chapter 1).
- Faulty master cylinder (Chapter 9).
- Air in hydraulic system (Chapter 9).
- Faulty vacuum servo unit (Chapter 9).

Brake pedal feels spongy when depressed
- Air in hydraulic system (Chapter 9).
- Deteriorated flexible rubber brake hoses (Chapter 9).
- Master cylinder mounting nuts loose (Chapter 9).
- Faulty master cylinder (Chapter 9).

Excessive brake pedal effort required to stop vehicle
- Faulty vacuum servo unit (Chapter 9).
- Disconnected, damaged or insecure brake servo vacuum hose (Chapter 9).
- Primary or secondary hydraulic circuit failure (Chapter 9).
- Seized brake caliper or wheel cylinder piston(s) (Chapter 9).
- Brake pads or brake shoes incorrectly fitted (Chapter 9).
- Incorrect grade of brake pads or brake shoes fitted (Chapter 1).
- Brake pads or brake shoe linings contaminated (Chapter 1).

Judder felt through brake pedal or steering wheel when braking
- Excessive run-out or distortion of front discs or rear drums/discs (Chapter 9).
- Brake pad or brake shoe linings worn (Chapter 1).
- Brake caliper or rear brake backplate mounting bolts loose (Chapter 9).
- Wear in suspension or steering components or mountings (Chapter 10).

Brakes binding
- Seized brake caliper or wheel cylinder piston(s) (Chapter 9).
- Incorrectly adjusted handbrake mechanism or linkage (Chapter 1).
- Faulty master cylinder (Chapter 9).

Rear wheels locking under normal braking
- Rear brake shoe linings contaminated (Chapter 1).
- Faulty dual proportioning valve (Chapter 9).

9 Suspension and steering systems

Note: *Before diagnosing suspension or steering faults, be sure that the trouble is not due to incorrect tyre pressures, mixtures of tyre types or binding brakes.*

Vehicle pulls to one side
- Defective tyre (Chapter 1).
- Excessive wear in suspension or steering components (Chapter 10).
- Incorrect front wheel alignment (Chapter 10).
- Accident damage to steering or suspension components (Chapter 10).

Wheel wobble and vibration
- Front roadwheels out of balance (vibration felt mainly through the steering wheel) (Chapter 10).
- Rear roadwheels out of balance (vibration felt throughout the vehicle) (Chapter 10).
- Roadwheels damaged or distorted (Chapter 1).
- Faulty or damaged tyre (Chapter 1).
- Worn steering or suspension joints, bushes or components (Chapter 10).
- Wheel bolts loose (Chapter 1).

Excessive pitching and/or rolling around corners or during braking
- Defective shock absorbers (Chapter 10).
- Broken or weak coil spring and/or suspension component (Chapter 10).
- Worn or damaged anti-roll bar or mountings (Chapter 10).

Wandering or general instability
- Incorrect front wheel alignment (Chapter 10).
- Worn steering or suspension joints, bushes or components (Chapter 10).
- Roadwheels out of balance (Chapter 1).
- Faulty or damaged tyre (Chapter 1).
- Wheel bolts loose (Chapter 1).
- Defective shock absorbers (Chapter 10).

Fault diagnosis

Excessively stiff steering
- Lack of steering gear lubricant (Chapter 10).
- Seized tie-rod end balljoint or suspension balljoint (Chapter 10).
- Broken or incorrectly adjusted power steering pump drivebelt (where fitted) (Chapter 1).
- Incorrect front wheel alignment (Chapter 10).
- Steering rack or column bent or damaged (Chapter 10).

Excessive play in steering
- Worn steering column universal joint(s) or intermediate coupling (Chapter 10).
- Worn steering tie-rod end balljoints (Chapter 10).
- Worn rack and pinion steering gear (Chapter 10).
- Worn steering or suspension joints, bushes or components (Chapter 10).

Lack of power assistance (where equipped)
- Broken or incorrectly adjusted power steering pump drivebelt (Chapter 1).
- Incorrect power steering fluid level (Chapter 1).
- Restriction in power steering fluid hoses (Chapter 10).
- Faulty power steering pump (Chapter 10).
- Faulty rack and pinion steering gear (Chapter 10).

Tyre wear excessive

Tyres worn on inside or outside edges
- Tyres underinflated (wear on both edges) (Chapter 1).
- Incorrect camber or castor angles (wear on one edge only) (Chapter 10).
- Worn steering or suspension joints, bushes or components (Chapter 10).
- Excessively hard cornering.
- Accident damage.

Tyre treads exhibit feathered edges
- Incorrect toe setting (Chapter 10).

Tyres worn in centre of tread
- Tyres overinflated (Chapter 1).

Tyres worn on inside and outside edges
- Tyres underinflated (Chapter 1).

Tyres worn unevenly
- Tyres out of balance (Chapter 1).
- Excessive wheel or tyre run-out (Chapter 1).
- Worn shock absorbers (Chapter 10).
- Faulty tyre (Chapter 1).

10 Electrical system

Note: *For problems associated with the starting system, refer to the faults listed under the 'Engine' heading earlier in this Section.*

Battery will not hold a charge for more than a few days
- Battery defective internally (Chapter 12).
- Battery electrolyte level low (Chapter 1).
- Battery terminal connections loose or corroded (Chapter 12).
- Alternator drivebelt worn or incorrectly adjusted (Chapter 1).
- Alternator not charging at correct output (Chapter 12).
- Alternator or voltage regulator faulty (Chapter 12).
- Short-circuit causing continual battery drain (Chapter 12).

Ignition warning light remains illuminated with engine running
- Alternator drivebelt broken, worn, or incorrectly adjusted (Chapter 1).
- Alternator brushes worn, sticking, or dirty (Chapter 12).
- Alternator brush springs weak or broken (Chapter 12).
- Internal fault in alternator or voltage regulator (Chapter 12).
- Broken, disconnected, or loose wiring in charging circuit (Chapter 12).

Ignition warning light fails to come on
- Warning light bulb blown (Chapter 12).
- Broken, disconnected, or loose wiring in warning light circuit (Chapter 12).
- Alternator faulty (Chapter 12).

Lights inoperative
- Bulb blown (Chapter 12).
- Corrosion of bulb or bulbholder contacts (Chapter 12).
- Blown fuse (Chapter 12).
- Faulty relay (Chapter 12).
- Broken, loose, or disconnected wiring (Chapter 12).
- Faulty switch (Chapter 12).

Instrument readings inaccurate or erratic

Instrument readings increase with engine speed
- Faulty voltage regulator (Chapter 12).

Fuel or temperature gauge give no reading
- Faulty gauge sender unit (Chapters 3 or 4).
- Wiring open-circuit (Chapter 12).
- Faulty gauge (Chapter 12).

Fuel or temperature gauges give continuous maximum reading
- Faulty gauge sender unit (Chapters 3 or 4).
- Wiring short-circuit (Chapter 12).
- Faulty gauge (Chapter 12).

Horn inoperative or unsatisfactory in operation

Horn operates all the time
- Horn push either earthed or stuck down (Chapter 12).
- Horn cable to horn push earthed (Chapter 12).

Horn fails to operate
- Blown fuse (Chapter 12).
- Cable or cable connections loose, broken or disconnected (Chapter 12).
- Faulty horn (Chapter 12).

Horn emits intermittent or unsatisfactory sound
- Cable connections loose (Chapter 12).
- Horn mountings loose (Chapter 12).
- Faulty horn (Chapter 12).

Windscreen/tailgate wipers inoperative or unsatisfactory in operation

Wipers fail to operate or operate very slowly
- Wiper blades stuck to screen or linkage seized or binding (Chapter 12).
- Blown fuse (Chapter 12).
- Cable or cable connections loose, broken or disconnected (Chapter 12).
- Faulty relay (Chapter 12).
- Faulty wiper motor (Chapter 12).

Wiper blades sweep over too large or too small an area of the glass
- Wiper arms incorrectly positioned on spindles (Chapter 1).
- Excessive wear of wiper linkage (Chapter 1).
- Wiper motor or linkage mountings loose or insecure (Chapter 12).

Wiper blades fail to clean the glass effectively
- Wiper blade rubbers worn or perished (Chapter 1).
- Wiper arm tension springs broken or arm pivots seized (Chapter 1).
- Insufficient windscreen washer additive to adequately remove road film (Chapter 1).

Windscreen/tailgate washers inoperative or unsatisfactory in operation

One or more washer jets inoperative
- Blocked washer jet (Chapter 1).
- Disconnected, kinked or restricted fluid hose (Chapter 12).
- Insufficient fluid in washer reservoir (Chapter 1).

Washer pump fails to operate
- Broken or disconnected wiring or connections (Chapter 12).
- Blown fuse (Chapter 12).
- Faulty washer switch (Chapter 12).
- Faulty washer pump (Chapter 12).

Washer pump runs for some time before fluid is emitted from jets
- Faulty one-way valve in fluid supply hose (Chapter 12).

Electric windows inoperative or unsatisfactory in operation

Window glass will only move in one direction
- Faulty switch (Chapter 12).

Window glass slow to move
- Incorrectly adjusted door glass guide channels (Chapter 12).
- Regulator seized or damaged, or in need of lubrication (Chapter 12).
- Door internal components or trim fouling regulator (Chapter 12).
- Faulty motor (Chapter 12).

Window glass fails to move
- Incorrectly adjusted door glass guide channels (Chapter 12).
- Blown fuse (Chapter 12).
- Faulty relay (Chapter 12).
- Broken or disconnected wiring or connections (Chapter 12).
- Faulty motor (Chapter 12).

Central locking system inoperative or unsatisfactory in operation

Complete system failure
- Blown fuse (Chapter 12).
- Faulty relay (Chapter 12).
- Broken or disconnected wiring or connections (Chapter 12).

Latch locks but will not unlock, or unlocks but will not lock
- Faulty master switch (Chapter 12).
- Broken or disconnected latch operating rods or levers (Chapter 12).
- Faulty relay (Chapter 12).

One solenoid/motor fails to operate
- Broken or disconnected wiring or connections (Chapter 12).
- Faulty solenoid/motor (Chapter 12).
- Broken, binding or disconnected latch operating rods or levers (Chapter 12).
- Fault in door latch (Chapter 12).

MOT test checks

Introduction

Motor vehicle testing has been compulsory in Great Britain since 1960 when the Motor Vehicle (Tests) Regulations were first introduced. At that time testing was only applicable to vehicles ten years old or older, and the test itself only covered lighting equipment, braking systems and steering gear. Current vehicle testing is far more extensive and, in the case of private cars, is now an annual inspection commencing three years after the date of first registration.

This Section is intended as a guide to getting your car through the MOT test. It lists all the relevant testable items, how to check them yourself, and what is likely to cause the vehicle to fail. Obviously it will not be possible to examine the vehicle to the same standard as the professional MOT tester who will be highly experienced in this work and will have all the necessary equipment available. However, working through the following checks will provide a good indication as to the condition of the vehicle and will enable you to identify any problem areas before submitting the vehicle for the test. Where a component is found to need repair or renewal, a cross reference is given to the relevant Chapter in the Manual where further information and the appropriate repair procedures will be found.

The following checks have been sub-divided into three categories as follows.

(a) Checks carried out from the driver's seat.
(b) Checks carried out with the car on the ground.
(c) Checks carried out with the car raised and with the wheels free to rotate.

In most cases the help of an assistant will be necessary to carry out these checks thoroughly.

Checks carried out from the driver's seat

Handbrake (Chapter 9)

Test the operation of the handbrake by pulling on the lever until the handbrake is in the normal fully-applied position. Ensure that the travel of the lever (the number of clicks of the ratchet) is not excessive before full resistance of the braking mechanism is felt. If so this would indicate incorrect adjustment of the rear brakes or incorrectly adjusted handbrake cables. With the handbrake fully applied, tap the lever sideways and make sure that it does not release which would indicate wear in the ratchet and pawl. Release the handbrake and move the lever from side to side to check for excessive wear in the pivot bearing. Check the security of the lever mountings and make sure that there is no corrosion of any part of the body structure within 30 cm (12 in) of the lever mounting. If the lever mountings cannot be readily seen from inside the vehicle, carry out this check later when working underneath.

Footbrake (Chapter 9)

Check that the brake pedal is sound without visible defects such as excessive wear of the pivot bushes or broken or damaged pedal pad. Check also for signs of fluid leaks on the pedal, floor or carpets which would indicate failed seals in the brake master cylinder. Depress the brake pedal slowly at first, then rapidly until sustained pressure can be held. Maintain this pressure and check that the pedal does not creep down to the floor which would again indicate problems with the master cylinder. Release the pedal, wait a few seconds then depress it once until firm resistance is felt. Check that this resistance occurs near the top of the pedal travel. If the pedal travels nearly to the floor before firm resistance is felt, this would indicate incorrect brake adjustment resulting in 'insufficient reserve travel' of the footbrake. If firm resistance cannot be felt, ie the pedal feels spongy, this would indicate that air is present in the hydraulic system which will necessitate complete bleeding of the system. Check that the servo unit is operating correctly by depressing the brake pedal several times to exhaust the vacuum. Keep the pedal depressed and start the engine. As soon as the engine starts, the brake pedal resistance will be felt to alter. If this is not the case, there may be a leak from the brake servo vacuum hose, or the servo unit itself may be faulty.

Steering wheel and column (Chapter 10)

Examine the steering wheel for fractures or looseness of the hub, spokes or rim. Move the steering wheel from side to side and then up and down, in relation to the steering column. Check that the steering wheel is not loose on the column, indicating wear in the column splines or a loose steering wheel retaining nut. Continue moving the steering wheel as before, but also turn it slightly from left to right. Check that there is no abnormal movement of the steering wheel, indicating excessive wear in the column upper support bearing, universal joint(s) or flexible coupling.

Electrical equipment (Chapter 12)

Switch on the ignition and operate the horn. The horn must operate and produce a clear sound audible to other road users. Note that a gong, siren or two-tone horn fitted as an alternative to the manufacturer's original equipment is not acceptable.

Check the operation of the windscreen washers and wipers. The washers must operate with adequate flow and pressure and with the jets adjusted so that the liquid strikes the windscreen near the top of the glass.

Operate the windscreen wipers in conjunction with the washers and check that the blades cover their designed sweep of the windscreen without smearing. The blades must effectively clean the glass so that the driver has an adequate view of the road ahead and to the front nearside and offside of the vehicle. If the screen smears or does not clean adequately, it is advisable to renew the wiper blades before the MOT test.

Depress the footbrake with the ignition switched on and have your assistant check that both rear stop lights operate, and are extinguished when the footbrake is released. If one stop light fails to operate it is likely that a bulb has blown or there is a poor electrical contact at, or near the bulbholder. If both stop lights fail to operate, check for a blown fuse, faulty stop light switch or possibly two blown bulbs. If the lights stay on when the brake pedal is released, it is possible that the switch is at fault.

Seat belts (Chapter 11)

Note: *The following checks are applicable to the seat belts provided for the driver's seat and front passenger's seat. Both seat belts must be of a type that will restrain the upper part of the body; lap belts are not acceptable.*

Carefully examine the seat belt webbing for cuts or any signs of serious fraying or deterioration. If the seat belt is of the retractable type, pull the belt all the way out and examine the full extent of the webbing.

Fasten and unfasten the belt ensuring that the locking mechanism holds securely and releases properly when intended. If the belt is of the retractable type, check also that the retracting mechanism operates correctly when the belt is released.

Check the security of all seat belt mountings and attachments which are accessible, without removing any trim or other components, from inside the car (photo). Any serious corrosion, fracture or distortion of the body structure within 30 cm (12 in) of any mounting point will cause the vehicle to fail. Certain anchorages will not be accessible, or even visible from inside the car and in this instance further checks should be carried out later, when working underneath. If any part of the seat belt mechanism is attached to the front seat, then the seat mountings are treated as anchorages and must also comply as above.

Checks carried out with the car on the ground

Electrical equipment (Chapter 12)

Switch on the side lights and check that both front and rear side lights are illuminated and that the lenses and reflectors are secure and

Check the security of all seat belt mountings

Check the flexible brake hoses for cracks or deterioration

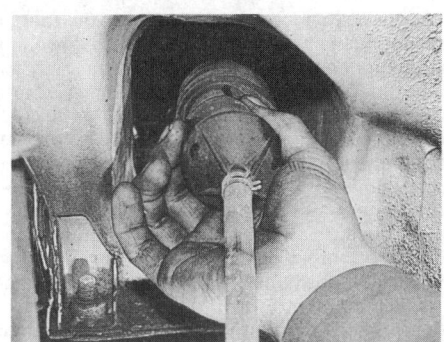
Examine steering gear rack rubber gaiters condition and security

undamaged. This is particularly important at the rear where a cracked or damaged lens would allow a white light to show to the rear, which is unacceptable. It is also worth noting that any lens that is excessively dirty, either inside or out, such that the light intensity is reduced, could also constitute a fail.

Switch on the headlamps and check that both dipped beam and main beam units are operating correctly and at the same light intensity. If either headlamp shows signs of dimness, this is usually attributable to a poor earth connection or severely corroded internal reflector. Inspect the headlamp lenses for cracks or stone damage. Any damage to the headlamp lens will normally constitute a fail, but this is very much down to the tester's discretion. Bear in mind that with all light units they must operate correctly when first switched on. It is not acceptable to tap a light unit to make it operate.

The headlamps must be aligned so as not to dazzle other road users when switched to dipped beam. This can only be accurately checked using optical beam setting equipment so if you have any doubts about the headlamp alignment, it is advisable to have this professionally checked and if necessary reset, before the MOT test.

With the ignition switched on, operate the direction indicators and check that they show a white or amber light to the front and red or amber light to the rear, that they flash at the rate of between one and two flashes per second and that the 'tell-tale' on the instrument panel also functions. Examine the lenses for cracks or damage as described previously.

Footbrake (Chapter 9)

From within the engine compartment examine the brake pipes for signs of leaks, corrosion, insecurity, chafing or other damage and check the master cylinder and servo unit for leaks, security of their mountings or excessive corrosion in the vicinity of the mountings.

Turn the steering as necessary so that the right-hand front brake flexible hose can be examined. Inspect the hose carefully for any sign of cracks or deterioration of the rubber. This will be most noticeable if the hose is bent in half and is particularly common where the rubber portion enters the metal end fitting (photo). Turn the steering onto full left then full right lock and ensure that the hose does not contact the wheel, tyre, or any part of the steering or suspension mechanism. While your assistant depresses the brake pedal firmly, check the hose for any bulges or fluid leaks under pressure. Now repeat these checks on the left-hand front hose. Should any damage or deterioration be noticed, renew the hose.

Steering mechanism and suspension (Chapter 10)

Have your assistant turn the steering wheel from side to side slightly, up to the point where the steering gear just begins to transmit this movement to the roadwheels. Check for excessive free play between the steering wheel and the steering gear which would indicate wear in the steering column joints, wear or insecurity of the steering column to steering gear coupling, or insecurity, incorrect adjustment, or wear in the steering gear itself. Generally speaking, free play greater than 1.3 cm (0.5 in) for vehicles with rack and pinion type steering or 7.6 cm (3.0 in) for vehicles with steering box mechanisms should be considered excessive.

Have your assistant turn the steering wheel more vigorously in each direction up to the point where the roadwheels just begin to turn. As this is done, carry out a complete examination of all the steering joints, linkages, fittings and attachments. Any component that shows signs of wear, damage, distortion, or insecurity should be renewed or attended to accordingly. On vehicles equipped with power steering also check that the power steering pump is secure, that the pump drivebelt is in satisfactory condition and correctly adjusted, that there are no fluid leaks or damaged hoses, and that the system operates correctly. Additional checks can be carried out later with the vehicle raised when there will be greater working clearance underneath.

Check that the vehicle is standing level and at approximately the correct ride height. Ensure that there is sufficient clearance between the suspension components and the bump stops to allow full suspension travel over bumps.

Shock absorbers (Chapter 10)

Depress each corner of the car in turn and then release it. If the shock absorbers are in good condition the corner of the car will rise and then settle in its normal position. If there is no noticeable damping effect from the shock absorber, and the car continues to rise and fall, then the shock absorber is defective.

Exhaust system (Chapter 1)

Start the engine and with your assistant holding a rag over the tailpipe, check the entire system for leaks which will appear as a rhythmic fluffing or hissing sound at the source of the leak. Check the effectiveness of the silencer by ensuring that the noise produced is of a level to be expected from a vehicle of similar type. Providing that the system is structurally sound, it is acceptable to cure a leak using a proprietary exhaust system repair kit or similar method.

Checks carried out with the car raised and with the wheels free to rotate

Jack up the front and rear of the car and securely support it on axle stands positioned at suitable load bearing points under the vehicle structure. Position the stands clear of the suspension assemblies and ensure that the wheels are clear of the ground and that the steering can be turned onto full right and left lock.

Steering mechanism (Chapter 10)

Examine the steering rack rubber gaiters for signs of splits, lubricant leakage or insecurity of the retaining clips (photo). If power steering is fitted, check for signs of deterioration, damage, chafing or leakage of the fluid hoses, pipes or connections. Also check for excessive stiffness or binding of the steering, a missing split pin or locking device or any severe corrosion of the body structure within 30 cm (12 in) of any steering component attachment point.

Have your assistant turn the steering onto full left then full right lock. Check that the steering turns smoothly without undue tightness or roughness and that no part of the steering mechanism, including a wheel or tyre, fouls any brake flexible or rigid hose or pipe, or any part of the body structure.

Front and rear suspension and wheel bearings (Chapter 10)

Starting at the front right-hand side of the vehicle, grasp the roadwheel at the 3 o'clock and 9 o'clock positions and shake it vigorously. Check for any free play at the wheel bearings, suspension ball joints, or suspension mountings, pivots and attachments. Check also for any serious deterioration of the rubber or metal casing of any mounting bushes, or any distortion, deformation or severe corrosion of any components (photo). Look for missing split pins, tab washers or other locking devices on any mounting or attachment, or any severe corrosion of the vehicle structure within 30 cm (12 in) of any suspension component attachment point. If any excess free play is suspected at a component pivot point, this can be confirmed by using a large screwdriver or similar tool and levering between the mounting and the component attachment. This will confirm whether the wear is in the pivot bush, its retaining bolt or in the mounting itself (the bolt holes can often become elongated). Now grasp the wheel at the 12 o'clock and 6 o'clock positions, shake it vigorously and repeat the previous inspection (photo). Rotate the wheel and check for roughness or tightness of the front wheel bearing such that imminent failure of the bearing is indicated. Carry out all the above checks at the other front wheel and then at both rear wheels. Note, however, that the condition of the rear wheel bearings is not actually part of the MOT test, but if they are at all suspect, it is likely that this will be brought to the owner's attention at the time of the test.

Roadsprings and shock absorbers (Chapter 10)

On vehicles with strut type suspension units, examine the strut assembly for signs of fluid leakage, corrosion or severe pitting of the piston rod or damage to the casing. Check also for security of the mounting points.

If coil springs are fitted check that the spring ends locate correctly in their spring seats, that there is no severe corrosion of the spring and that it is not cracked, broken or in any way damaged.

If the vehicle is fitted with leaf springs, check that all leaves are intact, that the axle is securely attached to each spring and that there is no wear or deterioration of the spring eye mountings, bushes, and shackles.

The same general checks apply to vehicles fitted with other suspension types, such as torsion bars, hydraulic displacer units etc. In all cases ensure that all mountings and attachments are secure, that there are no signs of excessive wear, corrosion, cracking, deformation or damage to any component or bush, and that there are no fluid leaks or damaged hoses or pipes (hydraulic types).

Inspect the shock absorbers for signs of fluid leakage, excessive wear of the mounting bushes or attachments or damage to the body of the unit.

Driveshafts (Chapter 8)

With the steering turned onto full lock, rotate each front wheel in turn and inspect the constant velocity joint gaiters for splits or damage (photo). Also check the gaiter is securely attached to its respective housings by clips or other methods of retention.

Continue turning the wheel and check that each driveshaft is straight with no sign of damage.

Braking system (Chapter 9)

If possible, without dismantling, check for wear of the brake pads and the condition of the discs. Ensure that the friction lining material has not worn excessively and that the discs are not fractured, pitted, scored or worn excessively.

Carefully examine all the rigid brake pipes underneath the car and the flexible hoses at the rear. Look for signs of excessive corrosion, chafing or insecurity of the pipes and for signs of bulging under pressure, chafing, splits or deterioration of the flexible hoses.

Look for signs of hydraulic fluid leaks at the brake calipers or on the brake backplates indicating failed hydraulic seals in the components concerned.

Slowly spin each wheel while your assistant depresses the footbrake then releases it. Ensure that each brake is operating and that the wheel is free to rotate when the pedal is released.

Examine the handbrake mechanism and check for signs of frayed or broken cables, excessive corrosion or wear or insecurity of the linkage (photo). Have your assistant operate the handbrake while you check that the mechanism works on each relevant wheel and releases fully without binding.

Exhaust system (Chapter 1)

Starting at the front, examine the exhaust system over its entire length checking for any damaged, broken or missing mountings, security of the pipe retaining clamps and condition of the system with regard to rust and corrosion (photo).

Check all rubber suspension mounting bushes (arrowed) for damage or deterioration

Shake the roadwheel vigorously to check for excess play in the wheel bearings and suspension components

Inspect constant velocity joint gaiters for splits or damage

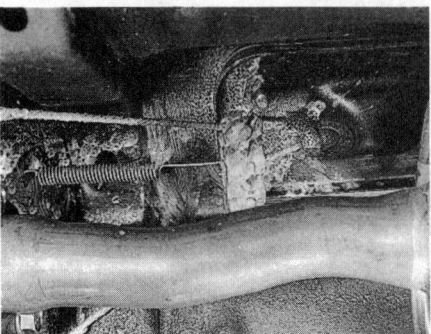

Check the handbrake mechanism for signs of frayed or broken cables or insecurity of the linkage

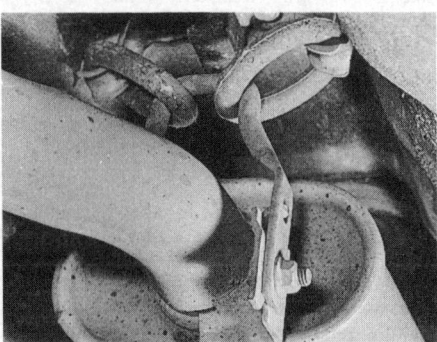

Check the condition of the exhaust system paying particular attention to the mountings

Wheels and tyres (Chapter 10)

Carefully examine each tyre in turn on both the inner and outer walls and over the whole of the tread area and check for signs of cuts, tears, lumps, bulges, separation of the tread and exposure of the ply or cord due to wear or other damage. Check also that the tyre bead is correctly seated on the wheel rim and that the tyre valve is sound and properly seated. Spin the wheel and check that it is not excessively distorted or damaged particularly at the bead rim. Check that the tyres are of the correct size for the car and that they are of the same size and type on each axle. They should also be inflated to the specified pressures.

Using a suitable gauge check the tyre tread depth. The current legal requirement states that the tread pattern must be visible over the whole tread area and must be of a minimum depth of 1.6 mm over at least three-quarters of the tread width. It is acceptable for some wear of the inside or outside edges of the tyre to be apparent but this wear must be in one even circumferential band and the tread must be visible. Any excessive wear of this nature may indicate incorrect front wheel alignment which should be checked before the tyre becomes excessively worn. See Chapters 1 and 10 for further information on tyre wear patterns and front wheel alignment.

Body corrosion

Check the condition of the entire vehicle structure for signs of corrosion in any load bearing areas. For the purpose of the MOT test all chassis box sections, side sills, subframes, crossmembers, pillars, suspension, steering, braking system and seat belt mountings and anchorages should all be considered as load bearing areas. As a general guide, any corrosion which has seriously reduced the metal thickness of a load bearing area to weaken it, is likely to cause the vehicle to fail. Should corrosion of this nature be encountered, professional repairs are likely to be needed.

Chapter 1 Routine maintenance and servicing

Contents

Lubricants, fluids and capacities
Maintenance schedule
Maintenance procedures

Bodywork	10	Engine	1
Braking system	8	Fuel and exhaust systems	3
Clutch	5	Ignition system	4
Cooling, heating and ventilation systems	2	Manual gearbox and automatic transmission	6
Driveshafts	7	Suspension and steering	9
Electrical system	11		

Specifications

Engine
Oil filter type .... Champion F110
Valve clearances (warm engine – between valve stem and rocker):
   Inlet .... 0.25 mm
   Exhaust .... 0.30 mm

Cooling system
Antifreeze mixtures: **Antifreeze**   **Water**
   Protection to –16°C (3°F) 35%   65%
   Protection to –35°C (–31°F) 50%   50%

Fuel and exhaust system
Air filter element type:
   1071 cc, 1323 cc, 1490 cc Saloon/Hatchback (except 1500 GT
   models) .... Champion W159
   1296 cc, 1490 cc Estate .... Champion W108
Fuel filter type:
   1071 cc, 1323 cc, 1490 cc Estate .... Champion L101
Idle speed – carburettor models:
   All manual gearbox models except 1500 GT .... 800 to 900 rpm
   1500 GT .... 850 to 950 rpm
   Automatic transmission models:
      Pre-September 1985 (selector in D position) .... 700 to 800 rpm
      September 1985 to July 1987 (selector in P position) ... 950 to 1050 rpm
      July 1987 on (selector in N position) .... 900 to 1000 rpm
Idle speed – fuel injected models .... 950 to 1050 rpm
Idle mixture CO content:
   Carburettor models .... 1.5 to 2.5%
   Fuel injected models .... 1.0 to 2.0%
Fuel octane requirement:
   Carburettor models except 1500 GT .... 91 RON unleaded or leaded
   1500 GT models .... 96 RON unleaded or leaded
   Fuel injected models .... 96 RON unleaded or leaded

Are your plugs trying to tell you something?

Normal.
Grey-brown deposits, lightly coated core nose. Plugs ideally suited to engine, and engine in good condition.

Heavy Deposits.
A build up of crusty deposits, light-grey sandy colour in appearance.
Fault: Often caused by worn valve guides, excessive use of upper cylinder lubricant, or idling for long periods.

Lead Glazing.
Plug insulator firing tip appears yellow or green/yellow and shiny in appearance.
Fault: Often caused by incorrect carburation, excessive idling followed by sharp acceleration. Also check ignition timing.

Carbon fouling.
Dry, black, sooty deposits.
Fault: over-rich fuel mixture.
Check: carburettor mixture settings, float level, choke operation, air filter.

Oil fouling.
Wet, oily deposits. Fault: worn bores/piston rings or valve guides; sometimes occurs (temporarily) during running-in period.

Overheating.
Electrodes have glazed appearance, core nose very white – few deposits. Fault: plug overheating. Check: plug value, ignition timing, fuel octane rating (too low) and fuel mixture (too weak).

Electrode damage.
Electrodes burned away; core nose has burned, glazed appearance. Fault: pre-ignition. Check: for correct heat range and as for 'overheating'.

Split core nose.
(May appear initially as a crack). Fault: detonation or wrong gap-setting technique.
Check: ignition timing, cooling system, fuel mixture (too weak).

WHY DOUBLE COPPER IS BETTER FOR YOUR ENGINE.

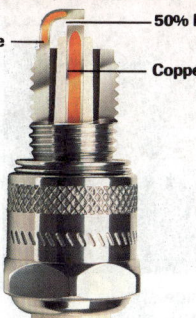

Unique Trapezoidal Copper Cored Earth Electrode — 50% Larger Spark Area — Copper Cored Centre Electrode

Champion Double Copper plugs are the first in the world to have copper core in both centre and earth electrode. This innovative design means that they run cooler by up to 100°C – giving greater efficiency and longer life. These double copper cores transfer heat away from the tip of the plug faster and more efficiently. Therefore, Double Copper runs at cooler temperatures than conventional plugs giving improved acceleration response and high speed performance with no fear of pre-ignition.

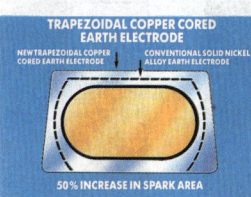
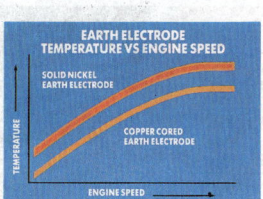

Champion Double Copper plugs also feature a unique trapezoidal earth electrode giving a 50% increase in spark area. This, together with the double copper cores, offers greatly reduced electrode wear, so the spark stays stronger for longer.

 FASTER COLD STARTING

 FOR UNLEADED OR LEADED FUEL

 ELECTRODES UP TO 100°C COOLER

 BETTER ACCELERATION RESPONSE

 LOWER EMISSIONS

 50% BIGGER SPARK AREA

 THE LONGER LIFE PLUG

Plug Tips/Hot and Cold.
Spark plugs must operate within well-defined temperature limits to avoid cold fouling at one extreme and overheating at the other.
Champion and the car manufacturers work out the best plugs for an engine to give optimum performance under all conditions, from freezing cold starts to sustained high speed motorway cruising.
Plugs are often referred to as hot or cold. With Champion, the higher the number on its body, the hotter the plug, and the lower the number the cooler the plug.

Plug Cleaning
Modern plug design and materials mean that Champion no longer recommends periodic plug cleaning. Certainly don't clean your plugs with a wire brush as this can cause metal conductive paths across the nose of the insulator so impairing its performance and resulting in loss of acceleration and reduced m.p.g.
However, if plugs are removed, always carefully clean the area where the plug seats in the cylinder head as grit and dirt can sometimes cause gas leakage.
Also wipe any traces of oil or grease from plug leads as this may lead to arcing.

This photographic sequence shows the steps taken to repair the dent and paintwork damage shown above. In general, the procedure for repairing a hole will be similar; where there are substantial differences, the procedure is clearly described and shown in a separate photograph.

First remove any trim around the dent, then hammer out the dent where access is possible. This will minimise filling. Here, after the large dent has been hammered out, the damaged area is being made slightly concave.

Next, remove all paint from the damaged area by rubbing with coarse abrasive paper or using a power drill fitted with a wire brush or abrasive pad. 'Feather' the edge of the boundary with good paintwork using a finer grade of abrasive paper.

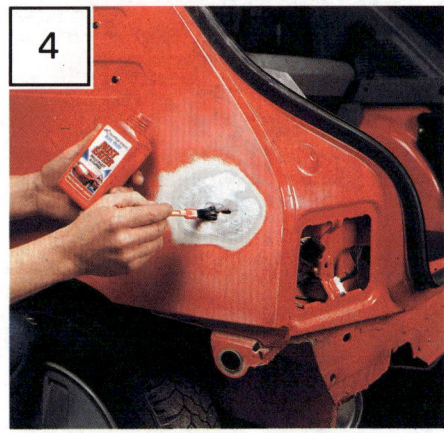

Where there are holes or other damage, the sheet metal should be cut away before proceeding further. The damaged area and any signs of rust should be treated with Turtle Wax Hi-Tech Rust Eater, which will also inhibit further rust formation.

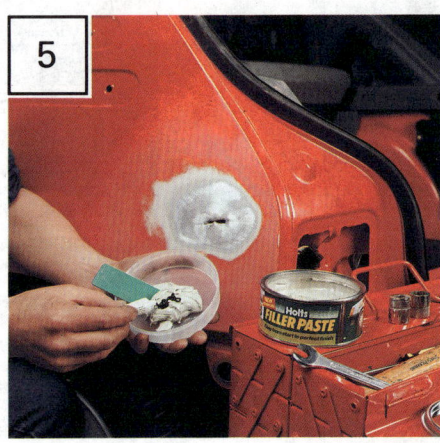

For a large dent or hole mix Holts Body Plus Resin and Hardener according to the manufacturer's instructions and apply around the edge of the repair. Press Glass Fibre Matting over the repair area and leave for 20-30 minutes to harden. Then ...

... brush more Holts Body Plus Resin and Hardener onto the matting and leave to harden. Repeat the sequence with two or three layers of matting, checking that the final layer is lower than the surrounding area. Apply Holts Body Plus Filler Paste as shown in Step 5B.

For a medium dent, mix Holts Body Plus Filler Paste and Hardener according to the manufacturer's instructions and apply it with a flexible applicator. Apply thin layers of filler at 20-minute intervals, until the filler surface is slightly proud of the surrounding bodywork.

For small dents and scratches use Holts No Mix Filler Paste straight from the tube. Apply it according to the instructions in thin layers, using the spatula provided. It will harden in minutes if applied outdoors and may then be used as its own knifing putty.

Use a plane or file for initial shaping. Then, using progressively finer grades of wet-and-dry paper, wrapped round a sanding block, and copious amounts of clean water, rub down the filler until glass smooth. 'Feather' the edges of adjoining paintwork.

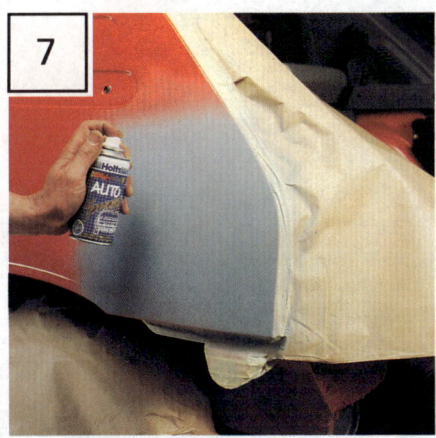

7 Protect adjoining areas before spraying the whole repair area and at least one inch of the surrounding sound paintwork with Holts Dupli-Color primer.

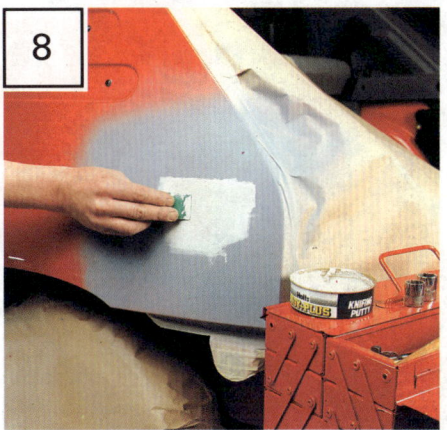

8 Fill any imperfections in the filler surface with a small amount of Holts Body Plus Knifing Putty. Using plenty of clean water, rub down the surface with a fine grade wet-and-dry paper – 400 grade is recommended – until it is really smooth.

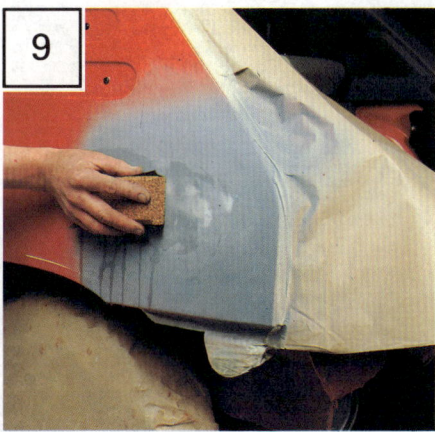

9 Carefully fill any remaining imperfections with knifing putty before applying the last coat of primer. Then rub down the surface with Holts Body Plus Rubbing Compound to ensure a really smooth surface.

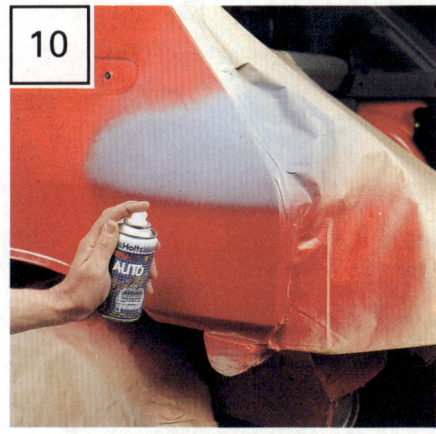

10 Protect surrounding areas from overspray before applying the topcoat in several thin layers. Agitate Holts Dupli-Color aerosol thoroughly. Start at the repair centre, spraying outwards with a side-to-side motion.

10A If the exact colour is not available off the shelf, local Holts Professional Spraymatch Centres will custom fill an aerosol to match perfectly.

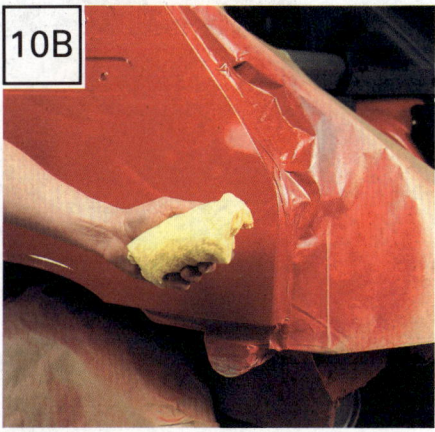

10B To identify whether a lacquer finish is required, rub a painted unrepaired part of the body with wax and a clean cloth.

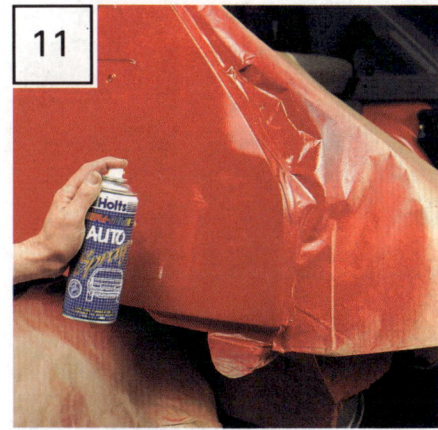

11 If *no* traces of paint appear on the cloth, spray Holts Dupli-Color clear lacquer over the repaired area to achieve the correct gloss level.

12 The paint will take about two weeks to harden fully. After this time it can be 'cut' with a mild cutting compound such as Turtle Wax Minute Cut prior to polishing with a final coating of Turtle Wax Extra.

14 When carrying out bodywork repairs, remember that the quality of the finished job is proportional to the time and effort expended.

HAYNES No1 for DIY

Haynes publish a wide variety of books besides the world famous range of *Haynes Owners Workshop Manuals*. They cover all sorts of DIY jobs. Specialist books such as the *Improve and Modify* series and the *Purchase and DIY Restoration Guides* give you all the information you require to carry out everything from minor modifications to complete restoration on a number of popular cars. In addition there are the publications dealing with specific tasks, such as the *Car Bodywork Repair Manual* and the *In-Car Entertainment Manual*. The *Household DIY* series gives clear step-by-step instructions on how to repair everyday household objects ranging from toasters to washing machines.

Whether it is under the bonnet or around the home there is a Haynes Manual that can help you save money. Available from motor accessory stores and bookshops or direct from the publisher.

Ignition system – general
Firing order 1-3-4-2
Location of No 1 cylinder .... Crankshaft pulley end

Contact breaker ignition system
Contact breaker points gap .... 0.45 to 0.55 mm
Dwell angle .... 49° to 55°
Ignition timing (vacuum hose connected):
   E1 engined models .... 3° to 5° BTDC at specified idle speed
   E3 and E5 engined models .... 5° to 7° BTDC at specified idle speed
Spark plugs:
   E1 and E5 engined models .... Champion RN11YCC or RN11YC
   E3 engined models .... Champion RN9YCC or RN9YC
Electrode gap .... 0.8 mm
Spark plug HT leads:
   All Saloon models .... Champion CLS08
   All Estate models .... Champion CLS13

Electronic ignition system
Ignition timing:
   Pre-September 1985 models (vacuum hose connected):
      E1 engined models .... 3° to 5° BTDC at specified idle speed
      E3 and E5 engined models .... 5° to 7° BTDC at specified idle speed
   September 1985 to July 1987 models:
      E series engined models (vacuum hose connected) ...... 3° to 5° BTDC at specified idle speed
      B6 engined models (vacuum hose disconnected) .... 5° to 7° BTDC at specified idle speed
   July 1987 onwards models (vacuum hose disconnected):
      B3 engined models .... 1° to 3° BTDC at specified idle speed
      B5 and B6 engined models .... 5° to 7° BTDC at specified idle speed
Spark plugs:
   E1 and E5 engined models .... Champion RN11YCC or RN11YC
   E3 and all B series engined models .... Champion RN9YCC or RN9YC
Electrode gap .... 0.8 mm
Spark plug HT leads:
   All Saloon models .... Champion CLS08
   All Estate models .... Champion CLS13

Clutch
Clutch pedal height:
   Pre-September 1985 models .... 225 to 230 mm
   September 1985 onwards models with cable operated clutch .... 215 to 220 mm
   Hydraulically operated clutch .... 229 to 234 mm
Clutch pedal free play:
   Pre-September 1985 models .... 11 to 17 mm
   September 1985 onwards models with cable operated clutch .... 9 to 15 mm
   Hydraulically operated clutch .... 0.6 to 3.0 mm
Minimum distance from pedal to floor (pedal depressed):
   Pre-September 1985 models .... 80 mm
   September 1985 onwards models with cable operated clutch .... 85 mm
   Hydraulically operated clutch .... 82 mm

Braking system
Brake pedal height:
   Pre-September 1985 models .... 215 to 220 mm
   September 1985 onwards models .... 219 to 224 mm
Brake pedal free play:
   Pre-September 1985 models .... 7 to 9 mm
   September 1985 onwards models .... 4 to 7 mm
Minimum distance from brake pedal to floor (pedal depressed) .... 83 mm
Minimum front brake pad lining thickness .... 1.0 mm
Minimum rear brake pad lining thickness .... 1.0 mm
Minimum rear brake shoe lining thickness .... 1.0 mm

Suspension and steering
Power steering drivebelt deflection:
   New belt .... 8 to 9 mm
   Used belt .... 9 to 10 mm
Tyre pressures (cold):
   Front:
      Pre-September 1985 models .... 1.8 bar (26 lbf/in^2)
      September 1985 models .... 2.0 bar (29 lbf/in^2)
   Rear (all models)* .... 1.8 bar (26 lbf/in^2)
*When fully loaded, increase the rear tyre pressure to 1.9 bar (28 lbf/in^2)

Chapter 1 Routine maintenance and servicing

Electrical system

Alternator drivebelt deflection:
- New belt:
  - E series engine .... 12 to 13 mm
  - B series engine .... 8 to 9 mm
- Used belt:
  - E series engine .... 13 to 14 mm
  - B series engine .... 9 to 10 mm

Wiper blade type:
- Front .... Champion X4503
- Rear .... Champion X3603

Wiper arm type:
- Front .... Champion CC A5
- Rear .... Champion CC A4

Torque wrench settings

	Nm	lbf ft
Fuel filter union bolt (fuel injected models)	25 to 35	18 to 25
Spark plugs	15 to 23	11 to 17
Automatic transmission oil pan bolts	5 to 8	4 to 6
Roadwheel nuts	90 to 110	65 to 80
Alternator pivot mounting bolt	19 to 31	14 to 22
Alternator adjusting arm bolt	43 to 61	32 to 45

Lubricants, fluids and capacities

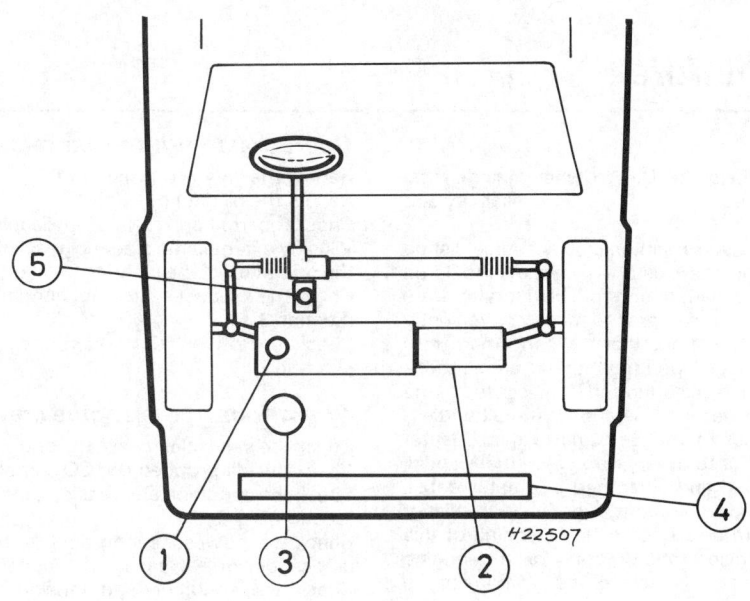

Lubricants and fluids

Component or system	Lubricant type/specification	Duckhams recommendation
1 Engine	Multigrade engine oil, viscosity 10W/40 to 20W/50	Duckhams QXR, Hypergrade, or 10W/40 Motor Oil
2 Manual gearbox Pre-Sept 1981 (approx) Sept 1981 on (approx)	ATF type F, (M2C33-F) Hypoid gear oil SAE 90 or 80W/90	Duckhams Uni-Matic or Q-Matic Duckhams Hypoid 90S
2 Automatic transmission Pre-July 1987 July 1987 on	ATF type F, (M2C33-F) Dexron type ATF	Duckhams Uni-Matic or Q-Matic Duckhams Uni-Matic or D-Matic
3 Power steering reservoir	ATF type F, (M2C33-F)	Duckhams Uni-Matic, Q-Matic or D-Matic
4 Cooling system	Ethylene glycol based antifreeze	Duckhams Universal Antifreeze and Summer Coolant
5 Brake fluid reservoir	Hydraulic fluid to SAE J1703, DOT 4	Duckhams Universal Brake and Clutch Fluid

Capacities

Engine oil
Oil and filter change:
    Pre-September 1985 models .... 3.7 litres
    September 1985 models onward .... 3.4 litres

Cooling system
Pre-September 1985 models .... 5.5 litres
September 1985 onward:
    Manual transmission models .... 5.0 litres
    Automatic transmission model .... 6.0 litres

Fuel tank:
Pre-September 1985 models .... 42 litres
September 1985 to July 1987 models .... 45 litres
July 1987 models onward .... 48 litres

Manual gearbox .... 3.2 litres

Automatic transmission .... 5.7 litres

Power-assisted steering reservoir .... 0.6 litres

Maintenance schedule

Introduction

This Chapter is designed to help the D.I.Y. owner maintain the Mazda 323 with the goals of maximum economy, safety, reliability and performance in mind.

On the following pages is a master maintenance schedule, listing the servicing requirements, and the intervals at which they should be carried out as recommended by the manufacturers. The operations are listed in the order in which the work can be most conveniently undertaken. For example, all the operations that are performed from within the engine compartment are grouped together, as are all those that require the car to be raised and supported for access to the suspension and underbody. Alongside each operation in the schedule is a reference which directs the user to the Sections in this Chapter covering maintenance procedures or to other Chapters in the Manual, where the operations are described and illustrated in greater detail. Specifications for all the maintenance operations, together with a list of lubricants, fluids and capacities are provided at the beginning of this Chapter. Refer to the accompanying photographs of the engine compartment and the underbody of the vehicle for the locations of the various components.

Servicing your vehicle in accordance with the mileage/time maintenance schedule and step-by-step procedures will result in a planned maintenance programme that should produce a long and reliable service life. Bear in mind that it is a comprehensive plan, so maintaining some items but not others at the specified intervals will not produce the same results.

The first step in this maintenance program is to prepare yourself before the actual work begins. Read through all the procedures to be undertaken then obtain all the parts, lubricants and any additional tools needed.

Every 250 miles (400 km) or weekly

Operations internally and externally
Visually examine the tyres for tread depth, and wear or damage (Section 9)
Check and if necessary adjust the tyre pressures (Section 9)

Operations in the engine compartment
Check the engine oil level (Section 1)
Check the engine coolant level (Section 2)
Check the screen washer fluid level (Section 11)
Check the battery electrolyte level (Section 11)

Every 6000 miles (10 000 km) or 6 months – whichever comes first

In addition to all the items listed above, carry out the following:

Operations internally and externally
Check the operation of all lights, indicators, instruments and windscreen washer system (Section 11)
Check the operation of the cooling system (Section 2)
Check and adjust the headlight beam alignment (Section 11)
Check the operation of the handbrake (Section 8)
Check the tightness of the wheel nuts (Section 9)
Check the brake pedal height and operation (Section 8)
Check the clutch pedal height and operation (Section 5)

Operations with the car raised and supported
Renew the engine oil (Section 1)
Renew the oil filter (Section 1)
Check the front and rear (as applicable) disc pads for wear (Section 8)
Visually examine the steering gear linkage and rubber gaiters and check the operation of the steering (Section 9)
Check the exhaust system for condition, leakage and security (Section 3)
Check the driveshafts and rubber gaiters for damage and leakage (Section 7)

Operations in the engine compartment
Check the valve clearances (B6 and all E series engines) (Section 1)
Check the idling speed and CO content (Section 3)
Check the condition and tension of the water pump/alternator drivebelt (Section 2)
Check the power steering fluid level and examine the fluid lines and unions for signs of leakage (Section 9)
Check the condition and tension of the power steering drivebelt (Section 9)
Check the condition of the spark plugs (Section 4)
Check and adjust the contact breaker points and ignition timing (contact breaker ignition system) (Section 4)
Check the ignition timing (electronic ignition system) (Section 4)
Check the condition of the distributor cap and rotor arm (Section 4)
Check the brake fluid level (Section 8)
Check the clutch fluid level or cable adjustment (as applicable) (Section 5)
Check the manual gearbox oil level (Section 6)

Every 12 000 miles (20 000 km) or 12 months – whichever comes first

In addition to the operations listed under the 6000 miles (10 000 km) heading, carry out the following:

Operations internally and externally
Check and if necessary adjust the front and rear wheel toe setting (Chapter 10, Section 26)
Check the operation of the braking system servo unit (Chapter 9, Section 3)

Operations with the car raised and supported
Check all chassis and body nuts and bolts for tightness (Section 10)
Visually examine the underbody, wheel arches and body panels for damage (Section 10)
Visually examine the brake pipes, hoses and unions (Chapter 9, Section 6)
Visually examine the fuel lines for signs of leakage (Section 3)
Check the rear drum brake shoes for wear (Section 8)
Check the operation of the front and rear suspension (Section 9)

Operations in the engine compartment
Renew the fuel filter (carburettor models) (Section 3)
Check the carburettor linkages and choke mechanism operation (carburettor models) (Section 3)
Check the operation of the throttle positioner system (B5 and all E series engined models equipped with manual gearbox (Section 3)
Check the operation of the dashpot and the throttle position switch (fuel injected models) (Section 3)
Check the automatic transmission fluid level (Section 6)
Visually examine the cooling system for leakage (Section 2)

Chapter 1 Routine maintenance and servicing

Every 24 000 miles (40 000 km) or 24 months – whichever comes first

In addition to the operations listed in the 12 000 mile service, carry out the following:

Operations with the car raised and supported
Grease the front and rear wheel bearings (Section 9)
Check the condition of the front suspension balljoints (Section 9)
Renew the manual gearbox oil (Section 6)

Operations in the engine compartment
Drain and flush the cooling system and refill with fresh coolant (Section 2)
Renew the air filter element (Section 3)
Check the cylinder head bolts for tightness (B6 and E series engines) (Section 1)
Check the inlet and exhaust manifold nuts and bolts for tightness (Section 3)
Renew the fuel filter (fuel injected models) (Section 3)
Renew the brake fluid (Section 8)

Every 60 000 miles (100 000 km)

In addition to all those operations listed under the 12 000 mile service, carry out the following:

Operations in the engine compartment
Renew the timing belt (B series engines) (Chapter 2)

Chapter 1 Routine maintenance and servicing

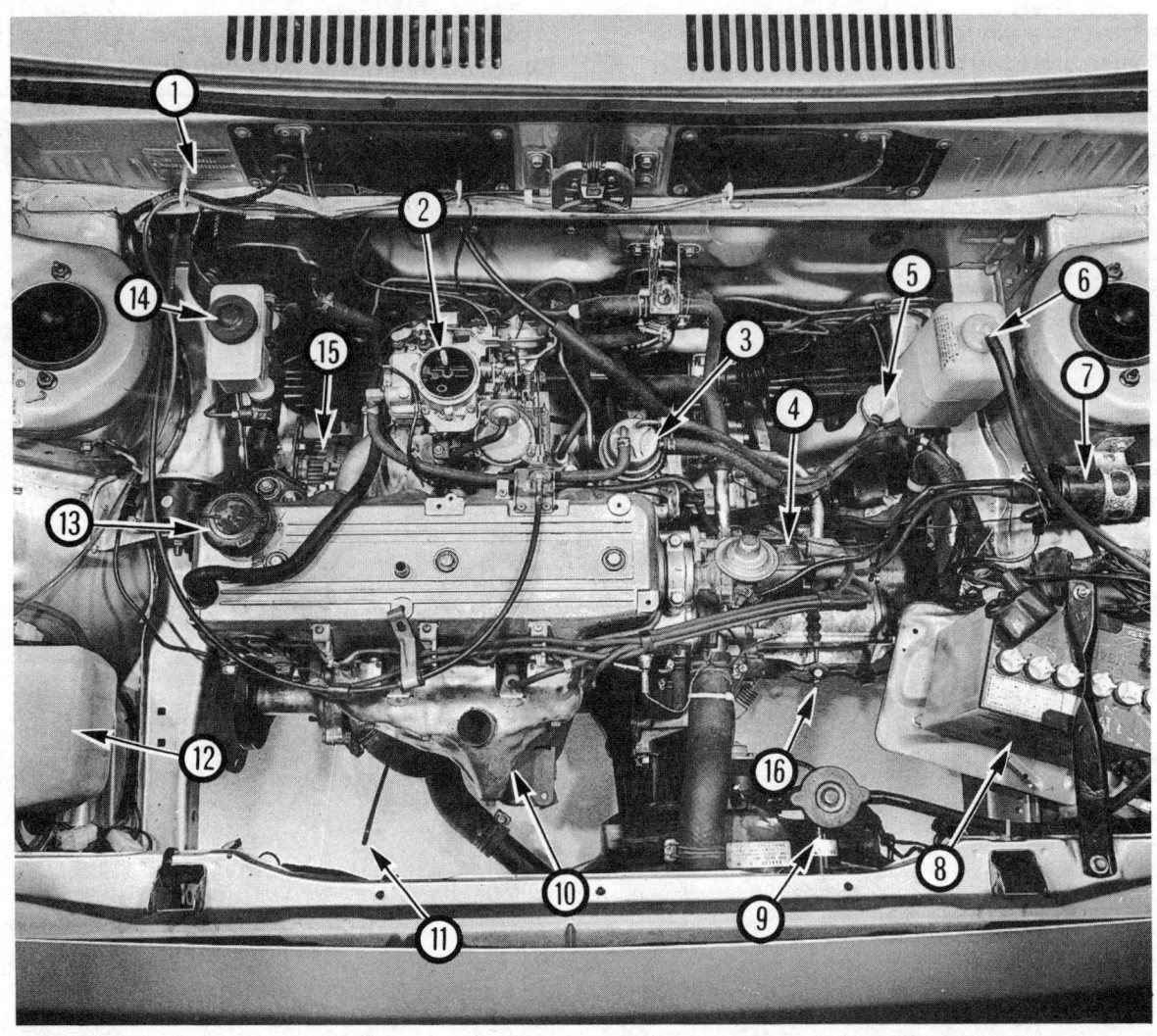

Engine and underbonnet component locations on pre-September 1985 models (air cleaner removed for clarity)

1 Vehicle identification number
2 Carburettor
3 Fuel pump
4 Distributor
5 Fuel filter
6 Cooling system expansion tank
7 Ignition coil
8 Battery
9 Radiator pressure cap
10 Exhaust manifold stove
11 Engine oil dipstick
12 Washer reservoir
13 Oil filler cap
14 Brake fluid reservoir
15 Alternator
16 Clutch cable adjuster

Chapter 1 Routine maintenance and servicing

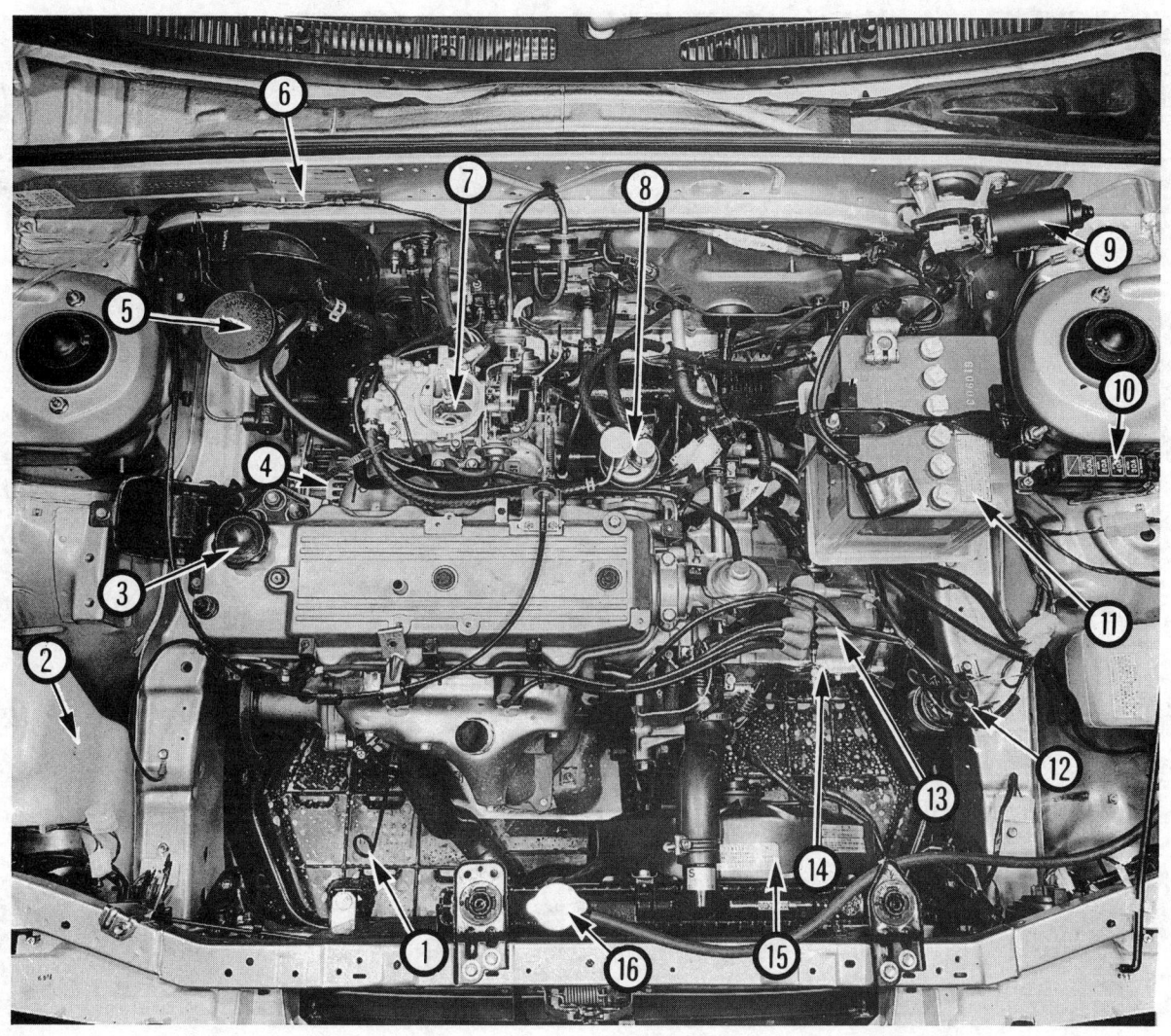

Engine and underbonnet component locations on September 1985 to July 1987 carburettor models (air cleaner removed for clarity)

1. Engine oil dipstick
2. Washer reservoir
3. Engine oil filler cap
4. Alternator
5. Brake fluid reservoir
6. Vehicle identification plate
7. Carburettor
8. Fuel pump
9. Windscreen wiper motor
10. Main fuse box
11. Battery
12. Ignition coil
13. Distributor
14. Clutch cable adjuster
15. Radiator cooling fan
16. Radiator pressure cap

40 Chapter 1 Routine maintenance and servicing

Engine and underbonnet components on July 1987 onwards carburettor models (air cleaner removed for clarity)

1. Engine oil dipstick
2. Washer reservoir
3. Anti-afterburn valve (where fitted)
4. Brake fluid reservoir
5. Vehicle identification plate
6. Carburettor
7. Distributor
8. Fuel filter
9. Windscreen wiper motor
10. Main fuse box
11. Cooling system expansion tank
12. Ignition coil
13. Battery
14. Clutch cable adjuster
15. Radiator cooling fan
16. Radiator pressure cap
17. Engine oil filler cap
18. Power steering fluid reservoir

Chapter 1 Routine maintenance and servicing

Engine and underbonnet components on 1600cc fuel injected model

1 Engine oil dipstick
2 Washer reservoir
3 Power steering idle-up solenoid
4 Brake fluid reservoir
5 Vehicle identification plate
6 Atmospheric pressure sensor
7 Fuel filter
8 Battery
9 Windscreen wiper motor
10 Main fuse box
11 Cooling system expansion tank
12 Ignition coil
13 Air cleaner and airflow meter assembly
14 Radiator cooling fan
15 Radiator pressure cap
16 Power steering fluid reservoir
17 Dashpot
18 Throttle position switch
19 Engine oil filler cap
20 Distributor

Front underbody view on pre-September 1985 models (undertrays removed)

1. Front towing hook
2. Front crossmember
3. Front lower suspension arm
4. Gearbox drain plug
5. Track rod outer balljoint
6. Steering gear track rod
7. Lower suspension arm front mounting bracket
8. Lower suspension arm rear mounting bracket
9. Exhaust rubber mounting
10. Gearchange rod
11. Exhaust intermediate pipe
12. Gearchange remote control housing
13. Exhaust system flexible joint
14. Driveshaft vibration damper
15. Balljoint retaining nuts
16. Engine oil drain plug
17. Exhaust front pipe
18. Radiator bottom hose
19. Radiator drain plug

Chapter 1 Routine maintenance and servicing 43

Front underbody view on September 1985 models onward (undertrays removed)

1 Radiator drain plug
2 Gearbox
3 Gearbox drain plug
4 Track rod
5 Brake pipes
6 Gearchange rod
7 Exhaust system
8 Steering gear gaiter
9 Front lower suspension arm
10 Driveshaft
11 Engine oil drain plug
12 Front anti-roll bar
13 Radiator bottom hose
14 Exhaust front pipe

44 Chapter 1 Routine maintenance and servicing

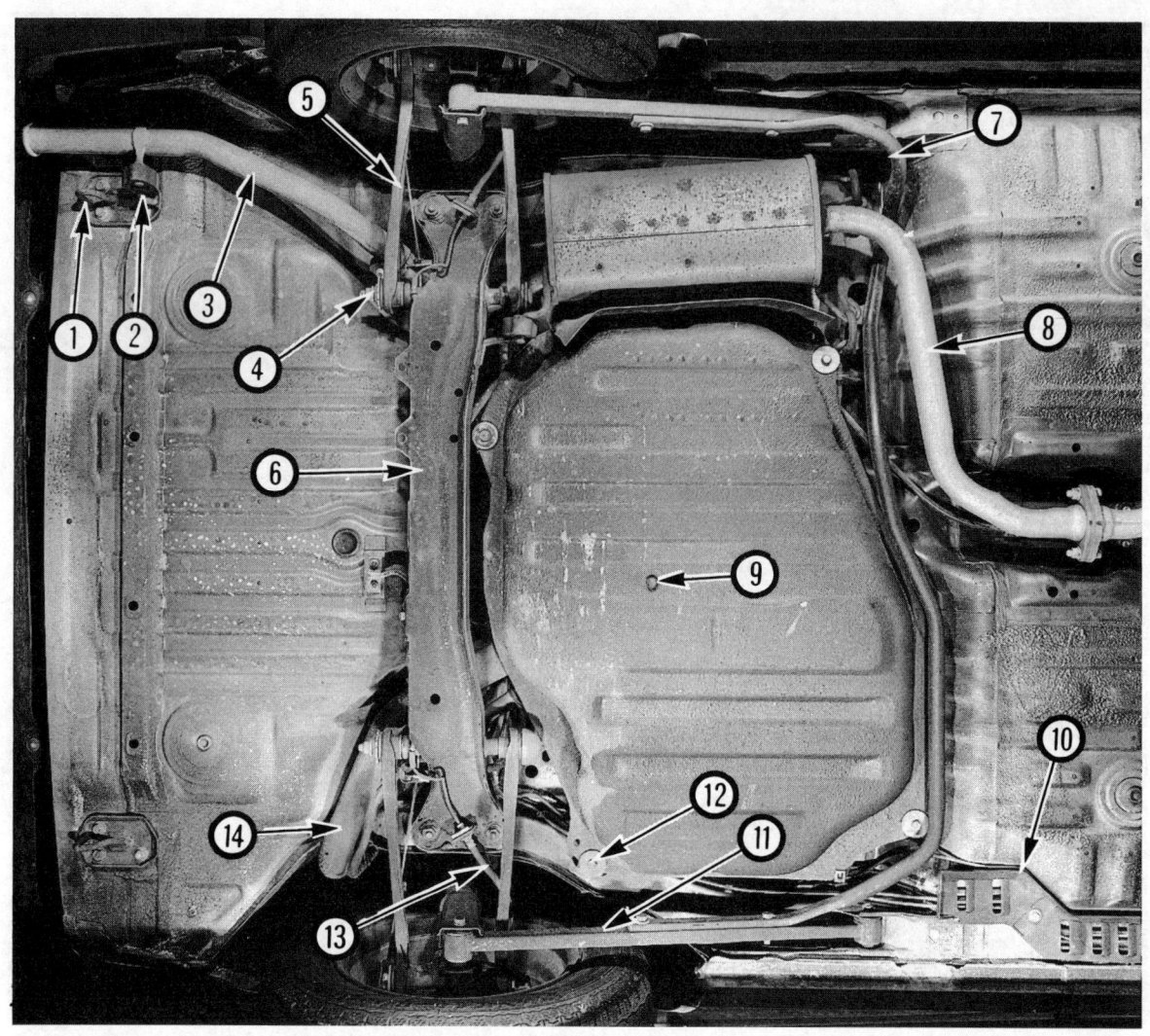

Rear underbody view on pre-September 1985 models

1. Rear towing hook
2. Exhaust mounting rubber
3. Exhaust tailpipe
4. Lateral link retaining nut and toe setting adjustment point
5. Lateral link
6. Rear crossmember
7. Rear anti-roll bar
8. Exhaust main silencer section
9. Fuel tank drain plug
10. Fuel and brake pipe cover
11. Trailing arm
12. Fuel tank mounting
13. Rear brake flexible hose
14. Fuel tank filler and vent hoses

Chapter 1 Routine maintenance and servicing 45

Rear underbody view on September 1985 onwards Saloon and Hatchback models

1 Rear towing hook
2 Exhaust mounting rubber
3 Exhaust tailpipe
4 Lateral link retaining nut and toe setting adjustment point
5 Lateral link
6 Handbrake cable
7 Exhaust main silencer section
8 Fuel tank drain plug
9 Rear crossmember
10 Rear brake flexible hose
11 Fuel tank filler and vent hoses
12 Rear anti-roll bar

Rear underbody view on Estate models

1. Fuel tank
2. Brake pipes
3. Trailing arm
4. Rear anti-roll bar
5. Spare wheel and carrier
6. Rear lateral link (adjustable)
7. Rear suspension strut
8. Rear lateral link (non-adjustable)
9. Exhaust system
10. Exhaust mounting rubber
11. Handbrake cable

Maintenance procedures

1 Engine

Engine oil level check

1 The engine oil level is checked with a dipstick that extends through a tube and into the sump at the bottom of the engine. The dipstick is located on the front face of the engine.

2 The oil level should be checked with the vehicle standing on level ground and before it is driven, or at least 5 minutes after the engine has been switched off. If the oil is checked immediately after driving the vehicle, some of the oil will remain in the upper engine components and oil galleries, resulting in an inaccurate reading on the dipstick.

3 Withdraw the dipstick from the tube and wipe all the oil from the end with a clean rag or paper towel. Insert the clean dipstick back into the tube as far as it will go, then withdraw it once more. Check that the oil level is between the upper (F) and lower (L) marks on the dipstick. If the level is towards the lower (L) mark, unscrew the oil filler cap on the front of the valve cover and add fresh oil until the level is on the upper (F) mark (photos).

4 Always maintain the level between the two dipstick marks. If the level is allowed to fall below the lower mark, oil starvation may result which could lead to severe engine damage. If the engine is overfilled by adding too much oil, this may result in oil fouled spark plugs, oil leaks or oil seal failures.

5 An oil can spout or funnel may help to reduce spillage when adding oil to the engine. Always use the correct grade and type of oil as shown in *Lubricants fluids and capacities*.

Engine oil and filter renewal

6 Frequent oil and filter changes are the most important preventative maintenance procedures that can be undertaken by the D.I.Y. owner. As engine oil ages, it becomes diluted and contaminated, which leads to premature engine wear.

7 Before starting this procedure, gather together all the necessary tools and materials. Also make sure that you have plenty of clean rags and newspapers handy to mop up any spills. Ideally, the engine oil should be warm as it will drain better and more built-up sludge will be removed with it. Take care however not to touch the exhaust or any other hot parts of the engine when working under the vehicle. To avoid any possibility of scalding and to protect yourself from possible skin irritants and other harmful contaminants in used engine oils, it is advisable to wear rubber gloves when carrying out this work. Access to the underside of the vehicle will be greatly improved if it can be raised on a lift, driven onto ramps or jacked up and supported on axle stands. Whichever method is chosen, the sump drain plug should be at the lowest point so ideally the car should be as level as possible since the drain plug is located in the centre of the sump (photo). Access to the drain plug is gained by removing the vehicle undertray.

8 Remove the undertray then position a suitable container beneath the drain plug. Clean the drain plug and the area around it, then slacken it half a turn using a suitable spanner. If possible, try to keep the plug pressed into the sump while unscrewing it by hand the last couple of turns. As the plug releases from the threads, move it away sharply so the stream of oil issuing from the sump runs into the container, not up your sleeve!

9 Allow some time for the old oil to drain, noting that it may be necessary to reposition the container as the oil flow slows to a trickle.

10 After all the oil has drained, wipe off the drain plug with a clean rag and renew the sealing washer. Clean the area around the drain plug opening then refit and tighten the plug securely.

11 Move the container into position under the oil filter which is located on rear of the cylinder block.

12 Using an oil filter removal tool, such as a strap wrench, slacken the filter initially (photo). Loosely wrap some rags around the oil filter, then unscrew it and immediately position it with its open end uppermost to prevent spillage of the oil. Remove the oil filter from the engine compartment and empty the oil into the container.

13 Use a clean rag to remove all oil, dirt and sludge from the filter sealing area on the engine. Check the old filter to make sure that the rubber sealing ring is not stuck to the engine. If it has, carefully remove it.

14 Apply a light coating of clean oil to the sealing ring on the new filter then screw it into position on the engine (photo). Tighten the filter firmly by hand only, do not use any tools. Wipe clean the exterior of the oil filter and refit the undertray.

15 Remove the old oil and all tools from under the car then, if applicable, lower the car to the ground.

16 Unscrew the oil filler cap on the valve cover and fill the engine with the specified quantity and grade of oil, as described earlier in this Section. Pour the oil in slowly otherwise it may overflow from the top of the valve cover. Check that the oil level is up to the maximum (F) mark on the dipstick.

17 Start the engine and run it for a few minutes while checking for leaks around the oil filter seal and the sump drain plug.

18 Switch off the engine and wait a few minutes for the oil to settle in the sump once more. With the new oil circulated and the filter now completely full, recheck the level on the dipstick and add more oil as necessary.

19 Dispose of the used engine oil safely with reference to *General repair procedures* in the preliminary Sections of this Manual.

Valve clearance adjustment (B6 and E series engines)

20 Warm the engine up to normal operating temperature then switch the engine off.

21 Remove the cylinder head cover as described in Chapter 2.

1.3A Engine oil level dipstick marks

1.3B Oil is added through the filler on the cylinder head cover

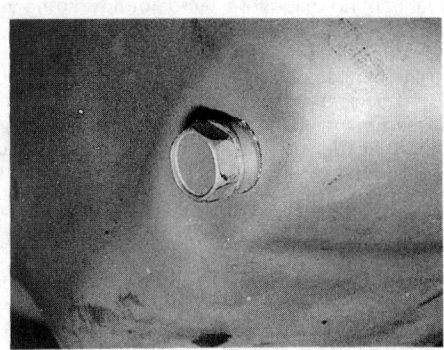

1.7 Oil drain plug location

Chapter 1 Routine maintenance and servicing

1.12 Using a strap wrench to remove the oil filter

1.14 Fit the new filter to the rear of the block and tighten using hand pressure only

1.25 Checking a valve clearance

22 Check that all cylinder head bolts are tightened to the specified torque as described in Chapter 2 (Part A or B as applicable).

23 With No 1 spark plug removed, place your finger over the plug hole and turn the crankshaft until compression can be felt building up in the cylinder. Continue turning the crankshaft until the timing marks on the crankshaft pulley and timing cover are aligned and the engine is at TDC with No 1 cylinder on compression.

24 With the engine in this position the following four valves can be adjusted:

No 1 cylinder exhaust
No 1 cylinder inlet
No 2 cylinder inlet
No 3 cylinder exhaust

25 Check the valve clearances are as stated in the Specifications at the start of this Chapter by inserting a feeler gauge of the correct thickness between the valve stem and the rocker adjusting screw (photo). If adjustment is necessary, slacken the adjusting screw locknut and turn the screw as necessary until the feeler blade is a light sliding fit. Once the correct clearance is obtained, hold the adjusting screw and securely tighten the locknut. Recheck the valve clearance and adjust if necessary.

26 Once all four clearances are as specified, turn the crankshaft one complete turn so that the engine is once again at TDC, but this time with No 4 cylinder on compression. With the engine in this position the remaining four valves can be adjusted:

No 2 cylinder exhaust
No 3 cylinder inlet
No 4 cylinder inlet
No 4 cylinder exhaust

27 Once all valve clearances are correct refit the cylinder head cover as described in Chapter 2.

General engine checks

28 Visually inspect the engine joint faces, gaskets and seals for any signs of water or oil leaks. Pay particular attention to the areas around the cylinder head gasket joint, valve cover joint, sump joint, and oil filter. Bear in mind that over a period of time some very slight seepage from these areas is to be expected but what you are really looking for is any indication of a serious leak. Should a leak be found, renew the offending gasket or oil seal by referring to the appropriate Chapters in this Manual.

29 Check the security and condition of all the engine related pipes and hoses. Ensure that all cable ties or securing clips are in place and in good condition. Clips which are broken or missing can lead to chafing of the hoses, pipes or wiring which could cause more serious problems in the future.

2 Cooling, heating and ventilation systems

Coolant level check

Warning: *DO NOT attempt to remove the radiator pressure cap when the engine is hot, as there is a very great risk of scalding.*

1 All vehicles covered by this manual are equipped with a pressurised cooling system. On later models an expansion tank is located on the left-hand side of the engine compartment. The expansion tank has only one hose which is connected directly to the radiator filler neck. As engine temperature increases, the coolant expands and travels through

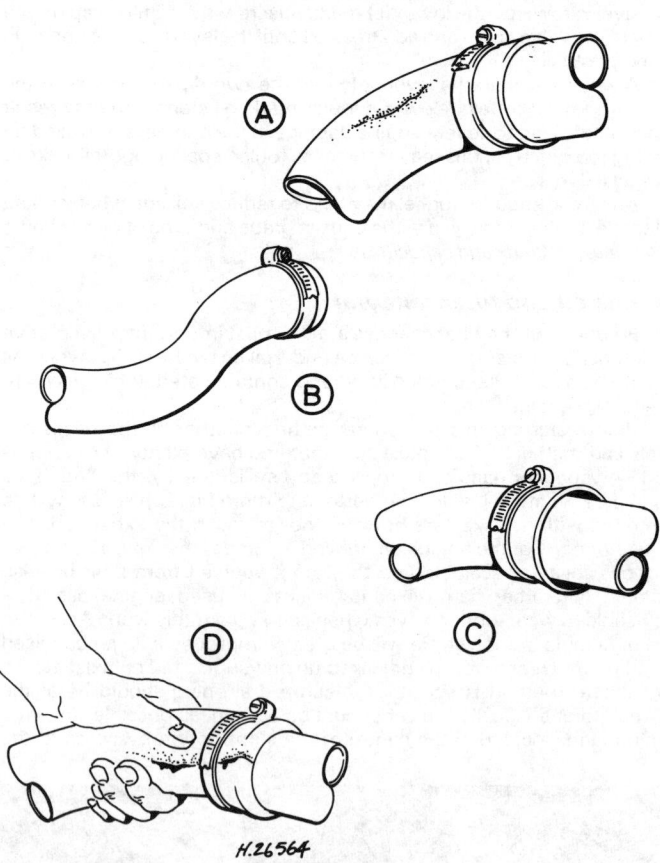

Fig. 1.1 Coolant hose inspection (Sec 2)

A Check for chafed or burned areas; these may lead to sudden and costly failure.
B A soft hose indicates inside deterioration, leading to contamination of the cooling system and clogging of the radiator.
C A hardened hose can fail at any time; tightening the clamps will not seal the joint or prevent leaks.
D A swollen hose or one with oil soaked ends indicates contamination from oil or grease. Cracks and breaks can be easily seen by squeezing the hose.

Chapter 1 Routine maintenance and servicing

2.2 Adding antifreeze to the cooling system – early models

2.3A On later models top up the expansion tank...

2.3B ...to the upper level mark (arrowed)

the hose to the expansion tank. As the engine cools, the coolant is automatically drawn back into the system to maintain the correct level.

2 To check the level on models without an expansion tank first ensure that the engine is cold, then remove the radiator pressure cap. The coolant level should be higher than the level gauges situated in the bottom of the filler neck. If this is not the case top up to the base of the filler neck using the specified water and antifreeze mixture (see below) (photo). Check the rubber seal on the radiator cap for signs of wear or damage and if necessary renew the cap. Fit the pressure cap to the radiator filler neck.

3 On models fitted with an expansion tank the coolant level is visible through the translucent material of the expansion tank. When the engine is cold, the coolant level should be between the upper (FULL) level mark and the lower (LOW) level mark on the side of the tank. When the engine is hot, the level may be slightly above the upper (FULL) mark. If topping up is necessary, remove the expansion tank cap and top up to the upper (FULL) level mark using the specified water and antifreeze mixture (see below) (photos). Refit the expansion tank cap.

4 With a sealed type cooling system, the addition of coolant should only be necessary at very infrequent intervals. If frequent topping up is required, it is likely there is a leak in the system. Check the radiator, all hoses and joint faces for any sign of staining or actual wetness, and rectify as necessary. If no leaks can be found, it is advisable to have the pressure cap and the entire system pressure tested by a dealer or suitably equipped garage as this will often show up a small leak not previously visible.

Coolant draining

Warning: *Wait until the engine is cold before starting this procedure. Do not allow antifreeze to come in contact with your skin or painted surfaces of the vehicle. Rinse off spills immediately with plenty of water.*

5 If the engine is cold, remove the radiator pressure cap. If it is not possible to wait until the engine is cold place a cloth over the radiator pressure cap and slowly turn the cap anti-clockwise until it reaches its first stop. Wait until the pressure has escaped then press the cap downwards and turn it further in an anti-clockwise direction. Release the cap slowly and, after making sure all the pressure in the system has been released, remove the cap.

6 Move the heater temperature control lever inside the car to the maximum heat setting.

7 Place a suitable container beneath the radiator drain plug situated on the bottom of the radiator (photo). Unscrew the drain plug and allow the coolant to drain into the container. On models manufactured before June 1981, there is also a cylinder block drain plug located on the front facing side of the block. Place a second container beneath the engine, unscrew the plug and allow the remaining coolant to drain.

8 If the system needs to be flushed after draining, refer to the following paragraphs, otherwise refit the drain plug(s) and tighten them securely.

System flushing

9 With time the cooling system may gradually lose its efficiency if the radiator matrix becomes choked with rust and scale deposits. If this is the case, the system must be flushed as follows. First drain the coolant as already described.

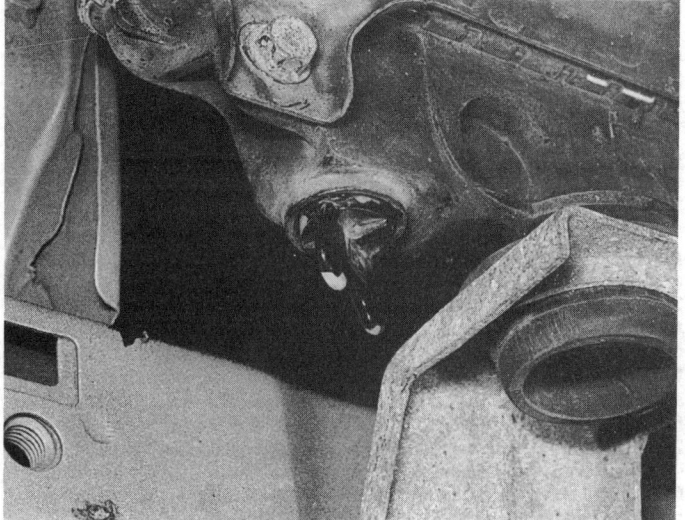

2.7 Cooling system drain plug location

10 Insert a garden hose in the radiator filler neck and allow the water to circulate through the radiator until it runs clear from the drain plug outlet.

11 In severe cases of contamination the radiator should be reverse-flushed. To do this, first remove it from the car, as described in Chapter 3, invert it and insert a hose in the bottom outlet. Continue flushing until clear water runs from the top hose outlet.

12 To flush the engine and the remainder of the system, remove the thermostat as described in Chapter 3. Insert the garden hose and allow the water to circulate through the engine until it runs clear from the radiator drain plug outlet.

13 If, after a reasonable period, the water still does not run clear, the radiator should be flushed with a good proprietary cleaning system such as Holts Radflush or Holts Speedflush. The regular renewal of corrosion inhibiting antifreeze should prevent severe contamination of the system. Note that as the radiator is of aluminium it is important not to use caustic soda or alkaline compounds to clean it.

Coolant filling

14 Refit the drain plug(s). If removed also refit the radiator and/or thermostat as described in Chapter 3. Ensure that all hoses are securely fitted.

15 Fill the system through the radiator filler neck with the appropriate mixture of water and antifreeze until the level is up to the base of the filler neck. Fill the system slowly to allow all trapped air to escape.

16 When the system is full, start the engine and allow it to idle with the radiator cap removed. If necessary, add coolant to maintain the correct level as any trapped air is expelled from the system.

17 When the engine reaches normal operating temperature depress the accelerator pedal two or three times then switch off the engine.
18 Allow the engine to cool then, if necessary, top up the coolant level and refit the radiator pressure cap. On models equipped with an expansion tank, add coolant until the level is up to the upper (FULL) level mark on the side of the tank.

Antifreeze mixture

19 The antifreeze should always be renewed at the specified intervals. This is necessary not only to maintain the antifreeze properties, but also to prevent corrosion which would otherwise occur as the corrosion inhibitors become progressively less effective.
20 Always use an ethylene-glycol based antifreeze which is suitable for use in mixed metal cooling systems. The percentage quantity of antifreeze and levels of protection afforded are indicated in the Specifications.
21 Before adding antifreeze, the cooling system should be completely drained, preferably flushed, and all hoses checked for condition and security.
22 After filling with antifreeze, a label should be attached to the radiator or expansion tank stating the type and concentration of antifreeze used and the date installed. Any subsequent topping up should be made with the same type and concentration of antifreeze.
23 Do not use engine antifreeze in the screen washer system, as it will cause damage to the vehicle paintwork. A screen wash such as Turtle Wax High Tech Screen Wash should be added to the washer system in the recommended quantities.

General cooling system checks

24 The engine should be cold for the cooling system checks, so perform the following procedure before driving the vehicle or after the engine been switched off for at least three hours.
25 Remove the radiator pressure cap (see above) and clean it thoroughly inside and out with a rag. Also clean the filler neck on the radiator. The presence of rust or corrosion in the filler neck indicates that the coolant should be changed. The coolant inside the expansion tank should be relatively clean and transparent. If it is rust coloured, drain and flush the system and refill with a fresh coolant mixture.
26 Carefully check the radiator and heater hoses along their entire length. Renew any hose which is cracked, swollen or deteriorated. Cracks will show up better if the hose is squeezed. Pay close attention to the hose clips that secure the hoses to the cooling system components. Hose clips can pinch and puncture hoses, resulting in cooling system leaks. If wire type hose clips are used, it may be a good idea to replace them with screw-type clips.
27 Inspect all the cooling system components (hoses, joint faces etc.) for leaks. A leak in the cooling system will usually show up as white or rust coloured deposits on the area adjoining the leak. Where any problems of this nature are found on system components, renew the component or gasket with reference to Chapter 3.
28 Clean the front of the radiator with a soft brush to remove all insects, leaves etc. imbedded in the radiator fins. Be extremely careful not to damage the radiator fins or cut your fingers on them.

Water pump drivebelt check, adjustment and renewal

29 The water pump is driven off the crankshaft pulley by the same belt which drives the alternator. Therefore the checking, adjustment and renewal procedures for the water pump drivebelt are the same as those given for the alternator drivebelt in Section 11 of this Chapter.

3 Fuel and exhaust systems

Warning: *Certain procedures in this Section require the removal of fuel lines and connections which may result in some fuel spillage. Before carrying out any operation on the fuel system refer to the precautions given in Safety First! at the beginning of this Manual and follow them implicitly. Petrol is a highly dangerous and volatile liquid and the precautions necessary when handling it cannot be overstressed.*

Air filter element renewal

1 On carburettor engined models, unscrew the wing nut then release the wire clips securing the top cover to the air cleaner body. Lift off the cover and remove the filter element (photo).
2 On fuel injected engined models, undo the clamp and disconnect the rubber hose from the airflow meter. Undo the bolts securing the air cleaner cover to the body and lift up the cover. Remove the filter element from the housing (photo).
3 Clean the inside of the air cleaner body and fit a new element.
4 Refit the air cleaner cover by a reverse of the removal procedure.

Checking the carburettor linkages and choke mechanism operation

5 Remove the air cleaner housing as described in Chapter 4.
6 Clean the external throttle and choke mechanism linkages using one of the special aerosol carburettor cleaners. Spray the carburettor cleaner down the carburettor bore whilst slowly operating the choke and throttle linkages, ensuring that the choke and throttle pivots are well lubricated. Once all traces of dirt and grease have been removed, dry the sprayed areas using compressed air.
7 Check that the accelerator and, where fitted, choke cable are correctly adjusted as described in Chapter 4.
8 Refit the air cleaner housing as described in Chapter 4.

Idle speed and CO content adjustment

9 Before carrying out the following adjustments, ensure that the spark plugs are in good condition and correctly gapped and that, where applicable, the contact breaker points and ignition timing settings are correct.
10 Make sure that all electrical components are switched off during the following procedure. If the electric cooling fan operates, wait until it has stopped before continuing.
11 Connect a tachometer to the engine in accordance with the manufacturer's instructions. The use of an exhaust gas analyser (CO meter) is also recommended to obtain an accurate setting.

Single carburettor models

12 Run the engine until it reaches normal operating temperature then allow it to idle. On pre-September 1985 models fitted with automatic

3.1 Removing the air cleaner filter element on carburettor models

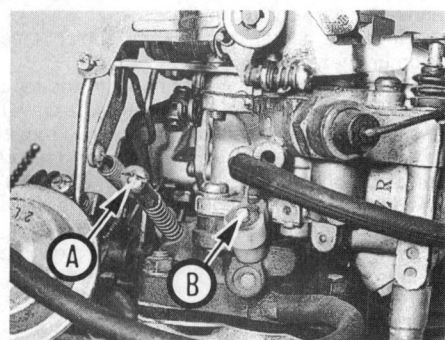

3.2 Removing the air cleaner filter element on fuel injected models

3.13A Throttle adjustment screw (A) and blind cap over mixture adjustment screw (B) on Hitachi carburettor

Chapter 1 Routine maintenance and servicing

3.13B Adjusting the idle speed on Aisan carburettor

3.14 Mixture screw on Aisan carburettor (blind cap removed)

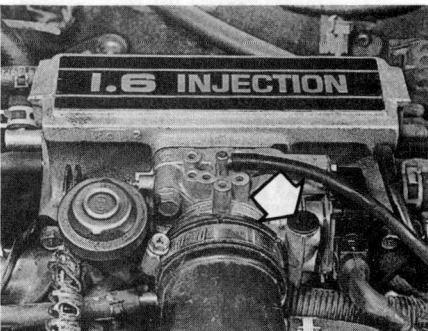

3.29A Idle speed screw is situated below rubber plug (arrowed) on the throttle housing on fuel injected models

Fig. 1.2 Disconnecting the linkage connecting rod at the throttle shaft lever on twin carburettor models (Sec 3)

Fig. 1.3 Removing the blind cap from the carburettor mixture adjustment screw tamperproof cover (Sec 3)

transmission set the selector lever to the D position, and on September 1985 onwards set the selector lever to the N position. Ensure that the handbrake is firmly applied and the wheels are chocked.

13 With the engine idling, turn the throttle adjustment screw until the specified idling speed is obtained (photos).

14 Remove the blind cap from the centre of the tamperproof cover over the mixture adjustment screw (photo).

15 Turn the mixture adjustment screw clockwise to weaken the mixture or anti-clockwise to richen the mixture until the CO reading is as given in the Specifications. If a CO meter is not being used weaken the mixture as described, then richen the mixture until the maximum engine speed is obtained consistent with even running.

16 If necessary, re-adjust the idling speed then check the CO reading again. Repeat as necessary until both the idling speed and CO reading are correct.

17 Fit a new tamperproof cap to the mixture adjustment screw.

Twin carburettor models

18 Run the engine until it reaches normal operating temperature then switch it off.

19 Disconnect the ball socket of the linkage connecting rod at the throttle shaft on the left-hand carburettor.

20 Remove the blind cap from the centre of the tamperproof cover over the mixture adjustment screw on each carburettor.

21 Starting with the left-hand carburettor, back off the throttle adjustment screw, then screw it in until it just touches the throttle lever. Now screw it in a further two complete turns. Repeat this operation for the right-hand carburettor.

22 Again starting with the left-hand carburettor, screw in the mixture adjustment screw until it makes light contact with its seat. Now back the screw off three complete turns and repeat the operation on the right-hand carburettor.

23 Start the engine and turn both throttle adjustment screws by equal amounts until the engine is idling at approximately 850 rpm.

24 Using a proprietary balancing meter, in accordance with the

Fig. 1.4 Adjusting the idle speed on twin carburettor models with balancing meter in use (Sec 3)

manufacturers instructions, balance the carburettors by altering the throttle adjustment screws until the airflow through both carburettors is the same and the engine is idling at the specified rpm. Once both the carburettors are balanced subsequent adjustment of the idling speed must be made by turning both the throttle adjustment screws by equal amounts.

25 Turn the mixture adjustment screw on the left-hand carburettor clockwise to weaken the mixture or anti-clockwise to richen the mixture until the CO reading is as given in the Specifications. If a CO meter is not being used weaken the mixture as described, then richen the mixture

3.29B Adjusting the idle speed on fuel injected models

3.31 Remove the rubber plug on the airflow meter to gain access to the bypass air adjusting screw

3.39 Checking dashpot operation

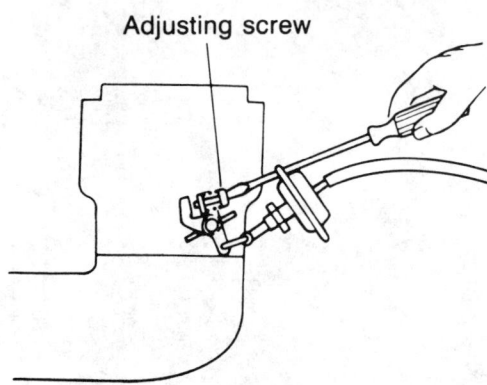
Fig. 1.5 Adjusting the throttle positioner system (Sec 3)

until the maximum engine speed is obtained consistent with even running. Repeat this operation on the right-hand carburettor then if necessary return the idling speed to the specified setting using the throttle adjustment screws. It will probably be necessary to repeat the mixture adjustment procedure two or three times until the engine is idling satisfactorily.
26 Make a final check that the carburettors are balanced then switch off the engine.
27 Refit the linkage connecting rod to the throttle shaft lever. If necessary, adjust the length of the rod so that it does not alter the throttle position as it is fitted.
28 Fit new tamperproof caps to the mixture adjustment screws.

Fuel injected models
Note: A special Mazda service tool, part number 49 HC27 140, is required to turn the bypass air adjusting screw. Without this tool it will not be possible to adjust the CO content.

29 Warm the engine up to normal operating temperature and allow the engine to idle. If necessary, remove the rubber plug and turn the idle speed (air adjusting) screw situated in the top of the throttle housing until the engine is idling at the specified speed (photos).
30 Run the engine at 2500–3000 rpm for two to three minutes then allow the engine to idle. Check the CO content is within the specified limits.
31 If adjustment is necessary remove the rubber plug from the top of the airflow meter housing to gain access to the bypass air adjusting screw (photo). Turn the screw clockwise to increase the CO content or anti-clockwise to decrease, until the reading is as given in the Specifications. If a CO meter is not being used weaken the mixture as described, then richen the mixture until the maximum engine speed is obtained consistent with even running.
32 If necessary, re-adjust the idling speed then check the CO reading again. Repeat as necessary until both the idling speed and CO reading are correct.
33 Refit the rubber plug to the airflow meter.

Checking the throttle positioner system (September 1985 onwards E series and B5 engined models with manual gearbox)

34 Check the idle speed and CO content as described above.
35 Disconnect the vacuum hose from the throttle positioner vacuum diaphragm mounted on the left-hand side of the carburettor. Disconnect the inlet manifold to throttle positioner solenoid hose from the solenoid and connect it directly to the vacuum diaphragm.
36 Start the engine and increase the engine speed to 2000 rpm using the throttle lever. Release the throttle lever and check that the idling speed is 1100 to 1200 rpm. If this is not the case turn the throttle positioner adjustment screw until the correct idle speed is obtained.
37 Once the idle speed is correct reconnect the hoses to their original positions.

Checking the dashpot and throttle position switch operation (fuel injected models)

Dashpot
38 The dashpot is located on the right-hand side of the throttle housing.
39 Push the dashpot rod into the dashpot, making sure that the rod enters the dashpot slowly (photo). Then release the rod making sure that the rod returns quickly. If this is not the case, the dashpot is faulty and must be renewed.
40 Warm the engine up to normal operating temperature and connect a tachometer to the engine in accordance with the manufacturer's instructions.
41 Allow the engine to idle then slowly increase the engine speed whilst closely observing the dashpot rod. The dashpot rod should come into contact with the throttle lever between 2600 and 3000 rpm.
42 If adjustment is necessary slacken the locknut and rotate the dashpot until it functions satisfactorily. Hold the dashpot and tighten the locknut securely.

Throttle position switch
43 The throttle position switch is located on the left-hand side of the throttle housing.
44 Disconnect the wiring connector from the switch and check the operation of the switch using a ohmmeter as follows.
45 Insert a 0.5 mm feeler gauge in between the throttle lever and the adjusting screw. With the throttle in this position there should be continuity between the bottom and centre switch terminal, and an open circuit between the top and centre terminal (photos).
46 Remove the 0.5 mm feeler gauge then insert a 0.7 mm feeler gauge in its place and recheck for continuity. With the throttle in this position there should be no continuity between either the top and centre terminal, or the bottom and centre switch terminal.
47 Remove the 0.7 mm feeler gauge and fully open the throttle lever. With the throttle fully open there should be continuity between the top

Chapter 1 Routine maintenance and servicing

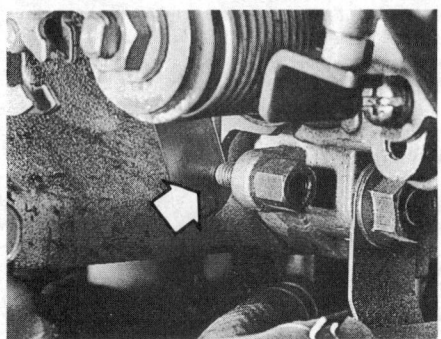
3.45A Insert a feeler gauge (arrowed) of the specified thickness between the throttle lever and adjusting screw...

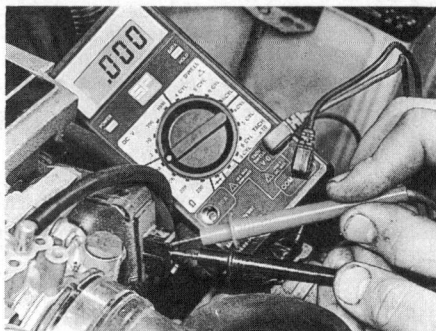
3.45B ...and check the resistances of throttle switch terminals

3.48 Adjustment is carried out by slackening the switch retaining screws and repositioning the switch

3.49A Fuel filter location on carburettor models

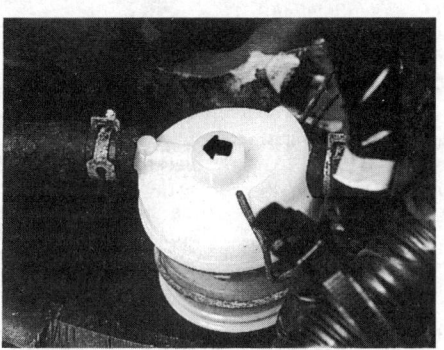
3.49B Filter must be installed with arrow (highlighted) on the top of the filter pointing in the direction of fuel flow

3.50 On fuel injected models the fuel filter is mounted on the engine compartment bulkhead

and centre terminal, and an open circuit between the bottom and centre terminal.
48 If adjustment is necessary, slacken the switch retaining screws and reposition the switch (photo). Once the switch is operating correctly tighten its retaining screws securely.

Fuel filter renewal

49 On carburettor models the fuel filter is mounted on the left-hand side of the engine compartment. Using a pair of pliers, release the hose clips and disconnect the hoses from the filter. Remove the filter from its retaining bracket. Fit a new filter into the retaining bracket ensuring that the arrow on the top of the filter is pointing in the direction of the fuel flow (towards the fuel pump) (photos). Refit the hoses and secure them in position with the retaining clips.
50 On fuel injected models the fuel filter is mounted on the engine compartment bulkhead (photo). Using a pair of pliers, release the clips securing the fuel hoses to the filter and disconnect the hoses. Undo the nuts securing the filter mounting bracket and remove the filter. Fit a new filter and tighten its retaining nuts securely. Where applicable, remove the upper hose union from the old filter and, positioning a new sealing washer on each side of the union, fit it to the new filter tightening the union bolt to the specified torque. Fit the hoses to the filter ensuring that they are held securely by their retaining clips.

General fuel system checks

51 The fuel system is most easily checked with the vehicle raised on a hoist or suitably supported on axle stands so the components underneath are readily visible and accessible.
52 If the smell of petrol is noticed while driving or after the vehicle has been parked in the sun, the system should be thoroughly inspected immediately.
53 Remove the petrol tank filler cap and check for damage, corrosion and an unbroken sealing imprint on the gasket. Renew the cap if necessary.
54 With the vehicle raised, inspect the petrol tank and filler neck for punctures, cracks and other damage. The connection between the filler neck and tank is especially critical. Sometimes a rubber filler neck or connecting hose will leak due to loose retaining clamps or deteriorated rubber.
55 Carefully check all rubber hoses and metal fuel lines leading away from the petrol tank. Check for loose connections, deteriorated hoses, crimped lines and other damage. Pay particular attention to the vent pipes and hoses which often loop up around the filler neck and can become blocked or crimped. Follow the lines to the front of the vehicle carefully inspecting them all the way. Renew damaged sections as necessary.
56 From within the engine compartment, check the security of all fuel hose attachments and inspect the fuel hoses and vacuum hoses for kinks, chafing and deterioration.
57 Check the operation of the throttle linkage and lubricate the linkage components with a few drops of light oil.

Exhaust system check

58 With the engine cold (at least an hour after the vehicle has been driven), check the complete exhaust system from the engine to the end of the tailpipe. Ideally the inspection should be carried out with the vehicle on a hoist to permit unrestricted access. If a hoist is not available, raise and support the vehicle safely on axle stands.
59 Check the exhaust pipes and connections for evidence of leaks, severe corrosion and damage. Make sure that all brackets and mountings are in good condition and tight. Leakage at any of the joints or in other parts of the system will usually show up as a black sooty stain in the vicinity of the leak. Holts Flexiwrap and Holts Gun Gum exhaust repair systems can be used for effective repairs to exhaust pipes and silencer boxes, including ends and bends. Holts Flexiwrap is an MOT approved permanent exhaust repair. Holts Firegum is suitable for the assembly of all exhaust system joints.
60 Rattles and other noises can often be traced to the exhaust system, especially the brackets and mountings. Try to move the pipes and silencers. If the components can come into contact with the body or suspension parts, secure the system with new mountings or if possible, separate the joints and twist the pipes as necessary to provide additional clearance.

61 Run the engine at idling speed then temporarily place a cloth rag over the rear end of the exhaust pipe and listen for any escape of exhaust gases that would indicate a leak.
62 On completion lower the car to the ground.
63 The inside of the exhaust tailpipe can be an indication of the engine's state-of-tune. If the pipe is black and sooty, the engine is in need of a tune-up, including a thorough fuel system inspection and adjustment.

4 Ignition system

Warning: *Voltages produced by an electronic ignition system are considerably higher than those produced by conventional systems. Extreme care must be taken when working on the system with the ignition switched on. Persons with surgically-implanted cardiac pacemaker devices should keep well clear of the ignition circuits, components and test equipment.*

Contact breaker points and condenser check, adjustment and renewal

1 Release the two spring clips and lift off the distributor cap. Pull the rotor arm off the shaft.
2 With the ignition switched off, use a screwdriver to open the contact breaker points, then visually check the points surfaces for pitting, roughness and discoloration. If the points have been arcing, there will be a build up of metal on the moving contact and a corresponding hole in the fixed contact, and if this is the case the points should be renewed.
3 Another method of checking the contact breaker points is by using a test meter available from most car accessory shops. If necessary, rotate the engine until the points are fully shut then connect the meter between the distributor LT wiring terminal and earth, and read off the condition of the points.
4 To remove the points, undo the two screws securing them to the distributor baseplate noting the location of the earth wire fitted under one of the screws.
5 Withdraw the points, undo the LT lead retaining screw noting the arrangement of the washer and insulator, then remove the contact breaker points from the distributor.
6 The purpose of the condenser, which is located externally on the side of the distributor body, is to ensure that, when the contact breaker points open, there is no sparking across them, which would cause wear of their faces and prevent the rapid collapse of the magnetic field in the coil. This would cause a reduction in coil HT voltage and ultimately lead to engine misfire.
7 If the engine becomes very difficult to start, or begins to miss after several miles of running, and the contact breaker points show signs of excessive burning, the condition of the condenser must be suspect. A further test can be made by separating the points by hand with the ignition switched on. If this is accompanied by a strong bright spark, it is indicative that the condenser has failed.
8 Without special test equipment, the only sure way to diagnose condenser trouble is to substitute a suspect unit with a new one and note if there is any improvement.
9 To remove the condenser, disconnect the wiring connector from the wiring harness. Unscrew the condenser retaining screw and remove the unit from the side of the distributor body.
10 Refitting of the condenser is a reversal of the removal procedure.
11 To fit the new contact breaker points, first check if there is any greasy deposit on them and if necessary clean them using methylated spirit.
12 Fit the points using a reversal of the removal procedure, then adjust them as follows. Turn the engine over using a socket or spanner on the crankshaft pulley bolt, until the heel of the contact breaker arm is on the peak of one of the four cam lobes.
13 With the points fully open, a feeler gauge equal to the contact breaker points gap, as given in the Specifications, should now just fit between the contact faces.
14 If the gap is too large or too small slacken the retaining screws and move the fixed point by inserting a flat bladed screwdriver through the hole in the breaker point plate and into the slot in the distributor baseplate.

15 Once the contact breaker point gap is correct tighten the retaining screws securely, then refit the rotor arm and distributor cap.
16 If a dwell meter is available, a far more accurate method of setting the contact breaker points is by measuring and setting the distributor dwell angle.
17 The dwell angle is the number of degrees of distributor shaft rotation during which the contact breaker points are closed (ie the period from when the points close after being opened by one cam lobe until they are opened again by the next cam lobe). The advantages of setting the points by this method are that any wear of the distributor shaft or cam lobes is taken into account, and also the inaccuracies of using a feeler gauge are eliminated.
18 To check and adjust the dwell angle, remove the distributor cap and rotor arm and connect one lead of the meter to the ignition coil positive (+) terminal and the other lead to the coil negative (–) terminal, or in accordance with the maker's instructions.
19 Whilst an assistant turns on the ignition and operates the starter observe the reading on the dwell meter scale. If the dwell angle is too small, the contact breaker gap is too wide, and if the dwell angle is excessive the gap is too small. **Note:** *Owing to machining tolerances, or wear in the distributor shaft or bushes, it is not uncommon for a contact breaker points gap correctly set with feeler gauges, to give a dwell angle outside the specified tolerances. If this is the case the dwell angle should be regarded as the preferred setting.*
20 If the dwell angle is not within the limits given in the Specifications, slacken the contact breaker point retaining screws and adjust the points as described in paragraph 14 whilst the engine is being turned over on the starter. Once the dwell angle is as specified tighten the contact breaker retaining screws securely. After completing the adjustment, switch off the engine and disconnect the dwell meter and refit the rotor arm and distributor cap.

Ignition timing check and adjustment

21 In order that the engine can run efficiently, it is necessary for a spark to occur at the spark plug and ignite the fuel/air mixture at the instant just before the piston on the compression stroke reaches the top of its travel. The precise instant at which the spark occurs is determined by the ignition timing, and this is quoted in degrees before top-dead centre (BTDC).
22 If the timing is being checked as a maintenance procedure refer to paragraph 32. If the distributor has been dismantled or renewed, or if its position on the engine has been altered, obtain an initial static setting as follows.

Static setting – conventional ignition system
23 First adjust the contact breaker point gap as described above.
24 Pull off the HT lead and remove No 1 spark plug (nearest the crankshaft pulley).
25 Place a finger over the plug hole and turn the engine in the normal direction of rotation (clockwise from the crankshaft pulley end) until pressure is felt in No 1 cylinder. This indicates that the piston is commencing its compression stroke. The engine can be turned using a socket or spanner on the crankshaft pulley bolt.
26 Continue turning the engine until the notch on the crankshaft pulley is aligned with the appropriate mark on the timing scale (photo). The scale is located just above the crankshaft pulley. The T on the scale indicates top dead centre (TDC) and the marks to the left are in increments of 2° BTDC. The marks to the right of the T are in increments of 2° ATDC.
27 Remove the distributor cap and check that the rotor arm is pointing towards the No 1 spark plug HT lead segment in the cap. If this is not the case, slacken the distributor clamp bolt and move the distributor body to the required position. Once the rotor arm position is correct tighten the clamp bolt securely.
28 Lift off the rotor arm then slacken the bolt securing the distributor clamp to the cylinder head. Turn the distributor body anti-clockwise slowly until the contact breaker points are closed, then slowly turn the distributor body clockwise until the points just open. Hold the distributor in this position and tighten the clamp bolt.
29 Refit the rotor arm, distributor cap, No 1 spark plug and HT lead.
30 It should now be possible to start the engine enabling the timing to be accurately checked with a timing light as follows.

Static setting – electronic ignition
31 On models with electronic ignition it should be possible to start the

Chapter 1 Routine maintenance and servicing

4.26 Ignition timing scale and crankshaft pulley notch on E series engine positioned at 6° BTDC

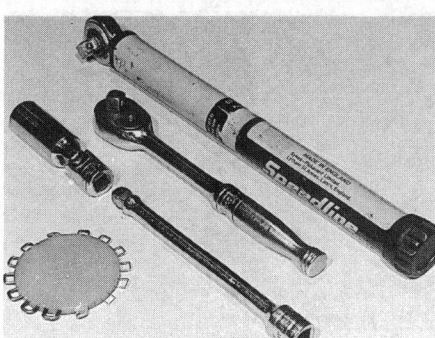

4.45 Tools required for removing, refitting and adjusting the spark plugs

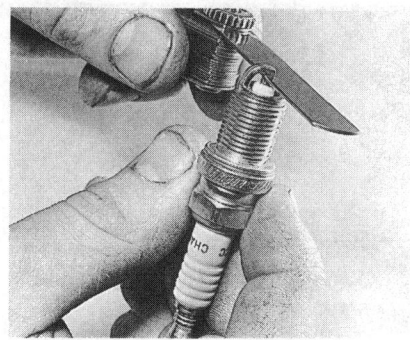

4.51 Measuring the spark plug gap with a feeler gauge

engine after checking the rotor arm position as described in paragraphs 24 to 27.

Dynamic setting

32 Refer to the Specifications at the beginning of this Chapter and note the specified ignition timing setting applicable to the engine being worked on. To make subsequent operations easier it is advisable to highlight the mark on the crankshaft pulley and the appropriate mark on the timing scale with white paint or chalk.
33 Connect the timing light in accordance with the manufacturer's instructions (usually interposed between the end of No 1 spark plug HT lead and No 1 spark plug terminal).
34 On all B series engines disconnect the vacuum advance pipe(s) from the distributor vacuum unit and plug the pipe ends.
35 Start the engine and leave it idling at the specified idling speed.
36 Point the timing light at the timing marks and they should appear to be stationary with the mark on the crankshaft pulley aligned with the appropriate mark on the scale.
37 If adjustment is necessary, (ie the crankshaft pulley mark does not line up with the appropriate mark) slacken the distributor clamp retaining nut and turn the distributor body clockwise to advance the timing, and anti-clockwise to retard it. Tighten the clamp nut when the setting is correct.
38 On E series engines disconnect the vacuum advance pipes from the distributor vacuum unit and plug the pipe ends.
39 On all models gradually increase the engine speed while still pointing the timing light at the marks. The mark on the flywheel should appear to advance further, indicating that the distributor centrifugal advance mechanism is functioning. If the mark remains stationary or moves in a jerky, erratic fashion, the centrifugal governor mechanism is suspect and the distributor should be removed and dismantled as described in Chapter 5.
40 Reconnect the vacuum pipe(s) to the distributor and check that the advance alters when the pipe is connected. If not, the vacuum unit on the distributor may be faulty and should be checked as described in Chapter 5.
41 After completing the checks and adjustments, switch off the engine and disconnect the timing light.

Spark plug inspection and renewal

42 The correct functioning of the spark plugs is vital for the correct running and efficiency of the engine. It is essential that the plugs fitted are appropriate for the engine, and the suitable type is specified at the beginning of this Chapter. If this type is used and the engine is in good condition, the spark plugs should not need attention between scheduled inspection intervals except for adjustment of the gaps. Spark plug cleaning is rarely necessary and should not be attempted unless specialised equipment is available as damage can easily be caused to the firing ends.
43 To remove the plugs, first open the bonnet and mark the HT leads one to four to correspond to the cylinder the lead serves (No 1 cylinder is at the crankshaft pulley end of the engine). Pull the HT leads from the plugs by gripping the end fitting, not the lead otherwise the lead connection may be fractured.
44 It is advisable to remove the dirt from the spark plug recesses using a clean brush, vacuum cleaner or compressed air before removing the plugs, to prevent the dirt dropping into the cylinders.
45 Unscrew the plugs using a spark plug spanner, suitable box spanner or a deep socket and extension bar (photo). Keep the socket in alignment with the spark plugs, otherwise if it is forcibly moved to either side, the porcelain top of the spark plug may be broken off. As each plug is removed, examine it as follows.
46 Examination of the spark plugs will give a good indication of the condition of the engine. If the insulator nose of the spark plug is clean and white, with no deposits, this is indicative of a weak mixture or too hot a plug (a hot plug transfers heat away from the electrode slowly, a cold plug transfers heat away quickly).
47 If the tip and insulator nose are covered with hard black-looking deposits, then this is indicative that the mixture is too rich. Should the plug be black and oily, then it is likely that the engine is fairly worn, as well as the mixture being too rich.
48 If the insulator nose is covered with light tan to greyish brown deposits, then the mixture is correct and it is likely that the engine is in good condition.
49 Examine the spark plug for signs of wear or damage such as worn electrodes or a cracked or chipped insulator nose and renew the plugs if necessary. If there are any traces of long brown tapering stains on the outside of the white porcelain insulator of the plug, then the plug must be renewed. Always renew the spark plugs as a set.
50 Whether the plugs are to be re-used or new ones are to be fitted, the spark plug gap is of considerable importance as if it is too large or too small, the size of the spark and its efficiency will be seriously impaired. For the best results the spark plug gap should be set in accordance with the Specifications at the beginning of this Chapter.
51 To set it, measure the gap with a feeler gauge, and then bend open, or close, the outer plug electrode until the correct gap is achieved (photo). The centre electrode should never be bent, as this may crack the insulation and cause plug failure, if nothing worse.
52 Special spark plug electrode gap adjusting tools are available from most motor accessory shops (photos).
53 Before fitting the spark plugs check that the threaded connector sleeves are tight and that the plug exterior surfaces and threads are clean.
54 It is very often difficult to insert spark plugs into their holes without cross-threading them. To avoid this possibility, fit a short length of rubber hose of suitable internal diameter over the end of the spark plug (photo). The flexible hose acts as a universal joint to help align the plug with the plug hole. Should the plug begin to cross-thread, the hose will slip on the spark plug, preventing thread damage to the aluminium cylinder head. Remove the rubber hose and tighten the plug to the specified torque using a spark plug socket and torque wrench. Refit the remaining spark plugs in the same manner.
55 Wipe the HT leads clean, then reconnect them in their correct order.

HT leads, distributor cap and rotor arm check and renewal

56 The spark plug HT leads should be checked whenever new spark plugs are installed in the engine.
57 Ensure that the leads are numbered before removing them to

4.52A Measuring the spark plug gap with a wire blade

4.52B Adjusting the spark plug gap using a special adjusting tool

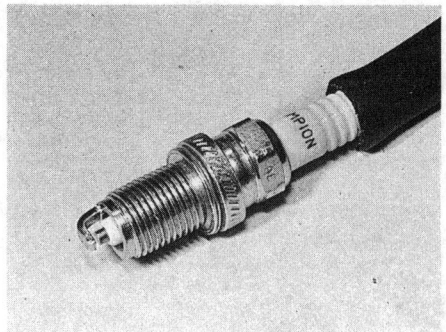

4.54 Using a short length of rubber hose to facilitate inserting the spark plugs

avoid confusion when refitting. Pull one HT lead from its plug by gripping the end fitting, not the lead, otherwise the lead connection may be fractured.

58 Check inside the end fitting for signs of corrosion, which will look like a white crusty powder. Push the end fitting back onto the spark plug ensuring that it is a tight fit on the plug. If it isn't, remove the lead again and use pliers to carefully crimp the metal connector inside the end fitting until it fits securely on the end of the spark plug.

59 Using a clean rag, wipe the entire length of the lead to remove any built-up dirt and grease. Once the lead is clean, check for burns, cracks and other damage. Do not bend the lead excessively or pull the lead lengthwise – the conductor inside might break.

60 Disconnect the other end of the lead from the distributor cap. Again, pull only on the end fitting. Check for corrosion and a tight fit in the same manner as the spark plug end. Refit the lead securely on completion.

61 Check the remaining HT leads one at a time, in the same way.

62 If new HT leads are required, purchase a set suitable for your specific vehicle and engine.

63 Remove the distributor cap, wipe it clean and carefully inspect it inside and out for signs of cracks, carbon tracks (tracking) and worn, burned or loose contacts. Similarly inspect the rotor arm. Renew these components if any defects are found. It is common practice to renew the cap and rotor arm whenever new HT leads are fitted. When fitting a new cap, remove the HT leads from the old cap one at a time and fit them to the new cap in the same location – do not simultaneously remove all the leads from the old cap or firing order confusion may occur.

64 Even with the ignition system in first class condition, some engines may still occasionally experience poor starting attributable to damp ignition components. To disperse moisture, Holts Wet Start can be very effective. Holts Damp Start should be used for providing a sealing coat to exclude moisture from the ignition system, and in extreme difficulty, Holts Cold Start will help to start a car when only a very poor spark occurs.

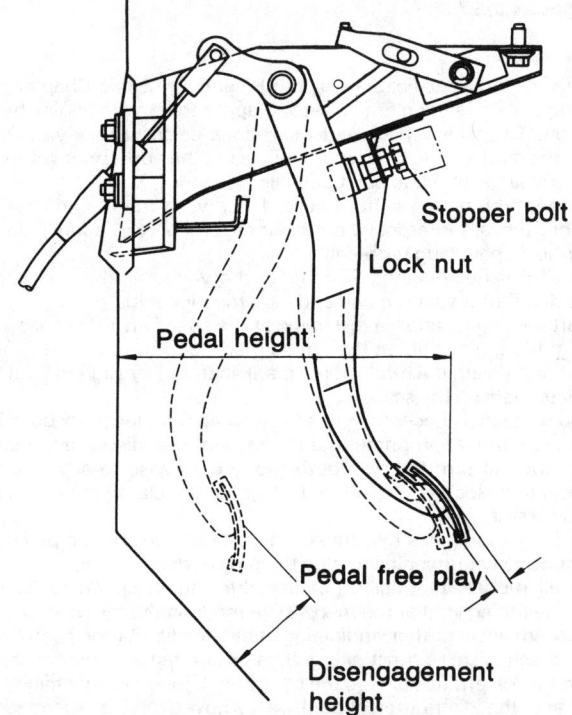

Fig. 1.6 Clutch pedal height and free play details (Sec 5)

5 Clutch

Clutch pedal and cable adjustment – cable operated clutch

1 Although adjustment is usually unnecessary, always check the clutch pedal height before adjusting the clutch cable.

2 Measure the distance from the centre of the clutch pedal pad upper surface to the engine compartment bulkhead (Fig. 1.6). Compare the figure obtained with the dimension given in the Specifications. Note the dimension does not take into account the thickness of sound deadening material or carpet affixed to the bulkhead and an allowance should be made for this.

3 If adjustment is necessary, remove the cover from under the facia, slacken the pedal stop locknut and turn the pedal stop bolt until the correct height is obtained. Tighten the locknut without moving the stop bolt and refit the cover to the facia.

4 Once the pedal height is correct, depress the pedal by hand and measure the distance the clutch pad travels from the pedal at rest position until firm resistance is met. This is the clutch pedal free play and should be as given in the Specifications.

5 If adjustment is necessary, open up the bonnet to gain access to the clutch operating lever situated on the top of the gearbox housing. Slacken the locknut (where fitted) on the clutch cable end and turn the adjuster nut until the distance from the release lever to the clutch cable roller is approximately 2 to 3 mm when all free play is removed from the cable and operating lever (photo). If necessary, tighten the adjuster locknut securely once the cable free play is correct.

6 Once the pedal free play is correct, check that when the pedal is depressed fully, the distance from the centre of the pedal pad upper surface to the floor is not less than the dimension given in the Specifications.

Clutch fluid level check and pedal adjustment – hydraulically operated clutch

7 The clutch fluid level is readily visible through the translucent material of the master cylinder reservoir. With the car on level ground the fluid level should be above the minimum (MIN) mark and preferably

Chapter 1 Routine maintenance and servicing

5.5 Adjusting the clutch cable free play

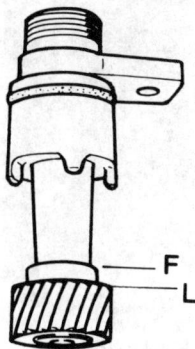

Fig. 1.7 Manual gearbox lubricant level markings on speedometer gear (Sec 6)

on or near the maximum (MAX) mark. Note that wear of the clutch friction plate lining causes the level of the clutch fluid to gradually fall. It is not therefore necessary to top up the level to compensate for this minimal drop, however the level must never be allowed to fall below the minimum mark.

8 If topping up is necessary, first wipe the area around the filler cap with a clean rag before removing the cap. When adding fluid, pour it carefully into the reservoir to avoid spilling it on surrounding painted surfaces. Be sure to use only the specified hydraulic fluid since mixing different types of fluid can cause damage to the system. See 'Lubricants fluids and capacities' at the beginning of this Chapter. **Warning:** Hydraulic fluid can harm your eyes and damage painted surfaces, so use extreme caution when handling and pouring it. Do not use fluid that has been standing open for some time as it absorbs moisture from the air. Excess moisture can cause a dangerous loss of braking effectiveness.

9 When adding fluid it is a good idea to inspect the reservoir for contamination. The system should be drained and refilled if deposits, dirt particles or contamination are seen in the fluid.

10 After filling the reservoir to the proper level, make sure that the cap is refitted securely to avoid leaks and the entry of foreign matter.

11 If the reservoir requires repeated replenishing to maintain the proper level, this is an indication of a hydraulic leak somewhere in the system which should be investigated immediately.

12 Once the fluid level is correct check and, if necessary adjust, the clutch pedal height as described in paragraphs 2 and 3.

13 Once the pedal height is correct check the pedal free play as described in paragraph 4. If adjustment is necessary, slacken the pedal pushrod locknut and turn the pushrod as necessary. Once the correct amount of free play is obtained tighten the pushrod locknut securely.

14 Once the pedal free play is correct, check that when the pedal is depressed fully, the distance from the centre of the pedal pad upper surface to the floor is as given in the Specifications.

6 Manual gearbox and automatic transmission

Manual gearbox oil level check

1 The gearbox oil level is checked using the speedometer drive gear ensuring that the vehicle is standing on level ground.

2 Working from inside the engine compartment, wipe clean the area around the speedometer cable connection on the top of the gearbox housing. Undo the speedometer drive gear retaining bolt and withdraw the cable and gear.

3 Wipe the gear clean, reinsert it fully, withdraw it once more and

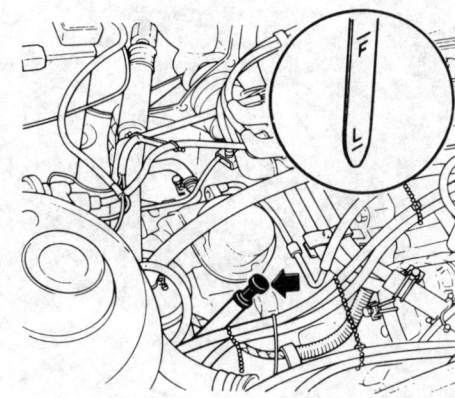

Fig. 1.8 Automatic transmission fluid level dipstick and markings (Sec 6)

observe the level. This should be between the upper edge of the gear teeth and the shoulder of the gear (Fig. 1.7). There is a diagrammatic representation of this cast into the gearbox housing just above the filler orifice. If topping up is necessary, add the specified lubricant through the filler hole until the level is correct, then refit the cable and gear and tighten the retaining bolt securely (photos).

Automatic transmission fluid level check

4 The automatic transmission fluid level should be checked when the engine is at normal operating temperature preferably after a short journey.

5 With the car standing on level ground and the engine running, apply the handbrake and slowly move the selector lever through all gear positions.

6 Return the selector lever to the P position and with the engine still idling, withdraw the dipstick from the filler tube and wipe it clean on a lint free cloth.

7 Reinsert the dipstick, withdraw it immediately and observe the fluid level. This should be between the L and F marks on the dipstick.

8 If topping up is necessary, add the specified fluid through the dipstick tube until the level is correct (photo). Use a funnel with a fine screen mesh to avoid spillage and ensure that any foreign matter is trapped. Take care not to overfill the transmission.

9 After topping up recheck the level again, refit the dipstick and switch the engine off.

Manual gearbox oil draining, renewal and filling

10 Position the car on level ground and place a suitable container beneath the gearbox drain plug, which is accessible through a hole in the crossmember (photo).

6.3A Speedometer drive level marking diagram cast on gearbox housing

6.3B Topping up the gearbox oil

6.8 Top up the automatic transmission fluid through the dipstick tube

6.10 Gearbox drain plug is accessible through hole in the crossmember

11 Using a socket and extension bar, unscrew the plug and allow the lubricant to drain into the container. Refit the plug once all the lubricant has drained.
12 From within the engine compartment, wipe clean the area around the speedometer cable attachment on the top of the gearbox housing. Undo the retaining bolt and withdraw the cable and drive gear.
13 Add the specified type and quantity of lubricant through the speedometer gear orifice until the level is between the upper edge of the gear teeth and the shoulder of the gear. Note that early models utilize automatic transmission fluid as a lubricant whereas later models are filled with gear oil. *Under no circumstances may the two types of lubricant be mixed or interchanged.*
14 When the gearbox has been filled to the correct level, refit the speedometer drive gear and cable assembly and tighten the retaining bolt securely.

Automatic transmission fluid draining, renewal and filling

15 Firmly apply the handbrake, then jack up the front of the car and support it on axle stands.
16 Wipe clean the area around the transmission drain plug located at the base of the final drive housing.
17 Place a suitable container beneath the drain plug and, working through the access hole in the crossmember, unscrew the plug and allow the fluid to drain. When all the fluid has drained, refit the plug.
18 If the transmission has been drained to allow for a repair operation to be carried out (ie driveshaft removal or transmission removal) then sufficient fluid will have been drained by this method to allow the work to proceed. However, if the transmission has been drained for fluid renewal then it will also be necessary to drain the small quantity of fluid remaining in the oil pan as follows.
19 Remove the vehicle undertray to gain access to the oil pan.
20 Wipe clean the area around the oil pan retaining bolts and the joint face.
21 Place a suitable container beneath the oil pan then undo all the bolts securing the oil pan to the transmission.
22 Carefully prise the oil pan downwards to break the seal. Once the pan is free lower the pan and tip the fluid into the container.
23 Thoroughly clean the oil pan in paraffin and remove all traces of old gasket from the pan and transmission sealing faces. Dry the pan with a lint free cloth.
24 Apply a small amount of jointing compound to both sides of the gasket and place the gasket onto the pan.
25 Refit the oil pan and tighten the bolts progressively and in a diagonal sequence to the specified torque.
26 Refit the undertray and lower the car to the ground.
27 Remove the transmission dipstick from the filler tube and add the specified transmission fluid, a small amount at a time until the level just starts to register on the dipstick. Use a funnel with a fine mesh screen to avoid spillage and ensure that any foreign matter is trapped. Take care not to overfill the transmission.
28 With the selector lever in the P position, start the engine and allow it to idle.
29 With the hand and footbrake applied slowly move the selector lever through each gear position then return it to the P position.
30 With the engine still idling check the fluid level on the dipstick and top up if necessary so that the level is between the L and F marks.
31 Refit the dipstick and drive the car on a short journey until the engine and transmission reach normal operating temperature.
32 With the car standing on level ground, the engine idling and the transmission in P, make a final check of the transmission fluid and top up if necessary.

7 Driveshafts

Driveshaft rubber gaiter and CV joint check

1 With the vehicle raised and securely supported on stands, turn the steering onto full lock then slowly rotate the roadwheel. Inspect the condition of the outer constant velocity (CV) joint rubber gaiters while squeezing the gaiters to open out the folds (photo). Check for signs of cracking, splits or deterioration of the rubber which may allow the grease to escape and lead to water and grit entry into the joint. Also check the security and condition of the retaining clips. Repeat these checks on the inner CV joints. If any damage or deterioration is found, the gaiters should be renewed as described in Chapter 8.
2 At the same time check the general condition of the outer CV joints themselves by first holding the driveshaft and attempting to rotate the wheels. Repeat this check on the right-hand inner joint by holding the inner joint yoke and attempting to rotate the driveshaft. The left-hand inner joint is concealed by a rubber gaiter which is bolted to the transmission casing, and it is not possible to hold the joint since the yoke

Chapter 1 Routine maintenance and servicing

7.1 Checking the condition of the driveshaft inner constant velocity (CV) joint gaiter

is an integral part of the differential sun wheel. However, one way to get round this problem is to have an assistant hold the right-hand wheel stationary with 4th gear selected, and then to attempt to rotate the left-hand driveshaft. If this method is used, beware of confusing wear in the left-hand CV joint with general wear in the transmission.

3 Any appreciable movement in the CV joint indicates wear in the joint, wear in the driveshaft splines or a loose driveshaft retaining nut.

8 Braking system

Hydraulic fluid level check

1 The brake fluid reservoir is located on the top of the brake master cylinder which is attached to the front of the vacuum servo unit.

2 The brake fluid inside the reservoir is readily visible. With the car on level ground the level should be above the minimum (MIN) mark and preferably on or near the maximum (MAX) mark (photo). Note that wear of the brake pad or brake shoe linings causes the level of the brake fluid to gradually fall, so that when the brake pads are renewed, the original level of the fluid is restored. It is not therefore necessary to top up the level to compensate for this minimal drop, however the level must never be allowed to fall below the minimum mark.

3 If topping up is necessary, first wipe the area around the filler cap with a clean rag before removing the cap. When adding fluid, pour it carefully into the reservoir to avoid spilling it on surrounding painted surfaces (photo). Be sure to use only the specified brake hydraulic fluid since mixing different types of fluid can cause damage to the system. See 'Lubricants fluids and capacities' at the beginning of this Chapter.

Warning: *Hydraulic fluid can harm your eyes and damage painted surfaces, so use extreme caution when handling and pouring it. Do not use fluid that has been standing open for some time as it absorbs moisture from the air. Excess moisture can cause a dangerous loss of braking effectiveness.*

4 When adding fluid it is a good idea to inspect the reservoir for contamination. The system should be drained and refilled if deposits, dirt particles or contamination are seen in the fluid.

5 After filling the reservoir to the proper level, make sure that the cap is refitted securely to avoid leaks and the entry of foreign matter.

6 If the reservoir requires repeated replenishing to maintain the proper level, this is an indication of a hydraulic leak somewhere in the system which should be investigated immediately.

Hydraulic fluid renewal

7 The procedure is similar to that for the bleeding of the hydraulic system as described in Chapter 9, except that the brake fluid reservoir should be emptied by syphoning, using a clean poultry baster or similar before starting, and allowance should be made for the old fluid to be removed from the circuit when bleeding a section of the circuit.

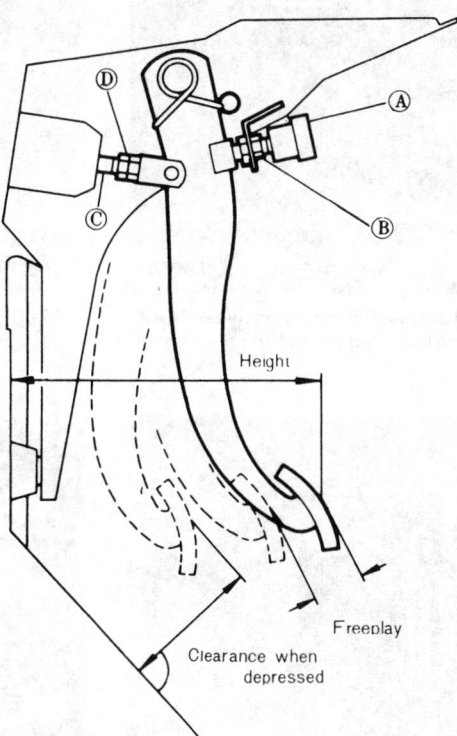

Fig. 1.9 Brake pedal adjustment details (Sec 8)

A Brake lamp switch
B Brake lamp switch locknut
C Vacuum servo pushrod
D Vacuum servo pushrod locknuts

Front brake disc pad wear check

8 Apply the handbrake, slacken the front wheel nuts then jack up the front of the of the car. Support the car on axle stands and remove the roadwheels.

9 Using a steel rule, check that the thickness of the brake pad linings is not less than the minimum thickness given in the Specifications. On pre-September 1985 models it will only be possible to view the centre area of the pads through the small aperture in the caliper, but on later models all of the pad area is visible (photo).

10 As a guide a wear indicator is provided on the pad. This is in the form of a groove running down the centre of the pads which can be seen from the front edge of the caliper. If the groove is not visible the pad is worn. On July 1987 models onwards a second wear indicator is also provided in the form of a metal tang which contacts the disc when the pad has worn to its service limit (photo). When the brakes are applied and the tang contacts the disc, a squealing noise will be omitted indicating that the pads need renewing.

11 If the friction material of any one pad is less than the minimum amount renew all the front pads as a set with reference to Chapter 9.

Rear brake disc pad wear check

12 Slacken the rear wheel nuts then chock the front wheels, jack up the rear of the car and support it on axle stands. Remove the rear roadwheels.

13 Using a steel rule, check that the thickness of the brake pad linings is not less than the minimum thickness given in the Specifications.

14 If any one pad thickness is less than the minimum amount, renew all the rear pads with reference to Chapter 9.

Rear brake shoe lining wear check

15 Remove the rear brake drums as described in Section 12 of Chapter 9.

Chapter 1 Routine maintenance and servicing

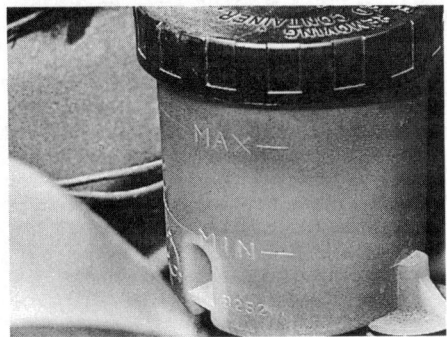

8.2 Brake fluid level must be between MAX and MIN markings on side of the reservoir

8.3 Topping up the brake fluid reservoir

8.9 Brake pad viewing aperture on pre-September 1985 models

8.10 Brake pad wear indicator tang on July 1987 onwards models

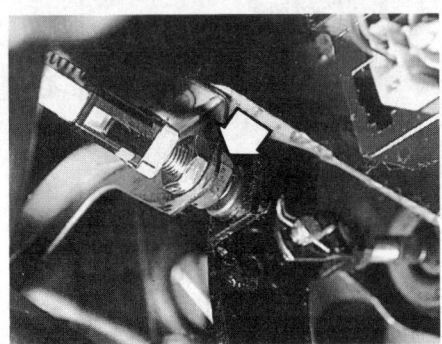

8.19 Stop lamp switch locknut (arrowed)

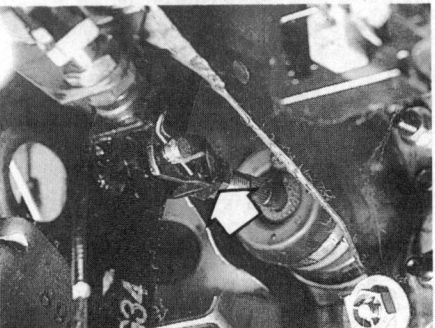

8.21 Brake pedal pushrod locknut (arrowed)

16 Check that each brake shoe lining thickness is not less than the thickness given in the Specifications. Also examine the lining for signs of contamination with brake fluid or grease.

17 If any one lining thickness is less than the minimum amount or the linings have been contaminated, renew all the rear brake shoes, as described in Chapter 9. If contamination is evident the cause must be traced and cured before new shoes are fitted.

Brake pedal adjustment

18 Measure the distance from the centre of the brake pedal pad upper surface to the engine compartment bulkhead (Fig. 1.9). Compare the figure obtained with the brake pedal height dimension given in the Specifications. Note the dimension does not take into account the thickness of sound deadening material or carpet affixed to the bulkhead and an allowance should be made for this.

19 If adjustment is necessary, remove the cover from under the facia, slacken the pedal stop lamp switch locknut and turn the switch until the correct height is obtained. Tighten the locknut without moving the switch (photo).

20 Once the pedal height is correct, depress the pedal by hand and measure the distance the brake pad travels from the pedal at rest position until firm resistance is met. This is the brake pedal free play and should be as given in the Specifications.

21 If adjustment is necessary, slacken the pedal pushrod locknut and screw the pushrod in or out, as necessary (photo). Once the correct amount of free play is obtained tighten the adjuster locknut securely.

22 Once the pedal free play is correct, check that when the pedal is depressed fully, the distance from the centre of the pedal pad upper surface to the floor is not less than the dimension given in the Specifications.

Handbrake check and adjustment

23 Before adjusting the handbrake cable, apply the foot brake several times whilst the vehicle is moving backwards. This will ensure that the automatic brake adjusters are fully adjusted.

24 From inside the vehicle, apply the handbrake in the normal way and count the number of clicks of the handbrake ratchet mechanism. On

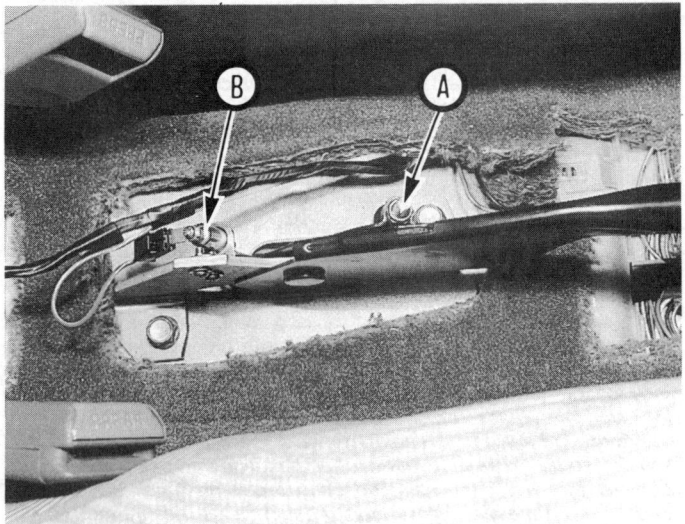

8.26 Handbrake lever cable adjusting nut (A) and warning lamp switch (B)

pre-September 1985 models, under a pull of approximately 10 kg (22 lb) the lever should travel between 5 and 9 clicks of the ratchet. On September 1985 onwards models, under a pull of 20 kg (44 lb) the lever should travel between 7 and 11 notches on models with rear drum brakes, and 9 to 15 notches on models with rear disc brakes. If the lever travel is not within the specified range adjust as follows.

25 Remove the centre console as described in Chapter 11.

26 Using a suitable spanner on the adjusting nut at the end of the cable, turn the nut as necessary until the correct amount of lever travel is obtained (photo).

27 Release the lever and make sure that the car is free to roll and that the rear brakes are not binding.

Chapter 1 Routine maintenance and servicing

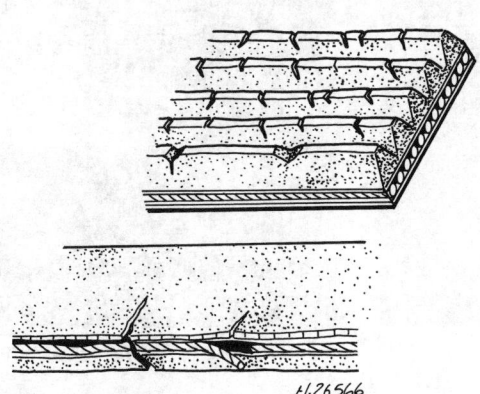

Fig. 1.10 Checking for power steering drivebelt wear – multi-ribbed type shown as fitted to later models (Sec 9)

28 Switch on the ignition and check that the warning light on the instrument panel illuminates when the handbrake lever is moved by one click of the ratchet. If not, slacken the switch retaining bolt and reposition the switch. Once the switch operation is correct tighten the retaining bolt securely.
29 Refit the centre console as described in Chapter 11.

9 Suspension and steering

Front suspension and steering check

1 Raise the front of the vehicle and securely support it on axle stands.
2 Visually inspect the balljoint dust covers and the steering rack and pinion gaiters for splits, chafing or deterioration (photo). Any wear of these components will cause loss of lubricant together with dirt and water entry resulting in rapid deterioration of the balljoints or steering gear.
3 On vehicles equipped with power steering, check the fluid hoses for chafing or deterioration and the pipe and hose unions for fluid leaks. Also check for signs of fluid leakage under pressure from the steering gear rubber gaiters which would indicate failed fluid seals within the steering gear.
4 Grasp the roadwheel at the 12 o'clock and 6 o'clock positions and try to rock it (photo). Very slight free play may be felt, but if the movement is appreciable further investigation is necessary to determine the source. Continue rocking the wheel while an assistant depresses the footbrake. If the movement is now eliminated or significantly reduced, it is likely that the hub bearings are at fault. If the free play is still evident with the footbrake depressed, then there is wear in the suspension joints or mountings.
5 Now grasp the wheel at the 9 o'clock and 3 o'clock positions and try to rock it as before. Any movement felt now may again be caused by wear in the hub bearings or the steering track-rod balljoints. If the outer balljoint is worn the visual movement will be obvious. If the inner joint is suspect it can be felt by placing a hand over the rack and pinion rubber gaiter and gripping the track-rod. If the wheel is now rocked, movement will be felt at the inner joint if wear has taken place.
6 Using a large screwdriver, or flat bar, check for wear in the suspension mounting bushes by levering between the relevant suspension component and its attachment point. Some movement is to be expected as the mountings are made of rubber, but excessive wear should be obvious. Also check the condition of any visible rubber bushes, looking for splits, cracks or contamination of the rubber.
7 With the car standing on its wheels, have an assistant turn the steering wheel back and forth about an eighth of a turn each way. There should be very little, if any lost movement between the steering wheel and roadwheels. If this is not the case, closely observe the joints and mountings previously described, but in addition check the steering column universal joints for wear and also check the rack and pinion steering gear itself.

Power steering fluid level check

8 The power steering reservoir is situated on top of the pump on the front right-hand side of the engine.
9 For the check the front wheels should be pointing straight ahead and the engine should be stopped. The vehicle must also be positioned on level ground.
10 Wipe clean the area around the reservoir filler cap to prevent the

Condition	Probable cause	Corrective action	Condition	Probable cause	Corrective action
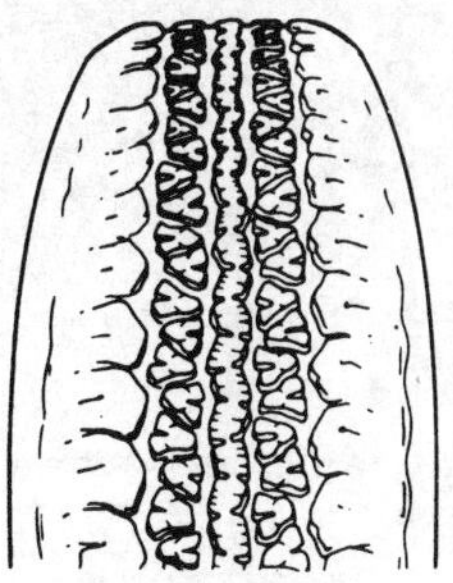 Shoulder wear	• Underinflation (wear on both sides) • Incorrect wheel camber (wear on one side) • Hard cornering	• Check and adjust pressure • Repair or renew suspension parts • Reduce speed	Feathered edge 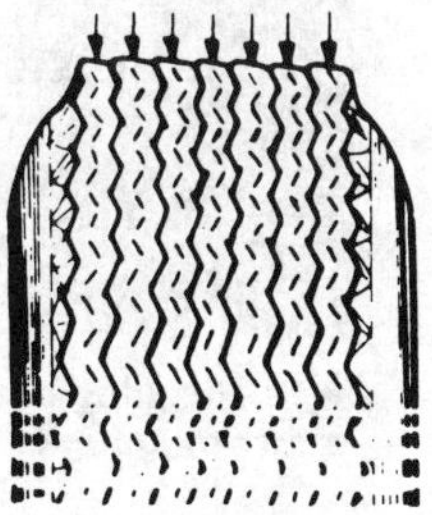 / Toe wear	• Incorrect toe setting	• Adjust front wheel alignment
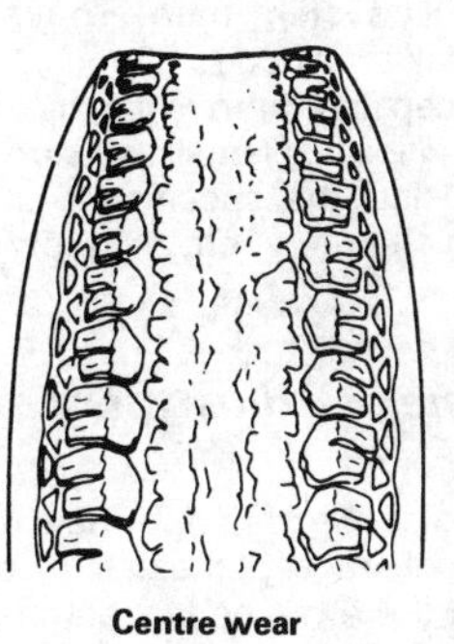 Centre wear	• Overinflation	• Measure and adjust pressure	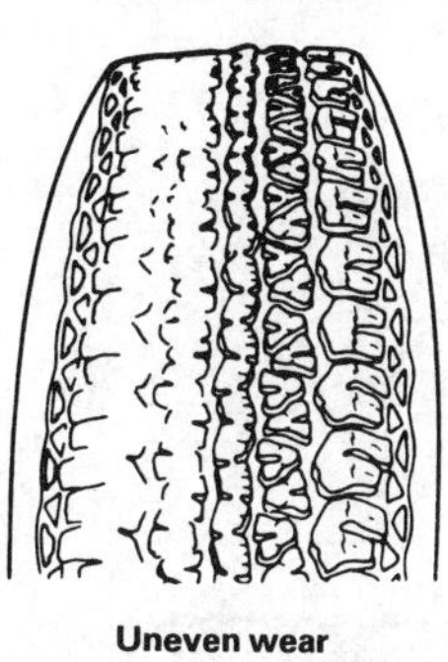 Uneven wear	• Incorrect camber or castor • Malfunctioning suspension • Unbalanced wheel • Out-of-round brake disc/drum	• Repair or renew suspension parts • Repair or renew suspension parts • Balance tyres • Machine or renew disc/drum

Fig. 1.11 Tyre wear patterns and causes (Sec 9)

Chapter 1 Routine maintenance and servicing

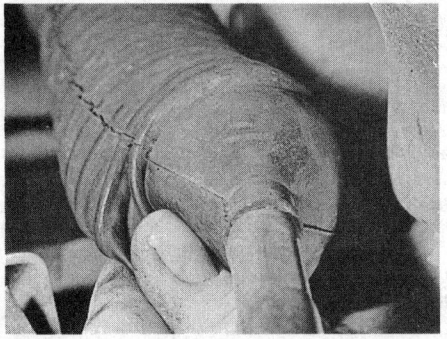

9.2 Checking the condition of the steering rack rubber gaiter

9.4 Checking for wear in the front suspension and hub bearings

9.11 Topping up the power steering reservoir

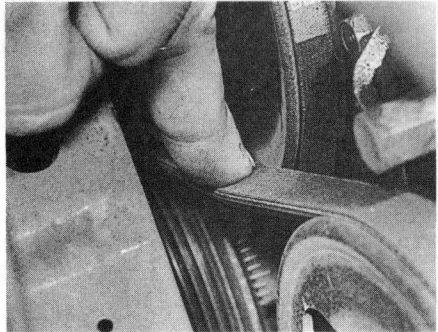

9.14 Checking power steering drive belt deflection

9.15A Slacken the power steering pump mounting bolt...

9.15B ...and adjuster locknut...

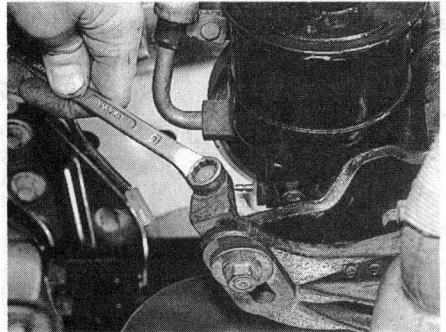

9.15C ...then adjust the drivebelt tension using the adjuster bolt

9.18 Checking the tyre tread depth with an indicator gauge

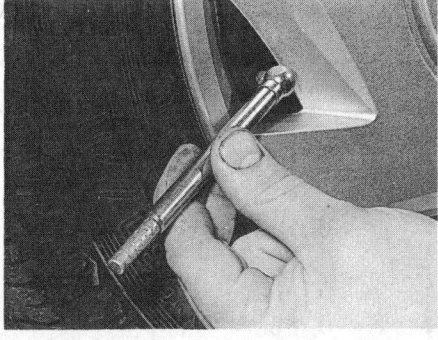

9.20 Checking the tyre pressures with a tyre pressure gauge

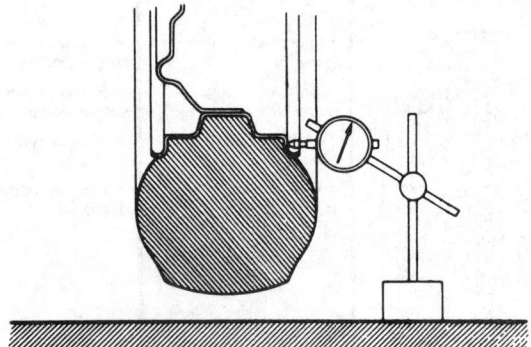

Fig. 1.12 Checking wheel run-out with a dial gauge and stand (Sec 9)

ingress of any foreign matter into the hydraulic system, then remove the cap from the reservoir. The cap has a dipstick attached to it.

11 Wipe clean the dipstick, then refit the cap and remove it again. The fluid level must be between the high (H) and low (L) level marks on the dipstick. If necessary top up the level using the specified type of lubricant, taking care not to overfill (photo). Securely refit the reservoir cap.

Power steering pump drivebelt check, adjustment and renewal

12 The power steering drivebelt is located on the front right-hand side of the engine. Due to its function and material makeup, the belt is prone to failure after a period of time and should therefore be inspected and adjusted periodically.

13 With the engine switched off, inspect the full length of the power steering pump drivebelt for cracks and separation of the belt plies. It will

Chapter 1 Routine maintenance and servicing

be necessary to turn the engine in order to move the belt from the pulleys so that the belt can be inspected thoroughly. Twist the belt between the pulleys so that both sides can be viewed. Also check for fraying, and glazing which gives the belt a shiny appearance. Check the pulleys for nicks, cracks, distortion and corrosion.

14 The tension of the belt is checked by pushing on it midway between the power steering pump and crankshaft pulleys (photo). The belt deflection should be as specified when a force of approximately 10 kg (22 lb) is applied.

15 If adjustment is necessary, loosen off the pump mounting bolt and adjuster locknut, then turn the pump adjusting bolt in the required direction until the drivebelt tension is correct. Securely tighten the pump mounting bolt and adjuster locknut (photos).

16 Run the engine for about five minutes, then recheck the tension.

17 To renew the belt, slacken the belt tension fully as described above. Slip the belt off the pulleys then fit the new belt ensuring that it is routed correctly. With the belt in position, adjust the tension as previously described.

Wheel and tyre maintenance and tyre pressure checks

18 The original tyres on this car are equipped with tread wear safety bands which will appear when the tread depth reaches approximately 1.6 mm. Tread wear can be monitored with a simple, inexpensive device known as a tread depth indicator gauge (photo).

19 Wheels and tyres should give no real problems in use provided that a close eye is kept on them with regard to excessive wear or damage. To this end, the following points should be noted.

20 Ensure that tyre pressures are checked regularly and maintained correctly (photo). Checking should be carried out with the tyres cold and not immediately after the vehicle has been in use. If the pressures are checked with the tyres hot, an apparently high reading will be obtained owing to heat expansion. Under no circumstances should an attempt be made to reduce the pressures to the quoted cold reading in this instance, or underinflation will result.

21 Note any abnormal tread wear with reference to Fig. 1.11. Tread pattern irregularities such as feathering, flat spots and more wear on one side than the other are indications of front wheel alignment and/or balance problems. If any of these conditions are noted, they should be rectified as soon as possible.

22 Underinflation will cause overheating of the tyre owing to excessive flexing of the casing, and the tread will not sit correctly on the road surface. This will cause a consequent loss of adhesion and excessive wear, not to mention the danger of sudden tyre failure due to heat build-up.

23 Overinflation will cause rapid wear of the centre part of the tyre tread coupled with reduced adhesion, harsher ride, and the danger of shock damage occurring in the tyre casing.

24 Regularly check the tyres for damage in the form of cuts or bulges, especially in the sidewalls. Remove any nails or stones embedded in the tread before they penetrate the tyre to cause deflation. If removal of a nail does reveal that the tyre has been punctured, refit the nail so that its point of penetration is marked. Then immediately change the wheel and have the tyre repaired by a tyre dealer. Do not drive on a tyre in such a condition. In many cases a puncture can be simply repaired by the use of an inner tube of the correct size and type. If in any doubt as to the possible consequences of any damage found, consult your local tyre dealer for advice.

25 Periodically remove the wheels and clean any dirt or mud from the inside and outside surfaces. Examine the wheel rims for signs of rusting, corrosion or other damage. Light alloy wheels are easily damaged by 'kerbing' whilst parking, and similarly steel wheels may become dented or buckled. Renewal of the wheel is very often the only course of remedial action possible.

26 The balance of each wheel and tyre assembly should be maintained to avoid excessive wear, not only to the tyres but also to the steering and suspension components. Wheel imbalance is normally signified by vibration through the vehicle's bodyshell, although in many cases it is particularly noticeable through the steering wheel. Conversely, it should be noted that wear or damage in suspension or steering components may cause excessive tyre wear. Out-of-round or out-of-true tyres, damaged wheels and wheel bearing wear/maladjustment also fall into this category. Balancing will not usually cure vibration caused by such wear.

27 Wheel balancing may be carried out with the wheel either on or off the vehicle. If balanced on the vehicle, ensure that the wheel-to-hub relationship is marked in some way prior to subsequent wheel removal so that it may be refitted in its original position.

28 General tyre wear is influenced to a large degree by driving style – harsh braking and acceleration or fast cornering will all produce more rapid tyre wear. Interchanging of tyres may result in more even wear, however it is worth bearing in mind that if this is completely effective, the added expense is incurred of replacing simultaneously a complete set of tyres, which may prove financially restrictive for many owners.

29 Front tyres may wear unevenly as a result of wheel misalignment. The front wheels should always be correctly aligned according to the settings specified by the vehicle manufacturer.

30 Legal restrictions apply to many aspects of tyre fitting and usage and in the UK this information is contained in the Motor Vehicle Construction and Use Regulations. It is suggested that a copy of these regulations is obtained from your local police if in doubt as to current legal requirements with regard to tyre type and condition, minimum tread depth, etc.

Regreasing the front and rear wheel bearings

31 At the intervals specified in the maintenance schedule, the front and rear hub assemblies should be removed and inspected as described in Chapter 10. On refitting, pack the bearings with fresh grease.

10 Bodywork

Underbody and general body check

1 With the car raised and supported on axle stands or over an inspection pit, thoroughly inspect the underbody and wheel arches for signs of damage and corrosion. In particular examine the bottom of the side sills and concealed areas where mud can collect. Where corrosion and rust is evident, press firmly on the panel by hand and check for possible repairs. If the panel is not seriously corroded, clean away the rust and apply a new coating of underseal. Refer to Chapter 11 for more details of body repairs.

2 Check all external body panels for damage and rectify where necessary.

11 Electrical system

Battery check and maintenance

Caution: *Before carrying out any work on the vehicle battery, read through the precautions given in Safety First! at the beginning of this manual.*

1 The battery is located in the engine compartment, on the left-hand side.

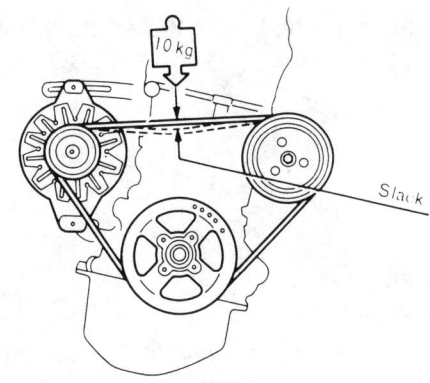

Fig. 1.13 Alternator drivebelt adjustment tension checking point – E series engines (Sec 11)

Chapter 1 Routine maintenance and servicing

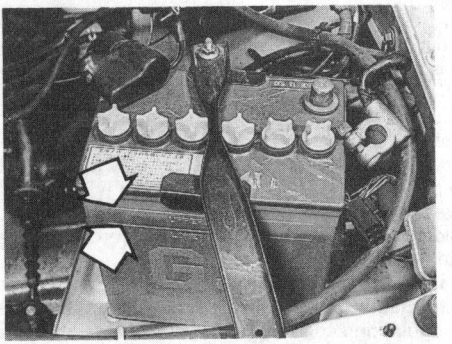

11.2 Electrolyte level must be maintained between the upper and lower level markings (arrowed) on the battery case

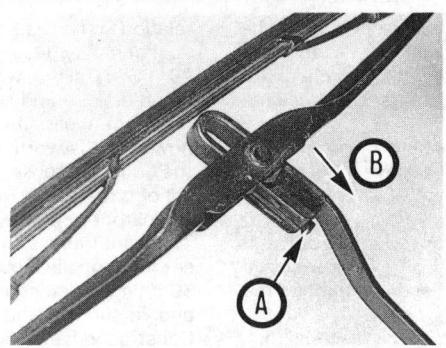

11.20 Windscreen wiper blade removal on later models – Press clip 'A' and withdraw blade from arm in direction of arrow 'B'

11.26A Front washer fluid reservoir location

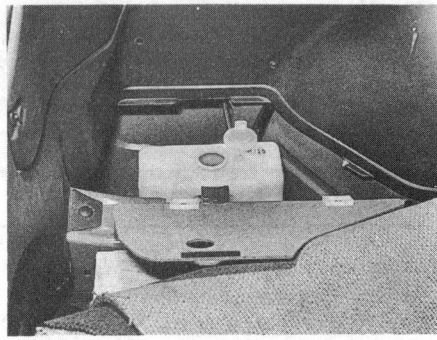

11.26B Tailgate washer reservoir location – Hatchback model shown

11.30 Headlamp adjusters on September 1985 onwards models – 'A' is for horizontal beam adjustment and 'B' is for vertical beam adjustment

2   Check that the level of the electrolyte is maintained between the upper and lower level markings on the side of the battery casing (photo).
3   If necessary, unscrew the six cell caps from the top of the battery and top up the level using only distilled or demineralised water. Once the level is correct securely refit the caps.
4   The exterior of the battery should be inspected periodically for damage such as a cracked case or cover.
5   Check the tightness of the battery cable clamps to ensure good electrical connections and check the entire length of each cable for cracks and frayed conductors.
6   If corrosion (visible as white, fluffy deposits) is evident, remove the cables from the battery terminals, clean them with a small wire brush then refit them. Corrosion can be kept to a minimum by applying a layer of petroleum jelly to the clamps and terminals after they are reconnected.

7   Make sure that the battery tray is in good condition and the retaining clamp is tight.
8   Corrosion on the tray, retaining clamp and the battery itself can be removed with a solution of water and baking soda. Thoroughly rinse all cleaned areas with plain water.
9   Any metal parts of the vehicle damaged by corrosion should be covered with a zinc-based primer then painted.
10 Further information on the battery, charging and jump starting can be found in Chapter 12 and in the preliminary sections of this Manual.

Alternator drivebelt check, adjustment and renewal

11 The alternator drivebelt is located on the rear right-hand side of the engine and also drives the water pump. Due to its function and material makeup, the belt is prone to failure after a period of time and should therefore be inspected and adjusted periodically.
12 With the engine switched off, inspect the full length of the alternator drivebelt for cracks and separation of the belt plies. It will be necessary to turn the engine in order to move the belt from the pulleys so that the belt can be inspected thoroughly. Twist the belt between the pulleys so that both sides can be viewed. Also check for fraying, and glazing which gives the belt a shiny appearance. Check the pulleys for nicks, cracks, distortion and corrosion.
13 The tension of the belt is checked by pushing on it midway between the alternator and water pump pulleys. The belt deflection should be as specified when a force of approximately 10 kg (22 lb) is applied.
14 If adjustment is necessary, slacken the alternator pivot mounting bolt and the adjusting arm bolt. Using a suitable bar or lever between the pulley end of the alternator and the cylinder block and lever the alternator out to tension the belt.
15 Once the alternator drivebelt tension is correct, hold the alternator in position then tighten the adjusting arm bolt followed by the pivot mounting bolt. Tighten both bolts to the specified torque setting.
16 Run the engine for about five minutes, then recheck the tension.
17 To renew the belt, slacken the belt tension fully as described above. If the car is also equipped with power steering, remove the

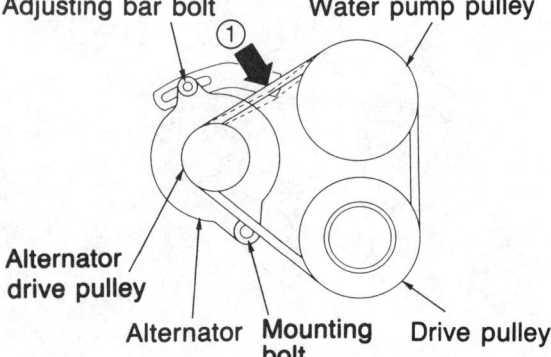

Fig. 1.14 Alternator drivebelt adjustment tension checking point (1) – B series engines (Sec 11)

Chapter 1 Routine maintenance and servicing

power steering pump drive belt as described in Section 9. Slip the belt off the pulleys then fit the new belt ensuring that it is routed correctly. With the belt in position, adjust the tension as previously described. Where necessary refit the power steering pump drivebelt as described in Section 9.

Windscreen/tailgate wiper blades and arms check and renewal

Wiper blades

18 The wiper blades should be renewed when they are deteriorated, cracked, or no longer clean the windscreen or tailgate glass effectively.
19 Lift the wiper arm away from the glass.
20 On early models release the spring retaining catch and separate the blade from the wiper arm. On later models release the catch on the arm, turn the blade through 90° and withdraw the blade from the arm fork (photo).
21 Insert the new blade into the arm, making sure that the spring retaining catch is properly engaged.

Wiper arms

22 Check the wiper arms for worn hinges and weak springs, and renew as necessary.
23 Make sure that the wiper is in its rest position, and note this position for correct refitting. If necessary, switch the wipers on and off in order to allow them to return to the 'park' position.
24 Lift up the hinged cover then unscrew the retaining nut and pull the arm from the spindle. If necessary use a screwdriver to prise off the arm, being careful not to damage the paintwork. On the tailgate wiper it will help if the arm is moved to its fully raised position before removing it from the spindle.
25 Fit the new arm using a reversal of the removal procedure.

Windscreen/headlamp and rear window/tailgate washer system check and adjustment

26 The windscreen/headlamp washer fluid reservoir is located in the right-hand front corner of the engine compartment. On Saloon and Hatchback models the rear window/tailgate washer fluid reservoir is located on the left-hand side of the boot, whereas on Estate models it is on the right-hand side (photos).
27 Check that the fluid level is at least up to the bottom of the filler neck and top up if necessary. When topping up the reservoir, a screen wash such as Turtle Wax High Tech Screen Wash should be added in the recommended quantities.
28 Check that the washer jets direct the fluid onto the middle of the windscreen/tailgate/headlamp and if necessary adjust the small sphere on the jet with a pin.

Headlamp beam alignment check

29 Accurate adjustment of the headlamp beam is only possible using optical beam setting equipment, and this work should therefore be carried out by a Mazda dealer or service station with the necessary facilities.
30 For reference, the headlamps on pre-September 1985 models are adjusted using the two screws (not the mounting screws) on the inner edge of each unit, and the headlamps on September 1985 onwards models are adjusted by means of the two adjusters on the rear of each unit (photo).

Chapter 2 Engine

Contents

Part A: E series engine – in-car engine repair procedures

Camshaft and rocker gear – inspection	6
Compression test – description and interpretation	2
Crankshaft oil seals – renewal	12
Cylinder head, camshaft and rocker gear – removal and refitting	5
Cylinder head cover – removal and refitting	4
Engine oil and filter – renewal	See Chapter 1
Engine/transmission mountings – renewal	11
Flywheel/driveplate – removal, inspection and refitting	10
General information	1
Locating top dead centre (TDC) for number one piston	3
Oil pump – removal, inspection and refitting	8
Sump – removal and refitting	7
Timing cover, chain and sprockets – removal, inspection and refitting	9
Valve clearance – adjustment	See Chapter 1

Part B: B series engine – in-car engine repair procedures

Camshaft – removal, inspection and refitting	20
Camshaft oil seal – renewal	21
Compression test – description and interpretation	14
Crankshaft oil seals – renewal	27
Cylinder head – removal and refitting	22
Cylinder head cover – removal and refitting	16
Engine oil and filter – renewal	See Chapter 1
Engine/transmission mountings – renewal	26
Flywheel/driveplate – removal, inspection and refitting	25
General information	13

Locating top dead centre (TDC) for number one piston	15
Oil pump – removal, inspection and refitting	24
Rocker gear – removal, inspection and refitting	17
Sump – removal and refitting	23
Timing belt – removal, inspection and refitting	18
Timing belt sprockets – removal, inspection and refitting	19
Valve clearance – adjustment (B6 engine only)	See Chapter 1

Part C: Engine removal and general engine overhaul procedures

Crankshaft – inspection	40
Crankshaft – refitting and main bearing running clearance check	44
Crankshaft – removal	37
Cylinder block/crankcase – cleaning and inspection	38
Cylinder head – dismantling	33
Cylinder head – reassembly	35
Cylinder head and valves – cleaning, inspection and renovation	34
Engine – initial start-up after overhaul	46
Engine – removal and refitting	31
Engine overhaul – dismantling sequence	32
Engine overhaul – general information	29
Engine overhaul – reassembly sequence	42
Engine removal – methods and precautions	30
General information	28
Main and big-end bearings – inspection	41
Piston/connecting rod assembly – inspection	39
Piston/connecting rod assembly – refitting and big-end bearing running clearance check	45
Piston/connecting rod assembly – removal	36
Piston rings – refitting	43

Specifications

Part A: E series engines

1100 cc engine

General

Type	Four cylinder in-line overhead camshaft
Designation	E1
Bore	70.0 mm
Stroke	69.6 mm
Capacity	1071 cc
Compression ratio	9.2 : 1
Compression test pressure:	
Standard	11.7 bars (170 lbf/in^2)
Limit	8.2 bars (119 lbf/in^2)
Maximum permissible difference between cylinders	1.9 bars (28 lbf/in^2)
Firing order	1–3–4–2 (No 1 cylinder at the crankshaft pulley end)

Crankshaft

Number of main bearings	5
Main bearing journal diameter	49.938 to 49.956 mm
Main bearing journal maximum permissible ovality	0.05 mm

Chapter 2 Engine

Crankshaft (continued)
Crankpin journal diameter	39.940 to 39.956 mm
Crankpin journal maximum permissible ovality	0.05 mm
Bearing undersizes available	0.25 mm, 0.5 mm, 0.75 mm
Maximum permissible crankshaft endfloat	0.04 mm
Crankshaft endfloat:	
Standard	0.10 to 0.15 mm
Limit	0.30 mm
Connecting rod big-end side clearance:	
Standard	0.110 to 0.262 mm
Limit	0.30 mm

Cylinder block and pistons
Cylinder bore diameter	70.00 to 70.019 mm
Maximum cylinder block face distortion	0.15 mm
Piston diameter	69.944 to 69.964 mm
Piston to bore clearance:	
Standard	0.036 to 0.075 mm
Limit	0.15 mm
Piston oversizes available	0.25 mm, 0.50 mm, 0.75 mm, 1.00 mm
Piston ring thickness:	
Top ring	1.17 to 1.19 mm
Second ring	1.47 to 1.49 mm
Piston ring to groove clearance:	
Top and second rings	0.03 to 0.07 mm
Limit	0.15 mm
Piston ring end gap:	
Top and second rings	0.20 to 0.40 mm
Oil control ring	0.30 to 0.90 mm
Limit	1.0 mm
Gudgeon pin diameter	19.976 to 19.988 mm
Gudgeon pin fit in piston	Hand push fit
Gudgeon pin fit in connecting rod	Interference

Camshaft
Camshaft journal diameter:	
Front and rear	41.949 to 41.965 mm
Centre	41.919 to 41.935 mm
Maximum permissible journal running clearance	0.15 mm
Maximum permissible camshaft endfloat	0.20 mm
Cam lobe height:	
Pre-September 1985 models:	
Standard	44.119 mm
Limit	43.919 mm
September 1985 models onward:	
Standard	43.733 mm
Limit	43.533 mm

Valves
Valve seat angle	45°
Head diameter:	
Inlet	35.9 to 36.1 mm
Exhaust	30.9 to 31.1 mm
Valve stem diameter:	
Inlet:	
Standard	8.030 to 8.045 mm
Limit	7.980 mm
Exhaust:	
Standard	8.025 to 8.045 mm
Limit	7.975 mm
Valve stem to guide clearance:	
Standard	0.018 to 0.053 mm
Limit	0.20 mm
Valve spring free length:	
Standard	43.3 mm
Limit	42.0 mm
Valve clearances (warm engine – between valve stem and rocker):	
Inlet	0.25 mm
Exhaust	0.30 mm
Valve timing:	
Inlet opens	15° BTDC
Inlet closes	44° ATDC
Exhaust opens	53° BBDC
Exhaust closes	6° ATDC

Rocker gear
Rocker shaft diameter .... 18.959 to 18.980 mm
Rocker arm inner diameter .... 19.000 to 19.027 mm
Rocker arm to shaft clearance:
    Standard .... 0.020 to 0.068 mm
    Limit .... 0.10 mm

Cylinder head
Height:
    Standard .... 90.5 mm
    Limit .... 90.3 mm
Maximum permissible warpage .... 0.15 mm

Lubrication system
Maximum oil pump outer rotor to body clearance .... 0.35 mm
Maximum rotor lobe clearance .... 0.25 mm
Maximum rotor endfloat .... 0.15 mm
Oil pressure at 3000 rpm .... 3.5 to 4.5 bars (50 to 64 lbf/in^2)

1300 cc engine
The engine specification is identical to the 1100 cc unit except for the following differences:

General
Designation E3
Bore .... 77.0 mm
Capacity .... 1296 cc

Cylinder block and pistons
Cylinder bore diameter .... 77.00 to 77.019 mm
Piston diameter .... 76.954 to 76.974 mm
Piston oversizes available .... 0.25 mm

Camshaft
Cam lobe height:
    Pre-September 1985 models:
        Standard .... 44.114 mm
        Limit .... 43.914 mm
    September 1985 models onward:
        Standard .... 43.731 mm
        Limit .... 43.531 mm

Valves
Valve timing:
    Inlet opens .... 15° BTDC
    Inlet closes .... 58° ABDC
    Exhaust opens .... 58° BBDC
    Exhaust closes .... 15° ATDC

1500 cc engine
The engine specification is identical to the 1300 cc unit except for the following differences:

General
Designation E5
Stroke .... 80.0 mm
Capacity .... 1490 cc
Compression ratio:
    1500 GT models 10.0:1
    All other models 9.0:1
Compression test pressure – 1500 GT models:
    Standard .... 13.7 bars (198 lbf/in^2)
    Limit .... 10.0 bars (145 lbf/in^2)

Cylinder block and pistons
Piston to bore clearance:
    Standard .... 0.026 to 0.065 mm
    Limit .... 0.15 mm
Piston oversizes available:
    Pre-September 1985 models .... 0.50 mm
    September 1985 models onward .... 0.25 mm

Camshaft
Cam lobe height – 1500 GT models:
    Standard .... 44.719 mm
    Limit .... 44.519 mm

Chapter 2 Engine 69

Valves
Valve spring free length – 1500 GT models:
  Inner spring:
    Standard .... 36.8 mm
    Limit .... 35.7 mm
  Outer spring:
    Standard .... 40.3 mm
    Limit .... 39.1 mm
Valve timing – 1500 GT models:
  Inlet opens .... 16° BTDC
  Inlet closes .... 59° ABDC
  Exhaust opens .... 59° BBDC
  Exhaust closes .... 16° ATDC

Torque wrench settings – all E series engines   Nm   lbf ft

	Nm	lbf ft
Cylinder head bolts:		
Pre-September 1985 models	78 to 82	57 to 60
September 1985 models onward	85 to 91	63 to 67
Main bearing cap bolts	65 to 71	48 to 52
Big-end bearing cap nuts	30 to 35	22 to 26
Camshaft sprocket nut	70 to 80	51 to 59
Crankshaft pulley bolt	110 to 120	81 to 88
Flywheel bolts (manual gearbox models)	88 to 94	65 to 69
Driveplate bolts (automatic transmission models)	69 to 83	51 to 61
Sump bolts	7 to 12	5 to 9
Timing cover bolts	19 to 26	14 to 19
Oil pump bolts	19 to 26	14 to 19
Engine/transmission mounting nuts and bolts	37 to 52	27 to 38
Engine/transmission mounting to crossmember nuts	32 to 47	23 to 35
Transmission front mounting (Pre-1982 models)	19 to 26	14 to 19

Part B: B series engines

1300 cc engine
General
Type .... Four cylinder in-line overhead camshaft
Designation B3
Bore .... 71.0 mm
Stroke .... 83.6 mm
Capacity .... 1323 cc
Compression ratio .... 9.4 : 1
Compression test pressure:
  Standard .... 13.7 bars (198 lbf/in^2)
  Limit .... 9.5 bars (138 lbf/in^2)
Maximum permissible difference between cylinders .... 1.9 bars (28 lbf/in^2)
Firing order .... 1–3–4–2 (No 1 cylinder at the crankshaft pulley end)

Crankshaft
Number of main bearings 5
Main bearing journal diameter .... 49.938 to 49.956 mm
Main bearing journal maximum permissible ovality .... 0.05 mm
Crankpin journal diameter .... 39.94 to 39.956 mm
Crankpin journal maximum permissible ovality .... 0.05 mm
Bearing undersizes available .... 0.25 mm, 0.5 mm, 0.75 mm
Maximum permissible crankshaft runout .... 0.04 mm
Crankshaft endfloat:
  Standard .... 0.10 to 0.15 mm
  Limit .... 0.30 mm
Connecting rod big-end side clearance:
  Standard .... 0.110 to 0.262 mm
  Limit .... 0.30 mm

Cylinder block and pistons
Cylinder bore diameter .... 71.000 to 71.019 mm
Maximum permissible cylinder ovality or taper .... 0.019 mm
Maximum cylinder block face distortion .... 0.15 mm
Piston diameter .... 70.954 to 70.974 mm
Piston to bore clearance:
  Standard .... 0.026 to 0.065 mm
  Limit .... 0.15 mm
Piston oversizes available .... 0.25 mm, 0.50 mm, 0.75 mm, 1.0 mm
Piston ring thickness:
  Top ring .... 1.17 to 1.19 mm
  Second ring .... 1.47 to 1.49 mm

Cylinder block and pistons (continued)
Piston ring to groove clearance:
    Top and second rings .... 0.030 to 0.065 mm
    Limit .... 0.15 mm
Piston ring end gap:
    Top and second rings .... 0.15 to 0.30 mm
    Oil control ring .... 0.20 to 0.70 mm
    Limit .... 1.0 mm
Gudgeon pin diameter .... 19.974 to 19.980 mm
Gudgeon pin fit in piston .... Hand push fit
Gudgeon pin fit in connecting rod .... Interference

Camshaft
Camshaft journal diameter:
    Front and rear .... 43.44 to 43.465 mm
    Centre .... 43.41 to 43.435 mm
Maximum permissible journal ovality .... 0.05 mm
Maximum permissible journal running clearance .... 0.15 mm
Camshaft endfloat:
    Standard .... 0.05 to 0.18 mm
    Limit .... 0.20 mm
Cam lobe height:
    Standard .... 36.378 to 36.528 mm
    Limit .... 36.23 mm

Valves
Valve seat angle .... 45°
Valve seat width .... 1.1 to 1.7 mm
Head diameter:
    Inlet .... 31.9 to 32.1 mm
    Exhaust .... 27.9 to 28.1 mm
Valve stem diameter:
    Inlet .... 6.970 to 6.985 mm
    Exhaust .... 6.965 to 6.980 mm
Valve stem to guide clearance:
    Inlet:
        Standard .... 0.025 to 0.060 mm
        Limit .... 0.20 mm
    Exhaust:
        Standard .... 0.030 to 0.065 mm
        Limit .... 0.20 mm
Valve timing:
    Inlet opens .... 14° BTDC
    Inlet closes .... 52° ABDC
    Exhaust opens .... 52° BBDC
    Exhaust closes .... 14° ATDC

Rocker gear
Rocker shaft outer diameter .... 17.959 to 17.980 mm
Rocker arm inner diameter .... 18.00 to 18.027 mm
Rocker arm to shaft clearance:
    Standard .... 0.020 to 0.068 mm
    Limit .... 0.10 mm

Cylinder head
Cylinder head height:
    Standard .... 107.4 to 107.6 mm
    Limit .... 107.2 mm
Maximum permissible warpage .... 0.15 mm

Lubrication system
Maximum oil pump outer rotor to body clearance .... 0.22 mm
Maximum rotor lobe clearance .... 0.20 mm
Maximum rotor endfloat .... 0.14 mm
Oil pressure at 3000 rpm .... 3.5 to 4.5 bars (50 to 64 lbf/in^2)

1500 cc engine
The engine specification is identical to the 1300 cc unit except for the following differences:

General
Designation B5
Bore .... 78.0 mm
Stroke .... 78.4 mm
Capacity .... 1498 cc

General (continued)
Compression ratio 9.1:1
Compression test pressure:
    Standard .... 13.5 bars (192 lbf/in^2)
    Limit .... 9.5 bars (135 lbf/in^2)

Cylinder block and pistons
Cylinder bore diameter .... 78.00 to 78.019 mm
Piston diameter .... 77.954 to 77.974 mm
Piston oversizes available .... 0.25 mm, 0.50 mm
Piston top and second ring thickness .... 1.47 to 1.49 mm

Valves
Head diameter:
    Inlet .... 37.9 to 38.1 mm
    Exhaust .... 31.9 to 32.1 mm

1600 cc engine
The engine specification is identical to the 1500 cc unit except for the following differences:

General
Designation B6
Stroke .... 83.6 mm
Capacity .... 1597 cc
Compression ratio 10.5:1
Compression test pressure:
    Standard .... 14.5 bars (206 lbf/in^2)
    Limit .... 10.0 bars (145 lbf/in^2)

Crankshaft
Main bearing journal undersize bearings available .... 0.25 mm, 0.50 mm
Crankpin journal diameter:
    Standard .... 44.94 to 44.956 mm
    Limit .... 44.89 mm
Bearing undersizes available .... 0.25 mm, 0.50 mm

Camshaft
Cam lobe height:
    Standard .... 36.811 to 36.911 mm
    Limit .... 36.66 mm

Torque wrench settings – all B series engines

	Nm	lbf ft
Cylinder head bolts	76 to 81	56 to 60
Main bearing cap bolts	54 to 59	40 to 43
Big-end cap nuts:		
B3 and B5 engines	29 to 34	21 to 25
B6 engines	47 to 52	35 to 38
Cylinder head cover bolts	5 to 9	4 to 7
Rocker gear bolts	22 to 28	16 to 21
Camshaft sprocket bolt	49 to 61	36 to 45
Camshaft thrust plate bolt	8 to 11	6 to 8
Crankshaft sprocket bolt	108 to 128	80 to 94
Crankshaft pulley bolts	12 to 17	9 to 13
Water pump pulley bolts	8 to 11	6 to 8
Flywheel/driveplate bolts	96 to 103	71 to 76
Sump bolts	6 to 9	4 to 7
Timing belt tensioner pulley bolt	19 to 26	14 to 19
Timing cover bolts	8 to 11	6 to 8
Oil pump bolts	19 to 26	14 to 19
Oil pump strainer bolts	8 to 11	6 to 8
Rear oil seal housing	8 to 11	6 to 8
Engine/transmission mounting nuts and bolts	37 to 52	27 to 38
Engine/transmission mounting to crossmember nuts	32 to 47	23 to 35

Part A: E series engine – in-car engine repair procedures

1 General information

How to use this Chapter

This Part of Chapter 2 is devoted to in-vehicle repair procedures for the earlier E series engine. Similar information covering the later B series engine will be found in Part B of this Chapter. All procedures concerning engine removal and refitting, and engine block/cylinder head overhaul for all engine types can be found in Part C of this Chapter.

Most of the operations included in this Part are based on the assumption that the engine is still installed in the vehicle. Therefore, if this information is being used during a complete engine overhaul, with the engine already removed, many of the steps included here will not apply.

Engine description

The E series engine is of the four-cylinder, in-line overhead camshaft type mounted transversely at the front of the car and available in 1071 cc (E1), 1296 cc (E3) and 1490 cc (E5) versions. Apart from the cylinder bore diameter, crankshaft stroke and minor detail differences, all three engines are identical in design and construction.

The crankshaft is supported within the cast iron cylinder block on five shell type main bearings. Thrustwashers are fitted at the rear main bearing to control crankshaft endfloat.

The connecting rods are attached to the crankshaft by horizontal split shell type big-end bearings, and to the pistons by interference fit gudgeon pins. The aluminium alloy pistons are of the slipper type and each is fitted with two compression rings and a three-piece oil control ring.

The camshaft is chain driven from the crankshaft and is carried in half bearings machined directly in the aluminium cylinder head. The inclined inlet and exhaust valves are actuated by rocker arms located on a rocker shaft assembly, the pedestals of which form the camshaft bearing upper halves. Valve clearances are adjusted via a screw and locknut arrangement.

Lubrication is by a rotor type oil pump, chain driven from the crankshaft and located in the crankcase. Engine oil is fed through an externally mounted full-flow oil filter to the engine oil gallery, and then to the crankshaft, camshaft and rocker shaft bearings. A pressure relief valve is incorporated in the oil pump.

A semi-closed crankcase ventilation system is employed and crankcase gases are drawn from the rocker cover via hoses to the air cleaner and inlet manifold.

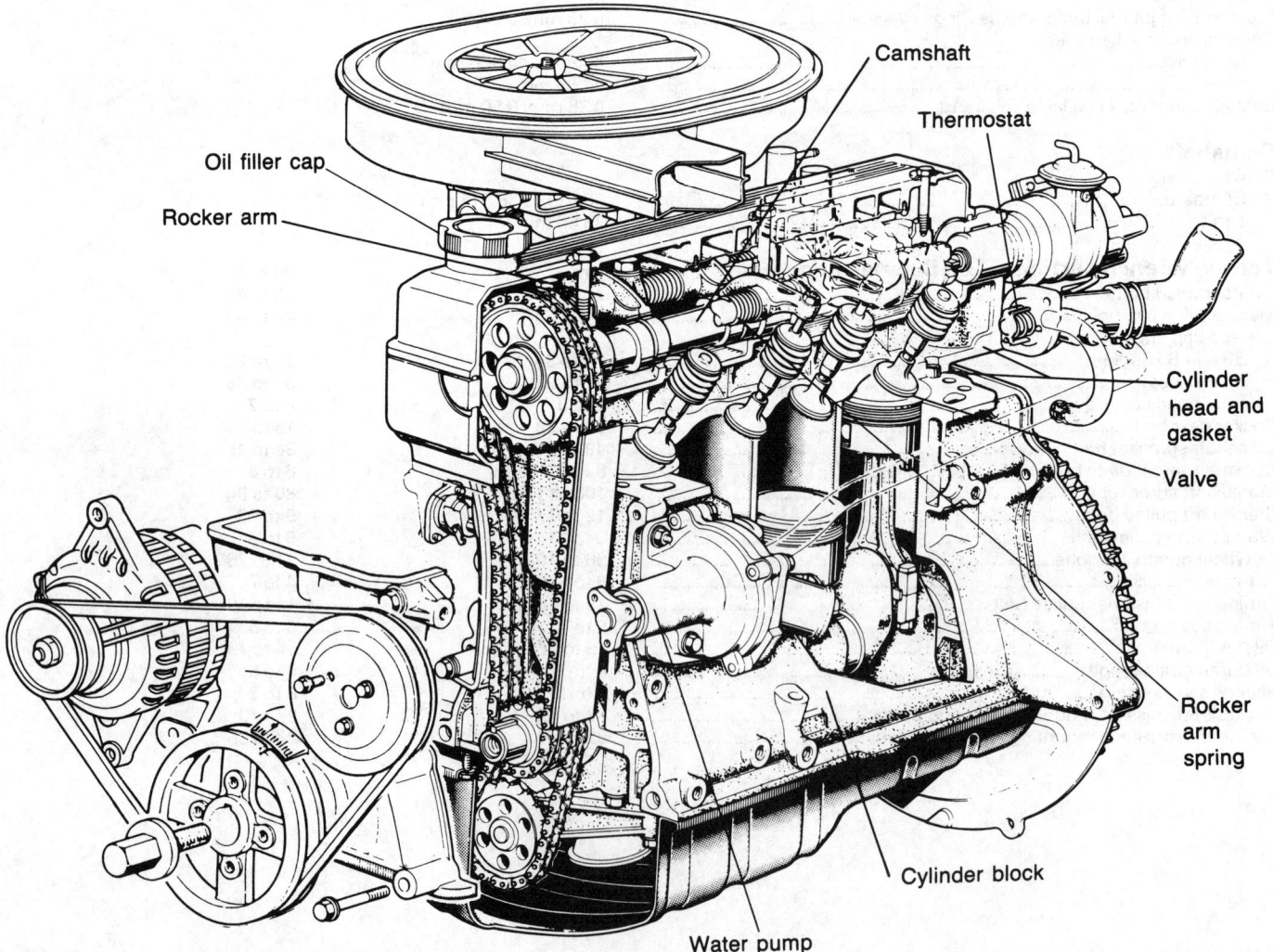

Fig. 2.1 Cutaway view of the E series engine (Sec 1)

Chapter 2 Part A: E series engine – in-car engine repair procedures

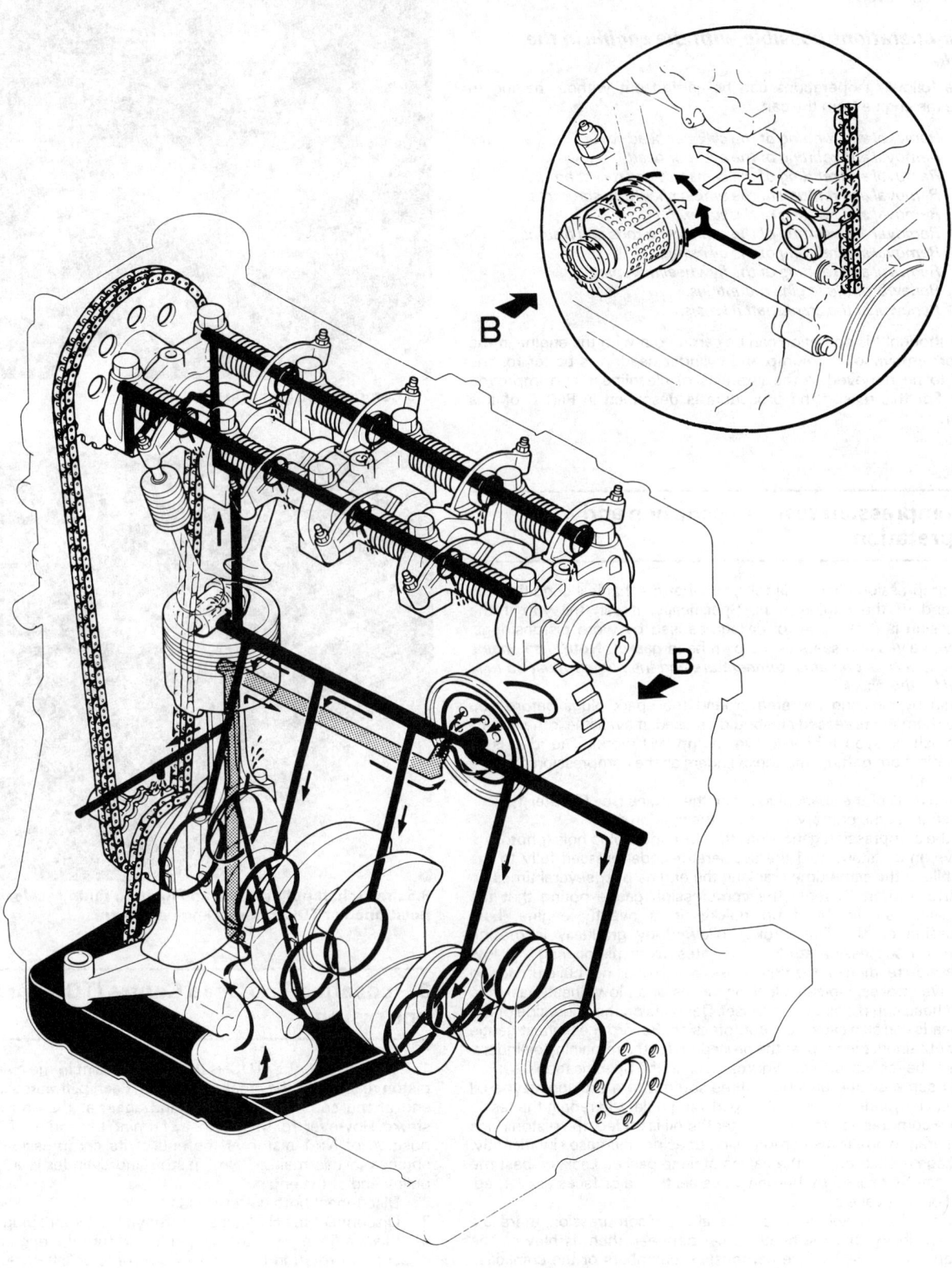

Fig. 2.2 Lubrication system of the E series engine (Sec 1)

The distributor rotor is driven direct from the left-hand end of the camshaft, and the fuel pump is also operated by the camshaft via an eccentric and lever.

Repair operations possible with the engine in the vehicle

The following operations can be carried out without having to remove the engine from the car:

(a) Removal and refitting of the cylinder head cover.
(b) Removal and refitting of the cylinder head.
(c) Removal and refitting of the camshaft and rocker arms.
(d) Removal and refitting of the timing chain and sprockets.
(e) Removal and refitting of the sump.
(f) Removal and refitting of the connecting rods and pistons*.
(g) Removal and refitting of the oil pump.
(h) Removal and refitting of the flywheel/driveplate.
(i) Renewal of the engine mountings.
(j) Renewal of the crankshaft oil seals.

* Although this operation can be carried out with the engine in the car after removal of the sump and cylinder head, it is better for the engine to be removed in the interests of cleanliness and improved access. For this reason the procedure is described in Part C of this Chapter.

2.5 Compression gauge in use

2 Compression test – description and interpretation

1 A compression check will tell you what mechanical condition the upper end of the engine is in. Specifically, it can tell you if the compression is down due to leakage caused by worn pistons/rings, defective valves and seats or a blown head gasket. **Note:** *The engine must be at normal operating temperature and the battery must be fully charged for this check.*
2 Begin by cleaning the area around the spark plugs before you remove them (compressed air should be used, if available, otherwise a small brush or even a bicycle tyre pump will work). The idea is to prevent dirt from getting into the cylinders as the compression check is being done.
3 Remove all of the spark plugs from the engine (see Chapter 1).
4 Disconnect the primary wire from the ignition coil.
5 Fit the compression gauge into the No 1 spark plug hole (photo).
6 Have an assistant hold the accelerator pedal pressed fully to the floor while at the same time cranking the engine over several times on the starter motor. Observe the compression gauge noting that the compression should build up quickly in a healthy engine. Low compression on the first stroke, followed by gradually increasing pressure on successive strokes, indicates worn piston rings. A low compression reading on the first stroke, which does not build up during successive strokes, indicates leaking valves or a blown head gasket (a cracked head could also be the cause). Deposits on the underside of the valve heads can also cause low compression. Record the highest gauge reading obtained, then repeat the procedure for the remaining cylinders. Compare the results with the figures given in the Specifications.
7 Add some engine oil (about three squirts from a plunger-type oil can) to each cylinder, through the spark plug hole, and repeat the test.
8 If the compression increases after the oil is added, the piston rings are definitely worn. If the compression does not increase significantly, the leakage is occurring at the valves or head gasket. Leakage past the valves may be caused by burned valve seats and/or faces or warped, cracked or bent valves.
9 If two adjacent cylinders have equally low compression, there is a strong possibility that the head gasket between them is blown. The appearance of coolant in the combustion chambers or the crankcase would verify this condition.
10 If one cylinder is about 20 percent lower than the other, and the engine has a slightly rough idle, a worn lobe on the camshaft could be the cause.
11 If the compression is unusually high, the combustion chambers are probably coated with carbon deposits. If this is the case, the cylinder head should be removed and decarbonised.

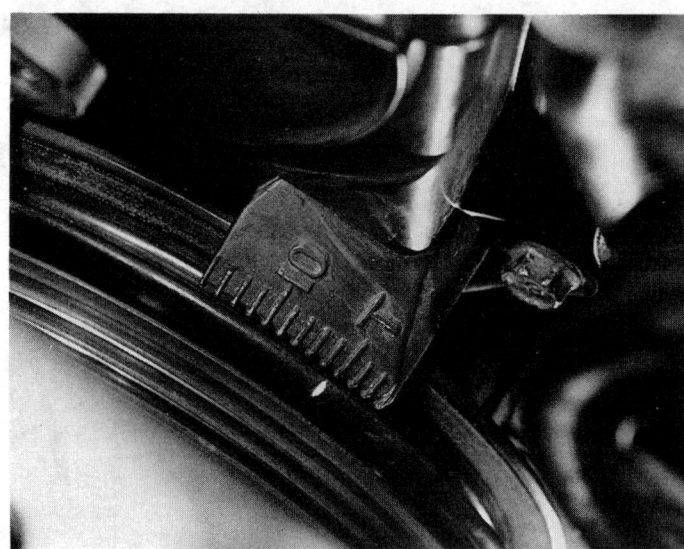

3.5 Crankshaft pulley notch and ignition timing scale with engine positioned at TDC (B series engine shown)

3 Locating Top Dead Centre (TDC) for number one piston

1 Top dead centre (TDC) is the highest point in the cylinder that each piston reaches as the crankshaft turns. Each piston reaches TDC at the end of the compression stroke and again at the end of the exhaust stroke. However, for the purpose of timing the engine, TDC refers to the position of No 1 piston at the end of its compression stroke. On all engines in this manual, No 1 piston and cylinder is at the crankshaft pulley end of the engine.
2 Disconnect both battery leads.
3 Disconnect the HT lead and remove No 1 spark plug.
4 Place a finger over the plug hole and turn the engine in the normal direction of rotation (clockwise from the crankshaft pulley end), until pressure is felt in No 1 cylinder. This indicates that the piston is commencing its compression stroke. The engine can be turned with a socket or spanner on the crankshaft pulley bolt.
5 Continue turning the engine until the notch on the crankshaft pulley is aligned with the T mark on the timing scale just above the pulley (photo). In this position the engine is at Top Dead Centre (TDC) with No 1 cylinder on compression.

Chapter 2 Part A: E series engine – in-car engine repair procedures

4.6 Fit the semi-circular grommet with the word OUT facing outwards

4.7A Locate the gasket in the cylinder head cover groove...

4.7B ...and refit the cylinder head cover to the engine

4 Cylinder head cover – removal and refitting

Removal

1 Remove the air cleaner assembly as described in Chapter 4.
2 Disconnect the accelerator cable from the carburettor, and release the cable from its mounting bracket and guide on the cylinder head cover.
3 Release the HT leads from their guides and disconnect the breather hose from the top of the cover.
4 Undo the three cylinder head cover retaining bolts and lift off the cover. Remove the semi-circular sealing grommet from the right-hand end of the cylinder head.

Refitting

5 Examine the cylinder head cover gasket, retaining bolt sealing washers and semi-circular sealing grommet for signs of damage or deterioration and renew as necessary.
6 Fit the semi-circular sealing grommet to the right-hand end of the cylinder head ensuring that the word OUT is facing outwards (photo).
7 Locate the gasket in the cylinder head cover groove and refit the cylinder head cover (photos). Refit the retaining bolts and sealing washers and tighten the bolts securely.
8 Reconnect the breather hose and refit the HT leads to their guides.
9 Referring to Chapter 4, reconnect the accelerator cable to the carburettor and refit the air cleaner assembly.

5 Cylinder head, camshaft and rocker gear – removal and refitting

Removal

1 Disconnect the battery negative lead.
2 Drain the cooling system as described in Chapter 1.
3 Remove the alternator as described in Chapter 12.
4 If the car is equipped with power steering, remove the power steering pump as described in Chapter 10. Leave the hydraulic hoses attached to the pump and position the pump clear of the head.
5 Refer to Chapter 4 and remove the air cleaner assembly, the exhaust manifold, and the fuel pump. Disconnect the accelerator and choke cable (if applicable) from the carburettor.
6 Remove the distributor as described in Chapter 5.
7 Slacken the retaining clips and disconnect the radiator top hose from the thermostat housing, and the coolant bypass hose from the outlet adjacent to the thermostat housing.
8 Disconnect the following wires and release them from any clips or guides. Identify them with adhesive tape if necessary to ensure correct refitting:

(a) Temperature gauge sender.
(b) Cooling fan thermostatic switch.
(c) Earth strap.
(d) Carburettor fuel cut-off valve solenoid.

9 Remove the cylinder head cover as described in Section 4.
10 Ensure that the engine is at TDC, with No 1 cylinder on compression, and note the position of the camshaft sprocket timing mark, which is next to the bright timing chain link (photo). The mark should be approximately in the 3 o'clock position when viewed from the right-hand end of the engine.
11 Using wire or a cable tie, secure the timing chain to the sprocket so that the chain to sprocket relationship is not lost.
12 Slacken and remove the small retaining bolt which is adjacent to the right-hand end of the camshaft (photo). For safety it is advisable to put some rag below the sprocket in case the bolt is dropped.
13 Slacken and remove the nut and washer securing the camshaft sprocket to the camshaft. Insert a stout screwdriver through one of the holes in the sprocket to prevent the camshaft rotating as the nut is slackened.
14 At the rear of the engine, beneath the engine mounting bracket, undo the two timing chain tensioner retaining bolts and withdraw the tensioner to relieve the tension on the chain. Note that due to a lack of clearance, it will be necessary to remove the right-hand engine mounting before the tensioner can be completely removed.
15 Slacken all the cylinder head bolts half a turn at a time in the order shown (photo). When all the tension has been relieved, remove the bolts.
16 Hold the two end rocker pedestals together and lift off the rocker gear as an assembly (photo).
17 Disengage the camshaft from the sprocket and lift it clear of the engine (photo). Pass a screwdriver through the sprocket to prevent it from falling down into the engine. Remove the Woodruff key from the end of the camshaft.
18 Release the cylinder head from its locating dowels by tapping it upwards using a hide or plastic mallet. When the head is free, lift it off the engine complete with the inlet manifold and carburettor, then remove the cylinder head gasket (photo).
19 If required, the right-hand engine mounting can be removed, with reference to Section 11, so that the timing chain tensioner can be withdrawn.

Refitting

20 Clean out all the bolt holes in the cylinder block using a cloth rag and screwdriver. Make sure that all oil is removed, otherwise there is a possibility of the block being cracked by hydraulic pressure when the bolts are tightened.
21 Make sure that the faces of the cylinder head and block are spotlessly clean, then lay a new cylinder head gasket on the block face.
22 Carefully lower the cylinder head onto the gasket and fit the small front retaining bolt, tightening it finger tight only at this stage.
23 Thoroughly lubricate the camshaft bearing journals in the cylinder head and refit the Woodruff key to its groove in the camshaft.
24 Align the Woodruff key with the slot in the camshaft sprocket and engage the camshaft with the sprocket (photo). Push the sprocket fully into place and lay the camshaft in its bearings.
25 Check that No 1 cylinder is still at TDC and that the timing mark on the sprocket is approximately in the 3 o'clock position, then remove the wire or cable tie from the timing chain.
26 Lubricate the camshaft bearing journals, then lower the rocker gear into position, ensuring that the arrows on the pedestals are all pointing towards the crankshaft pulley end of the engine (photo).

Chapter 2 Part A: E series engine – in-car engine repair procedures

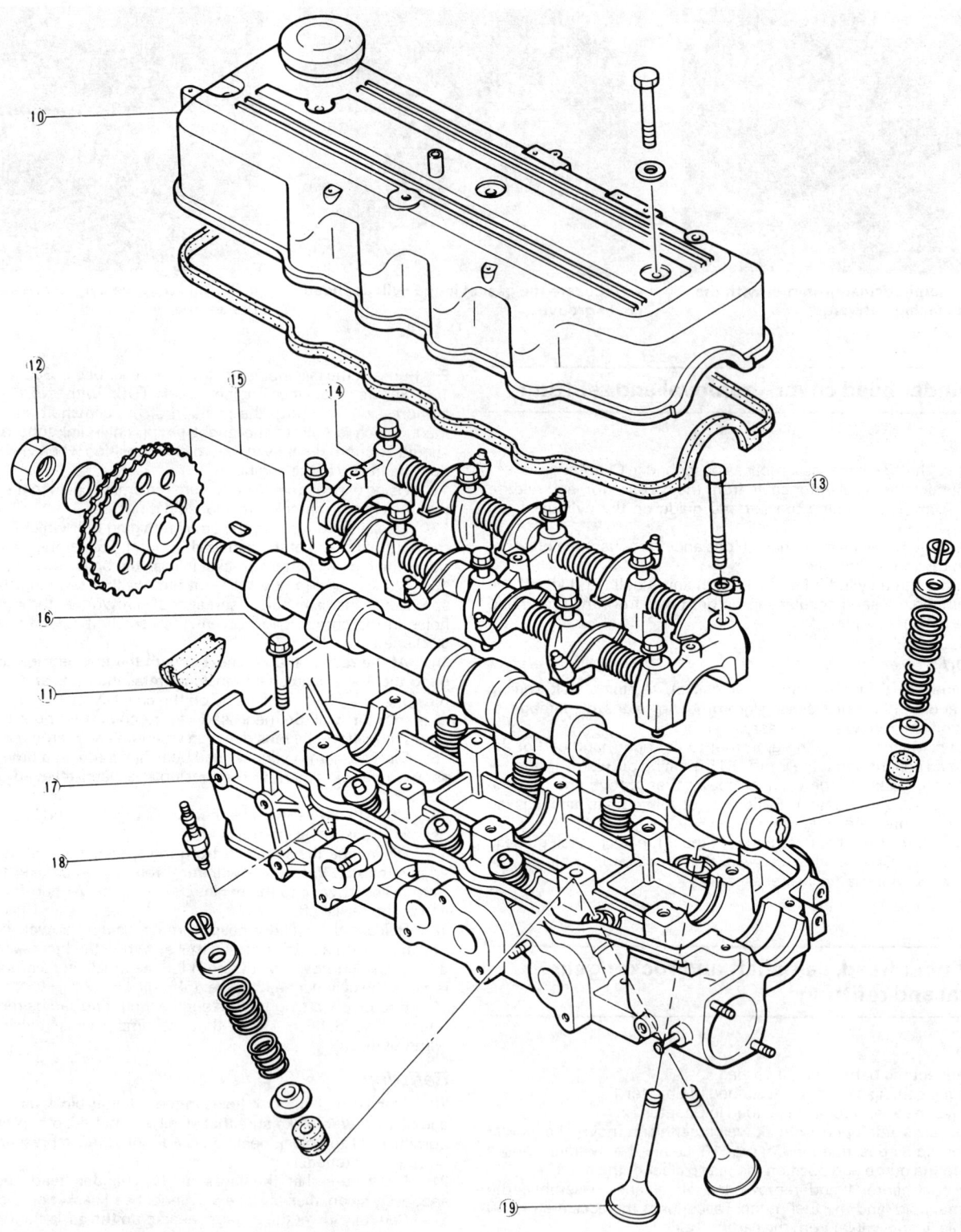

Fig. 2.3 Exploded view of the cylinder head components (Secs 5 and 6)

10 Cylinder head cover
11 Sealing grommet
12 Sprocket retaining nut
13 Cylinder head bolt
14 Rocker gear
15 Camshaft and sprocket
16 Cylinder head front retaining bolt
17 Cylinder head
18 Spark plug
19 Inlet valve

Chapter 2 Part A: E series engine – in-car engine repair procedures

5.8 Ensure camshaft sprocket timing mark and bright link (arrowed) are correctly positioned then secure the timing chain to the sprocket

5.10 Undo the small retaining bolt located next to the camshaft

5.13 Cylinder head bolt slackening sequence

5.14 Lift off the rocker gear assembly...

5.15 ...and remove the camshaft

5.16 Removing the cylinder head

5.22 Align Woodruff key with slot in the camshaft sprocket (arrowed)

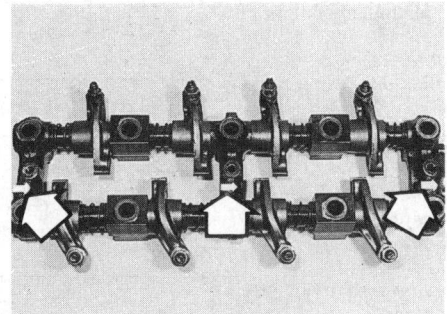

5.24 On refitting ensure all the arrows on the rocker pedestals (arrowed) point towards crankshaft pulley end of engine

5.26 Cylinder head bolt tightening sequence

5.28 Refit the camshaft sprocket retaining nut and washer

5.29 Checking camshaft endfloat

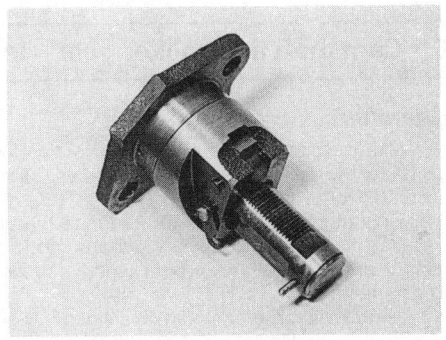

5.30A Push the tensioner plunger in...

5.30B ...and engage the catch over the tensioner plunger pin (arrowed)

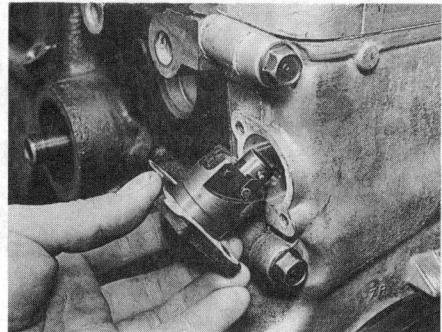

5.30C Fit the timing chain tensioner using a new gasket

27 Align the pedestal and rocker shaft holes then refit the cylinder head bolts.
28 Tighten the bolts to the specified torque in two stages in the sequence shown (photo).
29 Now tighten the small front retaining bolt securely.
30 Refit the camshaft sprocket washer and retaining nut and tighten the nut to the specified torque (photo). Prevent the camshaft from rotating by passing a stout screwdriver through one of the sprocket holes.
31 Check the camshaft endfloat by measuring the clearance between the sprocket and thrustplate using feeler gauges (photo). Renew the thrustplate if the clearance is more than specified.
32 Push the plunger fully into the timing chain tensioner body and lock it in position by engaging the small retaining catch on the tensioner body with the plunger pin. Apply jointing compound to a new tensioner gasket and fit the gasket to the tensioner. Refit the tensioner to the engine and tighten its retaining bolts securely (photos). Refit the rear engine mounting tightening the bolts to the specified torque.
33 Using a suitable spanner or socket, rotate the crankshaft pulley in a clockwise direction to release the timing chain tensioner and tension the chain.
34 Check the valve clearances as described in Chapter 1.
35 Refitting is now the reverse of the removal procedure noting the following points:

(a) Refer to Section 4 and refit the cylinder head cover.
(b) Refit the distributor as described in Chapter 5.
(c) Reconnect all the electrical connections ensuring that all wiring is correctly routed and secured by any relevant guides or clips.
(d) Reconnect all coolant hoses, tightening their retaining clips securely.
(e) Refit the exhaust manifold, fuel pump, air cleaner assembly, throttle and choke cables as described in Chapter 4.
(f) Where fitted, refit the power steering pump as described in Chapter 10, noting that it will not be necessary to bleed the pump.
(g) Refit the alternator as described in Chapter 12.
(h) Refill the cooling system as described in Chapter 1.

6 Camshaft and rocker gear – inspection

Camshaft
1 With the camshaft removed as described in the previous Section, examine the camshaft bearing surfaces, cam lobes and fuel pump eccentric for wear ridges and scoring. Renew the camshaft if any of these conditions are apparent. Examine the condition of the bearing surfaces both on the camshaft journals and in the cylinder head. If the head bearing surfaces are worn excessively, the cylinder head will need to be renewed.
2 Using a micrometer or vernier caliper, measure the diameter of each camshaft journal in at least two points and check the height of the cam lobe. If any of the measurements exceed the limits given in the Specifications, renew the camshaft.
3 Support the camshaft end journals on V-blocks and measure the camshaft runout at the centre journal using a dial gauge. If the runout exceeds the specified limit the camshaft should be renewed.

Rocker gear
4 If necessary, the rocker gear assembly can be dismantled by sliding the components off the shafts. Keep inlet and exhaust shaft components separate to avoid confusion and make a note of each components correct fitted position as it is removed to ensure it is positioned correctly on reassembly.
5 Examine the rocker arm bearing surfaces which contact the camshaft lobes for wear ridges and scoring. Renew any rocker arms on which these conditions are apparent.
6 If the rocker gear has been dismantled examine the rocker arm and shaft bearing surfaces for wear ridges and scoring. If the necessary measuring equipment is available, measure the internal diameter of the rocker arm and the outside diameter of the rocker shaft at the point where the rocker pivots and calculate the clearance. If the clearance exceeds the figure given in the Specifications at the start of this Chapter or there are obvious signs of wear the rocker arm and/or shaft must be renewed.

7 Sump – removal and refitting

Removal
1 Drain the engine oil as described in Chapter 1.
2 Apply the handbrake, then jack up the front of the car and support it on axle stands.
3 Remove the undertrays from the vehicle to gain access to the sump.

7.7 Refit the sump using a new gasket (engine removed for clarity)

Chapter 2 Part A: E series engine – in-car engine repair procedures

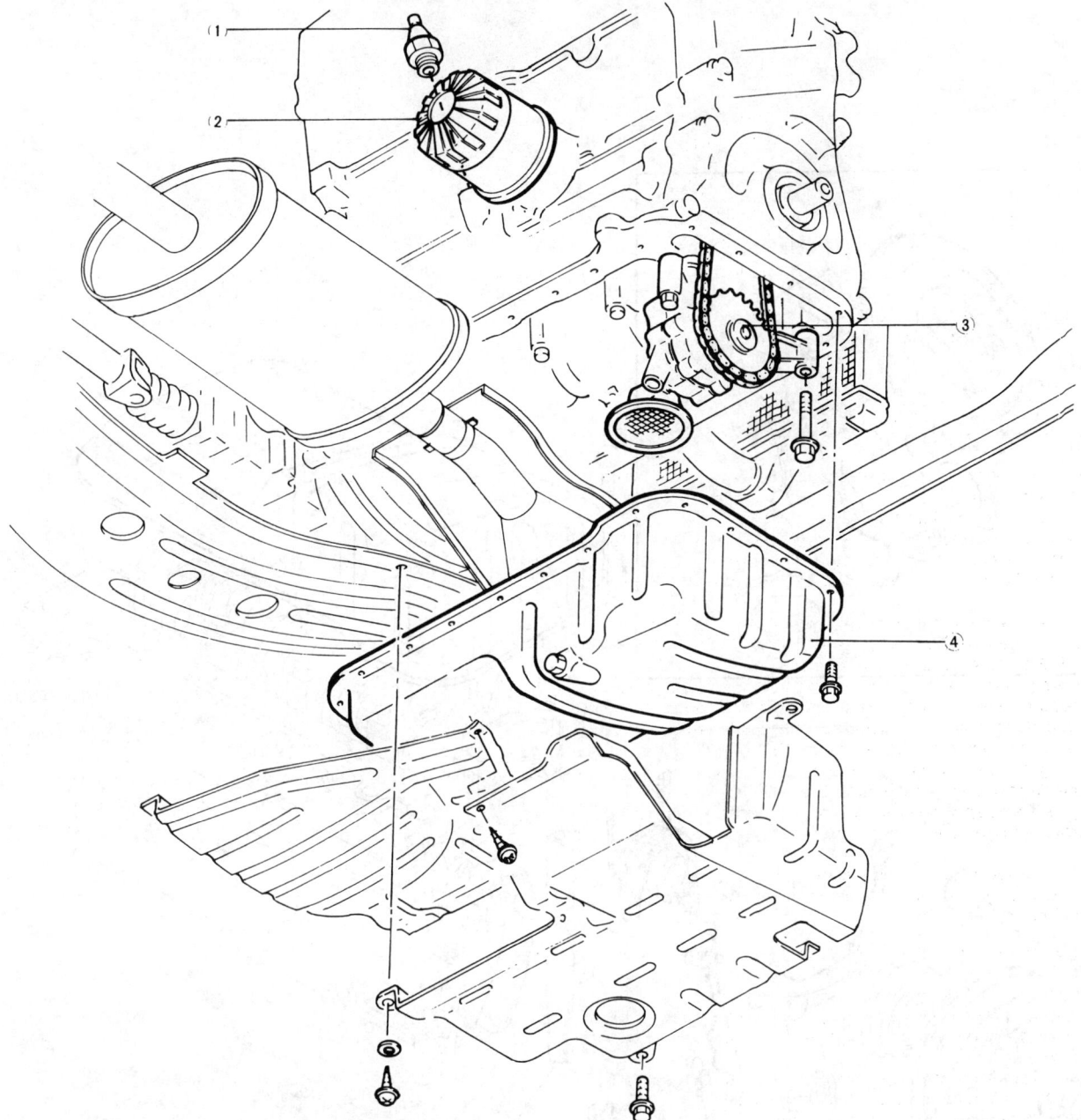

Fig. 2.4 Sump, oil pump and related components (Secs 7 and 8)

1 Oil pressure switch   2 Oil filter   3 Oil pump   4 Sump

4 Refer to Chapter 4 and remove the front exhaust pipe section.
5 Slacken and remove the bolts securing the sump to the crankcase. Tap the sump with a soft faced mallet to break the seal between the sump flange and crankcase and remove the sump.

Refitting

6 Ensure that the sump and cylinder block mating surfaces are perfectly clean and dry, with all traces of old gasket removed. If the ends of the timing cover gasket are protruding onto the sump face, carefully trim them with a sharp knife until flush.
7 Apply jointing compound to the new sump gasket and stick it in position on the sump face (photo).
8 Refit the sump and install all the sump retaining bolts. Tighten the bolts evenly and progressively in a diagonal sequence to the specified torque.
9 Refit the front exhaust section as described in Chapter 4.

10 Refit the undertrays and lower the car to the ground.
11 Refill the engine with oil as described in Chapter 1.

8 Oil pump – removal, inspection and refitting

Removal

1 Remove the sump as described in Section 7.
2 Undo the two oil pump retaining bolts and ease the pump away from the bottom of the crankcase.
3 Disengage the sprocket from the chain and remove the oil pump.
4 If necessary, the oil pump drive chain and sprocket can be removed once the timing chain and sprockets have been removed as described in Section 9.

Chapter 2 Part A: E series engine – in-car engine repair procedures

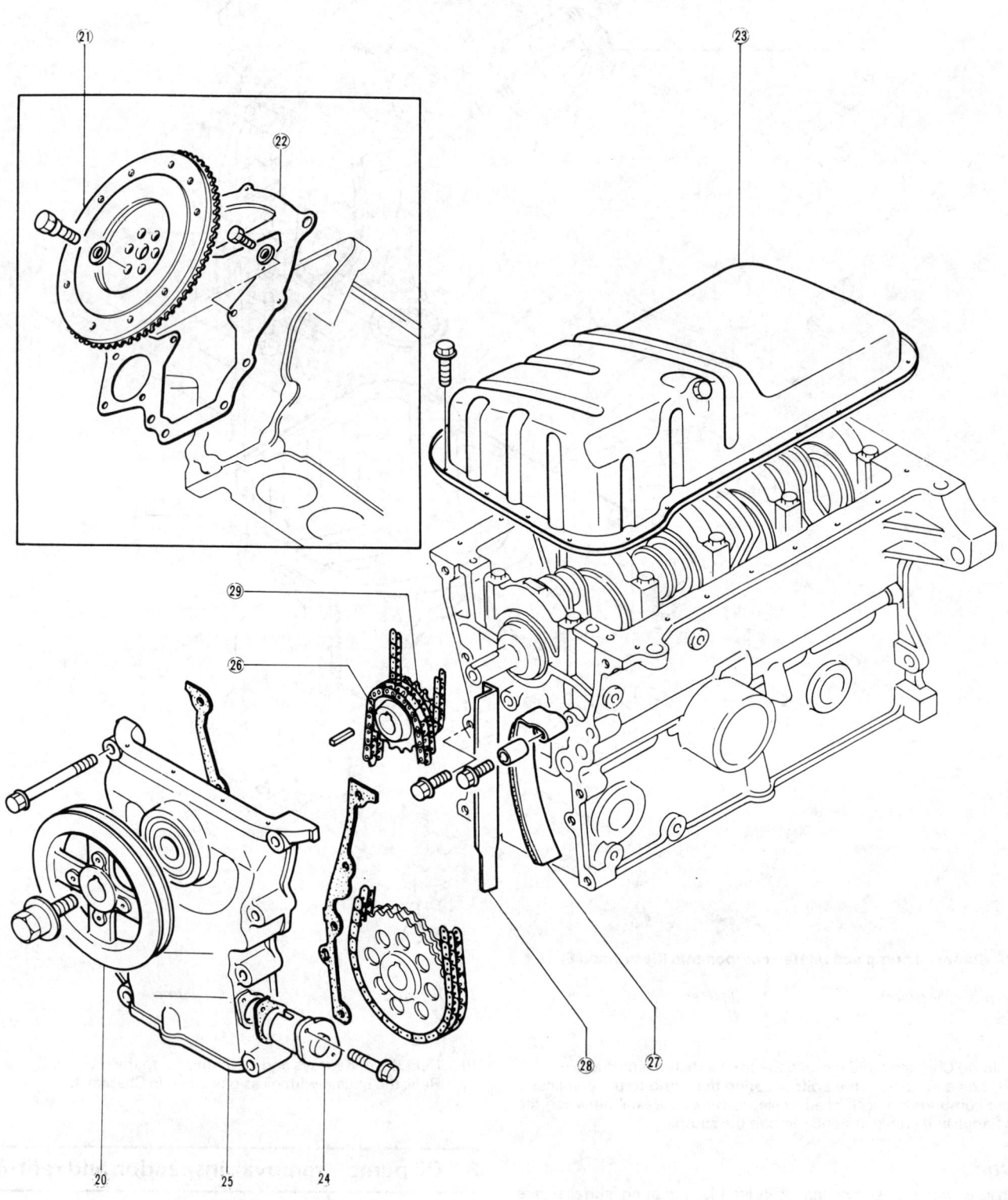

Fig. 2.5 Timing gear, flywheel and sump assemblies (Secs 7, 9 and 10)

- 20 Crankshaft pulley
- 21 Flywheel
- 22 Rear cover
- 23 Sump
- 24 Timing chain tensioner
- 25 Timing chain cover
- 26 Sprocket and chain
- 27 Tensioner arm
- 28 Chain guide
- 29 Oil pump drive sprocket

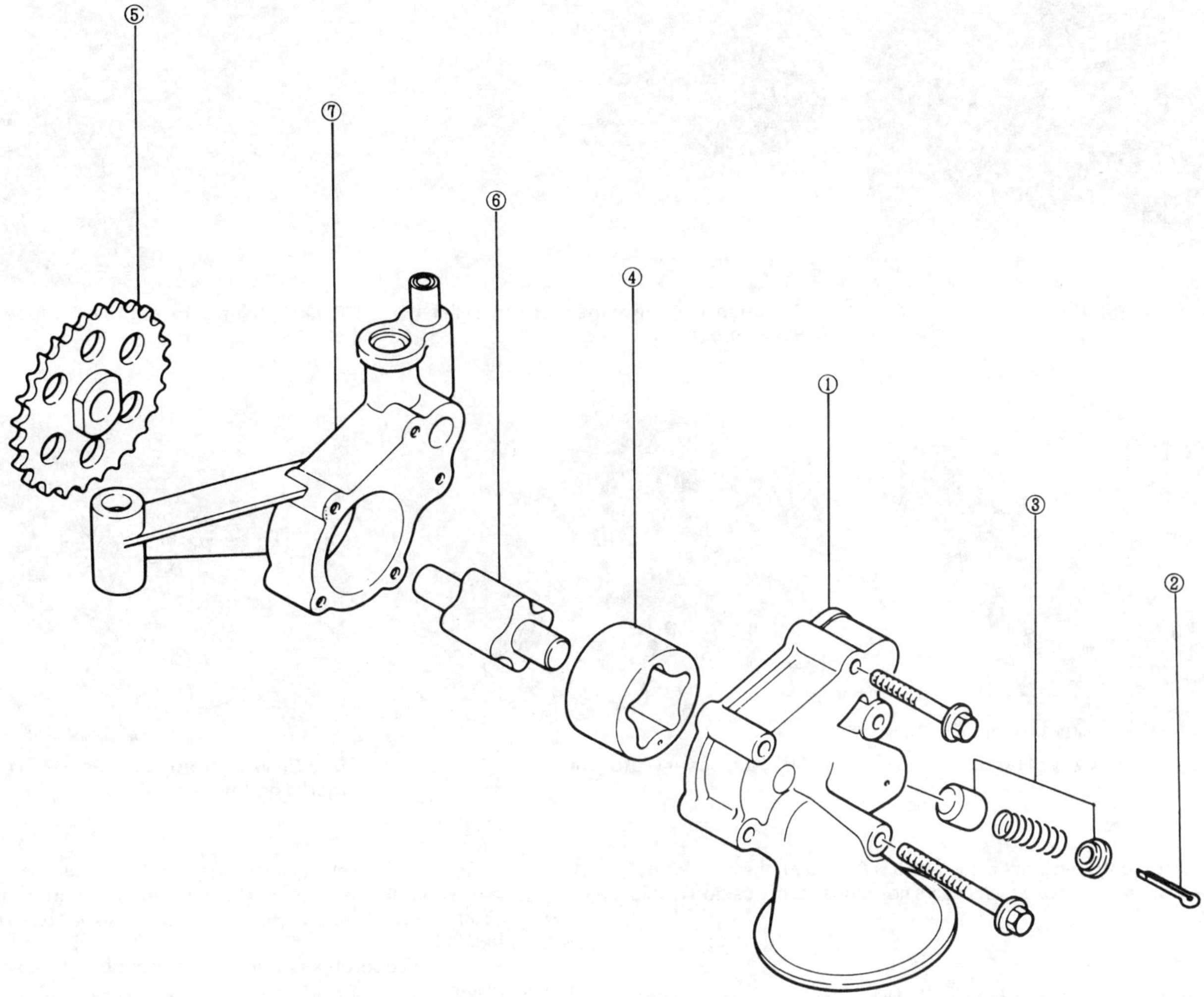

Chapter 2 Part A: E series engine – in-car engine repair procedures

Fig. 2.6 Exploded view of the oil pump (Sec 8)

1 Pump cover	3 Pressure relief valve components	4 Outer rotor	6 Inner rotor
2 Split pin		5 Sprocket	7 Pump body

Inspection

5 Undo the five bolts securing the pump cover to the body and lift off the cover (photo).

6 Extract the split pin from the pump cover and withdraw the pressure relief valve cap, spring and plunger (photos).

7 Withdraw the outer rotor from the pump body.

8 Clean the pump components and carefully examine the rotors, pump body and pressure relief valve body for signs of scoring or damage, renewing any component which is found to be worn. Note that to remove the inner rotor from the pump body it is first necessary to press off the pump sprocket.

9 If the components appear serviceable, refit the outer rotor and measure the clearance between the rotor lobes, the outer rotor and body and between the rotor and pump cover using feeler gauges. Also check for distortion of the pump cover using a straight edge (photos).

10 If any of the clearances exceed the limits given in the Specifications at the start of this Chapter, the pump assembly must be renewed.

11 Examine the oil pump drive chain and sprockets (if removed) for signs of wear or damage, and renew both sprockets and the chain as a set if necessary.

12 If the pump is satisfactory, reassemble the components in the reverse order of removal, lubricating them thoroughly with clean engine oil. Ensure that the inner and outer rotor are assembled with the identification marks on the same side (photo). If the pump sprocket was

8.5 Lift off the pump cover

Chapter 2 Part A: E series engine – in-car engine repair procedures

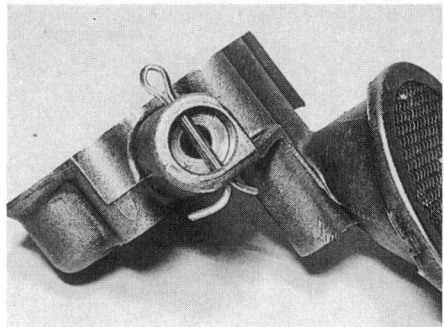

8.6A Extract the split pin...

8.6B ...then withdraw the pressure relief valve components

8.9A Checking the pump rotor lobe clearance...

8.9B ...outer rotor to body clearance...

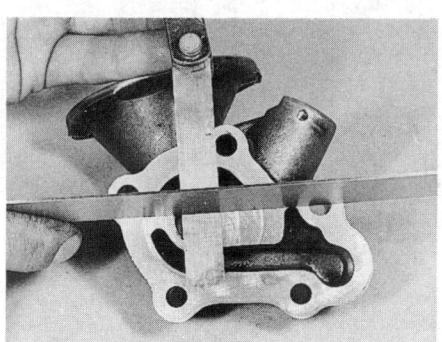

8.9C ...and cover distortion

8.12 Ensure pump is assembled with rotor marks on the same side

removed, press it onto the inner rotor until it is flush with the rotor shaft. Be sure to secure the pressure relief valve components in position using a new split pin.

Refitting

13 If removed, refit the oil pump drive chain and sprocket, and the timing chain components as described in Section 9.
14 Engage the pump sprocket with the chain and refit the pump to the bottom of the crankcase.
15 Refit the pump retaining bolts and tighten them to the specified torque.
16 Refit the sump as described in Section 7.

9 Timing cover, chain and sprockets – removal, inspection and refitting

Removal

1 Remove the cylinder head and sump as described in Sections 5 and 7 of this Chapter.
2 Remove the right-hand inner cover from under the wheel arch to gain access to the right-hand side of the engine.
3 Using a socket and extension bar, undo and remove the crankshaft pulley retaining bolt. Prevent the crankshaft from turning by placing a block of wood between the crankcase and one of the crankshaft webs.
4 Withdraw the pulley from the crankshaft.
5 Undo the retaining bolts and remove the timing chain cover from the crankcase. Support the camshaft sprocket and timing chain to prevent them falling down as the cover is removed.
6 Slacken and remove the timing chain tensioner blade retaining bolt and collar and withdraw the tensioner blade from the crankcase.
7 Undo the two bolts and remove the chain guide.
8 Slip the chain of the crankshaft sprocket and remove it complete with the camshaft sprocket.

9 Withdraw the timing chain sprocket from the crankshaft.
10 If necessary, the oil pump drive chain and sprocket can be removed from the crankshaft providing the oil pump has been removed as described in Section 8.
11 Recover the Woodruff key from the crankshaft groove and store it in a safe place.

Inspection

12 Examine all the teeth on the camshaft and crankshaft sprockets for signs of wear such as hooked, chipped or broken teeth, renewing them as a set if worn. Inspect the timing chain for wear ridges on the rollers, or

9.15 Fit Woodruff key to crankshaft groove ensuring the chamfered end is innermost

Chapter 2 Part A: E series engine – in-car engine repair procedures

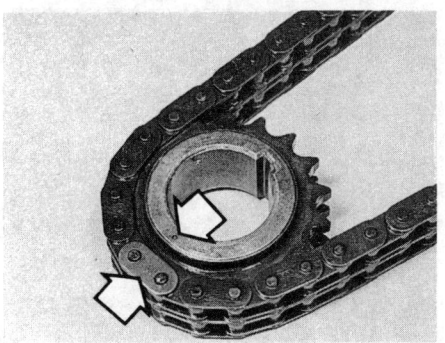

9.17A Crankshaft sprocket timing mark and bright chain link aligned (arrows)

9.17B Align camshaft sprocket timing mark and bright chain link (arrows) and secure chain to the sprocket

9.25 Tighten the crankshaft pulley bolt to the specified torque

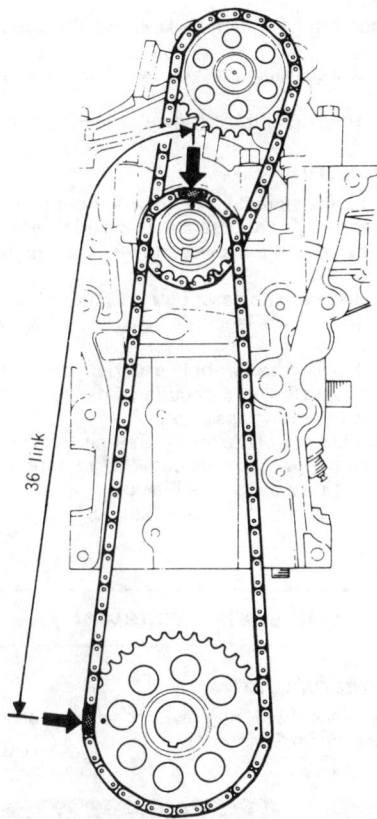

Fig. 2.7 Timing mark location and arrangement on sprockets and chain (Sec 9)

chain using wire or a cable tie to ensure that the chain sprocket relationship is not lost.
18 Position the crankshaft so that No 1 piston is at TDC and the crankshaft Woodruff key is facing the top of the cylinder block.
19 Hold the timing chain and sprockets together in the correct position and slide the crankshaft sprocket onto the crankshaft, ensuring that the sprocket timing marks are facing outwards.
20 Refit the timing chain tensioner blade and collar and tighten the retaining bolt securely.
21 Install the timing chain guide and tighten its retaining bolts securely.
22 Apply jointing compound to the timing chain cover gaskets and position them on the timing cover.
23 Refit the timing cover and progressively tighten the bolts to the specified torque in a diagonal sequence.
24 Slide the crankshaft pulley onto the end of the crankshaft.
25 Apply thread-locking compound to the shoulder of the crankshaft pulley bolt and tighten it to the specified torque setting (photo). Prevent the crankshaft from rotating by placing a block of wood between the crankcase and one of the crankshaft webs.
26 Refit the sump and cylinder head as described in Sections 5 and 7. Refit the inner cover to right-hand wheel arch.

10 Flywheel/driveplate – removal, inspection and refitting

Removal

1 Remove the gearbox or transmission as described in Chapter 7.
2 On manual gearbox models, remove the clutch as described in Chapter 6.
3 The flywheel/driveplate must now be held stationary while the retaining bolts are loosened. To do this, lock the crankshaft using a small strip of angle iron between the ring gear teeth and one of the locating dowels on the cylinder block flange (photo). Alternatively, if the sump has been removed, place a block of wood between the crankcase and crankshaft webs.
4 Mark the flywheel or driveplate in relation to the crankshaft, remove the retaining bolts and withdraw the unit.

Inspection

5 Examine the flywheel for scoring of the clutch face and for wear or chipping of the ring gear teeth. If the clutch face is scored, the flywheel may be machined until flat, but renewal is preferable. If the ring gear is worn or damaged it may be possible to renew it separately, but this job is best left to a Mazda dealer or engineering works. The temperature to which the new ring gear must be heated for installation is critical and, if not done accurately, the hardness of the teeth will be destroyed.
6 Check the torque converter driveplate carefully for signs of distortion or any hairline cracks around the bolt holes or radiating outwards from the centre and inspect the ring gear teeth for signs of wear or chipping. If any sign of wear or damage is found the driveplate must be renewed.

looseness of the rivets or side plates. If the chain has been in operation for some time, or if when held horizontally (rollers vertical) it takes on a deeply bowed appearance, renew the chain.
13 Check the timing chain tensioner for obvious signs of wear or damage. Check for free movement and spring return action of the plunger and ratchet. Renew the tensioner if it is suspect.
14 Inspect the tensioner blade and guide for signs of wear or damage and renew if necessary.

Refitting

15 Refit the Woodruff key in the crankshaft groove ensuring that the chamfered end of the key is innermost (photo).
16 Engage the oil pump drive chain with the sprocket and slide the oil pump drive sprocket onto the crankshaft so that its largest boss is on the outside (facing away from the crankcase).
17 Engage the crankshaft and camshaft sprockets with the timing chain so that the timing marks on the sprockets are aligned with the bright links on the chain (photos). Secure the camshaft sprocket to the

Chapter 2 Part A: E series engine – in-car engine repair procedures

10.3 Flywheel ring gear locked using a piece of angle iron

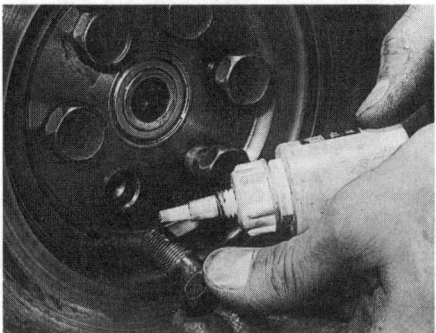

10.8A Apply sealant to flywheel retaining bolts...

10.8B ...and tighten them to the specified torque

Refitting

7 Clean the mating surfaces and fit the flywheel or driveplate aligning the marks made on dismantling.
8 Thoroughly clean the retaining bolts and apply fresh thread-locking compound. Fit the retaining bolts and tighten them to the specified torque in a diagonal sequence (photos). Lock the crankshaft using the method employed on removal.
9 On manual gearbox models, refit the clutch as described in Chapter 6.
10 Refit the gearbox or transmission as described in Chapter 7.

11 Engine/transmission mountings – renewal

1 Apply the handbrake, chock the rear wheels and jack up the front of the car. Support the front of the car securely on axle stands.
2 Remove the undertrays from the vehicle.

Front and rear mountings

3 Place a jack and interposed block of wood beneath the end of the transmission, and take the weight of the engine and transmission on the jack. Alternatively attach a hoist to the engine lifting eyes and take the weight of the engine on the hoist.
4 Undo the bolts securing the mountings to the transmission housing (photo).
5 From underneath the car undo the nuts and bolts (as applicable) securing the crossmember to the underbody.
6 Lower the crossmember at the rear to disengage the rear mounting, then disengage the front mounting and remove the crossmember from under the car.
7 Undo the retaining nuts and remove the mountings from the crossmember.
8 Refitting is the reverse of the removal sequence.

Right-hand mounting

9 Place a jack and interposed block of wood beneath the sump and take the weight of the engine and transmission on the jack. Alternatively attach a hoist to the engine lifting eyes and take the weight of the engine on the hoist.
10 Remove the alternator as described in Chapter 12.
11 Slacken the three nuts securing the mounting to the engine bracket (photo).
12 Undo the nut and through-bolt securing the mounting to the chassis bracket and undo the bolts securing the mounting bracket to the engine. Remove the mounting assembly.
13 Refitting is the reverse of the removal sequence, but do not tighten the mounting to engine bracket nuts until the mounting is in place and centralised on the engine and chassis bracket.
14 Refit the alternator as described in Chapter 12.

12 Crankshaft oil seals – renewal

Right-hand/front oil seal

1 Remove the power steering and/or alternator drivebelts (as applicable) as described in Chapter 1.

11.4 Front engine mounting viewed from underneath

11.11 Right-hand mounting viewed from above

Chapter 2 Part A: E series engine – in-car engine repair procedures

2 Remove the sump as described in Section 7.
3 Using a socket and extension bar, undo and remove the crankshaft pulley retaining bolt. Prevent the crankshaft from turning by placing a block of wood between the crankcase and one of the crankshaft webs.
4 Withdraw the pulley from the crankshaft.
5 The oil seal can then be removed by drilling two small holes diagonally opposite each other and inserting self-tapping screws in them. A pair of grips can then be used to pull out the oil seal, by pulling on each side in turn.
6 Wipe the oil seal seating clean, then dip the new seal in fresh engine oil and locate it over the crankshaft with its closed side facing outwards. Make sure that the oil seal lip is not damaged as it is located on the crankshaft.
7 Using a metal tube or socket, drive the oil seal squarely into the bore until it is flush with the timing chain cover.

8 Slide the crankshaft pulley onto the end of the crankshaft.
9 Apply thread-locking compound to the shoulder of the crankshaft pulley bolt and tighten it to the specified torque setting. Prevent the crankshaft from rotating by placing a block of wood between the crankcase and one of the crankshaft webs.
10 Refit the sump as described in Section 7, and install the power steering and alternator drivebelts (as applicable) as described in Chapter 1.

Left-hand/rear oil seal

11 Remove the flywheel/driveplate as described in Section 10.
12 Renew the oil seal as described in paragraphs 5 to 7.
13 Refit the flywheel/driveplate as described in Section 10.

Part B: B series engine – in-car engine repair procedures

13 General information

How to use this Chapter

This Part of Chapter 2 is devoted to in-vehicle repair procedures for the B series engines. Similar information covering the E series engine will be found in Part A. All procedures concerning engine removal and refitting, and engine block/cylinder head overhaul for all engine types can be found in Part C of this Chapter.

Most of the operations included in this Part are based on the assumption that the engine is still installed in the vehicle. Therefore, if this information is being used during a complete engine overhaul, with the engine already removed, many of the steps included here will not apply.

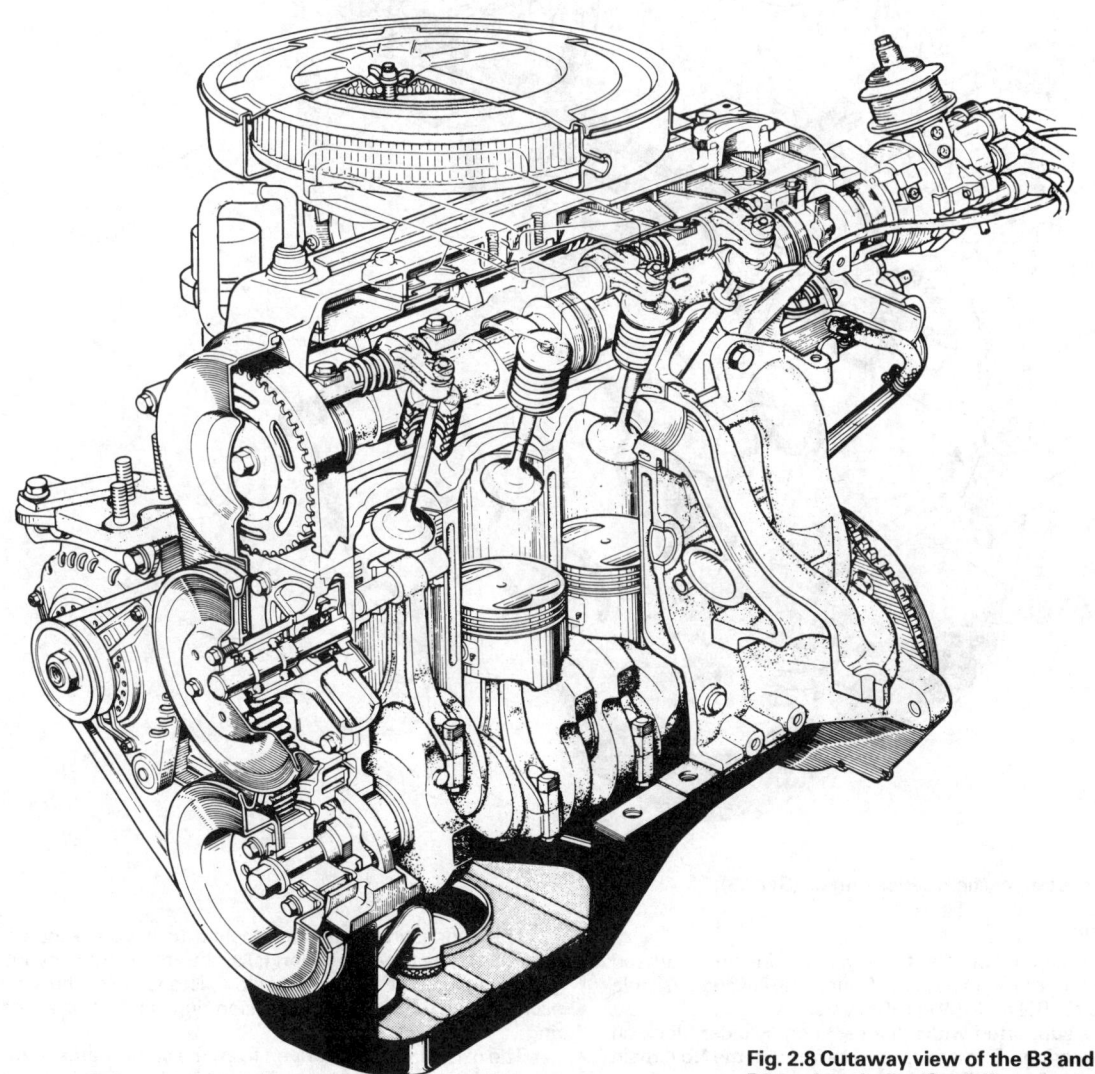

Fig. 2.8 Cutaway view of the B3 and B5 engine – B6 engine similar (Sec 13)

Chapter 2 Part B: B series engine – in-car engine repair procedures

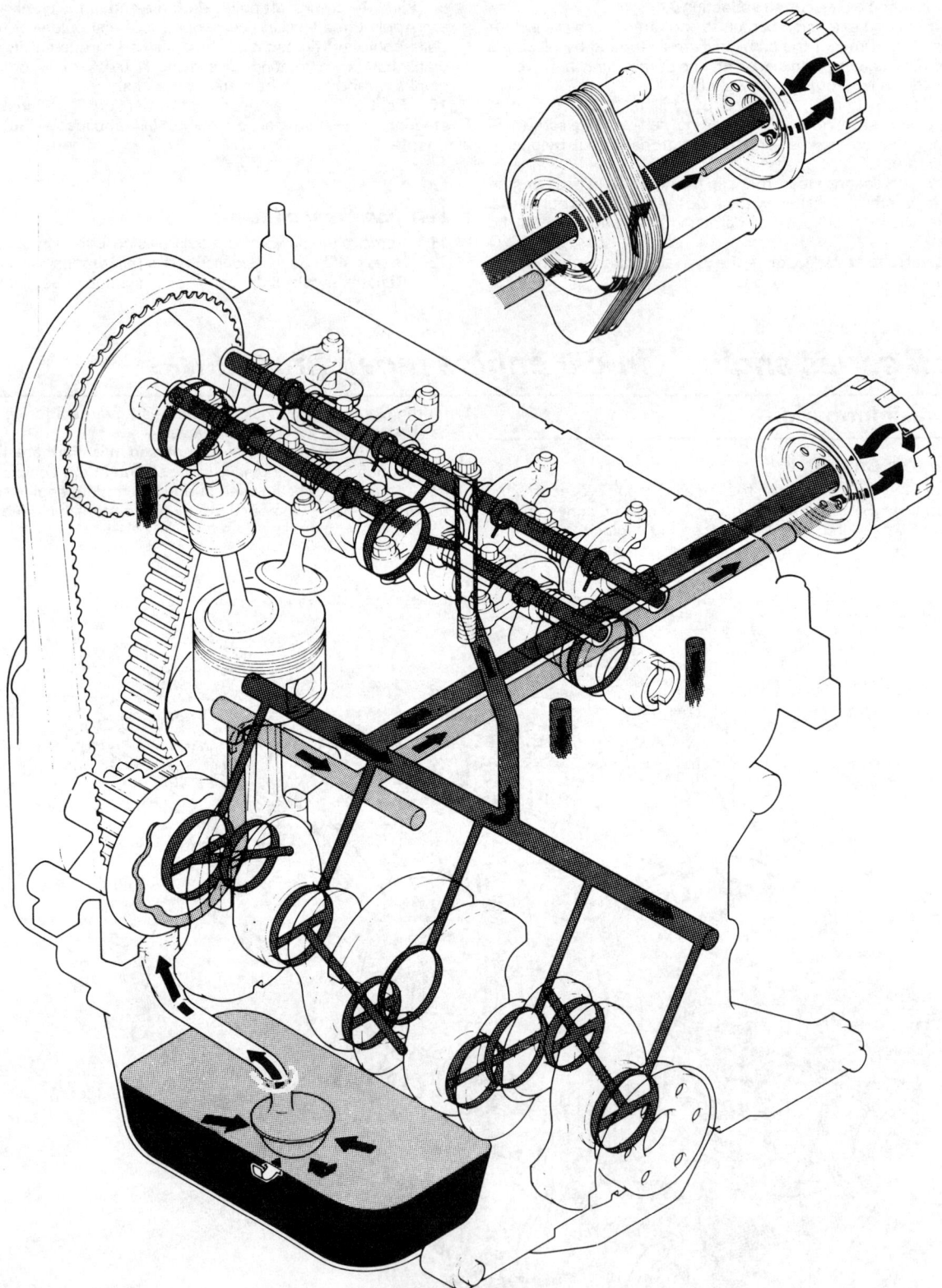

Fig. 2.9 Lubrication system of the B series engine (Sec 13)

Engine description

The B series engine is of the four-cylinder, in-line overhead camshaft type mounted transversely at the front of the car and available in 1323 cc (B3), 1498 cc (B5) and 1597 cc (B6) versions.

The crankshaft is supported within the cast iron cylinder block on five shell type main bearings. Thrustwashers are fitted to the No 4 main bearing to control crankshaft endfloat.

The connecting rods are attached to the crankshaft by horizontal split shell type big-end bearings, and to the pistons by interference fit gudgeon pins. The aluminium alloy pistons are of the slipper type and each is fitted with two compression rings and a three-piece oil control ring.

The camshaft is belt-driven from the crankshaft and runs directly in the aluminium cylinder head. The inclined inlet and exhaust valves are

16.3A Slacken the inlet duct retaining clamps...

16.3B ...and disconnect the hose from the cylinder head cover

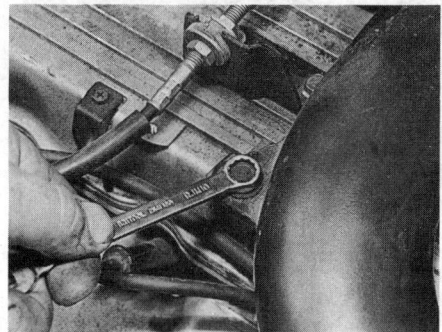

16.3C Undo the retaining bolt...

16.3D ...and remove the duct from the engine compartment

16.5 Slacken the accelerator cable locknuts and free the cable from the head cover

16.9A Fit the gasket to the cylinder head cover groove...

actuated by rocker arms located on a rocker shaft assembly which is bolted to the cylinder head. On B3 and B5 engines, the valve clearances are controlled by maintenance free 'Hydraulic Lash Adjusters' (HLA) which use the oil pressure to remove the clearance between the rocker arm and valve. On B6 engines, valve clearances are adjusted using a screw and locknut arrangement.

Lubrication is by a rotor type oil pump mounted on the right-hand end of the crankshaft. Engine oil is fed through an externally mounted full-flow oil filter to the engine oil gallery, and then to the crankshaft, camshaft and rocker shaft bearings. A pressure relief valve is incorporated in the oil pump. On B6 engines, a water cooled oil cooler is fitted between the oil filter and the crankcase.

A semi-closed crankcase ventilation system is employed and crankcase gases are drawn from the rocker cover via hoses to the air cleaner and inlet manifold.

The distributor rotor is driven direct from the left-hand end of the camshaft. On B3 and B5 engines, the fuel pump is also operated by the camshaft via an eccentric and lever.

Repair operations possible with the engine in the vehicle

The following operations can be carried out without having to remove the engine from the car:

(a) Removal and refitting of the cylinder head cover.
(b) Removal and refitting of the cylinder head.
(c) Removal and refitting of the camshaft and rocker gear.
(d) Removal and refitting of the sump.
(e) Removal and refitting of the connecting rods and pistons*.
(f) Removal and refitting of the oil pump.
(g) Removal and refitting of the flywheel/driveplate.
(h) Renewal of the engine mountings.
(i) Renewal of the crankshaft oil seals.
(j) Renewal of the camshaft oil seal.

* Although the pistons and connecting rods can be removed and refitted with the engine in the car, it is better to carry this work out with the engine removed in the interests of cleanliness and improved access. Refer to Part C for details.

14 Compression test – description and interpretation

Refer to Part A: Section 2.

15 Locating Top Dead Centre (TDC) for number one piston

Refer to Part A: Section 3.

16 Cylinder head cover – removal and refitting

Removal

1 Disconnect the battery leads.
2 On carburettor models, remove the air cleaner assembly as described in Chapter 4.
3 On fuel injected models, slacken the clamps securing the inlet duct to the airflow meter and throttle housing. Disconnect the hose linking the duct to the cylinder head cover from the cover, then undo the duct retaining bolt and remove the duct from the engine compartment (photos).

Chapter 2 Part B: B series engine – in-car engine repair procedures

16.9B ...and refit the cover to the engine

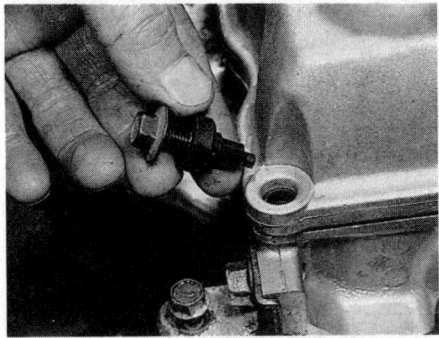
16.9C Refit the retaining bolts and sealing washers...

16.9D ...and tighten them to the specified torque

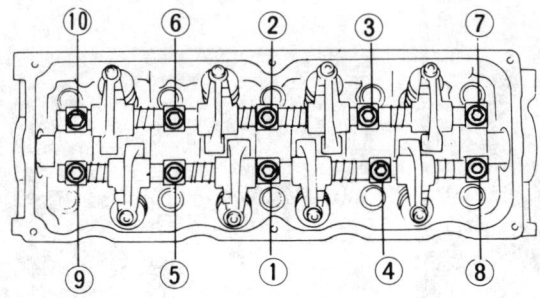
Fig. 2.10 Rocker shaft retaining bolt tightening sequence (Sec 17)

4 Undo the four bolts securing the upper timing belt cover to the right-hand end of the engine and remove the cover and gasket.
5 Disconnect the accelerator cable from the carburettor, or throttle housing, and release the cable from its mounting bracket and guide on the cylinder head cover (photo).
6 Release the HT leads from their guides and disconnect the breather hose from the top of the cover.
7 Undo the six cylinder head cover retaining bolts and remove the sealing washers. Lift off the cover and remove the gasket.

Refitting

8 Examine the cylinder head cover gasket and retaining bolt sealing washers for signs of damage or deterioration and renew as necessary.
9 Apply jointing compound to the groove of the cylinder head cover and position the gasket in the cover. Refit the cover to the cylinder head, then refit the retaining bolts and sealing washers and tighten them to the specified torque (photos).
10 Refit the upper timing belt gasket and cover and tighten its retaining bolts to the specified torque.
11 Reconnect the breather hose and refit the HT leads to their guides.
12 Referring to Chapter 4, reconnect the accelerator cable and, on B3 and B5 engines, refit the air cleaner assembly.
13 On B6 engines, refit the inlet duct and tighten its retaining clamps and bolt securely. Reconnect the hose to the cylinder head cover.
14 Reconnect the battery leads.

17 Rocker gear – removal, inspection and refitting

Removal

1 Remove the cylinder head cover as described in Section 16.
2 On B3 and B5 engines, check the operation of the 'Hydraulic Lash Adjusters' (HLA) by pushing down on the valve end of each rocker arm. If the rocker arm moves when pressure is applied, the HLA assembly is faulty and must be renewed. Make a note of all faulty HLA assemblies and renew them as described in paragraph 8 once the rocker gear has been removed.
3 Slacken all the rocker shaft retaining bolts half a turn at a time in the reverse order of the tightening sequence (Fig. 2.10) to relieve all the valve spring tension. **Note:** *Unscrew the bolts fully from the cylinder head but do not remove the bolts from the rocker shafts.*
4 Lift the inlet and exhaust rocker gear assemblies clear of the cylinder head ensuring that all the retaining bolts remain in position in the shaft (photo).
5 If necessary, the rocker gear assemblies can be dismantled by removing the retaining bolts and sliding the components off the shafts. Keep the inlet and exhaust assembly components separate and, to avoid confusion, make a note of each components correct fitted position as it is removed, to ensure it is positioned correctly on reassembly.

Inspection

6 Examine the rocker arm bearing surfaces which contact the camshaft lobes for wear ridges and scoring. Renew any rocker arms on which these conditions are apparent.
7 If the rocker gear has been dismantled examine the rocker arm and shaft bearing surfaces for wear ridges and scoring. If the necessary measuring equipment is available, measure the internal diameter of the rocker arm and the outside diameter of the rocker shaft at the point where the rocker pivots and calculate the clearance. If the clearance exceeds the figure given in the Specifications at the start of this Chapter

17.4 Ensure retaining bolts remain in position when removing rocker gear shaft assemblies

Chapter 2 Part B: B series engine – in-car engine repair procedures

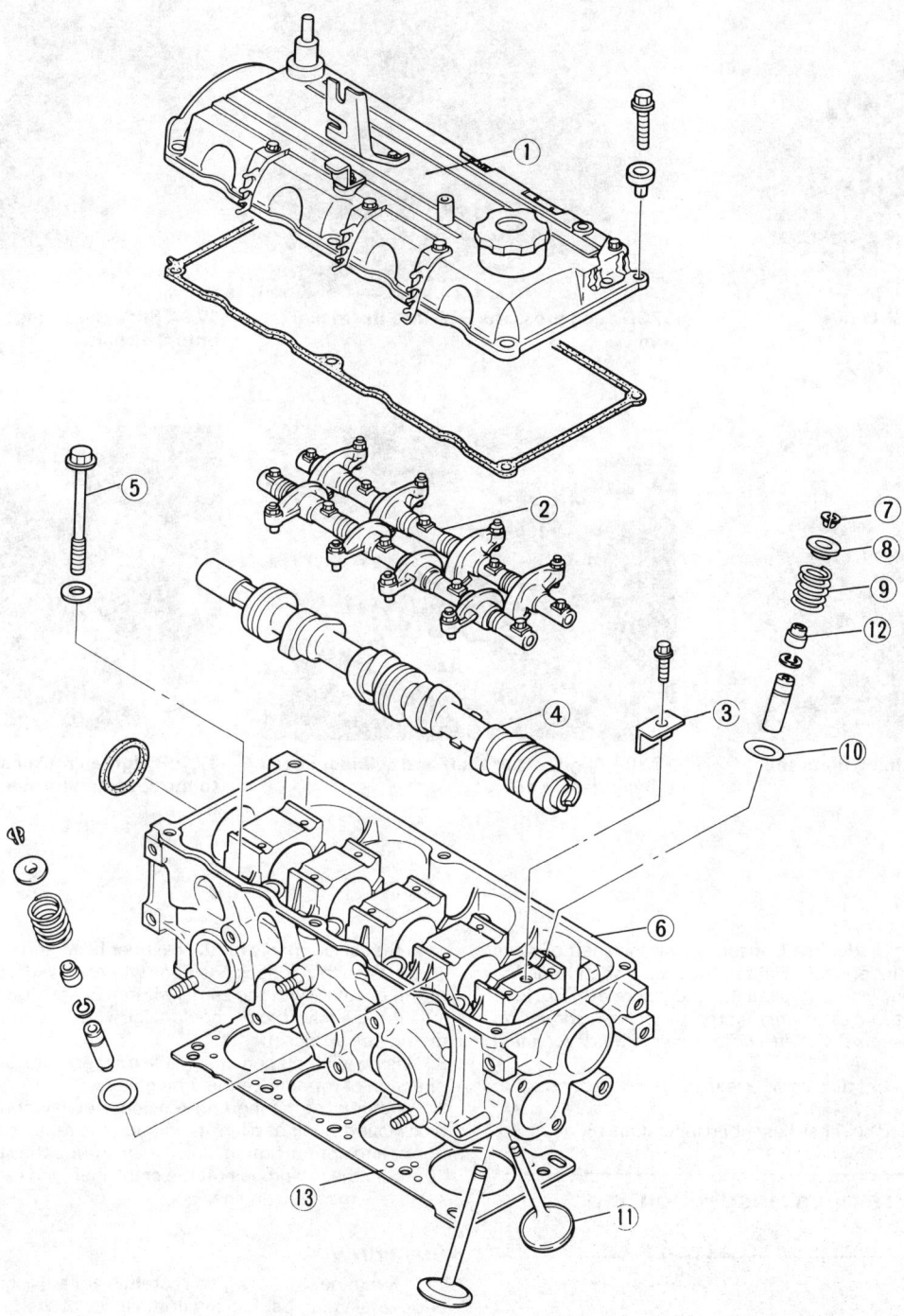

Fig. 2.11 Exploded view of the cylinder head components (Secs 17, 19, 20 and 22)

1 Cylinder head cover
2 Rocker gear
3 Thrust bearing
4 Camshaft
5 Cylinder head bolt
6 Cylinder head
7 Collets
8 Valve spring cap
9 Valve spring
10 Valve spring seat
11 Exhaust valve
12 Valve stem seal
13 Gasket

or there are obvious signs of wear the rocker arm and/or shaft must be renewed.

8 On B3 and B5, engines remove any faulty HLA assemblies from their respective rocker arms using a pair of pliers. Fill the rocker arm oil reservoir with clean engine oil and apply a smear of oil to the new HLA O-ring. Carefully press the new HLA assemblies into position in the rocker arms taking great care not to damage the O-ring (photos).

Refitting

9 If necessary, using the notes made on dismantling reassemble the rocker gear assemblies, noting that the rocker shaft oil holes must face downwards to align with the corresponding oil holes in the cylinder head (photos).

10 Lubricate the camshaft lobes with clean engine oil and refit the

17.8A Apply oil to the HLA O-ring (arrowed)...

17.8B ...and press the HLA into the rocker arm

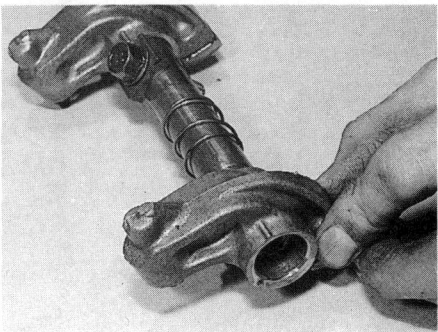
17.9A Slide the springs and rocker arms onto the shaft...

17.9B ...and refit the retaining bolts and plates

17.10A Align rocker shaft and cylinder head oilways (arrowed)

17.10B Tighten rocker shaft retaining bolts to the specified torque

rocker assemblies to the cylinder head, ensuring that the shaft oilways align with those in the cylinder head. Tighten the rocker shaft retaining bolts to the specified torque in two stages in the order shown in Fig. 2.10 (photos). **Note:** *Take great care to ensure that the rocker shaft springs do not become trapped between the shaft and cylinder head or retaining plates whilst tightening the bolts.*
11   On B6 engines check the valve clearances as described in Chapter 1.
12   Refit the cylinder head cover as described in Section 16.

18 Timing belt – removal, inspection and refitting

Removal

1   Firmly apply the handbrake, then jack up the front of the vehicle and support it on axle stands. Remove the right-hand roadwheel and the plastic cover from under the right-hand wheel arch to gain access to the side of the engine.
2   Slacken the three bolts securing the water pump pulley to the pump and remove the pulley.
3   Remove the power steering and/or alternator drivebelts (as applicable) as described in Chapter 1.
4   Undo the four bolts securing the upper timing belt cover to the engine and remove the cover and gasket.
5   Slacken the four crankshaft pulley retaining bolts and remove the bolts and retaining plate. Withdraw the pulley from the crankshaft end and remove the spacer (where fitted) and guide plate.
6   Undo the two lower timing belt cover retaining bolts and remove the cover and gasket.
7   Using a suitable spanner or socket, rotate the crankshaft in a clockwise direction until the two lines on the camshaft sprocket align with the raised marks on the cylinder head and cover, which are at the 12 o'clock and 3 o'clock positions when viewed from the right-hand end of the engine. With the camshaft in this position the raised arrow cast into the cylinder block should align with the notch on the crankshaft timing belt sprocket.
8   Using a felt tip pen or suitable marker, mark an arrow indicating the direction of rotation on the timing belt.
9   Unhook the timing belt tensioner pulley spring from the cylinder block using a pair of pliers, then undo the retaining bolt and remove the pulley and spring, noting which way around the spring is fitted.
10   Slip the timing belt off the crankshaft and camshaft sprockets and remove it from the engine.

Inspection

11   Examine the timing belt carefully for any signs of cracking, fraying or general wear, particularly at the roots of the teeth. Renew the belt if there is any sign of deterioration of this nature, or if there is any oil or grease contamination. The belt must, of course, be renewed if it has completed the mileage specified in Chapter 1.
12   Spin the tensioner by hand and check it for any roughness or tightness. Do not attempt to clean it with solvent as this may enter the bearing. If wear is evident, renew the tensioner. Renew the tensioner spring if there is any doubt about its condition.

Refitting

13   Disconnect the HT leads and remove the spark plugs.
14   Refit the tensioner pulley and spring, ensuring the spring is correctly located in the hole on the rear of the tensioner pulley, and install the retaining bolt and washer. Engage the spring with the peg on the cylinder block then pull on the tensioner pulley until the spring is fully extended. Tighten the retaining bolt securely to hold the tensioner pulley in position (photos).

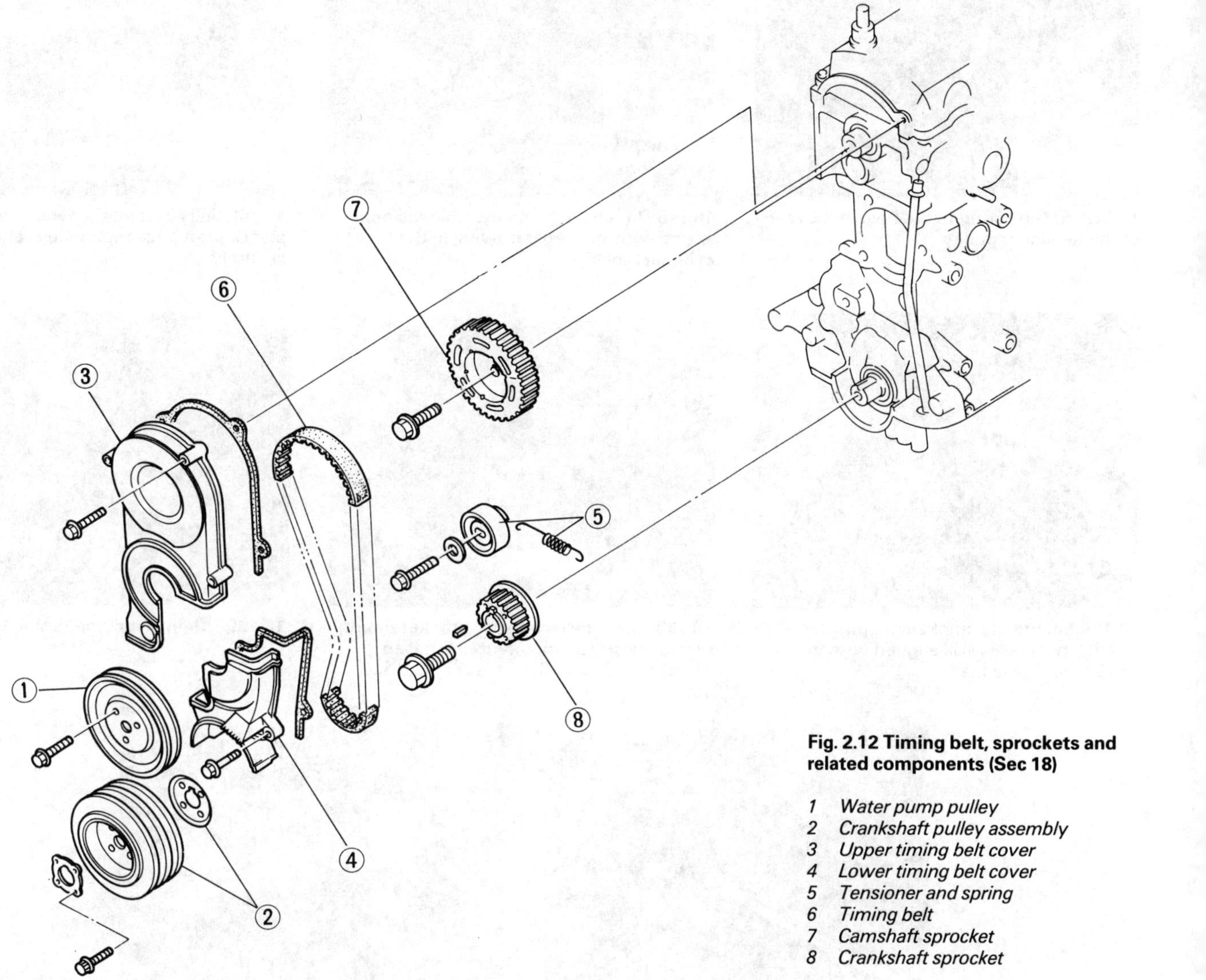

Chapter 2 Part B: B series engine – in-car engine repair procedures

Fig. 2.12 Timing belt, sprockets and related components (Sec 18)

1. Water pump pulley
2. Crankshaft pulley assembly
3. Upper timing belt cover
4. Lower timing belt cover
5. Tensioner and spring
6. Timing belt
7. Camshaft sprocket
8. Crankshaft sprocket

15 Ensure that the camshaft and crankshaft pulleys are still aligned with their respective timing marks as described in paragraph 7 then install the timing belt. If the belt is being re-used, ensure that the arrow made on removal is pointing in the normal direction of belt rotation. Ensure that all the slack is on the tensioner side of the belt then slacken the tensioner pulley retaining bolt to tension the timing belt (photos).

16 Using a suitable spanner or socket, turn the crankshaft through two complete turns in a clockwise direction until the arrow on the block is realigned with the mark on the crankshaft pulley and check that the camshaft sprocket timing marks are correctly aligned.

17 If the timing marks are not correctly positioned, pull the tensioner pulley away from the timing belt and tighten its retaining bolt securely. Remove the belt and repeat the procedures in paragraphs 15 and 16.

18 Once the timing marks are correctly positioned, tighten the tensioner pulley retaining bolt to the specified torque.

19 Apply a force of approximately 10 kg to the front edge of the timing belt at the point adjacent to the water pump and check that the timing belt deflection is 9 to 13 mm on B6 engines, and 12 to 13 mm on B3 and B5 engines (photo).

20 If the timing belt deflection is not within the specified limits the tensioner spring must be renewed. If spring renewal is necessary, unhook the spring from the pulley and block, using a pair of pliers, taking

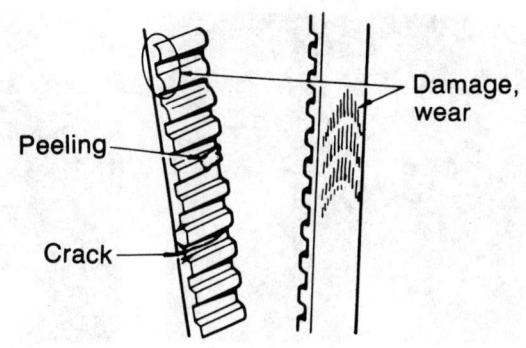

Fig. 2.13 Checking for timing belt wear (Sec 18)

Chapter 2 Part B: B series engine – in-car engine repair procedures

18.14A Fit the spring to the hole in the rear of the tensioner pulley

18.14B On refitting, engage the tensioner spring with the peg (arrowed) in the cylinder block

18.14C Fully extend the tensioner spring and tighten tensioner pulley retaining bolt securely

18.15A Ensure the crankshaft sprocket notch (arrowed) is still aligned with the raised crankcase mark...

18.15B ...and the camshaft sprocket timing marks are still aligned with the raised marks (arrows) on the cylinder head and cover...

18.15C ...then install the timing belt...

18.15D ...and release the tensioner pulley

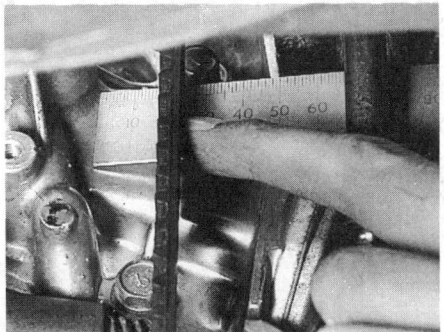

18.19 Checking timing belt deflection

18.21A Lower timing belt cover retaining bolts (arrowed)

18.21B Refitting the upper timing belt cover

18.23A Refit the guide plate...

Chapter 2 Part B: B series engine – in-car engine repair procedures

18.23B ...and spacer...

18.23C ...then install the crankshaft pulley and tighten its retaining screws securely (as applicable)

18.23D Refit the second spacer...

18.23E ...and crankshaft pulley (where fitted)

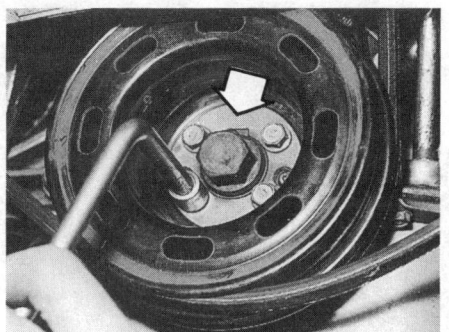

18.23F Refit the retaining plate (arrowed) and tighten the pulley retaining bolts

18.24 Refitting the water pump pulley

great care not to move the timing belt. Fit a new spring and repeat the procedures in paragraphs 16 to 19. If the timing belt deflection is still outside the specified limits renew the timing belt.
21 Once the timing belt is correctly tensioned, refit the lower and upper timing covers and gaskets and tighten their retaining bolts to the specified torque (photos).
22 Refit the spark plugs and reconnect the HT leads.
23 Refit the guide plate, ensuring that its concave surface is facing outwards, and spacer (where fitted) to the end of the crankshaft. Locate the crankshaft pulley in position and tighten its retaining screws (where fitted) securely. Refit the spacer and power steering pump pulley (where fitted) then install the retaining plate and bolts. Tighten the pulley retaining bolts to the specified torque (photos).
24 Refit the water pump pulley and tighten its retaining bolts by hand only (photo).
25 Refit the power steering pump and/or alternator drivebelts as described in Chapter 1, then tighten the water pump pulley retaining bolts to the specified torque.

26 Refit the cover to the right-hand wheel arch.
27 Refit the roadwheel and lower the car to the ground.

19 Timing belt sprockets – removal, inspection and refitting

Removal

1 Remove the timing belt as described in Section 18.

Camshaft sprocket

2 Undo the camshaft sprocket retaining bolt, whilst preventing the sprocket from turning by inserting a stout screwdriver through one of the holes in the camshaft sprocket and resting it against the cylinder

19.8A Ensure dowel pin is in position in the camshaft end...

19.8B ...and refit the sprocket

19.9A Tightening the sprocket retaining bolt to the specified torque whilst retaining the sprocket with either a screwdriver...

19.9B ...or open ended spanner if the cylinder head cover has been removed

19.11A Fit the Woodruff key with its chamfered end innermost...

19.11B ...and refit the crankshaft sprocket

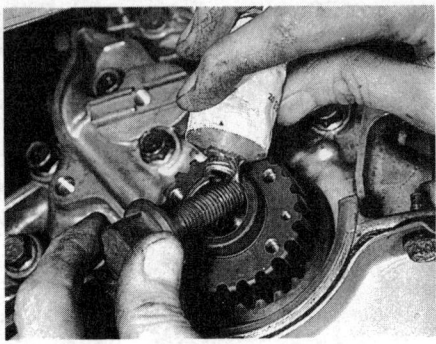

19.12A Apply thread-locking compound to the crankshaft pulley retaining bolt...

19.12B ...and tighten it to the specified torque

head. Alternatively, if the cylinder head cover has been removed, the camshaft can be held using a suitable open-ended spanner fitted to the flats on the right-hand end of the camshaft.

3   Remove the sprocket, then extract the dowel pin from the camshaft and store it safely.

Crankshaft sprocket

4   Remove the sump as described in Section 23.

5   Using a socket and extension bar, undo and remove the crankshaft sprocket retaining bolt. Prevent the crankshaft from turning by placing a block of wood between the crankcase and one of the crankshaft webs.

6   Extract the sprocket from the crankshaft end, then remove the Woodruff key from the crankshaft groove for safe keeping.

Inspection

7   Inspect the teeth of the sprockets for signs of nicks and damage. Also examine the water pump teeth. The teeth are not prone to wear and should normally last the life of the engine.

Refitting

Camshaft sprocket

8   Refit the dowel pin in the camshaft end and install the sprocket (photos). Ensure the pin is correctly located in the camshaft sprocket slot.

9   Refit the camshaft sprocket retaining bolt and tighten it to the specified torque, whilst preventing camshaft rotation using the method employed on removal (photos).

10 Refit the timing belt as described in Section 18.

Crankshaft sprocket

11 Refit the Woodruff key to the crankshaft groove, ensuring that its chamfered edge is innermost, and locate the sprocket on the end of the crankshaft (photos).

12 Apply a thread-locking compound to the shoulder of the crankshaft pulley bolt, then refit the bolt and tighten it to the specified torque, whilst preventing crankshaft rotation by placing a block of wood between the crankcase and one of the crankshaft webs (photos).

13 Refit the sump and timing belt as described in Sections 23 and 18 of this Chapter.

20 Camshaft – removal, inspection and refitting

Removal

1   Remove the rocker gear and camshaft sprocket as described in Sections 17 and 19 of this Chapter.

2   Remove the distributor as described in Chapter 5.

3   Remove the fuel pump as described in Chapter 4.

4   Undo the bolt securing the camshaft thrustplate to the left-hand side of the cylinder head and remove the thrustplate.

5   The camshaft can then be removed from the left-hand end (rear) of the cylinder head (photo). Lever the camshaft oil seal out of the cylinder head using a flat bladed screwdriver.

Inspection

6   Examine the camshaft bearing surfaces, cam lobes and fuel pump eccentric for wear ridges and scoring. Renew the camshaft if any of these conditions are apparent. Examine the condition of the bearing surfaces both on the camshaft journals and in the cylinder head. If the head bearing surfaces are worn excessively, the cylinder head will need to be renewed.

7   If the necessary measuring equipment, is available measure the outside diameter of each camshaft journal, and the internal diameter of the cylinder head bearing surfaces and calculate the camshaft journal oil clearance. Also measure the height of each cam lobe. If any of the measurements exceed the limits given in the Specifications, renew the camshaft and/or cylinder head.

8   Support the camshaft end journals on V-blocks and measure the runout at the centre journal using a dial gauge. If the runout exceeds the specified limit the camshaft should be renewed.

Chapter 2 Part B: B series engine – in-car engine repair procedures

20.5 Removing the camshaft

20.9 Lubricate the camshaft journals and refit the camshaft

20.10 Camshaft thrust bearing and retaining bolt

Refitting

9 Lubricate the camshaft journals with clean engine oil and insert the camshaft into the cylinder head (photo).
10 Refit the thrustplate and tighten its retaining bolt to the specified torque (photo).
11 Set up a dial gauge on one end of the camshaft and measure the endfloat whilst moving the camshaft to and fro (photo). If the endfloat exceeds the limit given in the Specifications renew the thrustplate.
12 Apply oil to the lip and outer edge of a new seal and position it on the camshaft. Using a hammer and suitable tubular drift, which bears only on the hard outer edge of the seal, tap the seal into position until it is flush with the cylinder head.
13 Refit the rocker gear and camshaft sprocket as described in Sections 17 and 19.
14 Refit the fuel pump as described in Chapter 4.
15 Refit the distributor as described in Chapter 5.

21 Camshaft oil seal – renewal

1 Remove the camshaft sprocket as described in Section 19.
2 The oil seal can then be removed by drilling two small holes diagonally opposite each other and inserting self-tapping screws in them. A pair of grips can then be used to pull out the oil seal, by pulling on each side in turn.
3 Wipe clean the oil seal seating, then dip the new seal in fresh engine oil and locate it over the camshaft with its closed side facing outwards. Make sure that the oil seal lip is not damaged as it is located on the camshaft.
4 Using a tubular drift, which bears only on the hard outer edge of the seal, tap the oil seal squarely into position until it is flush with the cylinder head (photo).
5 Refit the camshaft sprocket as described in Section 19.

22 Cylinder head – removal and refitting

Removal

1 Refer to Chapter 1 and drain the cooling system.
2 Remove the distributor as described in Chapter 5.
3 Remove the cylinder head cover and timing belt as described in Sections 16 and 18 of this Chapter.
4 Remove the inlet and exhaust manifolds as described in Chapter 4.
5 On B3 and B5 engines, make a note of the correct fitted positions of the fuel pump hoses. Take all the necessary precautions to prevent the risk of fire and disconnect the hoses from the fuel pump. Plug the hoses to minimise the loss of fuel.
6 Disconnect the following wires and release them from any clips or guides. Identify them with adhesive tape if necessary to ensure correct refitting (photo).

  (a) Temperature gauge sender.
  (b) Cooling fan thermostatic switch.
  (c) Earth strap(s).

7 Slacken the clips and disconnect the radiator top hose from the thermostat housing and the coolant bypass hose from the outlet adjacent to the housing (photo).
8 Slacken the cylinder head bolts half a turn at a time in the order shown in Fig. 2.14. When all the tension has been relieved, remove the bolts and washers.

20.11 Checking camshaft endfloat

21.4 Refitting the camshaft oil seal

Chapter 2 Part B: B series engine – in-car engine repair procedures

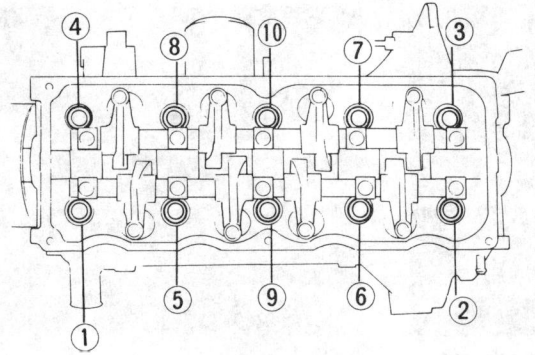

Fig. 2.14 Cylinder head bolt slackening sequence (Sec 22)

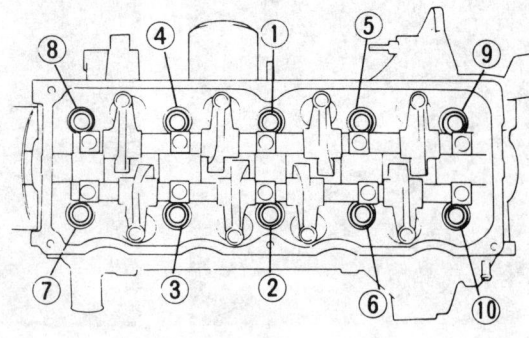

Fig. 2.15 Cylinder head bolt tightening sequence (Sec 22)

9 Release the cylinder head from its locating dowels by tapping it upwards using a hide or plastic mallet. When the head is free, lift it off the engine and remove the cylinder head gasket.

Refitting

10 Clean out all the bolt holes in the cylinder block using a cloth rag and screwdriver. Make sure that all oil is removed, otherwise there is a possibility of the block being cracked by hydraulic pressure when the bolts are tightened.
11 Make sure that the faces of the cylinder head and block are spotlessly clean, then check that No 1 piston is still at TDC.
12 Check that the location dowels are in position in the block.
13 Position the new gasket on the block and over the dowel. It can only be fitted one way round (photo).
14 Carefully lower the cylinder head onto the block and refit the cylinder head bolts and washers (photo).
15 Progressively tighten the cylinder head bolts to the specified torque in the sequence shown in Fig. 2.15 (photo).
16 Reconnect all the electrical connections ensuring that all wiring is correctly routed and secured by any relevant guides or clips.
17 Reconnect the radiator top hose to the thermostat housing and the coolant bypass hose tightening their retaining clips securely.
18 Reconnect the hoses to the fuel pump and secure them in position with their retaining clips.
19 Referring to Chapter 4, refit the inlet and exhaust manifolds.
20 Refit the timing belt as described in Section 19, then on B6 engines, check the valve clearances as described in Chapter 1.
21 Refit the cylinder head cover as described in Section 16.
22 Refit the distributor as described in Chapter 5.
23 Refill the cooling system as described in Chapter 1.

23 Sump – removal and refitting

Removal

1 Drain the engine oil as described in Chapter 1.

22.6 Disconnect the earth strap from the right-hand end of the cylinder head...

22.7 ...and the coolant hoses and electrical connections (arrowed) from the left-hand end

22.13 Fit a new gasket...

22.14 ...and refit the cylinder head

22.15 Tighten the cylinder head bolts to the specified torque setting

Chapter 2 Part B: B series engine – in-car engine repair procedures

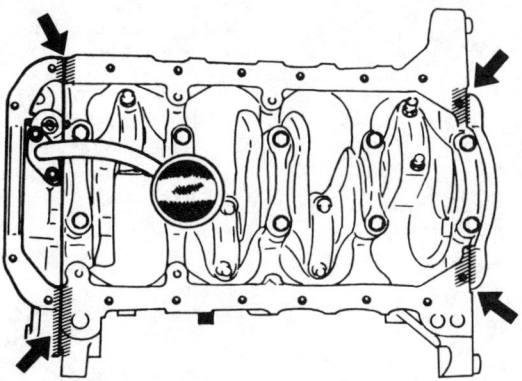

Fig. 2.16 Apply sealant to the shaded areas of the block sealing face (Sec 23)

compound to the cylinder block sealing face in the areas shown in Fig. 2.16 (photos).
8   Refit the sump, offer up the retaining plates and refit the retaining nuts and bolts (photo).
9   Tighten the sump retaining nuts and bolts to the specified torque setting in a diagonal sequence.
10   Refit the front exhaust section as described in Chapter 4.
11   Refit the undertrays and lower the car to the ground.
12   Refill the engine with oil as described in Chapter 1.

2   Apply the handbrake, then jack up the front of the car and support it on axle stands.
3   Remove the undertrays from the vehicle to gain access to the sump.
4   Refer to Chapter 4 and remove the front exhaust pipe section.
5   Slacken the nuts and bolts securing the sump to the crankcase and remove the four sump retaining plates.
6   Tap the sump with a soft faced mallet to break the seal between the sump flange and crankcase and remove the sump.

24 Oil pump – removal, inspection and refitting

Removal
1   Remove the crankshaft timing belt sprocket as described in Section 19.
2   From underneath the car, undo the two bolts securing the oil strainer to the pump and remove the strainer (photo).
3   Undo the bolts securing the oil pump assembly to the right-hand side of the crankcase. Carefully slide the pump housing off the end of the crankshaft and, if loose, remove the locating dowels for safe keeping.

Refitting
7   Apply jointing compound to the semi-circular areas of the sump and fit a new gasket onto the sump. Carefully cut away any protruding sections of the oil pump or seal housing gaskets and apply jointing

Inspection
4   Undo the six screws securing the cover to the pump housing and lift off the cover (photo).
5   Make reference marks on the inner and outer rotors using a felt tip pen and remove them from the cover.
6   Extract the split pin from the pump cover and remove the pressure relief valve cap, spring and plunger.
7   Clean the pump components and carefully examine the rotors,

23.7A Cut away any protruding sections of gasket...

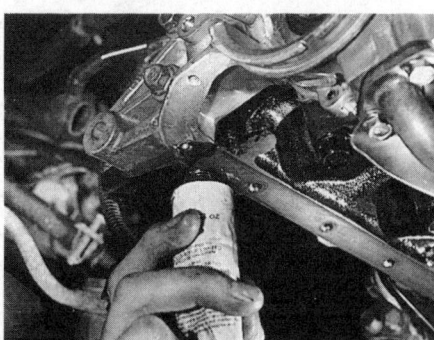

23.7B ...and apply sealant to the specified areas of the block sealing face

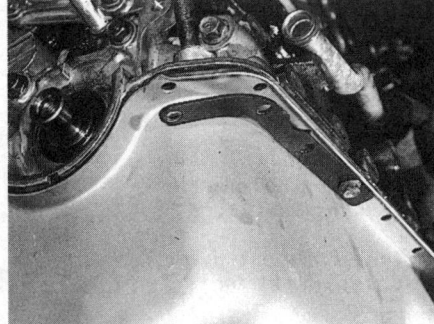

23.8 Refit the sump and install the retaining plates and bolts

24.2 Oil pump strainer retaining bolts

24.4 Removing the oil pump cover retaining screws

24.13 Ensure the locating dowels are in position (arrowed) and fit a new gasket

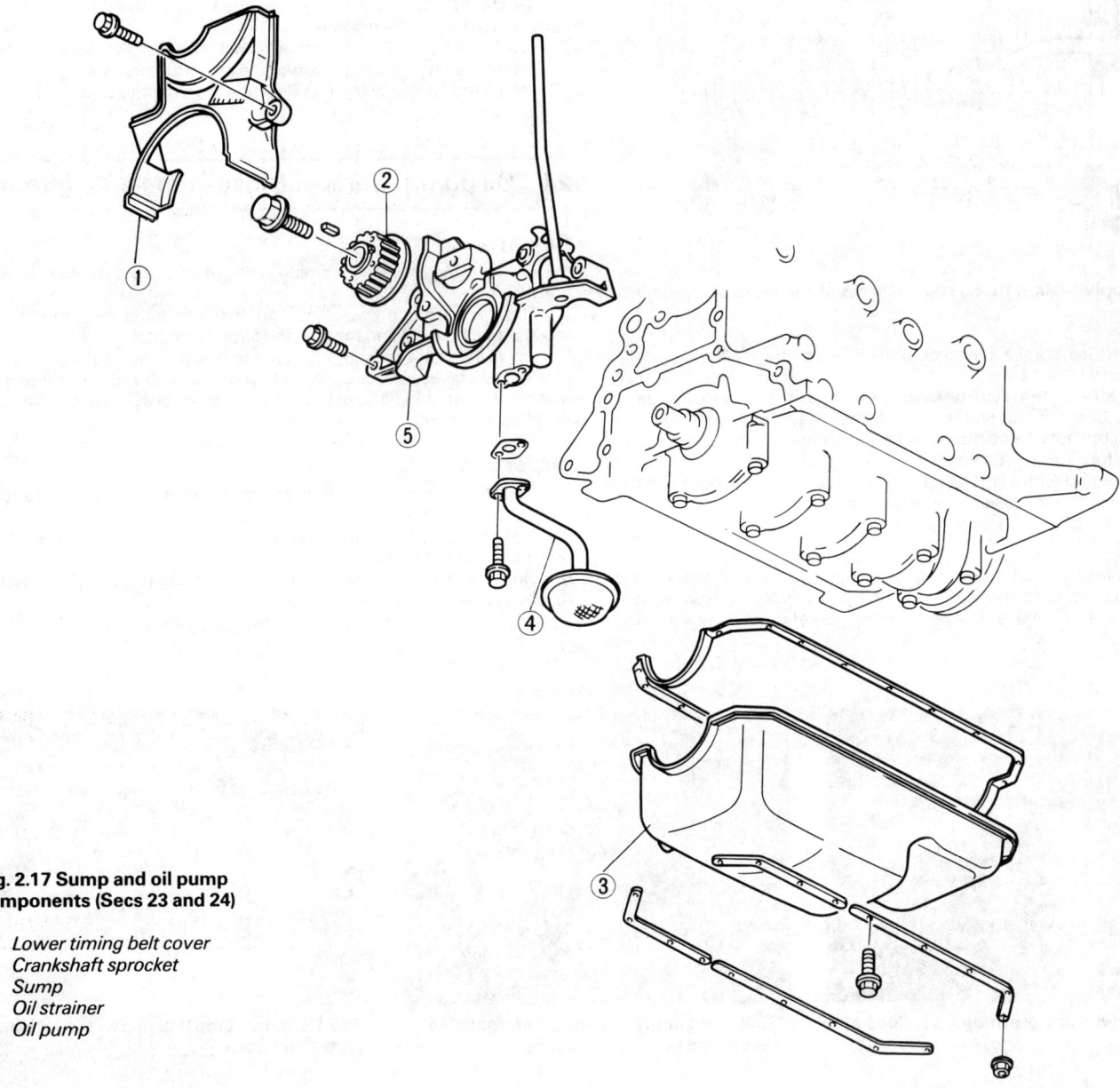

Fig. 2.17 Sump and oil pump components (Secs 23 and 24)

1 Lower timing belt cover
2 Crankshaft sprocket
3 Sump
4 Oil strainer
5 Oil pump

pump body and pressure relief valve body for signs of scoring or damage, renewing any component which is found to be worn. Note that the rotors must be renewed as a set.
8 Examine the oil seal for signs of damage or deterioration and renew if necessary. Lever the old seal out of position using a flat bladed screwdriver and tap the new seal in, using a hammer and suitable tubular drift which bears only on the hard outer edge of the seal.
9 If the components appear serviceable, refit the inner and outer rotors and measure the clearance between the rotor lobes, the outer rotor and body, and between the rotor and pump cover using feeler gauges. Also check for distortion of the pump cover using a straight edge.
10 If any of the clearances exceed the limits given in the Specifications at the start of this Chapter, the pump assembly must be renewed.
11 Clean the oil strainer gauze with a suitable solvent and examine it for signs of clogging or splitting. Renew the strainer if damaged.
12 If the pump is satisfactory reassemble the components in the reverse order of removal, lubricating them thoroughly with clean engine oil. Ensure that the inner and outer rotor are refitted in their original positions using the marks made on dismantling. Secure the pressure relief valve components in position using a new split pin. Apply locking compound to the pump cover retaining screws and tighten them securely.

Refitting

13 Fit the locating dowels to the crankcase or pump and fit a new gasket (photo).
14 Locate the oil pump assembly on the right-hand end of the crankshaft and refit the pump retaining bolts, tightening them to the specified torque.
15 Refit the strainer to the underside of the pump, positioning a new

Chapter 2 Part B: B series engine — in-car engine repair procedures

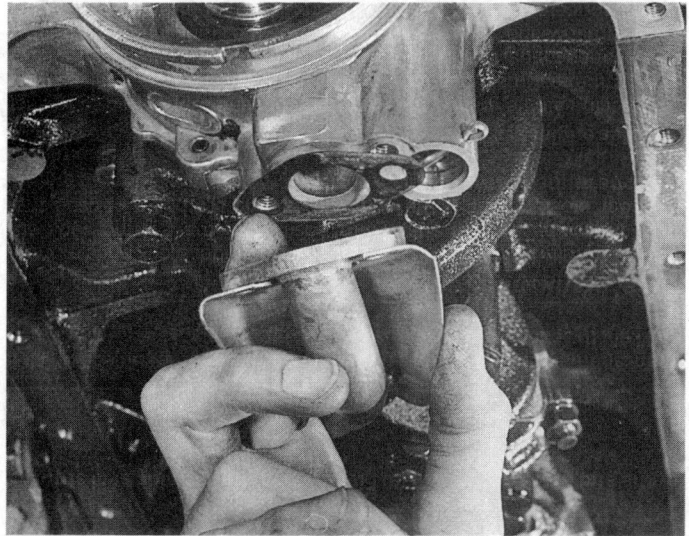

24.15 Refit the oil strainer using a new gasket

25.2 Fitting the flywheel pilot bearing

gasket between the pump and strainer, and tighten its retaining bolts securely (photo).
16 Refit the crankshaft timing belt sprocket as described in Section 19.

25 Flywheel/driveplate – removal, inspection and refitting

Removal
1 Remove the flywheel/driveplate as described in Part A, Section 10.

Inspection
2 Examine the flywheel for scoring of the clutch face and for wear or chipping of the ring gear teeth. If the clutch face is scored, the flywheel may be machined until flat, but renewal is preferable. If the ring gear teeth are damaged the flywheel must be renewed. Inspect the pilot bearing fitted to the centre of the flywheel for signs of roughness or free play and renew if necessary. Drift out the worn bearing and tap the new bearing into position using a hammer and tubular drift which bears only on the outer race of the bearing (photo).
3 Check the torque converter driveplate carefully for signs of distortion, or any hairline cracks around the bolt holes or radiating

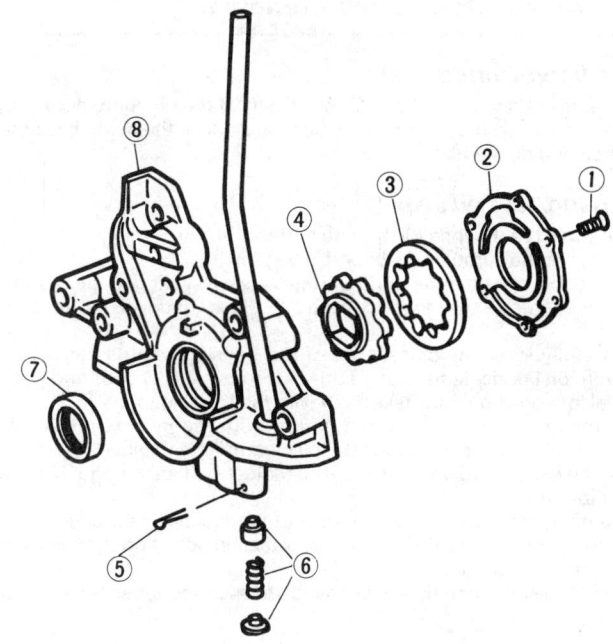

Fig. 2.18 Exploded view of the oil pump (Sec 24)

1 Screw
2 Pump cover
3 Outer rotor
4 Inner rotor
5 Split pin
6 Pressure relief valve components
7 Oil seal
8 Pump body

outwards from the centre, and inspect the ring gear teeth for signs of wear or chipping. If any sign of wear or damage is found the driveplate must be renewed.

Refitting
4 Refit the flywheel/driveplate as described in Part A, Section 10.

26 Engine/transmission mountings – renewal

Refer to Part A, Section 11.

27.5 Rear oil seal and housing

27 Crankshaft oil seals – renewal

Right-hand/front oil seal

1 The right-hand crankshaft oil seal is part of the oil pump assembly and can be renewed as described in Section 24 once the pump has been removed and dismantled.

Left-hand/rear oil seal

2 Remove the flywheel/driveplate as described in Section 25.
3 Remove the sump as described in Section 23.
4 Undo the four bolts securing the rear oil seal housing to the cylinder block and remove the housing from the block. Recover the locating dowels.
5 Carefully lever the oil seal out of the housing using a flat bladed screwdriver. Lubricate the outer edge of the new seal with oil and press the seal into position until its is flush with the housing (photo).
6 Remove all traces of the old gasket from the cylinder block and housing sealing faces and lubricate the oil seal lip with oil.
7 Fit the locating dowels to the cylinder block or housing, and fit a new gasket (photo).
8 Carefully ease the housing into position, taking great care not to damage the oil seal lip. Refit the cover retaining bolts and tighten them to the specified torque.
9 Refit the sump and flywheel/driveplate as described in Sections 23 and 25.

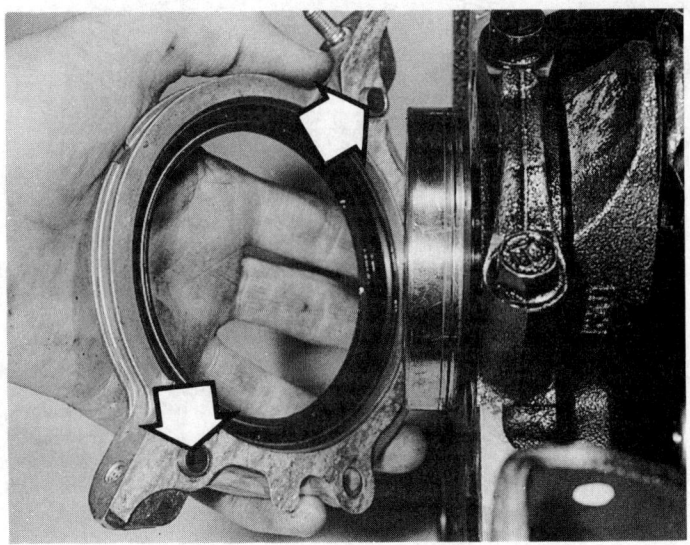

27.7 Ensure the locating dowels (arrowed) are in position and fit a new gasket

Part C: Engine removal and general engine overhaul procedures

28 General information

Included in this Part of Chapter 2 are the general overhaul procedures for the cylinder head, cylinder block/crankcase and internal engine components.

The information ranges from advice concerning preparation for an overhaul and the purchase of replacement parts, to detailed step-by-step procedures covering removal, inspection, renovation and refitting of internal engine parts.

The following Sections have been compiled based on the assumption that the engine has been removed from the vehicle. For information concerning in-vehicle engine repair, as well as the removal and refitting of the external components necessary for the overhaul, refer to Parts A and B of this Chapter and to Section 32 of this Part.

29 Engine overhaul – general information

It is not always easy to determine when, or if, an engine should be completely overhauled, as a number of factors must be considered.

High mileage is not necessarily an indication that an overhaul is needed, while low mileage does not preclude the need for an overhaul. Frequency of servicing is probably the most important consideration. An engine which has had regular and frequent oil and filter changes, as well as other required maintenance, will most likely give many thousands of miles of reliable service. Conversely, a neglected engine may require an overhaul very early in its life.

Excessive oil consumption is an indication that piston rings, valve seals and/or valve guides are in need of attention. Make sure that oil leaks are not responsible before deciding that the rings and/or guides are bad. Perform a cylinder compression check to determine the extent of the work required.

Check the oil pressure with a gauge fitted in place of the oil pressure sender, and compare it with the Specifications. If it is extremely low, the main and big-end bearings and/or the oil pump are probably worn out.

Loss of power, rough running, knocking or metallic engine noises, excessive valve gear noise and high fuel consumption may also point to the need for an overhaul, especially if they are all present at the same time. If a complete tune-up does not remedy the situation, major mechanical work is the only solution.

An engine overhaul involves restoring the internal parts to the specifications of a new engine. During an overhaul, the pistons and rings are replaced and the cylinder bores are reconditioned. New main bearings, connecting rod bearings and camshaft bearings are generally fitted, and if necessary, the crankshaft may be reground to restore the journals. The valves are also serviced as well, since they are usually in less than perfect condition at this point. While the engine is being overhauled, other components, such as the distributor, starter and alternator, can be overhauled as well. The end result should be a like-new engine that will give many trouble free miles. **Note:** *Critical cooling system components such as the hoses, drivebelts, thermostat and water pump MUST be renewed when an engine is overhauled. The radiator should be checked carefully to ensure that it is not clogged or leaking. Also it is a good idea to renew the oil pump whenever the engine is overhauled.*

Before beginning the engine overhaul, read through the entire procedure to familiarize yourself with the scope and requirements of the job. Overhauling an engine is not difficult if you follow all of the instructions carefully, have the necessary tools and equipment and pay close attention to all specifications; however, it can be time consuming. Plan on the vehicle being tied up for a minimum of two weeks, especially if parts must be taken to an engineering works for repair or reconditioning. Check on the availability of parts and make sure that any necessary special tools and equipment are obtained in advance. Most work can be done with typical hand tools, although a number of precision measuring tools are required for inspecting parts to determine if they must be renewed. Often the engineering works will handle the inspection of parts and offer advice concerning reconditioning and renewal. **Note:** *Always wait until the engine has been completely disassembled and all components, especially the engine block have been inspected before deciding what service and repair operations must be performed by an engineering works. Since the condition of the block will be the major factor to consider when determining whether to overhaul the original engine or buy a reconditioned unit, do not purchase parts or have overhaul work done on other components until the block has been thoroughly inspected.* As a general rule, time is the primary cost of an overhaul, so it does not pay to fit worn or sub-standard parts.

Chapter 2 Part C: Engine removal and general engine overhaul procedures

31.5 On carburettor models disconnect the fuel and return hoses from the fuel pump

31.8A Disconnect the HT leads...

31.8B ...and remove the distributor cap and leads as an assembly

As a final note, to ensure maximum life and minimum trouble from a reconditioned engine, everything must be assembled with care in a spotlessly clean environment.

30 Engine removal – methods and precautions

If you have decided that an engine must be removed for overhaul or major repair work, several preliminary steps should be taken.

Locating a suitable place to work is extremely important. Adequate work space, along with storage space for the vehicle, will be needed. If a shop or garage is not available, at the very least a flat, level, clean work surface is required.

Cleaning the engine compartment and engine before beginning the removal procedure will help keep tools clean and organized.

An engine hoist or A-frame will also be necessary. Make sure the equipment is rated in excess of the combined weight of the engine and transmission. Safety is of primary importance, considering the potential hazards involved in lifting the engine out of the vehicle.

If the engine is being removed by a novice, a helper should be available. Advice and aid from someone more experienced would also be helpful. There are many instances when one person cannot simultaneously perform all of the operations required when lifting the engine out of the vehicle.

Plan the operation ahead of time. Arrange for, or obtain all of the tools and equipment you'll need prior to beginning the job. Some of the equipment necessary to perform engine removal and installation safely and with relative ease are (in addition to an engine hoist) a heavy duty floor jack, complete sets of spanners and sockets as described in the front of this Manual, wooden blocks and plenty of rags and cleaning solvent for mopping up spilled oil, coolant and fuel. If the hoist must be hired, make sure that you arrange for it in advance, and perform all of the operations possible without it beforehand. This will save you money and time.

Plan for the vehicle to be out of use for quite a while. An engineering works will be required to perform some of the work which the do-it-yourselfer cannot accomplish without special equipment. These places often have a busy schedule, so it would be a good idea to consult them before removing the engine in order to accurately estimate the amount of time required to rebuild or repair components that may need work.

Always be extremely careful when removing and refitting the engine. Serious injury can result from careless actions. Plan ahead, take your time and a job of this nature, although major, can be accomplished successfully.

31 Engine – removal and refitting

Removal

1 Disconnect the battery leads and remove the battery from the car.
2 Remove the bonnet as described in Chapter 11.
3 Drain the engine oil, coolant and gearbox/transmission lubricant as described in Chapter 1.
4 Remove the air cleaner assembly as described in Chapter 4.
5 On all carburettor models, disconnect the fuel feed and return hoses from the fuel pump and plug the hoses to prevent fuel spillage (photo).
6 On models equipped with power steering, remove the pump as described in Chapter 10, however do not disconnect the hoses and position the pump clear of the engine unit.
7 Disconnect the brake servo vacuum hose from the inlet manifold.
8 Disconnect the HT leads from the spark plugs and ignition coil then remove the distributor cap and leads as an assembly (photos).
9 Referring to Chapter 3, remove the radiator. Disconnect the top hose from the thermostat housing and the bottom hose from the water pump housing and remove both hoses.
10 Disconnect the heater hoses from the coolant bypass pipe and the inlet manifold (photo).
11 Refer to Chapter 4 if necessary, and disconnect the choke and/or accelerator cable from the carburettor/throttle housing (as appropriate).
12 On manual gearbox models disconnect the clutch cable from the operating arm as described in Chapter 6.
13 Unscrew the speedometer cable retaining ring and disconnect the cable from the top of the gearbox/transmission housing.
14 Firmly apply the handbrake, chock the rear wheels and slacken the front wheel nuts. Jack up the front of the car and support it on axle stands. Remove the front roadwheels.
15 Remove both the undertray sections and the left and right-hand inner wheel arch covers to gain full access to the underside of the engine/transmission unit.
16 If an anti-roll bar is fitted, undo the two locknuts and remove the connecting link bolt securing the anti-roll bar to the lower suspension arm. Make a note of the fitted positions of the washers, rubber bushes and spacer for reference on reassembly.
17 Undo the nut and remove the pinch-bolt securing the lower suspension arm balljoint to the swivel hub.
18 Using a long stout bar, carefully lever the lower suspension arm down to release the balljoint from the swivel hub, whilst taking great care not to damage the balljoint rubber gaiter.
19 The inner constant velocity joint can be released from the gearbox by pulling the swivel hub firmly outwards. If this fails to release the inner joint, insert a suitable bar between the inner joint and the gearbox housing and carefully lever the joint out of position. Support the driveshaft at its inner end as it is removed to avoid damaging the oil seal.
20 Undo the nuts securing the front exhaust pipe section to the manifold, and the bolt securing the pipe to the bracket at the front of the engine. Separate the pipe from the manifold and recover the gasket.
21 Disconnect the wires from the following components and release them from any necessary guides or clips. If necessary, identify each wire with adhesive tape to ensure correct refitting (photos).

  (a) Ignition coil.
  (b) Cooling fan switch.
  (c) Temperature gauge sender.
  (d) Fuel cut-off valve solenoid (carburettor models).
  (e) Throttle position switch and injectors (fuel injection models).
  (f) Reversing lamp switch.
  (g) Alternator.
  (h) Starter motor and solenoid.
  (i) Oil pressure switch.
  (j) Starter inhibitor, neutral and kickdown solenoid switches (automatic transmission models).

31.10 Disconnect the heater hose from the inlet manifold

31.21A Disconnect the wiring from the ignition coil...

31.21B ...oil pressure switch...

31.21C ...and the various other electrical components

31.25A Front engine/transmission mounting to crossmember retaining nuts

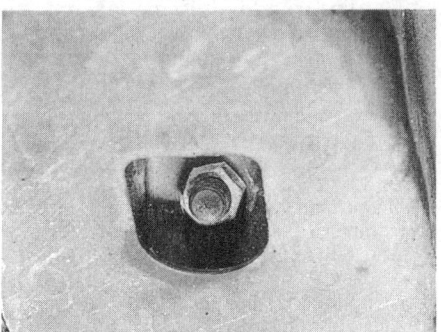
31.25B Rear engine/transmission mounting to crossmember retaining nut

31.26 Right-hand engine mounting through-bolt (arrowed)

31.29 Lifting the engine/transmission assembly

22 On automatic transmission models, disconnect and plug the fluid pipes at the transmission, and disconnect the vacuum hose from the diaphragm. Refer to Chapter 7 and disconnect the selector cable or linkage from the transmission.

23 On manual gearbox models, slacken the nut and remove the bolt securing the gearchange rod to the gearbox shift rod. Undo the nut then remove the washers and separate the remote control housing extension rod from the stud on the gearbox.

24 Attach a suitable hoist to the engine lifting brackets, then raise the hoist to just take the weight of the engine.

25 Undo the nuts securing the engine/transmission mountings to the crossmember (photos).

26 Undo the nut and remove the through-bolt from the right-hand engine mounting (photo).

27 On pre-1982 models remove the additional mounting from the front face of the gearbox.

28 Check that all pipes, cable, cable clips, hoses and other attachments have been removed and positioned well clear of the engine.

29 Lift the engine and transmission slightly and release the front and rear mountings from the crossmember. Continue lifting the unit carefully out of the engine compartment taking care not to damage any components on the surrounding panels (photo). As soon as the engine is high enough, move the hoist away from the car, swing the engine and transmission over the front body panel and lower the unit to the ground.

Refitting

30 Refitting is a reversal of removal, however note the following additional points.
   (a) Position the engine/transmission unit so that the mountings are not strained, twisted or in tension when the mounting nuts and bolts are tightened to the specified torque.
   (b) Renew the driveshaft retaining circlips and apply a smear of grease to the shaft splines prior to refitting
   (c) Refer to the applicable Chapters and Sections as for removal.
   (d) Tighten all nuts and bolts to the specified torque settings.
   (e) Refill the engine oil, coolant and transmission lubricant as described in Chapter 1.

Chapter 2 Part C: Engine removal and general engine overhaul procedures

32 Engine overhaul – dismantling sequence

1 It is much easier to disassemble and work on the engine if it is mounted on a portable engine stand. These stands can often be hired from a tool hire shop. Before the engine is mounted on a stand, the flywheel/driveplate should be removed from the engine so that the engine stand bolts can be tightened into the end of the cylinder block.

2 If a stand is not available, it is possible to disassemble the engine with it blocked up on a sturdy workbench or on the floor. Be extra careful not to tip or drop the engine when working without a stand.

3 If you are going to obtain a reconditioned engine, all external components must come off first in order to be transferred to the replacement engine (just as they will if you are doing a complete engine overhaul yourself). These components include:

(a) Alternator and brackets.
(b) Distributor and spark plugs.
(c) Thermostat and cover.
(d) Carburettor or throttle housing (as applicable).
(e) Inlet and exhaust manifolds.
(f) Oil filter.
(g) Fuel pump.
(h) Engine mountings.
(i) Flywheel/driveplate.

Note: *When removing the external components from the engine, pay close attention to details that may be helpful or important during refitting. Note the fitted position of gaskets, seals, spacers, pins, washers, bolts and other small items.*

4 If you are obtaining a short motor (which consists of the engine cylinder block, crankshaft, pistons and connecting rods all assembled), then the cylinder head, sump, oil pump, and timing chain/belt (as applicable) will have to be removed as well.

5 If you are planning a complete overhaul, the engine can be disassembled and the internal components removed in the following order.

(a) Timing belt and sprockets (B series engine).
(b) Timing chain and sprockets (E series engine).
(c) Cylinder head.
(d) Flywheel.
(e) Sump.
(f) Oil pump.
(g) Pistons.
(h) Crankshaft.

6 Before beginning the disassembly and overhaul procedures, make sure that you have all of the correct tools necessary. Refer to the introductory pages at the beginning of this Manual for further information.

33 Cylinder head – dismantling

Note: *New and reconditioned cylinder heads are available from the manufacturers and from engine overhaul specialists. Due to the fact that some specialist tools are required for the dismantling and inspection procedures, and new components may not be readily available, it may be more practical and economical for the home mechanic to purchase a reconditioned head rather than dismantle, inspect and recondition the original head.*

1 Using a valve spring compressor, compress each valve spring in turn until the split collets can be removed. Release the compressor and lift off the cap, spring(s) and spring seat.

2 If, when the valve spring compressor is screwed down, the valve spring cap refuses to free and expose the split collets, gently tap the top of the tool, directly over the cap with a light hammer. This will free the cap.

3 Withdraw the oil seal off the top of the valve guide, and then remove the valve through the combustion chamber.

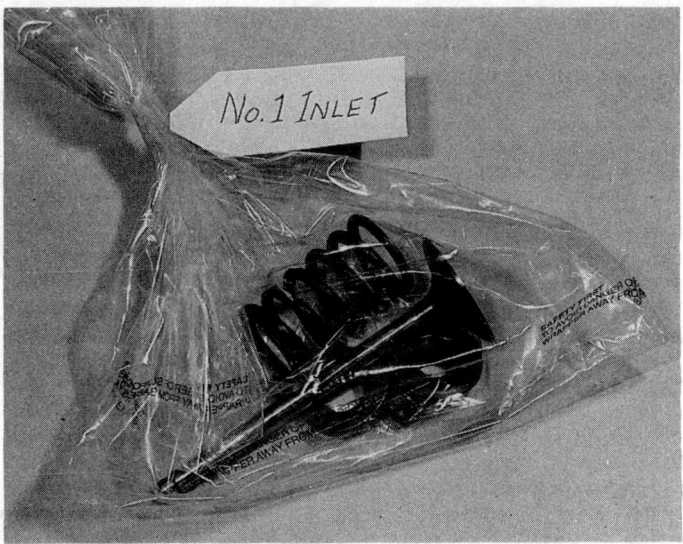

33.4 Store the valve components in a polythene bag after removal

4 It is essential that the valves are kept in their correct sequence unless they are so badly worn that they are to be renewed. If they are going to be kept and used again, place them in a labelled polythene bag or alternatively put them in a sheet of card having eight holes, numbered 1 to 4 inlet and 1 to 4 exhaust, corresponding to the relative fitted positions of the valves (photo). Note that No 1 cylinder is nearest to the crankshaft pulley end of the engine.

34 Cylinder head and valves – cleaning, inspection and renovation

1 Thorough cleaning of the cylinder head and valve components, followed by a detailed inspection, will enable you to decide how much valve service work must be carried out during the engine overhaul.

Cleaning

2 Scrape away all traces of old gasket material and sealing compound from the cylinder head.

3 Scrape away the carbon from the combustion chambers and ports, then wash the cylinder head thoroughly with paraffin or a suitable solvent.

4 Scrape off any heavy carbon deposits that may have formed on the valves, then use a power-operated wire brush to remove deposits from the valve heads and stems.

Inspection

Note: *Be sure to perform all the following inspection procedures before concluding that the services of a machine shop or engine overhaul specialist are required. Make a list of all items that require attention.*

Cylinder head

5 Inspect the head very carefully for cracks, evidence of coolant leakage and other damage. If cracks are found, a new cylinder head should be obtained.

6 Use a straight-edge and feeler blade to check that the cylinder head surface distortion does not exceed the limit given in the Specifications (photo). If it does, it may be possible to resurface it. This can be determined by measuring the height of the cylinder head. If the head height exceeds the specified minimum, have an engineering works grind the head surface until the distortion is within the specified limit or the minimum head height is reached. If when the head is machined to the specified minimum the distortion is not within the specified limit, the head must be renewed.

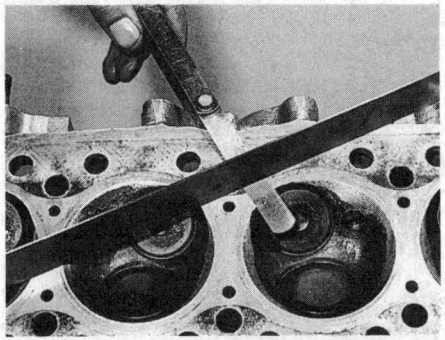
34.6 Checking the cylinder head for distortion

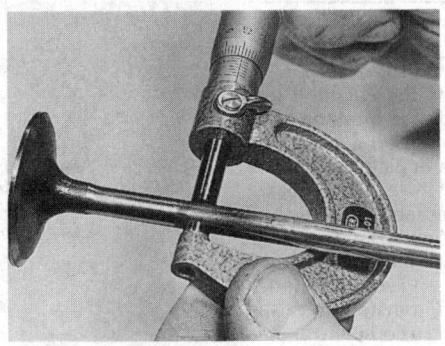
34.9 Checking the valve stem for excessive wear

34.12 Checking valve spring free length

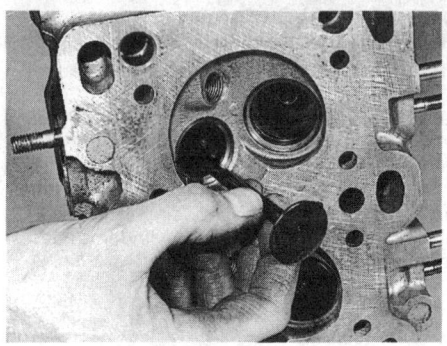
35.1 Lubricate the valve stem and insert it into its respective guide

35.2 Pressing a valve seal onto its guide

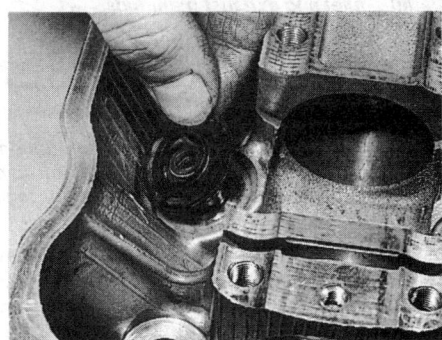
35.3A Refit the spring seat...

7   Examine the valve seats in each of the combustion chambers. If they are severely pitted, cracked or burned then they will need to be renewed or recut by an engine overhaul specialist. If they are only slightly pitted, this can be removed by grinding the valve heads and seats together with coarse then fine grinding paste as described below.
8   If the valve guides are worn, indicated by a side-to-side motion of the valve, new guides must be fitted. This work is best carried out by an engine overhaul specialist, however they may be renewed using a suitable mandrel making sure that they are at the correct height. A dial gauge may be used to determine the amount of side play of the valve.

Valves

9   Examine the heads of each valve for pitting, burning, cracks and general wear, and check the valve stem for scoring and wear ridges. Rotate the valve and check for any obvious indication that it is bent. Look for pits and excessive wear on the end of each valve stem. If the valve appears satisfactory at this stage, measure the valve stem diameter at several points using a micrometer (photo). Any significant difference in the readings obtained indicates wear of the valve stem. Should any of these conditions be apparent, the valve(s) must be renewed. If the valves are in satisfactory condition they should be ground (lapped) into their respective seats to ensure a smooth gas-tight seal.
10   Valve grinding is carried out as follows. Place the cylinder head upside down on a bench with a block of wood at each end to give clearance for the valve stems.
11   Smear a trace of coarse carborundum paste on the seat face and press a suction grinding tool onto the valve head. With a semi-rotary action, grind the valve head to its seat, lifting the valve occasionally to redistribute the grinding paste. When a dull matt even surface is produced on both the valve seat and the valve, wipe off the paste and repeat the process with fine carborundum paste. A light spring placed under the valve head will greatly ease this operation. When a smooth unbroken ring of light grey matt finish is produced on both the valve and seat, the grinding operation is complete. Be sure to remove all traces of grinding paste using paraffin or a suitable solvent before reassembly of the cylinder head.

Valve components

12   Examine the valve springs for signs of damage and discoloration, and also measure their free length using vernier calipers or by comparing the existing spring with a new component (photo).
13   Stand each spring on a flat surface and check it for squareness. If any of the springs are damaged, distorted or have lost their tension, obtain a complete new set of springs.

35 Cylinder head – reassembly

1   Lubricate the stems of the valves and insert them into their original locations (photo). If new valves are being fitted, insert them into the locations to which they have been ground.
2   Working on the first valve, dip the oil seal in engine oil then carefully locate it over the valve and onto the guide. Take care not to damage the seal as it is passed over the valve stem. Use a suitable socket or metal tube to press the seal firmly onto the guide (photo).
3   Locate the spring seat on the guide, followed by the spring and cap. Where applicable the spring should be fitted with its closest pitched coils towards the head (photos).
4   Compress the valve spring and locate the split collets in the recess in the valve stem (photo). Use a little grease to hold the collets in place. Note that the collets are different for the inlet and exhaust valves and must not be interchanged. Release the compressor, then repeat the procedure on the remaining valves.

Chapter 2 Part C: Engine removal and general engine overhaul procedures

35.3B ...followed by the spring, ensuring that its closest pitched coils are at the bottom...

35.3C ...and cap

35.4 Compress the valve spring and refit the collets

5 With all the valves installed, place the cylinder head flat on the bench and, using a hammer and interposed block of wood, tap the end of each valve stem to settle the components.

36 Piston/connecting rod assembly – removal

1 Remove the cylinder head and sump. Although not strictly necessary, it should be noted that access to the connecting rod assemblies will be greatly improved if on E series engines the oil pump is removed, and on B series engines the oil strainer is removed.
2 Rotate the crankshaft so that No 1 big-end cap (nearest the crankshaft pulley position) is at the lowest point of its travel. If the big-end cap and rod are not already numbered, mark them with a centre punch (photo). Mark both cap and rod in relation to the cylinder they operate in, i.e. one dot for No 1, two dots for No 2, noting that No 1 is nearest the crankshaft pulley end of the engine.
3 Before removing the big-end caps, use a feeler gauge to check the amount of side-play between the caps and the crankshaft webs (photo). If the clearance exceeds the limit given in the Specifications the connecting rod should be renewed.
4 Unscrew and remove the big-end bearing cap nuts and withdraw the cap, complete with shell bearing from the connecting rod. If only the bearing shells are being attended to, push the connecting rod up and off the crankpin and remove the upper bearing shell. Keep the bearing shells and cap together in their correct sequence if they are to be refitted.
5 Push the connecting rod up, and remove the piston and rod from the cylinder. Keep the big-end cap together with its rod so they do not become interchanged with any of the other assemblies.
6 Repeat the above operation on the remaining three piston and connecting rod assemblies.

37 Crankshaft – removal

1 Remove the connecting rods as described in the previous Section. On B series engines also remove the oil pump and rear oil seal housing as described in Sections 24 and 27 of Part B.
2 Identification numbers and an arrow should be visible on each main bearing cap. The caps are numbered 1 to 5 with No 1 being nearest to the crankshaft pulley and the arrows should all be pointing towards the pulley end. If no marks are visible, stamp the bearing caps with a centre punch as was done for the connecting rods and mark them in such a way as to indicate their fitted direction.
3 Before the crankshaft is removed, check the endfloat using a dial gauge in contact with the end of the crankshaft (photo). Push the crankshaft fully one way and then zero the gauge. Push the crankshaft fully the other way and check the endfloat. The result can be compared with the specified amount and will give an indication as to whether new thrustwashers are required.
4 If a dial gauge is not available, feeler gauges can be used (photo). First push the crankshaft fully towards the flywheel end of the engine, then slip the feeler gauge between the web of No 4 crankpin and the thrustwasher of the rear main bearing (E series engines), or between the web of No 3 crankpin and the thrustwasher of number 4 main bearing (B series engines).
5 Slacken and remove the main bearing cap bolts and withdraw the caps complete with bearing shells.
6 On E series engines, to remove the rear (No 5) cap, refit two of the

36.2 Big-end caps marked with a centre punch

36.3 Checking the big-end cap side play

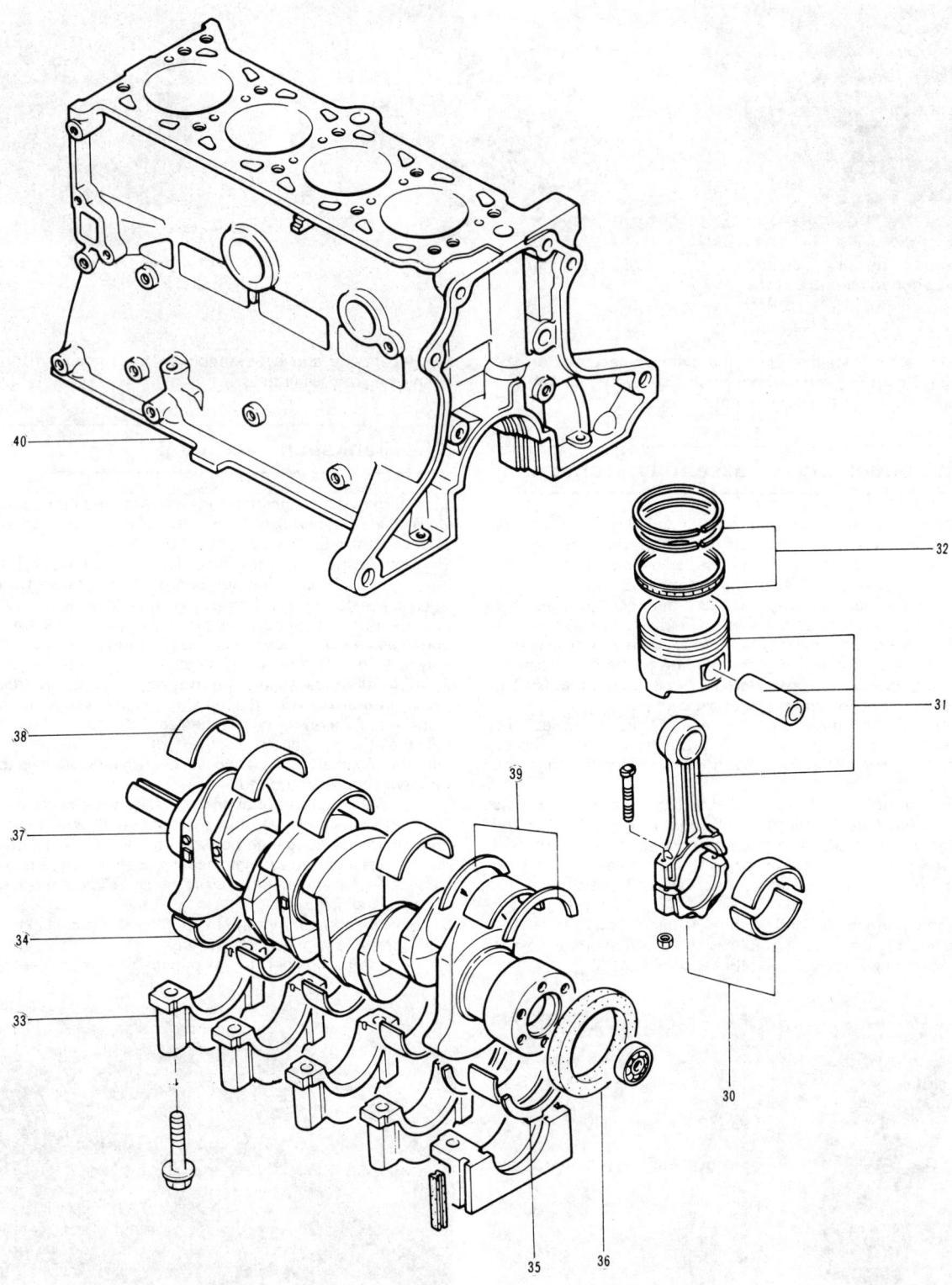

Fig. 2.19 E series engine crankshaft, piston and bearing assemblies – B series engine similar (Secs 36 and 37)

30 Connecting rod and big-end bearings
31 Piston and connecting rod assembly
32 Piston rings
33 Main bearing caps
34 Main bearing shells
35 Thrustwashers
36 Crankshaft rear oil seal
37 Crankshaft
38 Main bearing shells
39 Thrustwashers
40 Cylinder block

Chapter 2 Part C: Engine removal and general engine overhaul procedures

37.3 Checking the crankshaft endfloat with a dial gauge

37.4 Checking the crankshaft endfloat with a feeler gauge – B series engine shown

sump retaining bolts to the cap, then use the bolts to either pull or lever the cap out of position using pliers or screwdrivers. Recover the two oil seal strips from either side of the bearing cap and the thrustwasher halves.
7 Carefully lift the crankshaft out of the crankcase and, where necessary, slip the oil seal off the end boss.
8 Remove the thrustwasher upper halves from the rear (E series engine) or No 4 (B series engine) main bearing, then remove the bearing shell upper halves. Place each shell with its respective bearing cap.

38 Cylinder block/crankcase – cleaning and inspection

Cleaning
1 For complete cleaning, the core plugs should be removed (where fitted). Drill a small hole in them, then insert a self-tapping screw and pull out the plugs using a pair of grips or a slide hammer. Remove all external components and senders.
2 Scrape all traces of gasket from the cylinder block, taking care not to damage the head and sump mating faces.
3 Remove all oil gallery plugs where fitted. The plugs are usually very tight – they may have to be drilled out and the holes re-tapped. Use new plugs when the engine is reassembled.
4 If the block is extremely dirty, it should be steam cleaned.
5 After the block is returned, clean all oil holes and oil galleries one more time. Flush all internal passages with warm water until the water runs clear, dry the block thoroughly and wipe all machined surfaces with a light rust preventive oil. If you have access to compressed air, use it to speed the drying process and to blow out all the oil holes and galleries. **Warning:** *Wear eye protection when using compressed air!*
6 If the block is not very dirty, you can do an adequate cleaning job with hot soapy water and a stiff brush. Take plenty of time and do a thorough job. Regardless of the cleaning method used, be sure to clean all oil holes and galleries very thoroughly, dry the block completely and coat all machined surfaces with light oil.
7 The threaded holes in the block must be clean to ensure accurate torque readings during reassembly. Run the proper size tap into each of the holes to remove rust, corrosion, thread sealant or sludge and restore damaged threads. If possible, use compressed air to clear the holes of debris produced by this operation. Now is a good time to clean the threads on the head bolts and the main bearing cap bolts as well.
8 After coating the mating surfaces of the new core plugs with suitable sealant, refit them in the cylinder block. Make sure that they are driven in straight and seated properly or leakage could result. Special tools are available for this purpose, but a large socket, with an outside diameter that will just slip into the core plug will work just as well.
9 Apply suitable sealant to the new oil gallery plugs and insert them into the holes in the block. Tighten them securely.

10 If the engine is not going to be reassembled right away, cover it with a large plastic bag to keep it clean and prevent it rusting.

Inspection
11 Visually check the block for cracks, rust and corrosion. Look for stripped threads in the threaded holes. If there has been any history of internal water leakage, it may be worthwhile having an engine overhaul specialist check the block with special equipment. If defects are found, have the block repaired, if possible, or renewed.
12 Using a straight edge and feeler gauges, check the cylinder block top face for distortion. If the distortion exceeds the specified amount the face must be machined until flat noting that a maximum of 0.2 mm may be machined off of the block surface.
13 Check the cylinder bores for scuffing and scoring.
14 Measure the diameter of each cylinder at the top (just under the ridge area), centre and bottom of the cylinder bore, both parallel to the crankshaft axis and then at 90° to the crankshaft axis, so that a total of six measurements are taken. Compare these with the figures given in the Specifications. Repeat this procedure for the remaining cylinders.
15 If any of the measurements obtained exceed the specified limits remedial action must be taken.
16 If the cylinder walls are badly scuffed or scored, or if they are excessively out-of-round or tapered, have the cylinder block rebored. Oversize pistons will also be required.
17 If the cylinders are in reasonably good condition then it may only be necessary to renew the piston rings.
18 If this is the case, the bores should be honed in order to allow the new rings to bed in correctly and provide the best possible seal. The conventional type of hone has spring loaded stones and is used with a power drill. You will also need some paraffin or honing oil and rags. The hone should be moved up and down the cylinder to produce a crosshatch pattern and plenty of honing oil should be used. Ideally the crosshatch lines should intersect at approximately a 60° angle. Do not take off more material than is necessary to produce the required finish. If new pistons are being fitted, the piston manufacturers may specify a finish with a different angle, so their instructions should be followed. Do not withdraw the hone from the cylinder while it is still being turned, but stop it first. After honing a cylinder, wipe out all traces of the honing oil. If equipment of this type is not available, or if you are not sure whether you are competent to undertake the task yourself, an engine overhaul specialist will carry out the work at moderate cost.

39 Piston/connecting rod assembly – inspection

1 Examine the pistons for ovality, scoring and scratches, and for wear of the piston ring grooves. Use a micrometer to measure the pistons (photo).

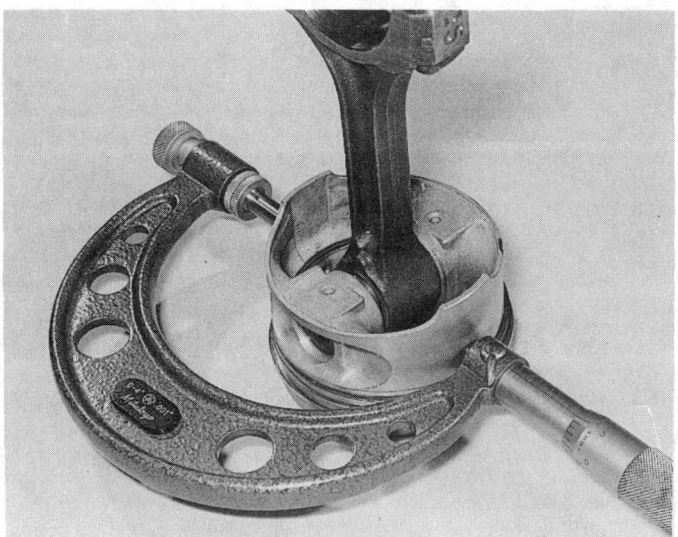

39.1 Measuring the pistons for ovality

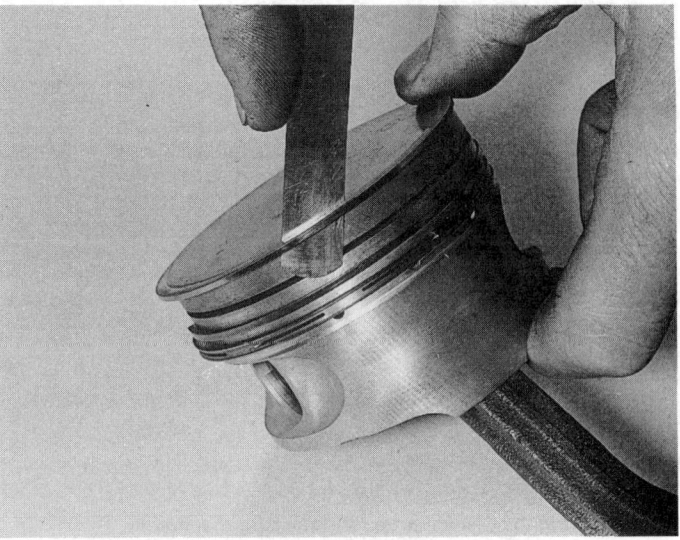

39.3 Using a feeler blade to remove the piston rings

40.3 Using a penny to check the crankshaft journals for scoring

40.5 Using a micrometer to measure the crankshaft journals

2 If the pistons or connecting rods are to be renewed, it is necessary to have this work carried out by a Mazda dealer, or suitable engine overhaul specialist, who will have the necessary tooling to remove the gudgeon pins.

3 If new rings are to be fitted to the original pistons, expand the old rings over the top of the pistons. The use of two or three old feeler blades will be helpful in preventing the rings dropping into empty grooves (photo).

40 Crankshaft – inspection

1 Clean the crankshaft and dry it with compressed air if available. **Warning:** *Wear eye protection when using compressed air!* Be sure to clean the oil holes with a pipe cleaner or similar probe.

2 Check the main and big-end bearing journals for uneven wear, scoring, pitting and cracking.

3 Rub a penny across each journal several times (photo). If a journal picks up copper from the penny, it is too rough and must be reground.

4 Remove all burrs from the crankshaft oil holes with a stone, file or scraper.

5 Using a micrometer, measure the diameter of the main and connecting rod journals and compare the results with the Specifications at the beginning of this Chapter (photo). By measuring the diameter at a number of points around each journal's circumference, you will be able to determine whether or not the journal is out-of-round. Take the measurement at each end of the journal, near the webs, to determine if the journal is tapered. If any of the measurements vary by more than 0.05 mm, the crankshaft will have to be reground and undersize bearings fitted.

6 Set up the crankshaft end journals in V-blocks and position a dial gauge on the centre main bearing journal. Slowly rotate the crankshaft and measure the runout. If the runout exceeds the specified limit the crankshaft must be renewed.

7 Check the oil seal journals as applicable at each end of the crankshaft for wear and damage. If the seal has worn an excessive groove in the journal, consult an engine overhaul specialist who will be able to advise whether a repair is possible, or whether a new crankshaft is necessary.

41 Main and big-end bearings – inspection

1 Even though the main and big-end bearings should be renewed during the engine overhaul, the old bearings should be retained for close examination, as they may reveal valuable information about the condition of the engine. The size of the bearing shells is stamped on the

Chapter 2 Part C: Engine removal and general engine overhaul procedures

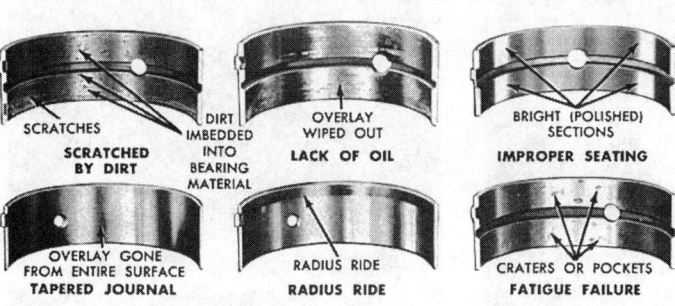

Fig. 2.20 Typical bearing failures (Sec 41)

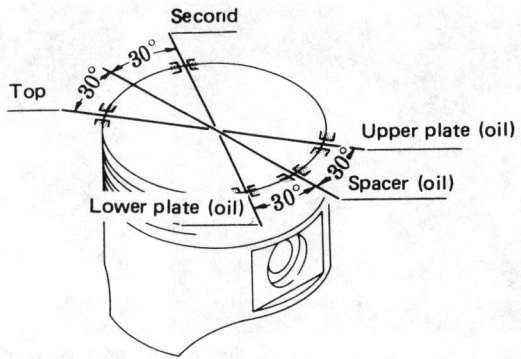

Fig. 2.21 Piston ring end gap spacing (Sec 43)

back metal and this information should be given to the supplier of the new shells.

2 Bearing failure occurs because of lack of lubrication, the presence of dirt or other foreign particles, overloading the engine, and corrosion. Regardless of the cause of bearing failure, it must be corrected before the engine is reassembled to prevent it from happening again.

3 When examining the bearings, remove them from the engine block, the main bearing caps, the connecting rods and the rod caps and lay them out on a clean surface in the same general position as their location in the engine. This will enable you to match any bearing problems with the corresponding crankshaft journal.

4 Dirt and other foreign particles get into the engine in a variety of ways. It may be left in the engine during assembly, or it may pass through filters or the crankcase ventilation system. It may get into the oil, and from there into the bearings. Metal chips from machining operations and normal engine wear are often present. Abrasives are sometimes left in engine components after reconditioning, especially when parts are not thoroughly cleaned using the proper cleaning methods. Whatever the source, these foreign objects often end up embedded in the soft bearing material and are easily recognized. Large particles will not embed in the bearing and will score or gouge the bearing and journal. The best prevention for this cause of bearing failure is to clean all parts thoroughly and keep everything spotlessly clean during engine assembly. Frequent and regular engine oil and filter changes are also recommended.

5 Lack of lubrication (or lubrication breakdown) has a number of interrelated causes. Excessive heat (which thins the oil), overloading (which squeezes the oil from the bearing face) and oil leakage (from excessive bearing clearances, worn oil pump or high engine speeds) all contribute to lubrication breakdown. Blocked oil passages, which usually are the result of misaligned oil holes in a bearing shell, will also oil starve a bearing and destroy it. When lack of lubrication is the cause of bearing failure, the bearing material is wiped or extruded from the steel backing of the bearing. Temperatures may increase to the point where the steel backing turns blue from overheating.

6 Driving habits can have a definite effect on bearing life. Full throttle, low speed operation (labouring the engine) puts very high loads on bearings, which tends to squeeze out the oil film. These loads cause the bearings to flex, which produces fine cracks in the bearing face (fatigue failure). Eventually the bearing material will loosen in pieces and tear away from the steel backing. Short trip driving leads to corrosion of bearings because insufficient engine heat is produced to drive off the condensed water and corrosive gases. These products collect in the engine oil, forming acid and sludge. As the oil is carried to the engine bearings, the acid attacks and corrodes the bearing material.

7 Incorrect bearing installation during engine assembly will lead to bearing failure as well. Tight fitting bearings leave insufficient bearing oil clearance and will result in oil starvation. Dirt or foreign particles trapped behind a bearing shell result in high spots on the bearing which lead to failure.

42 Engine overhaul – reassembly sequence

1 Before reassembly begins ensure that all new parts have been obtained and that all necessary tools are available. Read through the entire procedure to familiarise yourself with the work involved, and to ensure that all items necessary for reassembly of the engine are at hand. In addition to all normal tools and materials, a thread-locking compound will be needed. A tube of RTV sealing compound will also be required for the joint faces that are fitted without gaskets.

2 In order to save time and avoid problems, engine reassembly can be carried out in the following order.

  (a) Crankshaft.
  (b) Pistons.
  (c) Oil pump.
  (d) Timing chain and sprockets (E series engine).
  (e) Sump.
  (f) Flywheel.
  (g) Cylinder head.
  (h) Timing belt and sprockets (B series engine).
  (i) Engine external components.

43 Piston rings – refitting

1 Before fitting the new rings, ensure that the ring grooves in the piston are free of carbon by cleaning them using an old ring. Break the ring in half to do this.

2 Insert the new rings into the cylinder bore and use a feeler gauge to check that the end gaps are within the specified limits. Also check the ring to groove clearance in the piston using feeler gauges.

3 Install the new rings by fitting them over the top of the piston, starting with the oil control scraper ring. Note that the top and second compression rings must be fitted with the word TOP, or the letter R uppermost. On B series engines the top and second compression rings are different, the top ring has a rounded profile while the second ring has a flat profile with a ridge on the underside of the ring. Ensure the rings are correctly positioned on installation.

4 With all the rings in position, space the ring end gaps as shown in Fig. 2.21.

44 Crankshaft – refitting and main bearing running clearance check

Main bearing running clearance check

1 Clean the backs of the bearing shells and the bearing recesses in both the cylinder block and main bearing caps.

2 Press the bearing shells into the recesses in the cylinder block, noting if the original bearings are being re-used they must be refitted to their original positions in the block and cap.

3 Before the crankshaft can be permanently installed, the main bearing running clearance should be checked and this can be done in either of two ways. One method is to fit the main bearing caps to the cylinder block, with bearing shells in place. With the cap retaining bolts tightened to the specified torque, measure the internal diameter of each assembled pair of bearing shells using a vernier dial indicator or internal micrometer. If the diameter of each corresponding crankshaft journal is

44.5 Thread of Plastigage (arrowed) placed on a crankshaft main journal

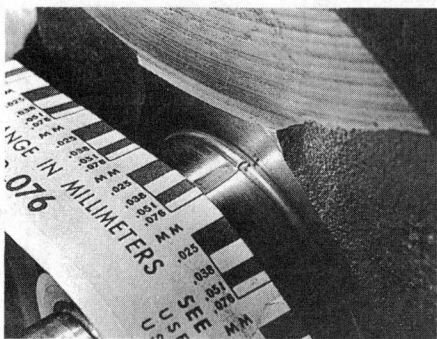
44.9 Measuring the Plastigage width with the special gauge

44.14A Fit the thrustwasher upper halves to the crankcase...

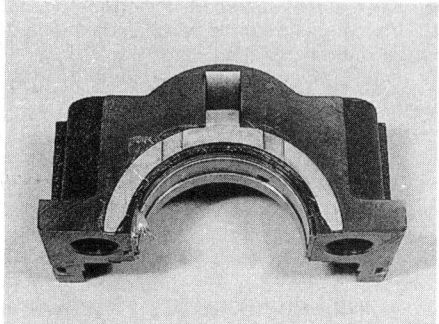
44.14B ...and the lower halves to the main bearing cap ensuring the oilway grooves are facing outwards

44.15 Slip the oil seal over the end of the crankshaft

44.20 Fit new bearing cap side seal strips

measured and then subtracted from the bearing internal diameter, the result will be the main bearing running clearance. The second (and more accurate) method is to use an American product known as Plastigage. This consists of a fine thread of perfectly round plastic which is compressed between the bearing cap and the journal. When the cap is removed, the plastic is deformed and can be measured with a special card gauge supplied with the kit. The running clearance is determined from this gauge. Plastigage is sometimes difficult to obtain in this country but enquiries at one of the larger specialist chains of quality motor factors should produce the name of a stockist in your area. The procedure for using Plastigage is as follows.

4   With the upper main bearing shells in place, carefully lay the crankshaft in position. Do not use any lubricant; the crankshaft journals and bearing shells must be perfectly clean and dry.
5   Cut several pieces of the appropriate size Plastigage (they should be slightly shorter than the width of the main bearings) and place one piece on each crankshaft journal axis (photo).
6   With the bearing shells in position in the caps, fit the caps to their numbered or previously noted locations. Take care not to disturb the Plastigage.
7   Starting with the centre main bearing and working outward, tighten the main bearing cap bolts progressively to their specified torque setting. Don't rotate the crankshaft at any time during this operation.
8   Remove the bolts and carefully lift off the main bearing caps, keeping them in order. Don't disturb the Plastigage or rotate the crankshaft. If any of the bearing caps are difficult to remove, tap them from side-to-side with a soft faced mallet.
9   Compare the width of the crushed Plastigage on each journal to the scale printed on the Plastigage envelope to obtain the main bearing running clearance (photo).
10   If the clearance is not as specified, the bearing shells may be the wrong size (or excessively worn if the original shells are being re-used). Before deciding that different size shells are needed, make sure that no dirt or oil was trapped between the bearing shells and the caps or block when the clearance was measured. If the Plastigage was wider at one end than at the other, the journal may be tapered.
11   Carefully scrape away all traces of the Plastigage material from the crankshaft and bearing shells using a fingernail or other object which is unlikely to score the shells.

Final crankshaft refitting

12   Carefully lift the crankshaft out of the cylinder block once more.
13   Press the bearing shells into position in the bearing caps and cylinder block, noting if the original bearings are being re-used they must be refitted to their original positions.

E series engines

14   Using a little grease, stick the thrustwasher upper halves to each side of the rear main bearing, and the thrustwasher lower halves (with the locating tangs) to the bearing cap. Ensure that the oilway grooves on each thrustwasher face outwards (towards the crankshaft) (photos).
15   Lubricate the lips of the new crankshaft rear oil seal and carefully slip it over the crankshaft rear journal (photo). Do this carefully as the seal lips are very delicate. Ensure that the open side of the seal faces the engine.
16   Liberally lubricate each bearing shell in the cylinder block and lower the crankshaft into position. Check that the rear oil seal is positioned correctly.
17   Lubricate the bearing shells, then fit the main bearing caps in their numbered order ensuring that the arrows on the caps all point towards the pulley end of the crankshaft. Alternatively use the marks made on removal to position the caps correctly.
18   Fit the main bearing cap bolts and tighten them progressively to the specified torque.
19   Check that the crankshaft turns freely without any tight spots, then check the endfloat with reference to Section 37.
20   Lubricate the rear bearing cap side seal strips with a little grease, and push them fully in to their grooves ensuring that the groove in the seal is at 90° to the crankshaft axis (photo).

B series engines

21   Using a little grease, stick the thrustwasher halves to each side of crankcase No 4 main bearing. Ensure that the oilway grooves on each

Chapter 2 Part C: Engine removal and general engine overhaul procedures

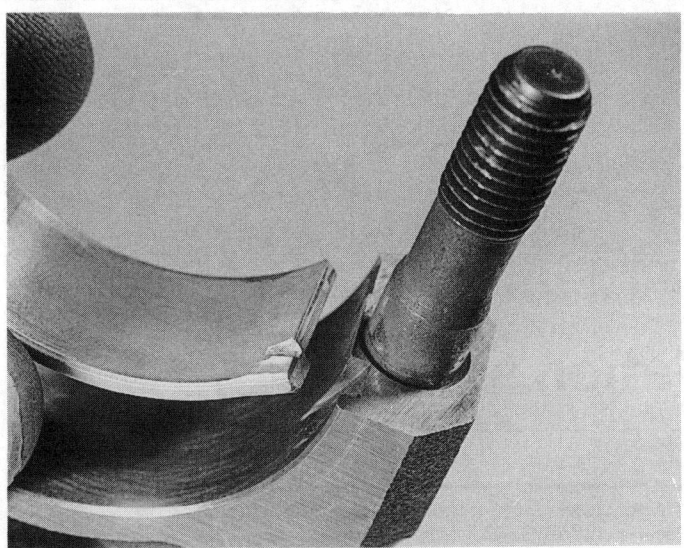

45.2 Ensure the bearing shell tab locates with the cutout in the connecting rod

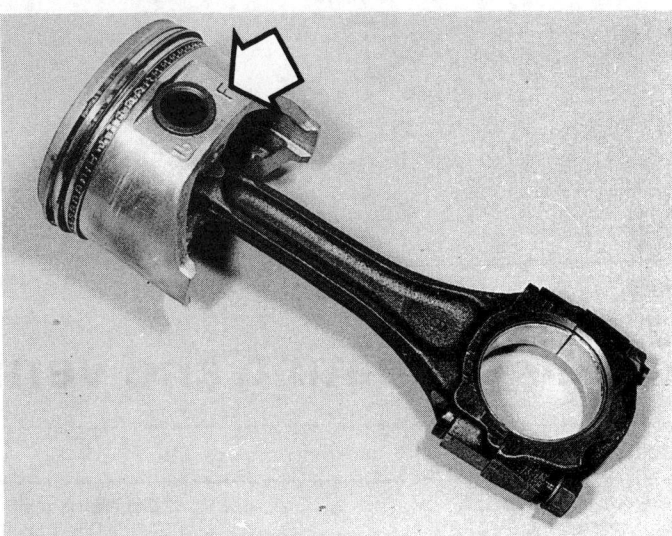

45.4 Piston/connecting rod assembly must be installed with F mark (arrowed) on piston facing the crankshaft pulley end of the engine

thrustwasher face outwards (towards the crankshaft).
22 Liberally lubricate each bearing shell in the cylinder block and lower the crankshaft into position.
23 Refit the bearing caps as described above in paragraphs 17 to 19.

45 Piston/connecting rod assembly – refitting and big-end bearing running clearance check

1 Clean the backs of the big-end bearing shells and the recesses in the connecting rods and big-end caps. If new shells are being fitted, ensure that all traces of the protective grease are cleaned off using paraffin. Wipe the shells and connecting rods dry with a lint-free cloth.
2 Press the big-end bearing shells into the connecting rods and caps in their correct positions. Make sure that the location tabs are engaged with the cut-outs in the connecting rods (photo).

Big-end bearing running clearance check

3 Lubricate No 1 piston and rings and check that the ring gaps are still arranged as described in Section 43.
4 Fit a ring compressor to No 1 piston then insert the piston and connecting rod into No 1 cylinder, noting that the letter F on the side of the piston must be facing the crankshaft pulley end of the engine (photo). With No 1 crankpin at its lowest point, drive the piston carefully into the cylinder with the wooden handle of a hammer and at the same time guide the connecting rod onto the crankpin.
5 To measure the big-end bearing running clearance, refer to the information contained in Section 44 as the same general procedures apply. If the Plastigage method is being used, ensure that the crankpin journal and the big-end bearing shells are clean and dry then engage the connecting rod with the crankpin. Lay the Plastigage strip on the crankpin, fit the bearing cap in its previously noted position, using the marks made on removal, then tighten the nuts to the specified torque. Do not rotate the crankshaft or connecting rod during this operation. Remove the cap and check the running clearance by measuring the Plastigage as previously described.

6 Repeat the foregoing procedures on the remaining piston/connecting rod assemblies.

Final connecting rod fitting

7 Having checked the running clearance of all the crankpin journals and taken any corrective action necessary, clean off all traces of Plastigage from the bearing shells and crankpin.
8 Liberally lubricate the crankpin journals and big-end bearing shells and refit the bearing caps once more, ensuring correct positioning as previously described. Tighten the bearing cap bolts to the specified torque and turn the crankshaft each time to make sure that it is free before moving on to the next assembly.

46 Engine – initial start-up after overhaul

1 With the engine refitted in the vehicle, double-check the engine oil and coolant levels.
2 With the spark plugs removed and the ignition system disabled by disconnecting the coil LT wire, crank the engine over on the starter until the oil pressure light goes out.
3 Refit the spark plugs and connect all the HT leads.
4 Start the engine, noting that this may take a little longer than usual due to the fuel pump and carburettor being empty.
5 While the engine is idling, check for fuel, water and oil leaks. Don't be alarmed if there are some odd smells and smoke from parts getting hot and burning off oil deposits.
6 Keep the engine idling until hot water is felt circulating through the top hose, then switch it off.
7 After a few minutes, recheck the oil and water levels and top up as necessary.
8 If new pistons, rings or crankshaft bearings have been fitted, the engine must be run-in for the first 500 miles (800 km). Do not operate the engine at full throttle or allow it to labour in any gear during this period. It is recommended that the oil and filter be changed at the end of this period.

Chapter 3
Cooling, heating and ventilation systems

Contents

Antifreeze mixture .... See Chapter 1	Heater blower unit – removal and refitting 11
Coolant draining .... See Chapter 1	Heater control cables – adjustment 14
Coolant filling .... See Chapter 1	Heater control panel – removal and refitting 13
Coolant level check .... See Chapter 1	Heater coolant valve (pre-September 1985 models) – removal and refitting 12
Coolant level sensor (July 1987 onwards models) – removal, testing and refitting 7	Heater matrix – removal and refitting 10
Coolant system flushing .... See Chapter 1	Heater unit – removal and refitting 9
Electric cooling fan – removal, testing and refitting 4	Logic type heater control system – general 15
Electric cooling fan thermostatic switch – removal, testing and refitting 5	Radiator – removal, inspection, cleaning and refitting 2
General cooling system checks .... See Chapter 1	Temperature gauge sender unit – removal and refitting 6
General information 1	Thermostat – removal, testing and refitting 3
	Water pump – removal and refitting 8

Specifications

System type .... Pressurised, pump-assisted with front mounted radiator and electric cooling fan

Thermostat
Type .... Wax
Opening temperature:
  Single stage .... 88°C ± 1.5°C
  Dual stage:
    Sub valve .... 85°C
    Main valve .... 88°C
Fully open temperature .... 100°C
Lift height:
  Single stage .... 8.0 mm minimum
  Dual stage:
    Sub valve .... 1.5 mm minimum
    Main valve .... 8.0 mm minimum
Valve closing temperature:
  Single stage .... NA
  Dual stage:
    Sub valve .... 80°C
    Main valve .... 83°C

Radiator cooling fan
Fan motor current consumption:
  Pre-September 1985 models:
    270 mm diameter fan .... 9.5 amp (maximum)
    250 mm diameter fan .... 6.5 amp (maximum)
  September 1985 models onward .... 6.1 to 7.3 amp

Torque wrench settings

	Nm	lbf ft
Cooling fan temperature switch	30 to 40	22 to 30
Temperature gauge sender unit	5 to 10	4 to 7
Thermostat housing	19 to 31	14 to 22
Water pump:		
Pre-September 1985 models	19 to 31	14 to 22
September 1985 models onward	19 to 26	14 to 19

Chapter 3 Cooling, heating and ventilation systems

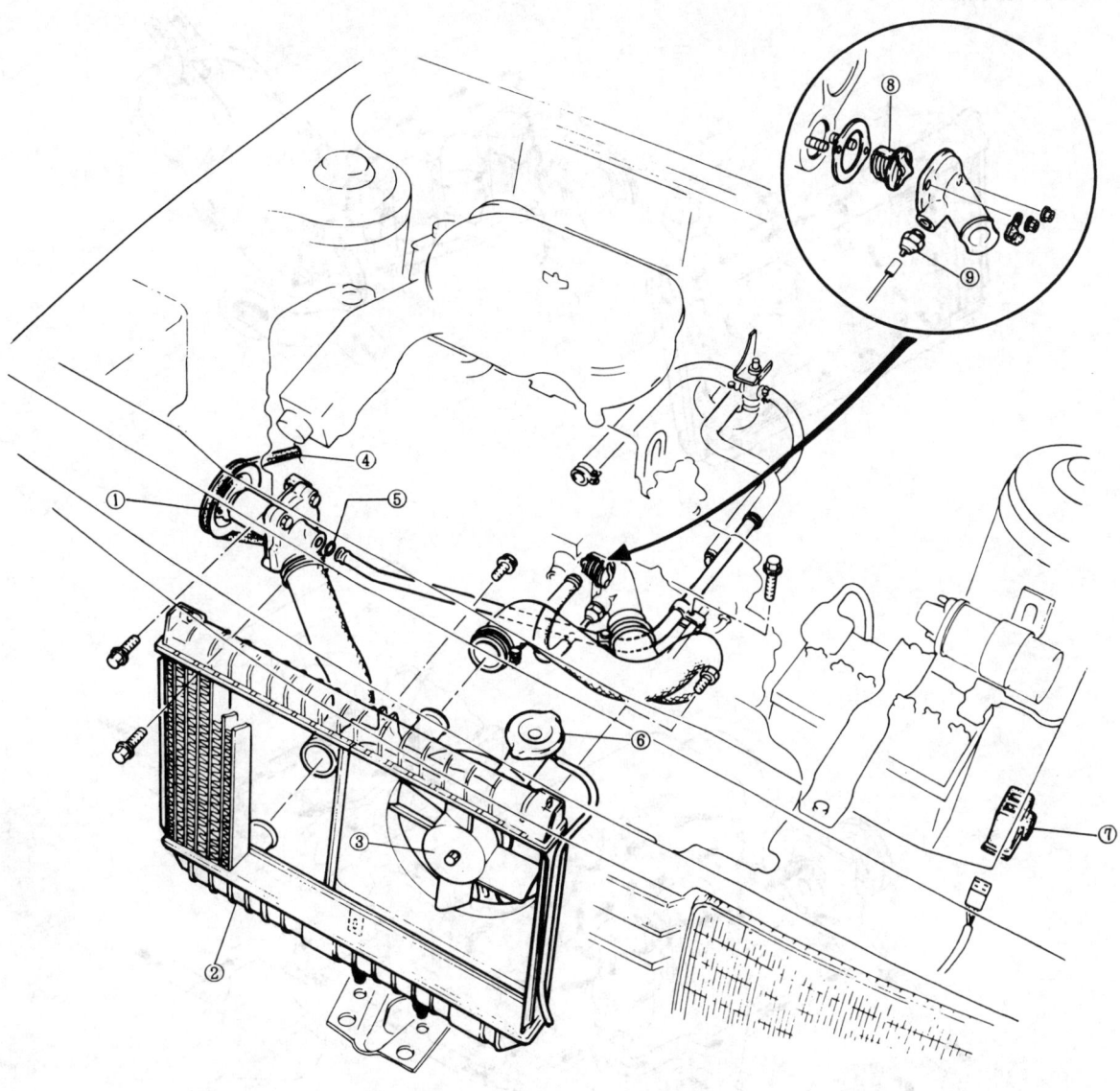

Fig. 3.1 Cooling system component layout – E series engine (Sec 1)

1 Water pump	3 Cooling fan	5 Bypass pipe O-ring	8 Thermostat
2 Radiator	4 Drivebelt	6 Radiator pressure cap	9 Cooling fan temperature switch
		7 Cooling fan relay	

1 General information

The cooling system is of pressurised type consisting of a belt-driven pump, aluminium crossflow radiator, electric cooling fan, thermostat and on later models, a radiator expansion tank.

The system functions as follows. Cold coolant in the bottom of the radiator passes through the bottom hose to the water pump where it is pumped around the cylinder block and head passages. After cooling the cylinder bores, combustion surfaces and valve seats, the coolant reaches the underside of the thermostat, which is initially closed. The coolant passes through the heater and inlet manifold and is returned to the water pump.

When the engine is cold the coolant circulates only through the cylinder block, cylinder head, heater and inlet manifold. When the coolant reaches a predetermined temperature, the thermostat opens and the coolant passes through the top hose to the radiator. As the coolant circulates through the radiator it is cooled by the inrush of air when the car is in forward motion. Airflow is supplemented by the action of the electric cooling fan when necessary. Upon reaching the bottom of the radiator, the coolant is now cooled and the cycle is repeated.

The electric cooling fan mounted behind the radiator is controlled by a thermostatic switch located in the thermostat housing. At a predetermined coolant temperature the switch contacts close, thus actuating the fan via a relay.

Fig. 3.2 Cooling system coolant flow diagram – E series engine (Sec 1)

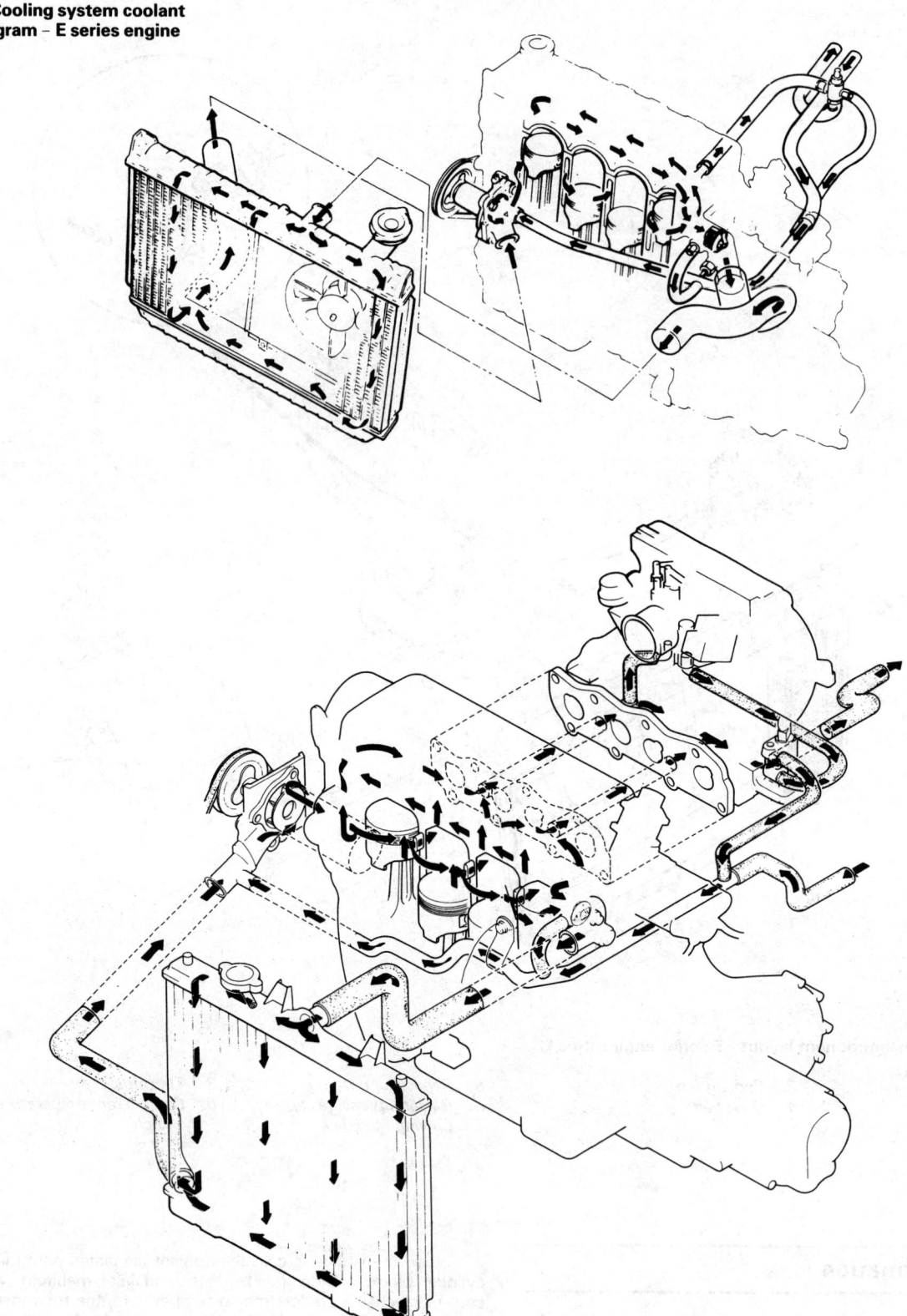

Fig. 3.3 Cooling system coolant flow diagram – B series engine (Sec 1)

Chapter 3 Cooling, heating and ventilation systems

Fig. 3.4 Radiator and cooling fan attachments – later models shown (Secs 2 and 4)

1 Cowling
2 Cooling fan
3 Cooling fan motor
4 Top and bottom hoses
5 Expansion tank hose
6 Radiator

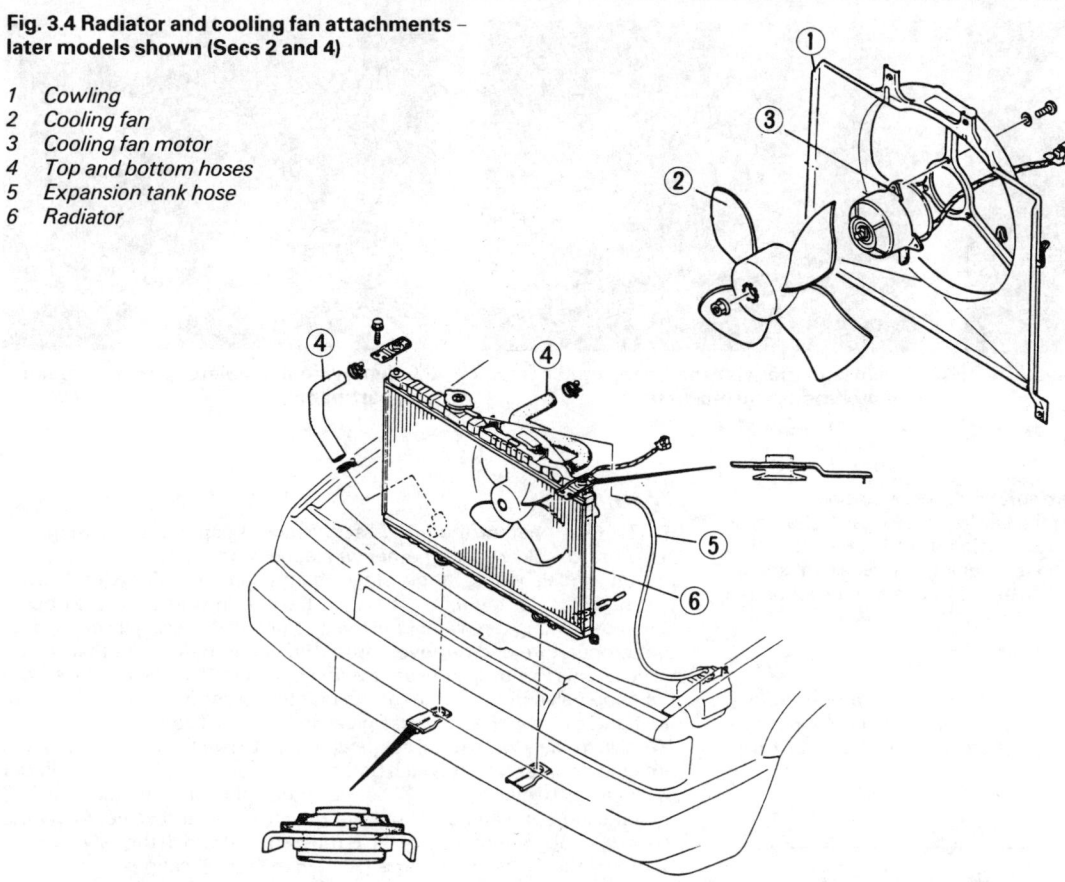

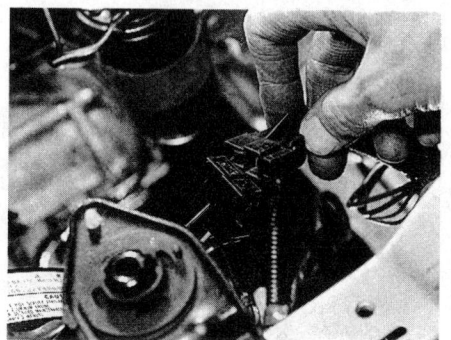

2.3 Disconnecting the radiator cooling fan switch wiring connector

2.4A Disconnect the top hose...

2.4B ...bottom hose...

On later models a radiator expansion tank is incorporated in the system. When the engine is at normal operating temperature the coolant expands and some of it is displaced into the expansion tank. This coolant collects in the tank and is returned to the radiator when the system cools.

2 Radiator – removal, inspection, cleaning and refitting

Removal

1 Disconnect the battery negative terminal.
2 Drain the cooling system as described in Chapter 1.
3 Disconnect the radiator cooling fan wiring at the block connector and, if necessary, at the temperature switch in the thermostat housing (photo).
4 Slacken the retaining clips and detach the radiator top and bottom hoses. On later models detach the expansion tank hose from the radiator filler neck (photos).
5 On early models undo the two upper radiator retaining bolts, and on later models undo the retaining bolts and remove the radiator upper mounting brackets (photo).
6 Lift the radiator up and out of the engine compartment (photo).

Inspection and cleaning

7 Radiator repair is best left to a specialist, but minor leaks may be sealed using a radiator sealant such as Holts Radweld. Clear the radiator

Chapter 3 Cooling, heating and ventilation systems

2.4C and expansion tank hose from the radiator (if applicable)

2.5 On later models remove the upper radiator mounting brackets

2.6 Removing the radiator from the engine compartment

matrix of flies and small leaves with a soft brush, or by hosing.
8 If the radiator is to be left out of the car for more than 48 hours, all traces of coolant should be flushed out using clean water through a garden hose inserted in the top hose opening. In cases of severe contamination, or blockage, reverse flush the radiator as described in Chapter 1.

Refitting

9 Refitting is a reversal of removal, but check the radiator mounting bushes for signs of damage or deterioration and if necessary renew them. On completion refill the cooling system with reference to Chapter 1.

3 Thermostat – removal, testing and refitting

Removal

1 Disconnect the battery negative terminal.
2 Drain the cooling system as described in Chapter 1.
3 Slacken the retaining clip and disconnect the radiator top hose from the thermostat housing (photo).
4 Disconnect the electrical lead from the cooling fan temperature switch (photo).
5 On pre-September 1985 models, undo the two nuts securing the thermostat housing to the cylinder head, noting the position of the clip which is fitted under the front nut.
6 On September 1985 models onward, slacken and remove the two bolts securing the thermostat housing to the cylinder head (photo).
7 Withdraw the housing and gasket then lift out the thermostat.

Testing

8 To test whether the unit is serviceable, suspend it on string in a saucepan of cold water together with a thermometer.
9 On models with a single stage thermostat, heat the water slowly and note the temperature at which the thermostat begins to open. Continue heating the water until the thermostat valve is fully open, note the temperature and remove it from the water. Before the thermostat valve begins to close, measure the lift height which is the distance from the open valve in the centre of the unit to the sealing flange. Compare the figures obtained with those given in the Specifications.
10 On models with a dual stage thermostat, heat the water slowly and note the temperature at which both the smaller sub valve and the larger main valves begin to open. Continue heating the water to above 100°C then measure the lift height of both the main and sub valve. Allow the water to cool slowly and note the temperature at which the valves close. Compare the figures with those given in the Specifications.
11 If the thermostat does not perform as specified then it must be discarded and a new unit fitted. Under no circumstances should the car be used without a thermostat, as uneven cooling of the cylinder walls and head passages may occur, causing distortion and possible seizure of the engine internal components.

Refitting

12 The thermostat is refitted by a reversal of the removal procedure bearing in mind the following points (photos).

(a) Position the thermostat in its seating in the cylinder head so that the 'jiggle pin' is uppermost.
(b) Use a new gasket when refitting the thermostat housing and tighten the housing mounting nuts or bolts to the specified torque.
(c) On completion refill the cooling system as described in Chapter 1.

Fig. 3.5 Thermostat removal – early models shown (Sec 3)

1 Top hose
2 Thermostat housing
3 Thermostat

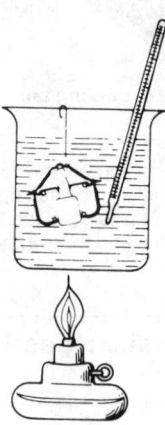

Fig. 3.6 Testing the thermostat (Sec 3)

Chapter 3 Cooling, heating and ventilation systems

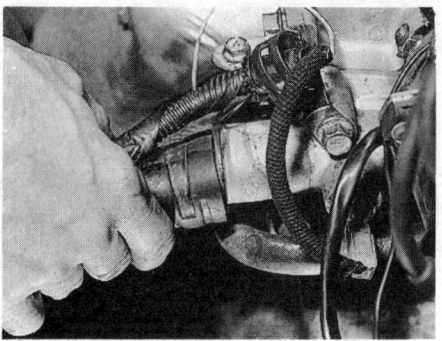

3.3 Disconnect the radiator top hose from the thermostat housing...

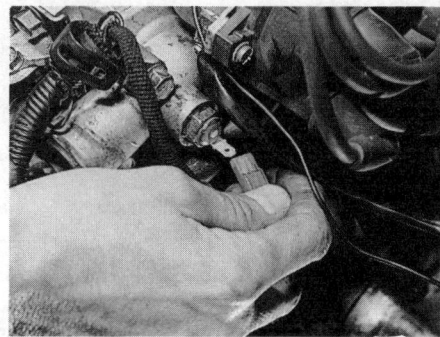

3.4 ...and the fan switch wiring connector

3.6 Removing the thermostat housing bolts – September 1985 models onward

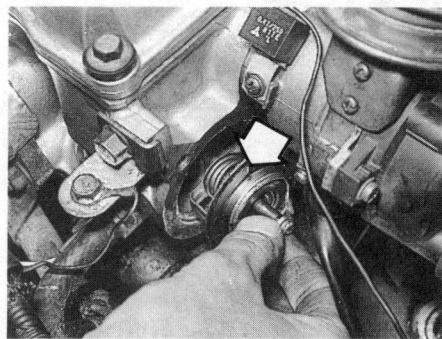

3.12A Install the thermostat with the jiggle pin (arrowed) uppermost...

3.12B ...and refit the housing using a new gasket

4.4A Undo the bolts securing the cooling fan cowling to the radiator...

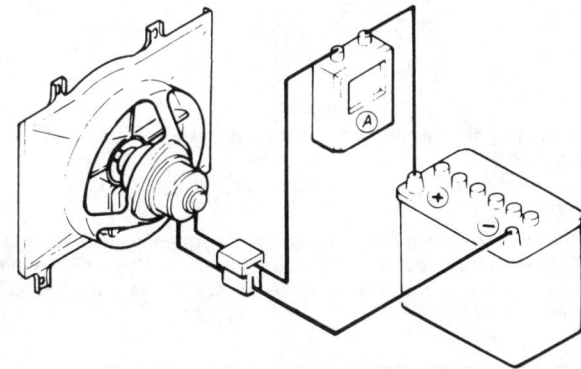

Fig. 3.7 Testing the cooling fan motor (Sec 4)

assembly to the back of the radiator and lift the fan and cowling assembly out of position (photos).

Testing

5 To test the motor a fully charged 12 V battery, an ammeter and an auxiliary wire are required.
6 Connect the positive lead of the ammeter to the battery positive terminal and the meter negative lead to the horizontal terminal of the cooling fan wiring connector (Fig. 3.7).
7 Using the auxiliary wire connect the battery negative terminal to the vertical terminal of the cooling fan wiring connector.
8 With the battery connected as described, the motor should run normally and the current consumption should not exceed the figure given in the Specifications at the start of this Chapter.
9 If the current consumption is excessive, or if the motor does not run at all, then the motor is faulty and must be renewed.

4 Electric cooling fan – removal, testing and refitting

Removal

1 Disconnect the battery negative terminal.
2 Drain the cooling system as described in Chapter 1 and disconnect the top hose from the radiator.
3 Disconnect the cooling fan wiring at the cable connector and, if necessary, at the temperature switch in the thermostat housing.
4 Undo the four retaining bolts securing the cooling fan and cowling

4.4B ...and lift the fan assembly out of the engine compartment

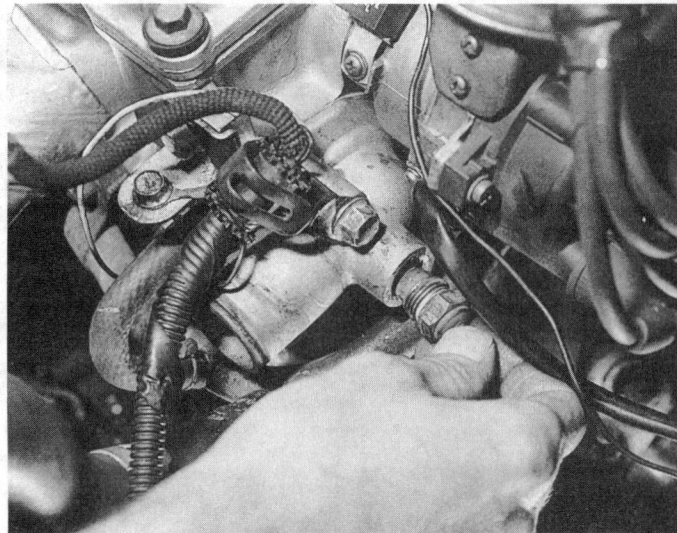

5.3 Removing the cooling fan thermostatic switch

ohmmeter. Above 97°C the switch should open and an infinite resistance reading (indicating an open circuit) obtained on the ohmmeter.

6 If the switch does not function as described it is faulty and must be renewed.

Refitting

7 Renew the switch O-ring (where fitted) then refit the switch to the thermostat housing and tighten it to the specified torque.
8 Reconnect the switch wire and the battery negative terminal.
9 Refill the cooling system as described in Chapter 1.

6 Temperature gauge sender unit – removal and refitting

Removal

1 The coolant temperature gauge sender unit is located on the left-hand side of the front face of the cylinder head and can be removed as follows.
2 Disconnect the battery negative terminal.
3 Drain the coolant as described in Chapter 1.
4 Disconnect the wire from, and unscrew the sender unit from the cylinder head (photos).

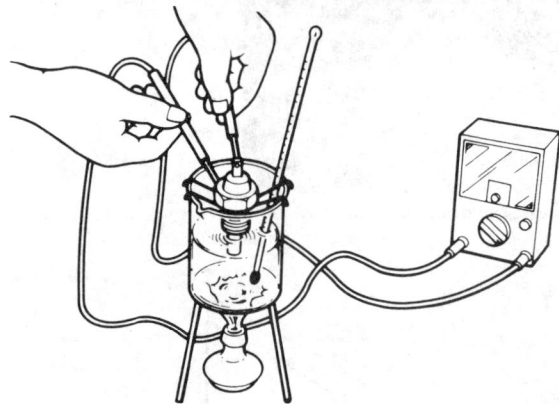

Fig. 3.8 Testing the cooling system temperature switches and senders (Sec 5)

Refitting

10 Refitting is the reverse of the removal sequence ensuring that the cooling fan wiring is correctly routed and secured by any necessary guides or clamps. On completion refill the cooling system as described in Chapter 1.

5 Electric cooling fan thermostatic switch – removal, testing and refitting

Note: *Ensure the ignition switch is switched off, or disconnect the battery negative terminal, before disconnecting the wire from the cooling fan switch. If not the fan will operate the instant the switch is disconnected.*

Removal

1 Disconnect the battery negative terminal.
2 Drain the cooling system as described in Chapter 1.
3 Disconnect the wire from the temperature switch and unscrew the switch from its location in the thermostat housing (photo).

Testing

4 Referring to Fig. 3.8, suspend the switch in a saucepan or suitable vessel together with a thermometer. Fill the vessel with water so that the switch probe is completely submerged. Connect an ohmmeter between the switch terminal and the switch body.
5 Heat the water and note the operation of the switch. Below 90°C the switch should be closed and a reading of 0 ohms obtained on the

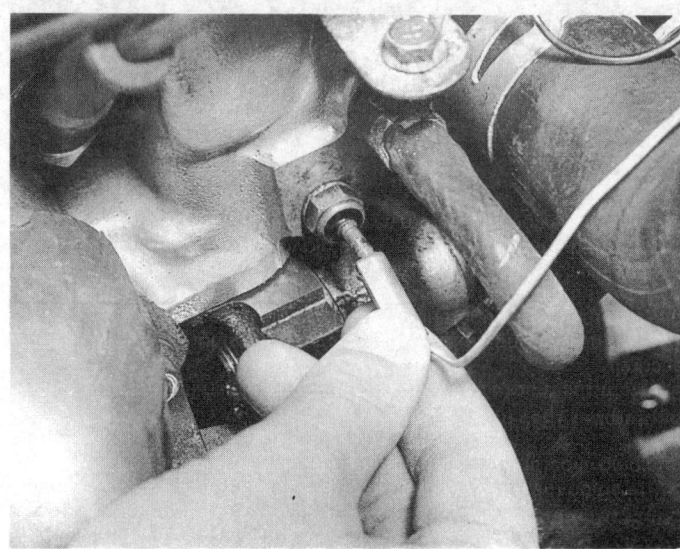

6.4A Disconnect the wire...

Chapter 3 Cooling, heating and ventilation systems

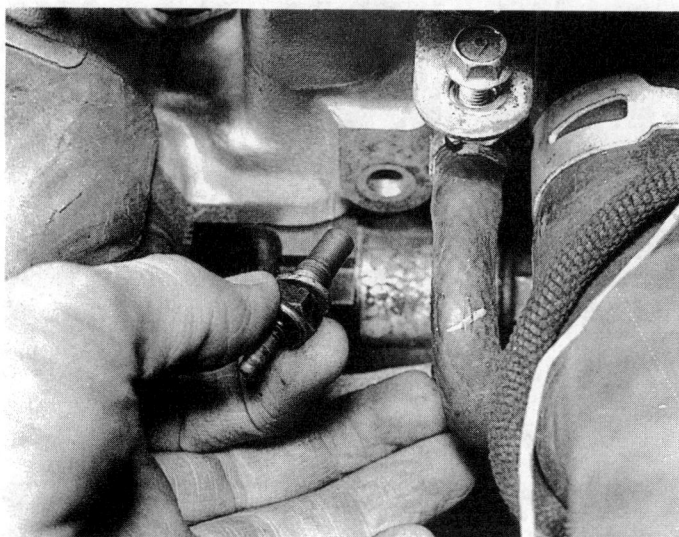

6.4B ...and unscrew the temperature gauge sender

Testing

5 Referring to Fig. 3.8, suspend the sender unit in a saucepan or suitable vessel together with a thermometer. Fill the vessel with water so that the sender unit probe is completely submerged. Connect an ohmmeter between the sender unit terminal and the unit body.
6 Heat the water to 80°C and note the resistance reading obtained. If the sender unit is functioning correctly this should be 49.3 to 57.7 ohms. If the reading obtained differs greatly from this the sender unit is faulty and must be renewed.

Refitting

7 Renew the sender unit sealing washer, then screw the unit into the cylinder head and tighten to the specified torque.
8 Reconnect the sender unit wire and the battery negative terminal.
9 Refill the cooling system as described in Chapter 1.

7 Coolant level sensor (July 1987 models onward) – removal, testing and refitting

Removal

1 Disconnect the coolant level sensor wiring connector and unscrew the sensor from the top of the radiator. Be prepared for some coolant loss and mop up any split coolant.

Testing

2 Fit a suitable bolt to the sensor thread in the top of the radiator, or plug the hole with a suitable bung.

3 Reconnect the coolant level sensor wire to the sensor and start the engine.
4 With the sensor probe not earthed, the instrument panel warning lamp should be illuminated. Earth the sensor probe on the engine and check that the warning lamp goes out.
5 If the lamp does not go out when the sensor probe is earthed the sensor is faulty and must be renewed.

Refitting

6 Refit the sensor to the radiator and tighten it securely.
7 Reconnect the sensor wiring connector and check the coolant level as described in Chapter 1.

8 Water pump – removal and refitting

Removal

E series engines

1 Disconnect the battery negative terminal.
2 Referring to Chapter 1, drain the cooling system and disconnect the water pump drivebelt.
3 Apply the handbrake, then jack up the front of the car and support it on axle stands.
4 Undo the retaining bolts and remove the undertray from the vehicle to gain access to the water pump.
5 Slacken the hose clip and detach the radiator bottom hose from the pump.
6 Undo the exhaust manifold nut which retains the bypass pipe support bracket and carefully ease the pipe away from the pump. Remove the O-ring seal from the end of the pipe.
7 Undo the nut and two bolts securing the pump to the cylinder block, and the bolt securing the dipstick tube to the pump body.
8 Remove the pump from the engine.

B series engines

9 Drain the cooling system as described in Chapter 1.
10 Remove the timing belt and camshaft sprocket as described in Chapter 2, Part B.
11 Disconnect the bottom hose from the water pump bypass pipe union, then undo the two bolts securing the union to the water pump housing. Separate the union from the pump and bypass pipe and remove it from the engine.
12 Undo the four bolts securing the water pump to the cylinder block and lift the pump out of position.

Refitting

E series engines

13 Fit a new gasket in position over the water pump mounting stud and install the water pump (photos). Tighten the pump mounting bolts and nut to the specified torque.
14 Fit a new O-ring to the bypass pipe. Apply a smear of grease to the O-ring and push the pipe firmly into the pump (photo).

8.13A Fit a new gasket over the mounting stud...

8.13B ...then refit the water pump

8.14 Push the bypass pipe (arrowed) firmly into its pump location...

Chapter 3 Cooling, heating and ventilation systems

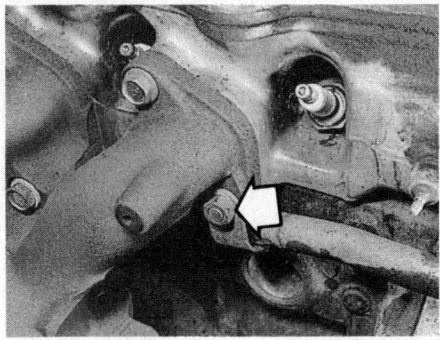

8.15 ...and refit the exhaust manifold nut (arrowed) which retains the pipe support bracket

8.20A Refit the water pump using a new gasket

8.20B Water pump mounting bolts (arrowed)

8.21 Fit a new O-ring to the bypass pipe (arrowed) and position a new gasket on the water pump housing

8.22A Push the pipe union firmly onto the bypass pipe and refit the retaining bolts

8.22B Reconnect the hose and secure it in position with the retaining clip

15 Secure the bypass pipe in position by tightening its support bracket bolt securely (photo).
16 Refit the radiator bottom hose and the undertray.
17 Lower the car to the ground.
18 Refit the water pump drivebelt, adjusting the tension as described in Chapter 1, and refill the cooling system with reference to Chapter 1.
19 Reconnect the battery negative terminal.

B series engines

20 Renew the gasket and refit the water pump to the cylinder block. Tighten the pump mounting bolts to the specified torque (photos).
21 Fit a new O-ring to the bypass pipe and push the pipe union firmly onto the pipe (photo).
22 Refit the bypass pipe union to the water pump using a new gasket and tighten its retaining bolts securely. Reconnect the bottom hose to the union ensuring that it is retained securely by the retaining clip (photos).
23 Refit the camshaft sprocket and timing belt as described in Part B of Chapter 2.
24 Refill the cooling system as described in Chapter 1.

9 Heater unit – removal and refitting

Removal

1 Disconnect the battery negative terminal.
2 Drain the cooling system as described in Chapter 1.

Pre-September 1985 models

3 Remove the facia as described in Chapter 11.
4 From within the engine compartment, slacken the clamps and disconnect the heater hoses at the matrix outlets.

5 Release the retaining clips and disconnect the control cables from the heater unit.
6 Disconnect the blower motor wiring at the connector.
7 Disconnect the air and demister ducts from the heater unit.
8 Undo the bolts securing the heater unit in position and ease it away from the bulkhead. Tip the unit back, so that the matrix outlets are uppermost, to prevent coolant spillage from the heater core as the unit is removed from the vehicle.

September 1985 models onward

9 With reference to Chapter 11, remove the right and left-hand lower

9.10 Heater unit control cable connection (arrowed)

Chapter 3 Cooling, heating and ventilation systems

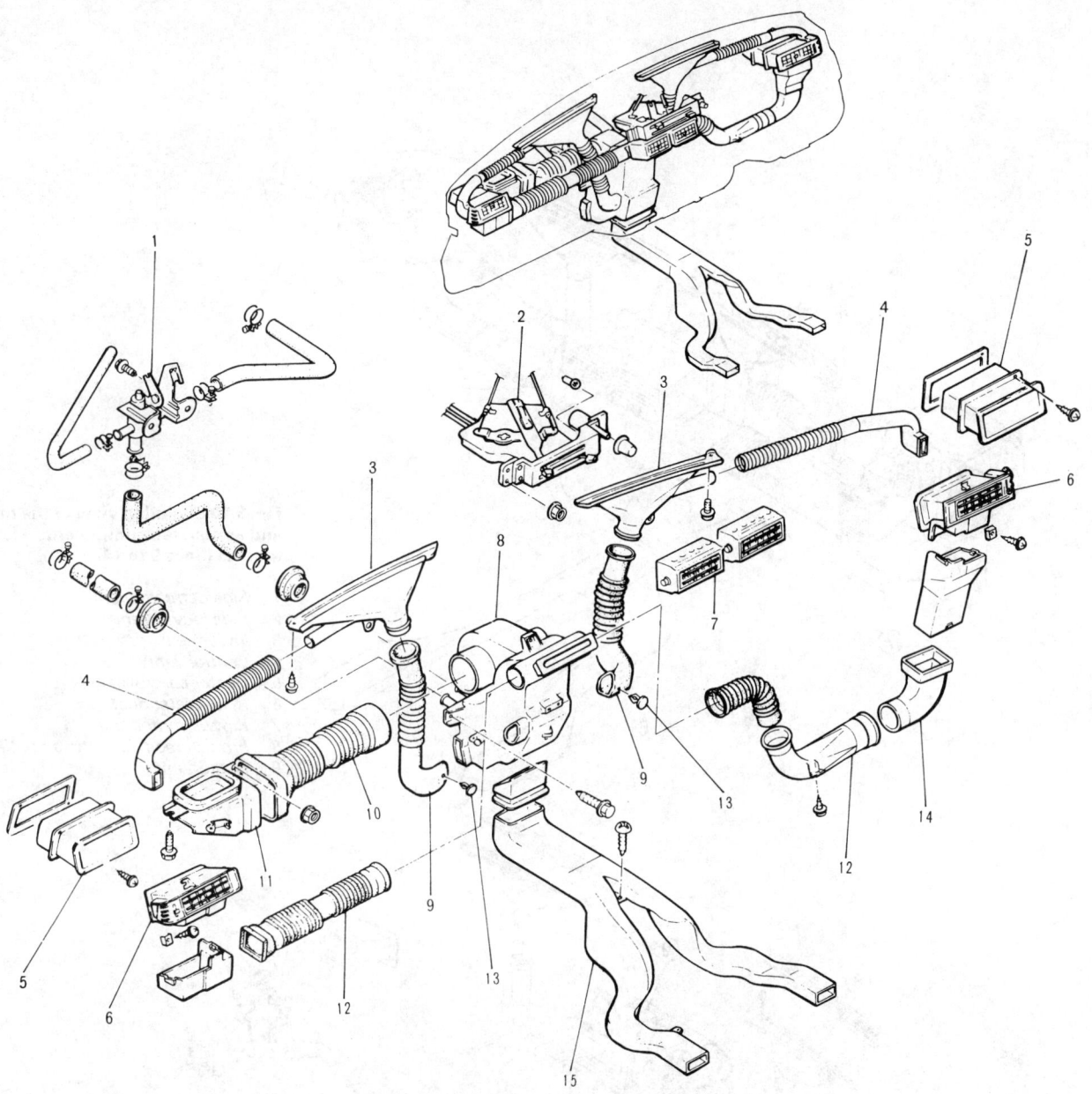

Fig. 3.9 Exploded view of the heater and associated components – early models (Secs 9 to 14)

1. Coolant valve
2. Heater controls
3. Demister nozzles
4. Side demister hoses
5. Ventilator ducts
6. Side ventilator grilles
7. Centre ventilator grille
8. Heater unit
9. Demister hoses
10. Duct
11. Air intake
12. Boost ventilator ducts
13. Retaining buttons
14. Elbow joint
15. Rear compartment duct

facia trim panels, the glovebox, the front console unit and side covers, the lower cover and carpet trim.

10 Disconnect the heater control cables from the heater unit (photo).

11 Undo the nuts securing the coolant inlet and outlet pipes to the matrix, and detach the pipes noting the sealing O-rings. Be prepared for a certain amount of coolant spillage as the pipes are disconnected.

12 Disconnect the airflow ducts from the heater unit and where necessary, remove the ducts completely to gain the necessary space for heater unit removal.

13 Undo the heater unit retaining nuts and bolts and withdraw the unit. Check that all the necessary items are disconnected and take great care not to spill any coolant remaining in the matrix over the carpets or upholstery.

Refitting

14 The heater unit is refitted by a reversal of the removal procedure, bearing in mind the following points.

(a) On September 1985 models onward renew the matrix coolant pipe O-rings.
(b) Adjust the heater cables as described in Section 14.
(c) On completion, refill the cooling system as described in Chapter 1.

Chapter 3 Cooling, heating and ventilation systems

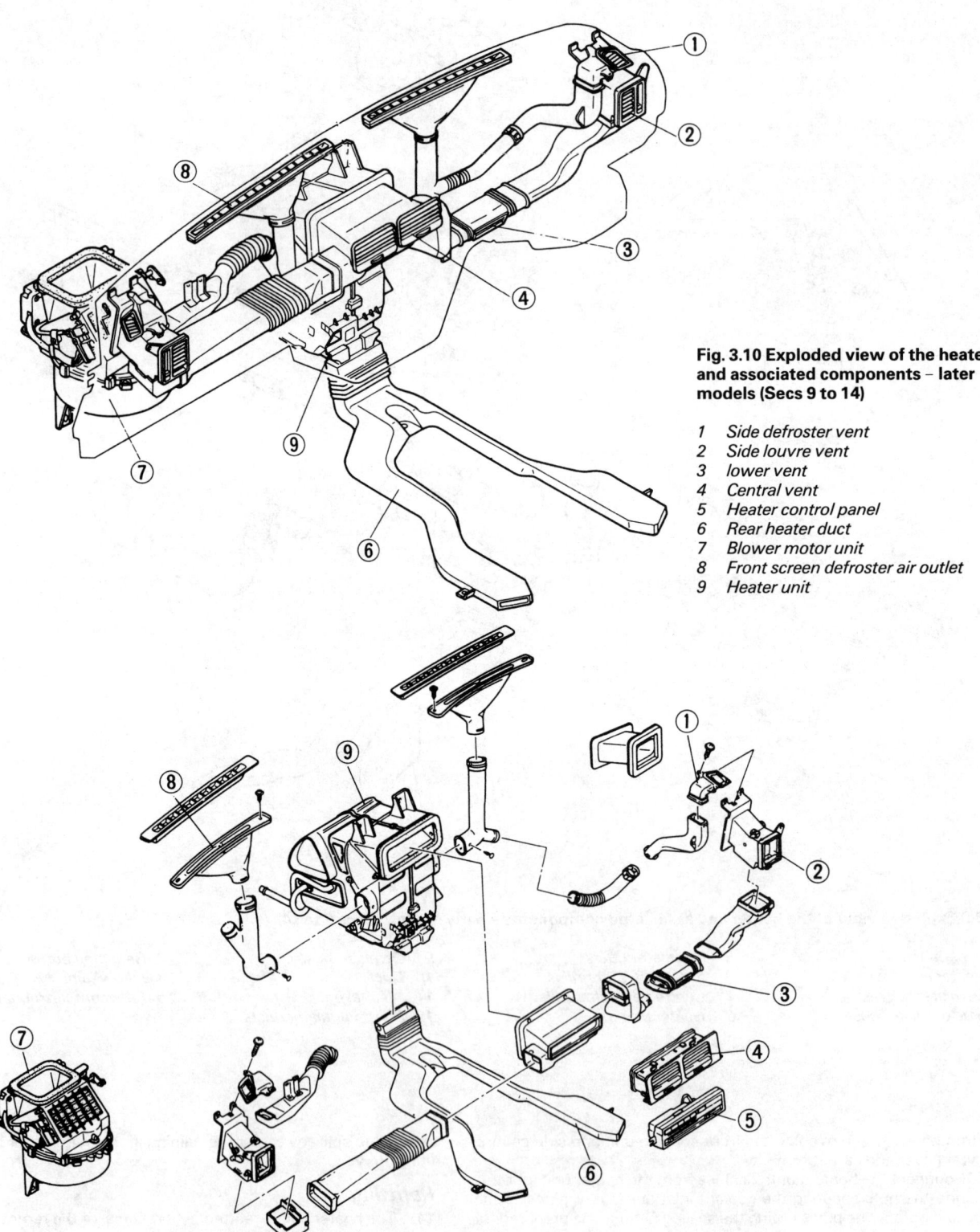

Fig. 3.10 Exploded view of the heater and associated components – later models (Secs 9 to 14)

1 Side defroster vent
2 Side louvre vent
3 lower vent
4 Central vent
5 Heater control panel
6 Rear heater duct
7 Blower motor unit
8 Front screen defroster air outlet
9 Heater unit

Chapter 3 Cooling, heating and ventilation systems

10 Heater matrix – removal and refitting

Removal

Pre-September 1985 models
1 Remove the heater unit as described in Section 9.
2 Prise out the retaining clips and remove the heater matrix from the rear of the heater unit.

September 1985 models onward
3 Drain the cooling system as described in Chapter 1.
4 Referring to Chapter 11, remove the glovebox and facia undercovers.
5 Disconnect the mode control cable, then undo the screw securing the pivot control lever and move the lever and cable out of the way.
6 Undo the nuts securing the inlet and outlet coolant pipes to the matrix and detach the pipes noting the sealing O-rings. Be prepared for a certain amount of coolant spillage as the pipes are disconnected.
7 Undo the two matrix retaining screws and withdraw the matrix from the heater unit, taking great care not to spill any remaining coolant over the carpets or upholstery.

Refitting

Pre-September 1985 models
8 Position the matrix in the heater unit, securing it in position with its retaining clips, and refit the heater unit as described in Section 9.

September 1985 models onward
9 Refit the matrix to the heater unit and tighten its retaining screws securely.
10 Fit new O-rings to the coolant inlet and outlet pipes and refit the pipes to the matrix, tightening their retaining nuts securely.
11 Refit the pivot lever and reconnect the mode control cable. Adjust the cable as described in Section 14.
12 Refit the facia undercovers and glovebox.
13 Refill the cooling system as described in Chapter 1.

11 Heater blower unit – removal and refitting

Removal

Pre-September 1985 models
1 Disconnect the battery negative terminal.
2 Remove the facia undercover from under the steering wheel.
3 Disconnect the heater motor wiring connector, undo the three retaining screws and withdraw the motor from the heater unit. For greater access remove the air ducts from under the facia.

September 1985 models onward
4 Disconnect the battery negative terminal.
5 Referring to Chapter 11, remove the facia undercover, the glovebox and the black metal upper panel from inside the glovebox compartment.
6 Disconnect the wiring connectors to the heater blower (photo).
7 Remove the duct between the heater unit and blower motor (photo).
8 Undo the blower unit mounting nuts, disconnect the fresh air recirculation control wire and remove the heater blower unit (photos). Note the following points when removing the blower motor from vehicles that are fitted with a 'Logic' type heater control.

   (a) *Removal of the instrument panel will ease withdrawal of the unit.*
   (b) *Set the 'REC/FRESH' air control selector to the 'REC' position to ease removal of the upper retaining nuts.*

Refitting

Pre-September 1985 models
9 Refitting is a reversal of the removal sequence.

September 1985 models onward
10 Refitting is a reversal of the removal procedure. Adjust the 'REC/FRESH' air control wire as described in Section 14 on installation.

11.6 Blower unit wiring connector

11.7 Blower motor to heater unit duct (arrowed)

11.8A Undo the mounting nuts...

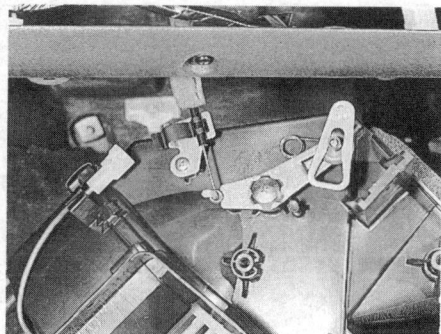

11.8B ...then disconnect the fresh air recirculation control wire...

11.8C ...and remove the blower unit

Chapter 3 Cooling, heating and ventilation systems

12.3 Heater coolant valve is mounted on the engine compartment bulkhead

13.4 Removing heater control lever grub screws – pre-September 1985 models

12 Heater coolant valve (pre-September 1985 models) – removal and refitting

Removal

1 Drain the cooling system as described in Chapter 1.
2 Move the heater air temperature control lever fully to the left.
3 From within the engine compartment, release the control cable retaining clip from the coolant valve which is mounted on the engine compartment bulkhead, and disconnect the cable (photo).
4 Slacken the clamps and disconnect the three coolant hoses from the valve.
5 Undo the two bolts securing the coolant valve to the bulkhead and remove the valve from the engine compartment.

Refitting

6 Refitting is a reverse of the removal sequence. On completion refill the cooling system as described in Chapter 1.

13 Heater control panel – removal and refitting

Removal

Pre-September 1985 models

1 Disconnect the battery negative terminal.
2 Pull the knob off the heater blower control.

3 Undo the four screws securing the upper part of the instrument panel shroud to the facia. Pull the bottom part of the shroud out of its retaining clips and remove it.
4 Undo the grub screws securing the knobs to the heater control levers and remove the knobs (photo).
5 Undo the screws securing the heater control faceplate and remove the faceplate.
6 Undo the heater control retaining nuts and withdraw the control assembly from the facia. Release the control cable retaining clips, disconnect the cables and remove the heater control assembly.

September 1985 models onward

7 Disconnect the battery negative terminal.
8 Undo the two retaining screws and remove the central vent panel (photos).
9 Remove both the facia undercovers for access to the heater unit and blower motor operating cables.
10 Disconnect the cables from the heater unit and blower motor, then undo the control unit retaining screws and withdraw the unit and cables (photo). Disconnect the wiring plug from the rear of the unit as it is removed.

Refitting

11 Refitting is the reverse of the removal procedure. Adjust the control cables as described in the following Section.

14 Heater control cables – adjustment

1 Before carrying out any adjustments, disconnect the battery

13.8A On September 1985 models onward undo the retaining screws...

13.8B ...and remove the central vent panel

13.10 Heater control panel retaining screws (arrowed)

Chapter 3 Cooling, heating and ventilation systems

negative terminal and remove the necessary panels to gain access to the heater unit.

Pre-September 1985 models
Air control adjustment cable
2 Move the control lever fully to the left.
3 Release the control cable retaining clip at the air flap on the lower left-hand side of the heater.
4 Move the flap fully clockwise, hold it in position and refit the control cable retaining clip.
5 A second control cable from the air control lever operates a flap valve on the air intake to open or close the intake to outside air.
6 To adjust the air intake valve flap, pull the control lever out and release the retaining clip at the air intake valve clip.
7 Move the flap fully anti-clockwise, hold it in this position and refit the control cable retaining clip.

Air temperature control adjustment cable
8 Move the air temperature control lever fully to the right.
9 Release the control cable retaining clip at the air flap on the lower right-hand side of the heater.
10 Move the flap fully clockwise, hold it in this position and refit the control cable retaining clip.
11 Now move the air temperature control lever fully to the left.
12 From within the engine compartment release the control cable retaining clip from the coolant valve mounted on the bulkhead. Move the valve lever fully downwards, hold it in this position and refit the cable retaining clip.

September 1985 models onward (lever type controls)
Mode control cable
13 Position the mode control lever at the 'DEF' position.
14 Release the cable retaining clip, then at the heater unit, pivot the control lever fully downwards to the stop position. Hold the lever and secure the cable in position with the retaining clip.
15 Disengage the connecting rod and pivot the connecting rod arm fully anti-clockwise onto its stop. Hold the arm in this position then refit the rod and secure it in position with the retaining clip.
16 Check that the mode control cable is operating correctly by setting the fan speed control to the number 4 position and turning the fan on.

Air mix door control cable
17 Move the temperature control lever fully to the left.
18 Release the cable retaining clip, then pivot the heater unit air mix control cable lever fully clockwise to the stop position. Secure the cable in position with the retaining clip.
19 Disengage the connecting rod, then pivot the air mix door lever clockwise to the stop position. Refit the connecting rod to the air mix door lever and secure it in position with its retaining clip.
20 Check that the air mix door moves smoothly from the fully open to fully closed position while operating the control lever.

Recirculating/fresh air (REC/FRESH) control cable
21 Set the lever to the fresh air inlet position.
22 Remove the control cable retaining clip, then push the control lever on the blower unit to its extreme stop position. Hold the lever in this

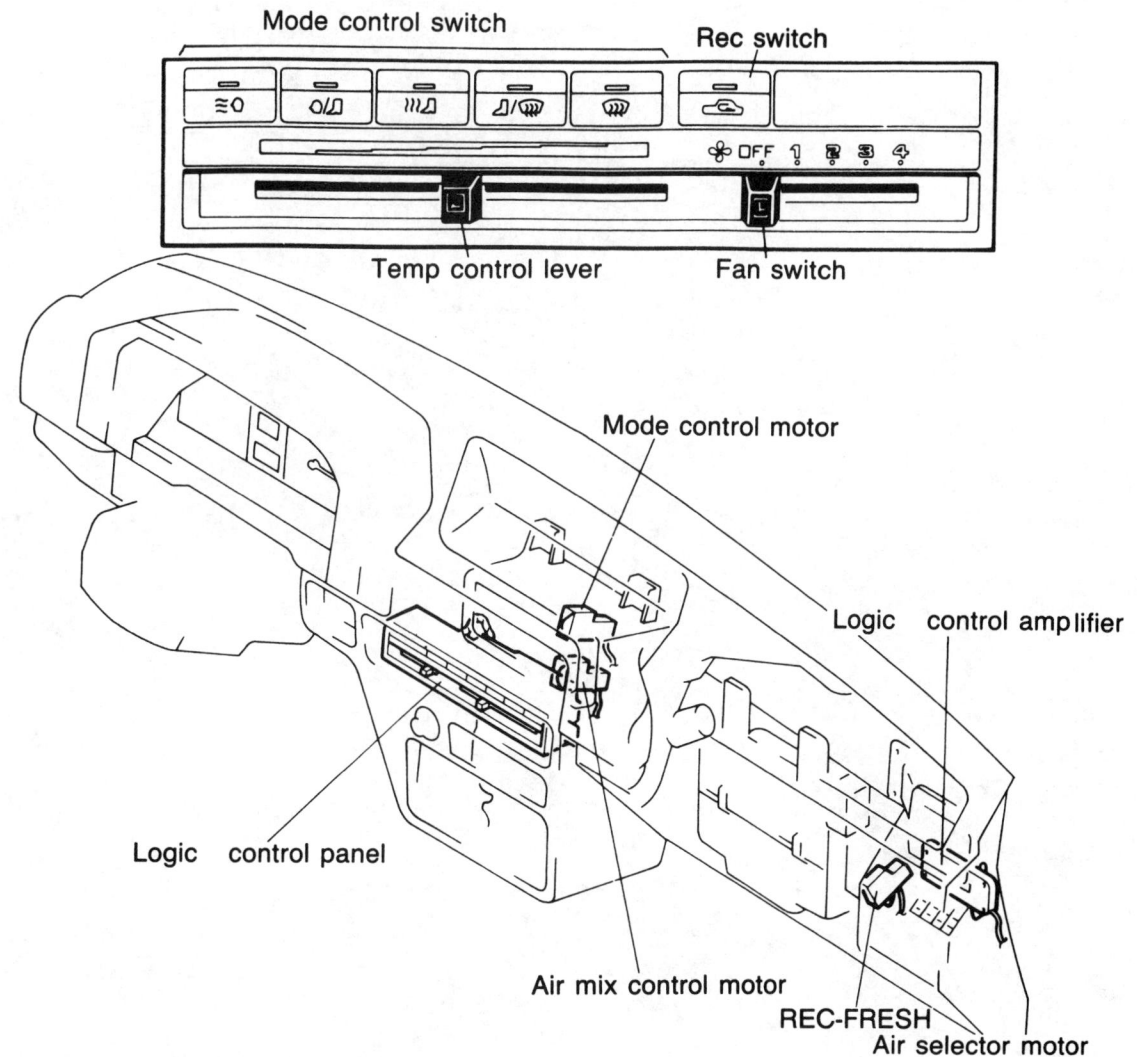

Fig. 3.11 Logic control type heater controls – left-hand drive shown (Sec 15)

position and refit the control cable retaining clip.
23 Operate the blower motor and check that the recirculating/fresh air control operation is satisfactory.

15 Logic type heater control system – general

1 The 'logic' type heater control panel (where fitted) and the associated components of the system are shown in Fig. 3.11. The function of each is as follows.

(a) *Logic control panel: This is a one touch push button control panel.*
(b) *Logic control amplifier: This controls the air mixer valve motor and protects it in the event of a malfunction.*
(c) *Mode control motor: This controls the air outlet during each mode, operating the ventilation and defrost valve.*
(d) *Air mixture control motor: This controls the outgoing air temperature by operating the air mixer valve in accordance with the signal from the logic control amplifier.*
(e) *REC/FRESH air selector motor: This motor operates the valve for the REC/FRESH shift by the selector lever.*

2 If a malfunction occurs in the 'Logic' type heater system, the checking procedures are considerable and require the use of a special 'Logicon checker' unit. For this reason if the system malfunctions it will be necessary to take the car to a suitably equipped Mazda dealer for fault diagnosis and repair.

Chapter 4
Fuel, exhaust and emission control systems

Contents

Part A: Carburettor engines
Accelerator cable – removal, refitting and adjustment 6
Accelerator pedal – removal and refitting 7
Air cleaner assembly – removal and refitting 2
Air cleaner filter element – renewal .... See Chapter 1
Carburettor – fault diagnosis, overhaul and adjustments 12
Carburettor – general information 10
Carburettor – on-car adjustments 13
Carburettor – removal and refitting 11
Choke cable (Hitachi carburettor) – removal, refitting and
 adjustment 8
Economy drive indicator system – general information 15
Exhaust manifold – removal and refitting 18
Exhaust system – check .... See Chapter 1
Exhaust system – general information, removal and refitting 19
Fuel filter – renewal .... See Chapter 1
Fuel gauge sender unit – removal and refitting 4
Fuel pump – testing, removal and refitting 3
Fuel tank – removal and refitting 5
General fuel system checks .... See Chapter 1
General information and precautions 1
Idle speed and CO content – adjustment .... See Chapter 1
Inlet manifold – removal and refitting 17
PTC heater system (Aisan carburettor) – testing 16
Shutter valve control system (B3 and B5 engines) – general
 information and testing 14
Unleaded petrol – general information and usage 9

Part B: Fuel injected engines
Accelerator cable – removal, refitting and adjustment 22
Accelerator pedal – removal and refitting 23
Air cleaner filter element – renewal .... See Chapter 1
Air cleaner housing assembly – removal and refitting 21
Economy drive indicator system – general information 34
Exhaust manifold – removal and refitting 36
Exhaust system – check .... See Chapter 1
Exhaust system – general information, removal and refitting 37
Fuel filter – renewal .... See Chapter 1
Fuel injection system – general information and fault diagnosis 25
Fuel injection system components – testing 31
Fuel injection system components – removal and refitting 32
Fuel pump and fuel gauge sender unit assembly – removal and
 refitting 27
Fuel system pressure tests 26
Fuel tank – removal and refitting 28
General fuel system checks .... See Chapter 1
General information and precautions 20
Idle speed and CO content – adjustment .... See Chapter 1
Idle up system – general information and adjustment 33
Inlet manifold – removal and refitting 35
Surge tank – removal and refitting 30
Throttle housing – removal and refitting 29
Unleaded petrol – general information and usage 24

Part C: Emission control systems
Emission control system components – testing and renewal 39
General information 38

Specifications

Part A: Carburettor engines

Fuel grade
Fuel octane requirement:
    All models except 1500 GT .... 91 RON unleaded or leaded
    1500 GT models .... 96 RON unleaded or leaded

Fuel pump
Type .... Mechanical, operated by eccentric on camshaft
Delivery pressure .... 0.20 to 0.31 bars (2.9 to 4.5 lbf/in^2)

Carburettor (general)
Type:
  Pre-September 1985 models:
    All models except 1500 GT .... Hitachi dual throat downdraught
    1500 GT models .... Twin Hitachi dual throat downdraught
  September 1985 models onward .... Aisan dual throat downdraught
Choke type:
  Pre-September 1985 models .... Manual
  September 1985 models onward .... Automatic

Hitachi carburettor data (except 1500 GT models)

	E1 engined models	E3 and E5 engined models
Throat diameter:		
Primary	26 mm	26 mm
Secondary	30 mm	30 mm
Main jet:		
Primary	90	106
Secondary	145	160
Main air bleed:		
Primary	60	80
Secondary	100	100
Slow running jet:		
Primary	48	48
Secondary	120	130
Slow air bleed:		
Primary:		
No 1	150	170
No 2	110	110
Secondary	190	130
Power jet	40	40

Hitachi carburettor data (1500 GT models)

Throat diameter:	
Primary	26 mm
Secondary	30 mm
Main jet:	
Primary	90
Secondary	135
Main air bleed:	
Primary	60
Secondary	80
Slow running jet:	
Primary	46
Secondary	130
Slow air bleed:	
Primary:	
No 1	170
No 2	100
Secondary	150
Power jet	40

Aisan carburettor data (E series engine)

	E1 engined models	E3 and E5 engined models
Throat diameter:		
Primary	28 mm	28 mm
Secondary	32 mm	32 mm
Venturi diameter:		
Primary	20 x 11 mm	22 x 11 mm
Secondary	25 x 11 mm	27 x 11 mm
Main nozzle:		
Primary	2.40	2.40
Secondary	2.40	2.40
Main jet:		
Primary	0.79	0.93
Secondary	1.13	1.29
Main air bleed:		
Primary	0.60	0.55
Secondary	0.70	0.70
Slow jet:		
Primary	0.50	0.50
Secondary	0.80	0.90
Slow air bleed:		
Primary (No 1)	1.60	1.60
Secondary (No 2)	0.50	0.50
Power jet	0.40	0.45

Aisan carburettor data (B series engine)

Throat diameter:	
Primary	28 mm
Secondary	32 mm
Venturi diameter:	
Primary	22 x 11 mm
Secondary	27 x 11 mm
Main nozzle:	
Primary	2.40
Secondary	2.40

Chapter 4 Fuel, exhaust and emission control systems

Aisan carburettor data (B series engine) (continued)

Main jet:
- Primary 0.91
- Secondary 1.56

Main air bleed:
- Primary 0.70
- Secondary 0.70

Slow jet:
- Primary 0.51
- Secondary 0.80

Slow air bleed:
- Primary (No 1) 1.40
- Secondary (No 2) 0.50

Power jet 0.50

Hitachi carburettor adjustment data

Choke vacuum diaphragm adjustment clearance:
- All models except 1500 GT .... 1.35 to 1.65 mm
- 1500 GT models .... 1.45 to 1.75 mm

Secondary fuel cut-off valve clearance .... 2.0 mm
Float height .... 11.0 mm
Float opening clearance .... 1.3 to 1.7 mm

Fast idle clearance:
- All models except 1500 GT .... 1.1 mm
- 1500 GT models .... 0.6 mm

Aisan carburettor adjustment data

Float height settings:
- Float to sealing face clearance (L):
  - E series engined models .... 45 to 49 mm
  - B series engined models .... 46 to 48 mm
- Float-to-sealing face clearance (H):
  - E series engined models .... 4 to 5 mm
  - B series engined models .... 6 to 7 mm

Choke breaker diaphragm adjustment choke valve-to-bore clearance:
- E1 engined models .... 0.85 to 1.35 mm
- E3 and E5 engined models .... 1.30 to 1.80 mm
- B3 and B5 engined models .... 1.05 to 1.24 mm

Primary throttle valve-to-bore clearance (fast idle cam on 3rd position):
- E1 engined models .... 0.23 to 0.53 mm
- E3 and E5 engined models .... 0.35 to 0.65 mm
- B3 and B5 engined models .... 1.05 to 1.24 mm

Choke valve-to-bore clearance (fast idle cam on 2nd position) ... 0.67 to 1.17 mm
Choke valve clearance (throttle valve fully open) .... 1.55 to 2.05 mm
Primary throttle valve-to-bore clearance (secondary throttle valve just open) .... 5.4 to 6.4 mm

Torque wrench settings

	Nm	lbf ft
Inlet manifold to cylinder head nuts and bolts	19 to 26	14 to 19
Exhaust manifold to cylinder head nuts and bolts	16 to 27	12 to 20

Part B: Fuel injected models

Fuel grade

Fuel octane requirement .... 96 RON unleaded or leaded

Fuel pump

Type .... Electric
Fuel pump operating test pressure .... 4.5 to 6.0 bars (65.0 to 87.0 lbf/in^2)

Fuel pressure regulator test pressure:
- Vacuum hose connected .... 2.0 to 2.2 bars (29.0 to 31.9 lbf/in^2)
- Vacuum hose disconnected .... 2.5 to 2.9 bars (36.3 to 42.0 lbf/in^2)

Fuel injection system component resistances

Fuel injectors .... 12 to 16 ohms
Auxiliary air valve .... 30 to 50 ohms

Torque wrench settings

	Nm	lbf ft
Inlet manifold to cylinder head nuts and bolts	19 to 26	14 to 19
Exhaust manifold to cylinder head nuts and bolts	16 to 27	12 to 20

Part A: Carburettor engines

1 General information and precautions

The fuel system consists of a centrally mounted fuel tank, mechanical fuel pump and a dual throat downdraught carburettor. Twin carburettors are used on 1500 GT models.

The mechanical fuel pump is operated by an eccentric on the camshaft and is mounted on the rear facing side of the cylinder head. A disposable in-line filter is located between the tank and the pump.

The air cleaner contains a paper element and incorporates an air temperature control valve, which is either manually operated via a lever on the front of the air cleaner assembly, or automatically by bi-metallic type valve. The valve allows either warm air from the exhaust manifold stove or cold air from the air cleaner intake to enter the air cleaner according to the position of the valve flap.

The carburettor(s) are mounted on a cast aluminium inlet manifold. Early models employ a Hitachi carburettor on which the choke is cable controlled via a control lever on the facia. Later models use an Aisan carburettor which has a fully automatic choke system. On later models the inlet manifold incorporates a heater, which heats the fuel/air mixture when the engine temperature is cold to improve fuel atomization.

The exhaust system consists of four or five sections secured by flanges, joints or push fit with clamps, and a cast iron exhaust manifold. A spring loaded semi ball and socket joint is used to connect the exhaust front pipe to the front silencer thus catering for engine and exhaust system movement. The system is suspended throughout its length on rubber ring or block type mountings.

Warning: *Many of the procedures in this Chapter require the removal of fuel lines and connections which may result in some fuel spillage. Before carrying out any operation on the fuel system refer to the precautions given in Safety First! at the beginning of this Manual and follow them implicitly. Petrol is a highly dangerous and volatile liquid and the precautions necessary when handling it cannot be overstressed.*

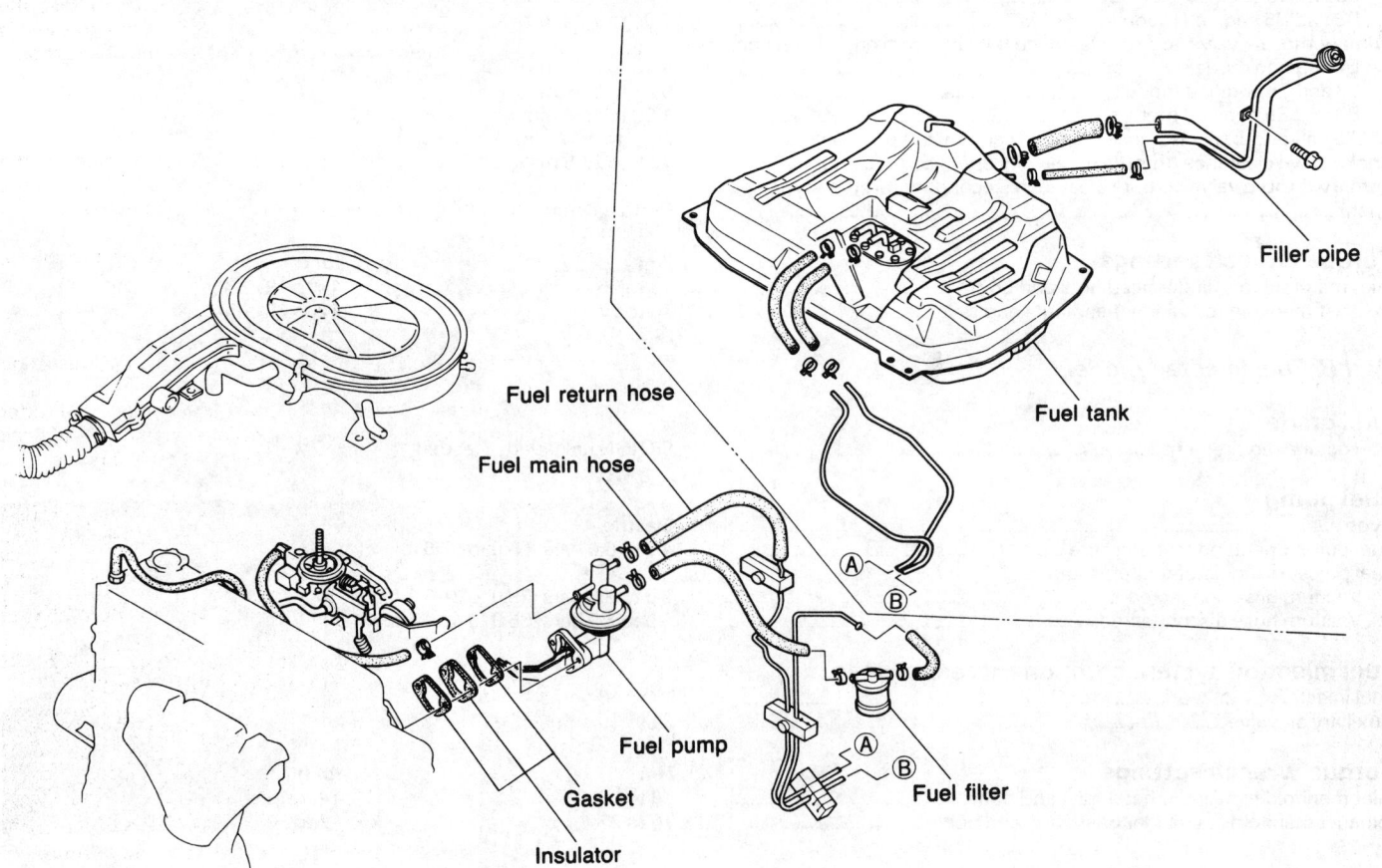

Fig. 4.1 Fuel system layout (Sec 1)

Chapter 4 Part A: Carburettor engines

2.2A Air cleaner assembly retaining bolt...

2.2B ...and nut

2 Air cleaner assembly – removal and refitting

Removal
1 Disconnect the breather hose from the top of the cylinder head cover, and the air cleaner warm air intake duct from the exhaust manifold stove.
2 Undo the nut and bolts securing the air cleaner assembly to the cylinder head cover and the rear support bracket (photos).
3 Undo the top cover wing nut(s) and lift the air cleaner upwards off the carburettor and away from the engine.

Refitting
4 Refitting is the reverse of the removal sequence.

3 Fuel pump – testing, removal and refitting

Note: *Refer to the warning note in Section 1 before carrying out the following operation.*

Testing
1 To test the fuel pump on the engine, temporarily disconnect the outlet pipe which leads to the carburettor, and hold a wad of rag over the pump outlet while an assistant spins the engine on the starter. Regular spurts of fuel should be ejected as the engine turns.
2 The pump can also be tested after removing it. With the pump outlet pipe disconnected, but the inlet pipe still connected, hold a wad of rag by the outlet. Operate the pump lever by hand and if the pump is in a satisfactory condition a strong jet of fuel should be ejected.
3 If a suitable pressure gauge is available, a more accurate test may be carried out. Before connecting the gauge to the fuel system, run the engine at idling speed for several minutes in order to completely fill the carburettor float chamber. With the engine switched off, disconnect the fuel supply pipe at the carburettor end, then connect the pressure gauge to it. Using a hose clamp, pinch the return pipe leading to the fuel tank, alternatively disconnect the hose and plug the pump return pipe union. Start the engine and allow it to idle. Check that the pump pressure is as given in the Specifications. Check the return pipe for obstruction by removing the clamp from the return hose and checking that the pressure then drops to between 0.01 and 0.02 bars – if the pressure is higher than this, blow through the return pipe to clear the obstruction.

Removal
4 Disconnect the battery negative lead.
5 Note the location of the fuel inlet, outlet and return pipes at the pump then, using pliers, release the retaining clips and disconnect the three hoses. Plug the hose ends with a suitable screw or bolt to minimise fuel spillage.
6 Undo the two pump retaining bolts then withdraw the pump, insulating block and gaskets from the cylinder head.

Refitting
7 Remove all traces of old gasket from the pump flange, insulating block and cylinder head sealing face.
8 Position a new gasket on each side of the insulating block, then offer up the pump and insulating block to the cylinder head and refit the retaining bolts. Tighten the pump retaining bolts securely (photos).

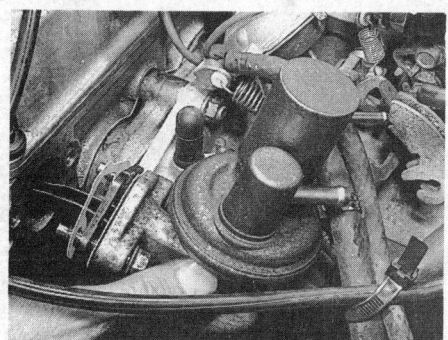

3.8A Position a new gasket on each side of the insulator block and refit the fuel pump

3.8B Tighten the pump mounting bolts securely

3.9 Reconnect the fuel hoses and secure them in position with the retaining clips

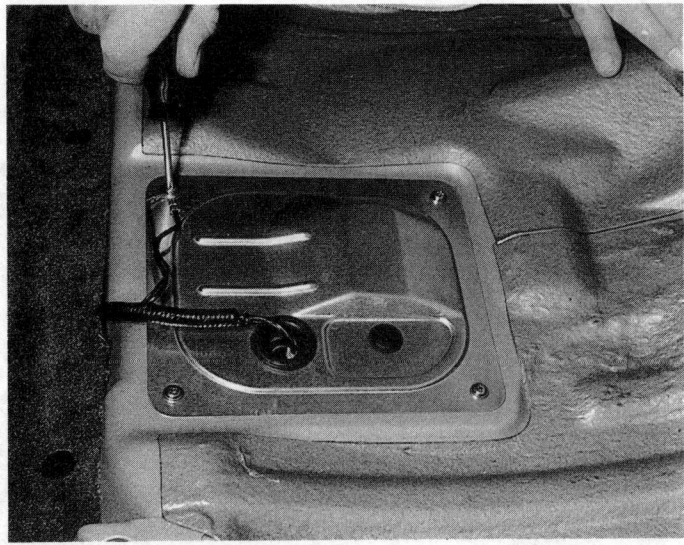

4.3 Remove the fuel tank cover...

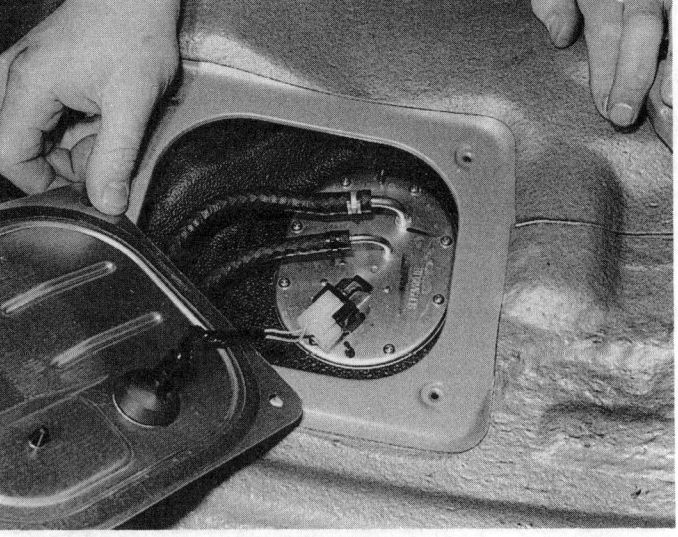

4.4 ...to gain access to the fuel gauge sender unit

9 Connect the fuel inlet, outlet and return pipes to the pump and secure them in position with the retaining clips (photo).
10 Reconnect the battery negative lead, start the engine and check for fuel leaks.

4 Fuel gauge sender unit – removal and refitting

Note: *Refer to the warning note in Section 1 before carrying out the following operation.*

Removal

1 Disconnect the battery negative lead.
2 Remove the rear seat cushion as described in Chapter 11.
3 Undo the four screws securing the fuel tank cover to the floor panel and lift off the cover (photo).
4 Disconnect the wiring and, where necessary, the fuel hose(s) from the fuel gauge sender unit (photo).
5 Undo the screws securing the sender unit to the tank then remove the assembly, taking care not to damage the float arm. Remove the gasket.

Refitting

6 Refitting is a reversal of removal, but use a new gasket if the old one is damaged or shows signs of deterioration.

5 Fuel tank – removal and refitting

Note: *Refer to the warning note in Section 1 before carrying out the following operation.*

Removal

1 On models where a drain plug is not provided on the fuel tank, it is preferable to carry out the removal operation when the tank is nearly empty. Before proceeding, disconnect the battery negative lead and then syphon or hand pump the remaining fuel from the tank. Where a drain plug is fitted to the underside of the tank, place a suitable container beneath the tank then remove the drain plug and allow the fuel to drain. Once the tank is empty refit the drain plug and tighten it securely.
2 From inside the car, remove the rear seat cushion as described in Chapter 11.
3 Undo the four screws securing the fuel tank cover to the floor panel and lift off the cover.
4 Disconnect the fuel gauge sender unit wiring connector and, where necessary, the fuel tank feed and return hoses.
5 Chock the front wheels, jack up the rear of the car and support it on axle stands.
6 Make a note of the correct fitted positions of the filler and breather hose positions, then loosen the clips and disconnect them from the tank unit. Release any cable or pipe retaining clips from the sides of the fuel tank.
7 Support the tank on a jack with interposed block of wood then undo the tank retaining bolts.
8 Slowly lower the jack and tank. As access improves, disconnect any remaining breather pipes then remove the tank from under the car.
9 If the tank is contaminated with sediment or water, remove the sender unit as described in Section 4 and swill the tank out with clean fuel. If the tank is damaged or leaks, it should be repaired by a specialist or alternatively renewed.

Refitting

10 Refitting is a reversal of the removal procedure, ensuring that all fuel and breather hoses are securely retained and do not become trapped as the tank is lifted into position.

6 Accelerator cable – removal, refitting and adjustment

Removal

1 Remove the air cleaner assembly as described in Section 2.
2 Working in the engine compartment, open the throttle linkage fully

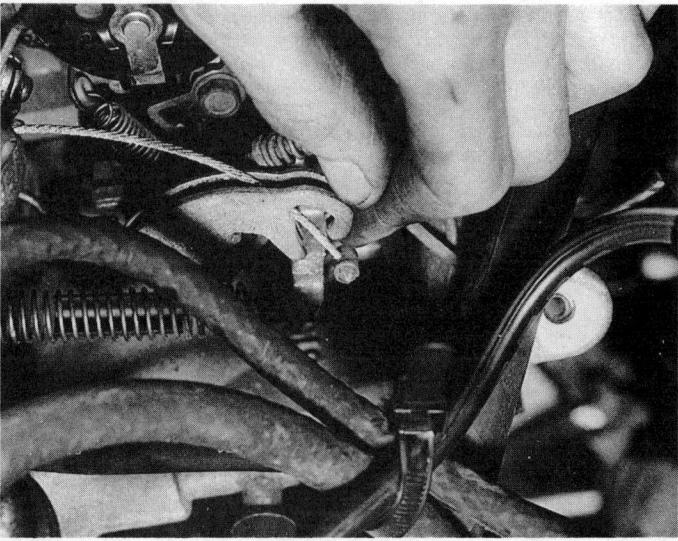

6.2 Disconnecting the accelerator cable from the carburettor

6.3A Slacken the accelerator cable locknuts...

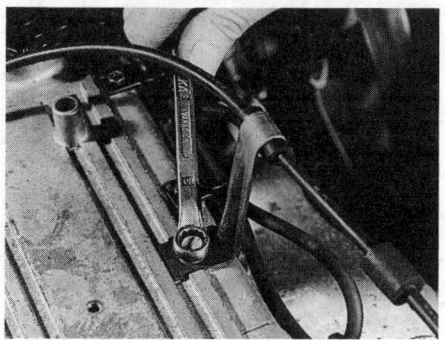
6.3B ...and remove the support bracket retaining bolt

6.6 Accelerator pedal stop bolt and locknut

8.4 Slacken the choke cable clamp screw (arrowed)...

8.5 ...and disconnect the inner cable (arrowed) from the choke linkage

8.6 Undo the choke knob grub screw using a small screwdriver

by hand and disconnect the cable end from the linkage lever (photo).
3 Slacken the locknuts securing the cable to the bracket on the cylinder head cover and undo the bolt securing the cable support bracket to the cylinder head cover (where fitted) (photos).
4 From inside the car remove the cover under the facia.
5 Slip the cable end out of the slot on the accelerator pedal, then withdraw the cable through the bulkhead grommet into the engine compartment.

Refitting and adjustment

6 Refitting is a reversal of the removal sequence. Adjust the cable by means of the locknuts on the cylinder head cover bracket so that there is a small amount of slack in the cable with the throttle linkage lever on its stop. Have an assistant fully depress the accelerator pedal and check that the throttle linkage on the carburettor opens fully. If not, working from inside the car, slacken the accelerator pedal stop bolt locknut and screw the stop bolt in or out as necessary until the throttle linkage opens fully (photo). Hold the stop bolt in this position and tighten the locknut securely.

7 Accelerator pedal – removal and refitting

Removal

1 Remove the cover from under the facia.
2 Slip the accelerator cable out of the slot on the upper end of the pedal.
3 Unhook the pedal return spring to release the tension, and remove the retaining clip and bush from the pedal pivot pin.
4 Withdraw the accelerator pedal from its mounting bracket and recover the return spring and bush from the pedal pivot pin.

Refitting

5 Refitting is a reversal of removal. On completion check the accelerator cable adjustment as described in Section 6.

8 Choke cable (Hitachi carburettor) – removal, refitting and adjustment

Removal

1 Disconnect the battery negative lead.
2 Remove the air cleaner assembly as described in Section 2.
3 For greater access, refer to Chapter 12 and remove the instrument panel.
4 Working in the engine compartment, undo the clamp screw and release the choke outer cable from the carburettor bracket (photo).
5 Close the choke linkage by hand and disconnect the inner cable from the linkage lever (photo).
6 From inside the car, using a small screwdriver, undo the grub screw securing the choke cable knob and remove the knob (photo).
7 Unscrew the knurled retaining ring and push the choke cable through the facia panel.
8 Working through the instrument panel aperture, pull the cable through the bulkhead grommet and remove it from the car.

Refitting

9 Refitting is the reverse of the removal sequence. Adjust the position of the outer cable in the carburettor bracket so that the choke linkage closes fully when the knob is pulled out, and opens fully when it is pushed in.

9 Unleaded petrol – general information and usage

All the models covered in this Section can run on either leaded or unleaded fuel providing it meets the minimum octane rating requirement. On 1500 GT models the fuel must have a minimum octane rating of 96 RON, and on all other models it must have a minimum rating of 91 RON. No adjustments are necessary to the ignition timing.

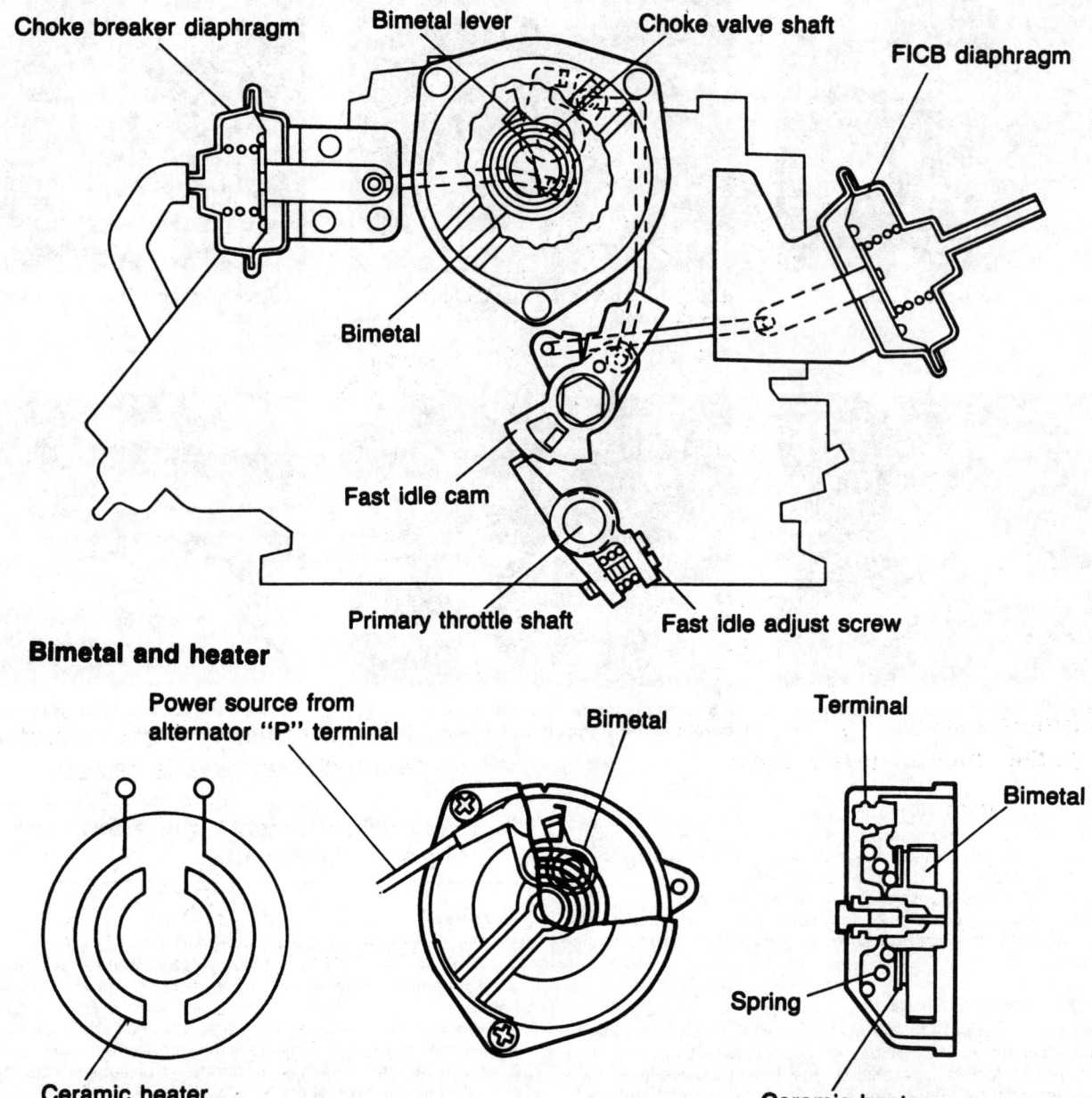

Fig. 4.2 Choke system components – Aisan carburettor (Sec 10)

10 Carburettor – general information

Pre-September 1985 models

Pre-September 1985 models are fitted with dual throat downdraught Hitachi carburettors. Twin carburettors are used on 1500 GT models whereas all other models use a single carburettor. On all models fitted with a manual gearbox, a throttle positioner system is fitted to prevent engine stalling and afterburning. On models equipped with power steering a fast idle system is also fitted to increase the idle speed when the power steering pump is operating.

September 1985 onward E series engined models

On all these models a single dual throat downdraught Aisan carburettor is fitted. All carburettors are equipped with a fully automatic choke system of which the main components are a choke valve, a bi-metal heater, a fast idle cam, a choke breaker diaphragm and a fast idle cam breaker (FICB) diaphragm and functions as follows.

The bi-metal heater strip automatically positions the choke valve correctly, then the fast idle cam opens the throttle valve relative to the choke valve opening and provides the correct mixture. When the engine is started the choke breaker diaphragm opens the choke valve a sufficient amount to provide the correct mixture ratio. In the event of the engine flooding with fuel when cold, a choke unloader opens the choke partially to weaken the mixture.

When the engine temperature reaches 40°C, the coolant temperature sensor switch causes the three-way solenoid valve to open, allowing vacuum from the inlet manifold to be directed to the FICB diaphragm. This then shuts down the fast idle cam to reduce the engine speed. On E1 and E3 engines a delay valve is incorporated in the circuit to delay the operation of the FICB diaphragm by approximately 10 to 15 seconds.

Models fitted with a manual gearbox are also fitted with a throttle positioner system to prevent stalling. Models equipped with power

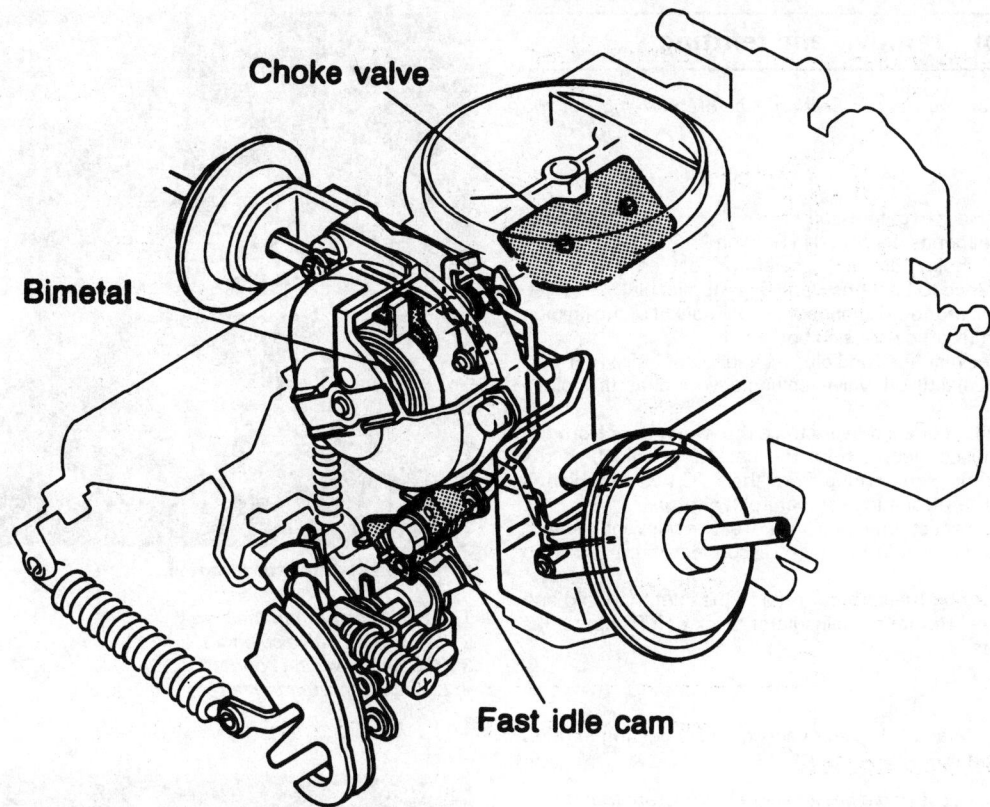

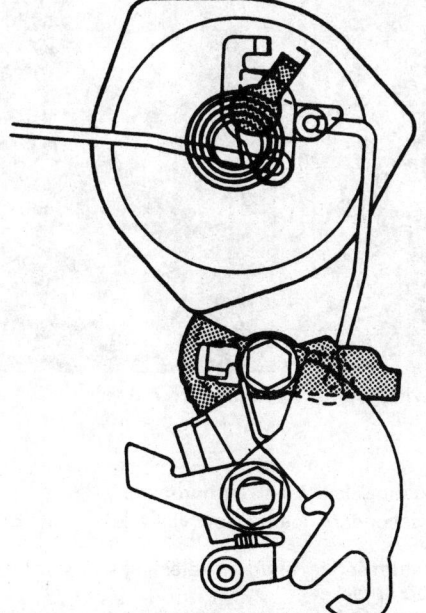

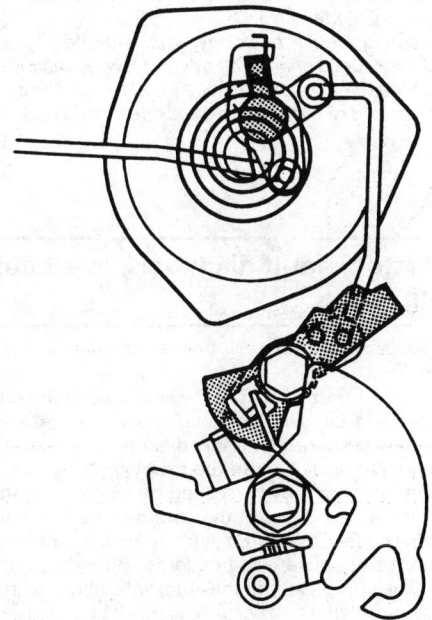

Fig. 4.3 Choke system and fast idle cams – Aisan carburettor (Sec 10)

steering incorporate an idle up system which increases the engine idle speed when the power steering pump is operating. Both systems are similar to those fitted to the earlier carburettors.

July 1987 onward B series engined models

The carburettor fitted to these models is identical to that fitted to the September 1985 onward E series engined models. However the throttle positioner system is only fitted to B5 engined models which are equipped with a manual gearbox.

A slight modification to the choke system is that the fast idle cam breaker diaphragm (FICB) is now controlled by a thermostatically operated valve which is fitted to the underside of the inlet manifold. The valve opens when the inlet manifold temperature reaches 50°C allowing the vacuum from the inlet manifold to reach the FICB diaphragm.

11 Carburettor – removal and refitting

Note: *Refer to the warning note in Section 1 before carrying out the following operation.*

Removal

1 Disconnect the battery negative lead.
2 Remove the air cleaner as described in Section 2.
3 Disconnect the choke and/or accelerator cables from the carburettor with reference to Sections 6 and 8 (as applicable).
4 On twin carburettor models disconnect the ball socket of the linkage connecting rods at the throttle levers on both shafts.
5 Disconnect the fuel inlet hose and plug its end to minimise fuel loss.
6 Disconnect the fuel cut-off valve solenoid wiring at the block connector.
7 On September 1985 models onward trace the wiring back from the bi-metal heater and disconnect it from the back (P terminal) of the alternator. Also disconnect the wiring from the PTC heater which is fitted between the carburettor and inlet manifold (photo).
8 Make a note of the correct fitted positions of all the relevant vacuum hoses, to use as a guide on refitting, then disconnect them from the carburettor(s).
9 Undo the nuts securing the carburettor(s) to the inlet manifold and lift off the carburettor(s). Remove the insulator block or PTC heater (as applicable) and gaskets.

Refitting

10 Refitting is the reverse of the removal sequence, bearing in mind the following points (photo).

   (a) Ensure that all traces of old gasket are removed from the carburettor and inlet manifold sealing faces.
   (b) Position a new gasket on either side of the insulator block or PTC heater on refitting.
   (c) Adjust the choke and/or accelerator cables as described in Sections 6 and 8 (as applicable).
   (d) On completion adjust the idle speed and mixture settings as described in Chapter 1.
   (e) On September 1985 models onwards, with the engine idling at the specified speed check that the fuel level is maintained at the centre of the window on the right-hand side of the carburettor body. If not, check the float height as described in Section 12.

12 Carburettor – fault diagnosis, overhaul and adjustments

Fault diagnosis

1 Faults with the carburettor are usually associated with dirt entering the float chamber and blocking the jets, causing a weak mixture or power failure within a certain engine speed range. If this is the case, then a thorough clean will normally cure the problem. If the carburettor is well worn, uneven running may be caused by air entering through the throttle valve spindle bearings. All the carburettors fitted to the pre-September 1985 models are fitted with manually-operated chokes which do not normally cause any problems. However, although not common, due to the complexity of the automatic choke system on the later carburettors, problems can occur which will be extremely difficult to trace. If a fault is suspected in the automatic choke system operation, the vehicle should be taken to a Mazda dealer for inspection.

Overhaul and adjustments

2 The following paragraphs describe cleaning and adjustment procedures which can be carried out by the home mechanic after the carburettor has been removed from the inlet manifold. If the carburettor is worn or damaged, it should either be renewed or overhauled by a specialist who will be able to restore the carburettor to its original calibration.

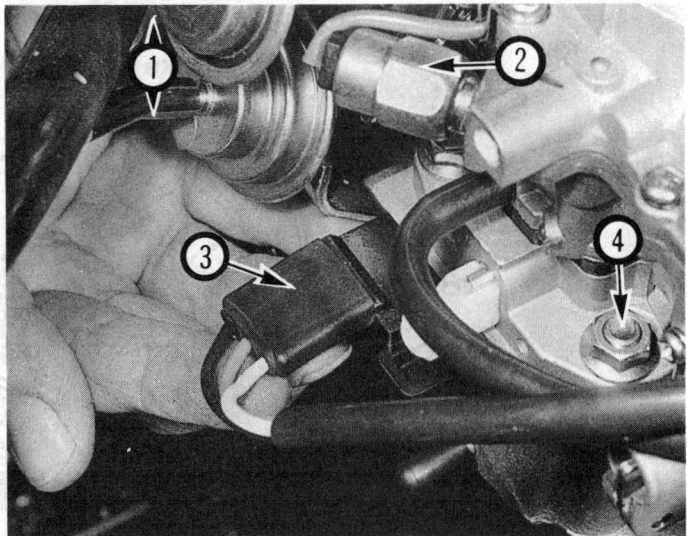

11.7 Aisan carburettor removal

1 Vacuum diaphragm hoses
2 Fuel cut-off valve solenoid
3 PTC heater wiring connector
4 Carburettor retaining nut

11.10 Fuel level viewing window – Aisan carburettor

Pre-September 1985 models (Hitachi carburettor)

3 Disconnect the secondary fuel cut-off valve lever rod from the throttle linkage.
4 Disconnect the return spring, undo the retaining screw and remove the choke cable support bracket.
5 Extract the split pin, and remove the small washers securing the choke vacuum diaphragm connecting rod to the choke valve linkage.
6 Disconnect the vacuum hose, undo the two screws and lift off the choke vacuum diaphragm and mounting bracket.
7 Undo the retaining screws securing the carburettor top cover to the main body. Carefully lift off the top cover, disengage the throttle linkage connecting rod and recover the gasket.
8 The various jets and carburettor components are shown in Fig. 4.4. Each component should be removed and identified for position, then the float chamber can be cleaned of any sediment. Clean the main body and the cover thoroughly with fuel, and blow through the carburettor internal channels and jets using air from an air line or foot pump. **Note:**

Fig. 4.4 Exploded view of the Hitachi carburettor (Sec 12)

1. Fuel cut-off valve lever pivot
2. Choke cable support bracket
3. Choke vacuum diaphragm
4. Top cover
5. Gasket
6. Accelerator pump piston
7. Rubber boot
8. Float and pivot pin
9. Fuel needle valve
10. Spring
11. Accelerator pump inlet ball
12. Accelerator pump discharge weight
13. Rubber boot
14. Secondary fuel cut-off valve cap, spring and ball
15. Fuel cut-off solenoid valve
16. Primary slow running jet
17. Primary main air bleed
18. Power jet
19. Secondary main air bleed
20. Secondary slow running jet
21. Main diaphragm
22. Main body
23. Gasket
24. Insulator block

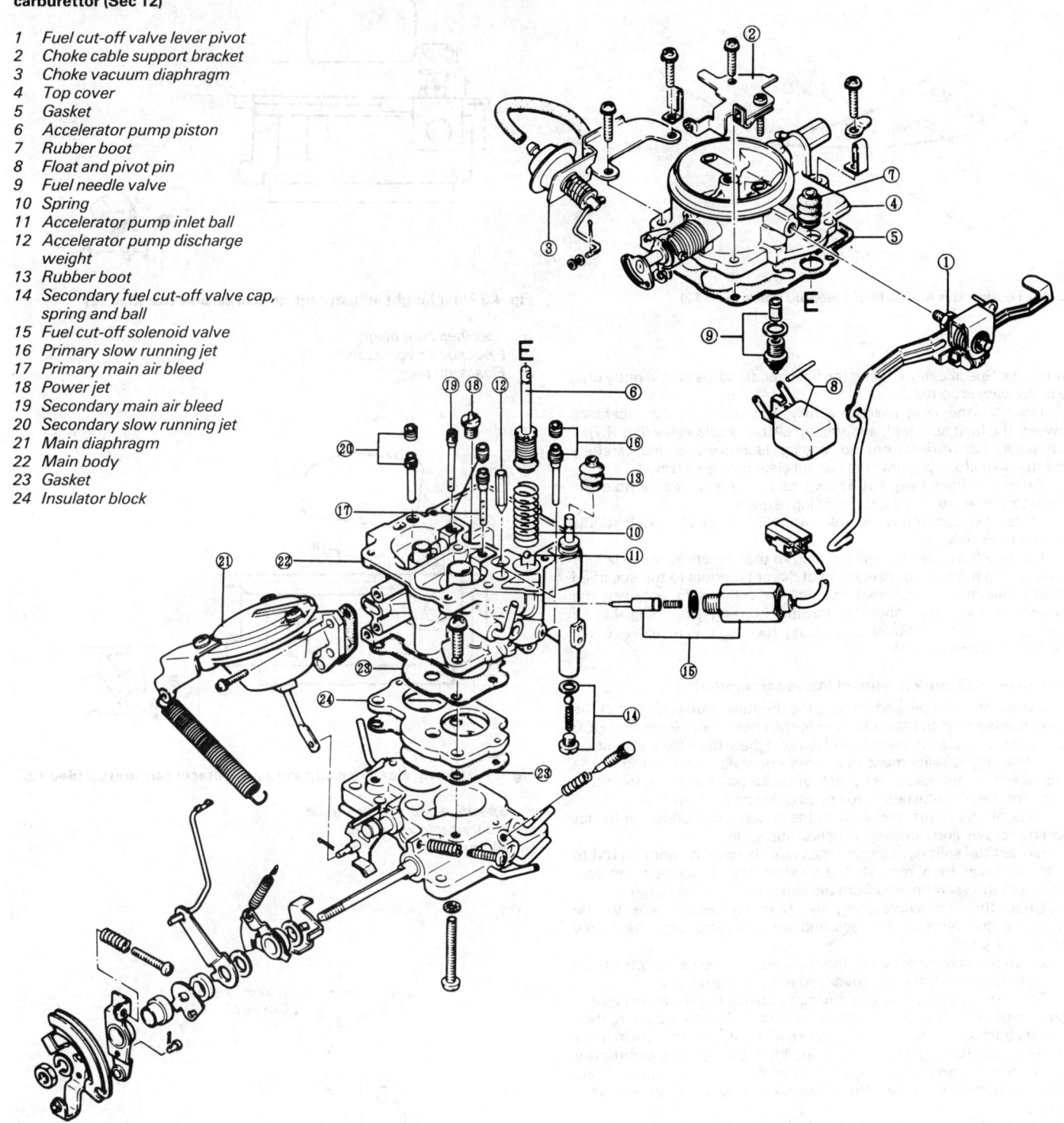

Do not probe the jet or carburettor channels with wire as they are easily enlarged.

9 Check the operation of the fuel cut-off solenoid valve by connecting it across a 12 volt battery as shown in Fig. 4.5. If the solenoid is functioning correctly, the plunger should be pulled into the valve body when the voltage is applied and return when the voltage is disconnected. If this is not the case renew the solenoid.

10 Before refitting the top cover check the float height adjustment as follows.

11 With the float and fuel valve in place, invert the top cover and allow the float to close the needle valve under its own weight. Measure the distance between the float upper face and top cover sealing surface. This measurement must be taken without the top cover gasket in place (Fig. 4.6). If this measurement differs from the float height dimension

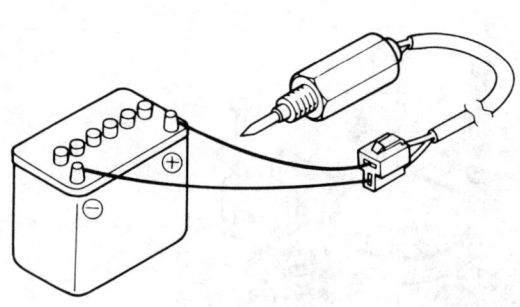

Fig. 4.5 Testing the fuel cut-off solenoid valve (Sec 12)

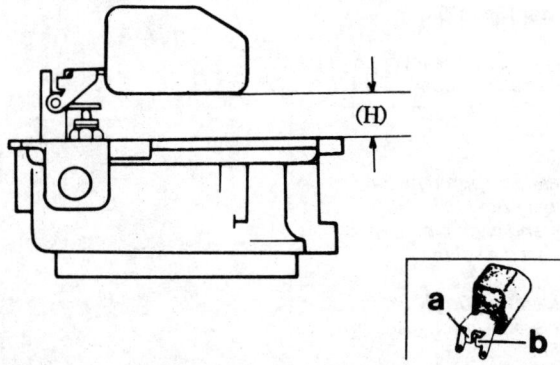

Fig. 4.6 Float height adjustment Hitachi carburettor (Sec 12)

H Specified float height
a Float needle valve tang
b Float stop tang

given in the Specifications, bend the float needle valve tang slightly until the dimension is correct.

12 Now lift the float assembly fully and measure the clearance between the float seat tang and the top of the needle valve (Fig. 4.7). If the measurement differs from the opening clearance specified, carefully bend the float stop tang until the correct clearance is obtained.

13 Once the float height is known to be correct, reassemble the carburettor by reversing the dismantling sequence.

14 Once the carburettor is fully assembled check the fast idle adjustment as follows.

15 Operate the choke linkage by hand so that the choke valve is fully closed. Using a drill bit or gauge rod of diameter equal to the specified fast idle clearance, check that the drill bit will just fit between the primary throttle valve and the throttle housing bore (Fig. 4.8). If necessary bend the linkage connecting rod slightly to achieve the specified clearance.

September 1985 models onward (Aisan carburettor)

16 Extract the split pin and remove the washers from each end of the rods which connect the fast idle cam to the choke valve lever and FICB diaphragm. Disconnect the rods and remove them from the carburettor.

17 If the original alignment marks are not visible on the top of the bi-metal heater and case, use a dab of white paint to mark the fitted position of the bi-metal heater to use as guide on reassembly.

18 Unhook the return spring from the choke lever, undo the heater retaining screws and remove the heater and spring.

19 Extract the split pin securing the choke breaker diaphragm rod to the choke lever then remove the washer and disconnect the rod. Disconnect the vacuum pipe from the choke breaker diaphragm.

20 Undo the screws securing the bi-metal heater case to the carburettor and remove the case and vacuum diaphragm assembly from the carburettor.

21 Undo the screws securing the top cover to the carburettor main body. Carefully lift off the top cover and remove the gasket.

22 The various jets and carburettor components are shown in Fig. 4.9. Each component should be removed and identified for position, then the float chamber can be cleaned of any sediment. Clean the main body and the cover thoroughly with fuel, and blow through the carburettor internal channels and jets using air from an air line or foot pump. Note: *Do not probe the jet or carburettor channels with wire as they are easily enlarged.*

23 Check the operation of the fuel cut-off solenoid valve by connecting a 12 volt battery across its block connector terminals (photo). If the solenoid is functioning correctly, the plunger should be pulled into the valve body when the voltage is applied and return when the voltage is disconnected. If this is not the case renew the solenoid.

24 On carburettors fitted to B3 and B5 engines, check the operation of the open vent solenoid valve. Connect the battery positive (+) terminal to the solenoid wire and connect the negative (−) to the solenoid body. With the voltage applied it should be possible to blow through the solenoid union which is normally connected to the inlet manifold, but not through the breather pipe, and vice versa when the voltage is disconnected. If this is not the case renew the solenoid.

25 Before refitting the top cover check the float height as follows.

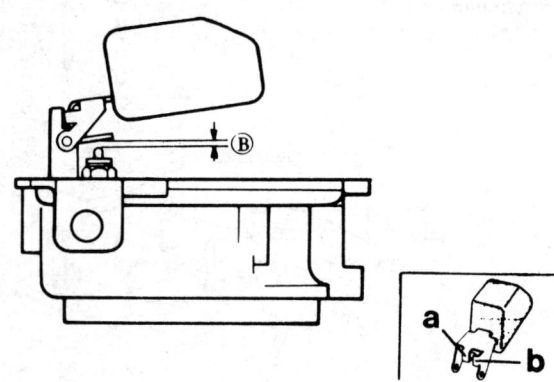

Fig. 4.7 Opening clearance adjustment – Hitachi carburettor (Sec 12)

B Specified opening clearance
a Float needle valve tang
b Float stop tang

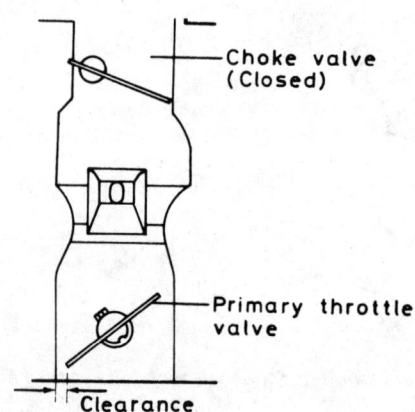

Fig. 4.8 Fast idle clearance checking point – Hitachi carburettor (Sec 12)

Chapter 4 Part A: Carburettor engines

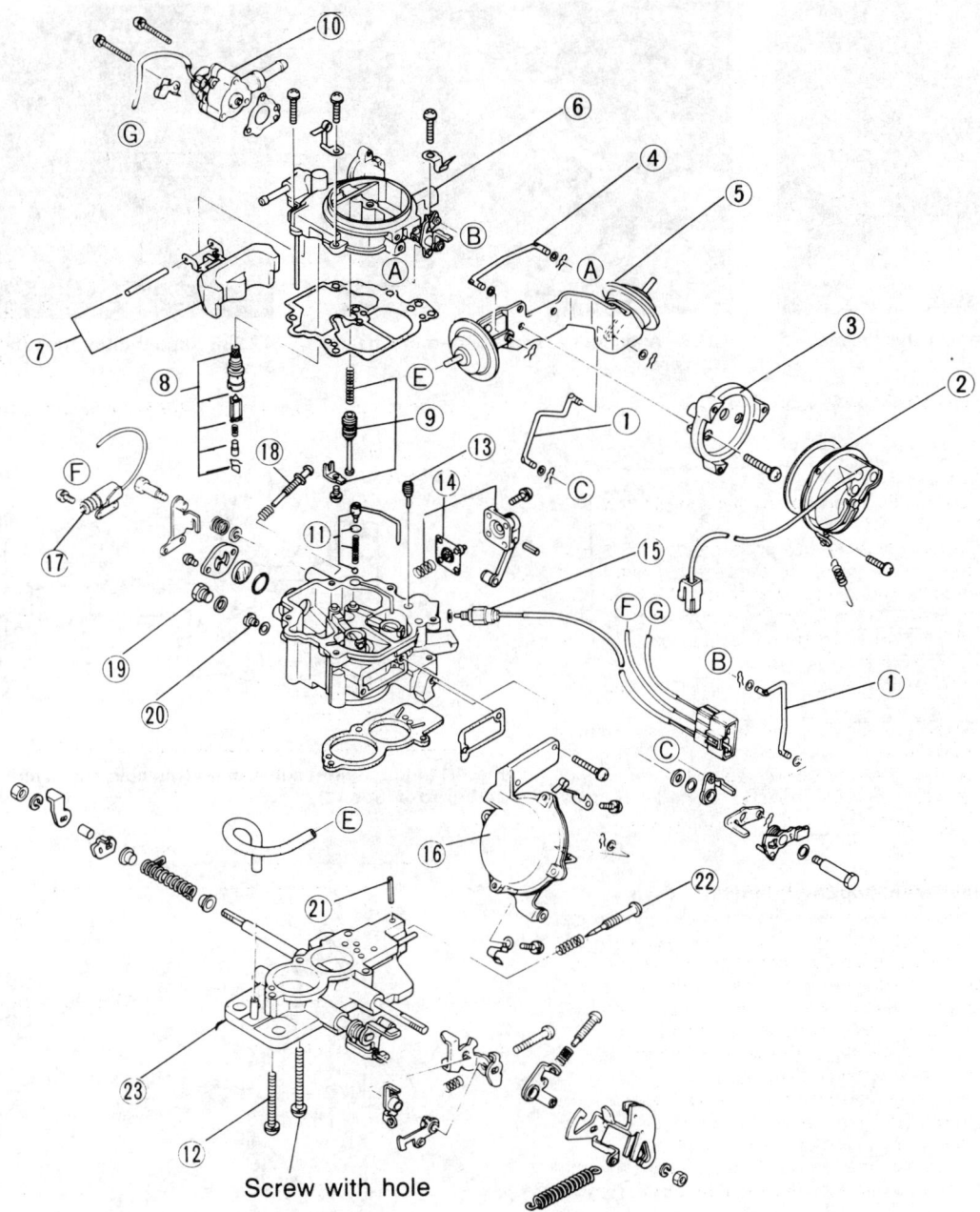

Fig. 4.9 Exploded view of the Aisan carburettor (Sec 12)

1. Rod
2. Bi-metal and heater
3. Bracket
4. Rod
5. Diaphragm and bracket
6. Top cover
7. Float
8. Needle valve
9. Vacuum piston
10. Open vent solenoid valve (where fitted)
11. Pump nozzle and check ball
12. Screw
13. Primary slow jet
14. Accelerator pump
15. Fuel cut-off valve solenoid
16. Secondary diaphragm
17. Idle switch (where fitted)
18. Throttle adjust screw
19. Plug
20. Main jet
21. Spring pin (where fitted)
22. Mixture adjust screw
23. Throttle valve housing

Letters indicate connections between components

26 Hold the top cover horizontally, with the float assembly on the bottom. Measure the distance between the bottom of the float assembly and the sealing surface of the top cover noting that, if necessary, the gasket must first be removed (Fig. 4.10). If this distance (L) is not within the limits given in the Specifications, carefully bend the float stop tang (A) until the correct height is achieved.

27 Invert the top cover and let the float pivot downwards under its own weight. Measure the distance (H) between the top cover sealing face and the float at the point shown in Fig. 4.11, and check that it is as given in the Specifications. If adjustment is necessary, carefully bend the float needle valve tang until the specified distance is obtained.

28 Once the float level is correct, reassemble the carburettor by reversing the dismantling sequence. When fitting the bi-metal heater ensure the angled tang of the heater engages with the choke lever. Align the marks and tighten the bi-metal heater retaining screws securely (photos). Carry out the following adjustments before refitting the carburettor.

29 Connect a vacuum pump and gauge to the choke breaker

12.23 Fuel cut-off solenoid valve location

12.28A Align index marks of bi-metal heater and case...

12.28B ...and tighten the retaining screws securely

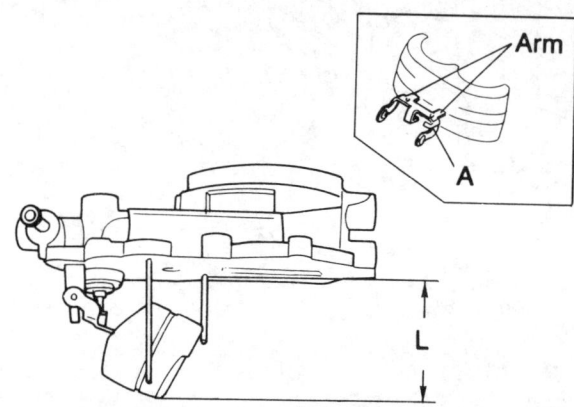

Fig. 4.10 Float height adjustment (top cover upright) – Aisan carburettor (Sec 12)

A Float stop tang
L Specified float height

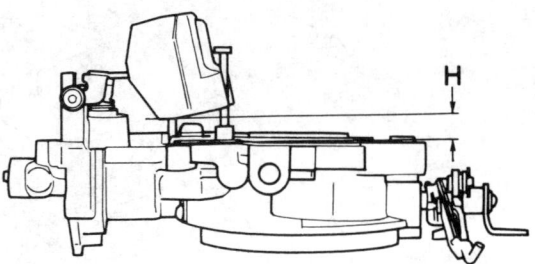

Fig. 4.11 Float height adjustment (top cover inverted) – Aisan carburettor (Sec 12)

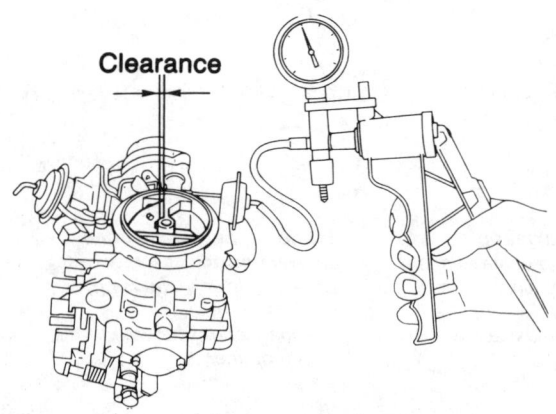

Fig. 4.12 Checking the choke breaker diaphragm clearance – Aisan carburettor (Sec 12)

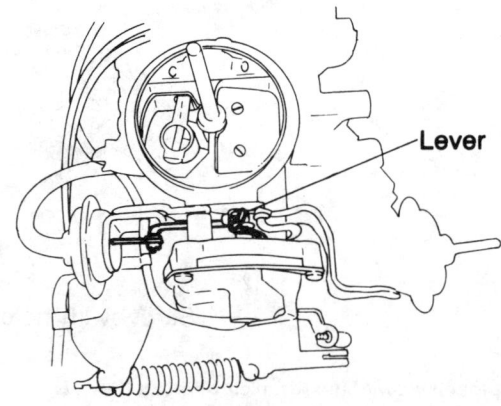

Fig. 4.13 Choke breaker diaphragm adjustment lever – Aisan carburettor (Sec 12)

diaphragm as shown in Fig. 4.12. Check that the choke valve is shut, then apply a vacuum of 400 mm Hg (15.7 in Hg). Insert a gauge rod, or drill, with a diameter equal to the specified clearance, between the choke valve and bore. If the clearance is not as specified, carefully bend the lever (Fig. 4.13) using a pair of pointed nose pliers. **Note:** *If access to a vacuum pump and gauge cannot be gained, choke breaker diaphragm adjustment should be entrusted to a Mazda dealer.*

30 Move the fast idle cam onto the third step position, then check the primary throttle-to-carburettor bore clearance using a drill or gauge rod of the diameter of the specified clearance. If adjustment is necessary, turn the adjuster screw as required (Fig. 4.14).
31 Move the fast idle cam onto the second step position and measure the specified choke valve-to-bore clearance (Fig. 4.15). If adjustment is necessary, carefully bend the operating lever to suit.
32 Fully open the primary throttle valve and measure the choke valve clearance (Fig. 4.16). The clearance should be as specified.
33 Open the primary throttle valve whilst observing the secondary throttle valve movement. When the primary throttle valve is opened

Chapter 4 Part A: Carburettor engines

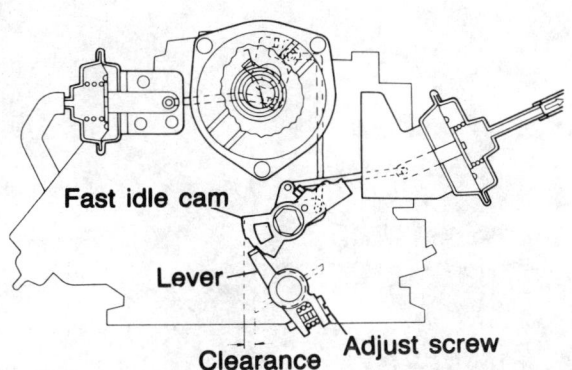

Fig. 4.14 Checking the primary throttle-to-carburettor bore clearance (fast idle cam on third step) – Aisan carburettor (Sec 12)

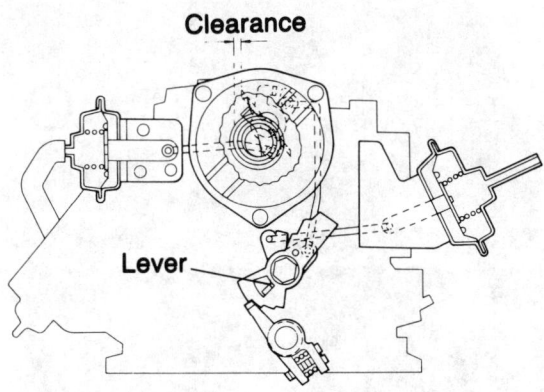

Fig. 4.15 Checking the choke valve-to-bore clearance (fast idle cam on second step) – Aisan carburettor (Sec 12)

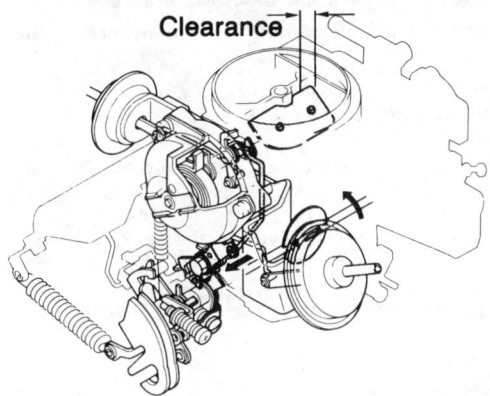

Fig. 4.16 Checking the choke valve clearance – Aisan carburettor (Sec 12)

approximately 47° to 53°, the secondary throttle valve should just start to open. Hold the primary throttle valve in this position and measure the primary throttle valve-to-bore clearance (Fig. 4.17) using a gauge rod or drill with the same diameter as the specified clearance. If adjustment is necessary, bend the tab to suit using a pair of pointed nose pliers.

13 Carburettor – on-car adjustments

Hitachi carburettor

Throttle positioner system

Note: *Full adjustment of the throttle positioner system requires the use of a vacuum gauge. If access to a suitable gauge cannot be gained, have the system adjusted by a Mazda dealer.*

1 On all models equipped with a manual gearbox a throttle positioner system is fitted to the carburettor. This system prevents the engine stalling and reduces harmful exhaust emissions by preventing afterburning. The main components are a vacuum diaphragm unit, which varies the position of the throttle valve when the throttle is closed, and a vacuum valve, which is connected to the inlet manifold and controls the diaphragm unit.
2 Prior to adjustment, ensure that the idle speed and CO content are correct as described in Chapter 1. Leave the tachometer connected.
3 With the engine stopped, disconnect the manifold supply vacuum hose at the T-piece connector, and the small vacuum hose from the T-piece to the throttle positioner servo diaphragm, at the diaphragm. Attach the manifold supply vacuum hose directly to the diaphragm.
4 Start the engine and check that the idle speed is now 1400 to 1600 rpm. If adjustment is necessary, turn the adjusting screw on the throttle linkage as necessary (photo).

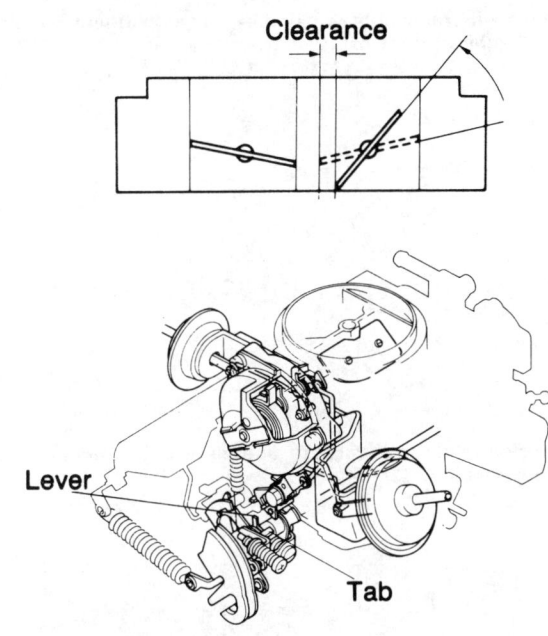

Fig. 4.17 Checking the secondary throttle valve opening clearance – Aisan carburettor (Sec 12)

5 Switch off the engine and reconnect the vacuum hoses to their original positions.
6 Using a length of suitable pipe, connect the vacuum gauge into the vacuum supply hose between the manifold and vacuum control valve.
7 Start the engine, increase the engine speed to 3000 rpm and quickly release the accelerator. Observe the reading on the vacuum gauge which should be initially high as the accelerator is released, then fall to approximately 77.3 to 79.9 kPa (22.8 to 23.6 in Hg). This reading should be maintained for one to two seconds and then fall to a slightly lower reading which will be steady as the engine idles. The critical period is the one to two second intermediate reading and if this is not as specified adjust the vacuum control valve as follows.
8 Remove the rubber cap from the centre of the valve and turn the adjusting screw as necessary while carrying out the foregoing test. Turning the adjusting screw clockwise will decrease the vacuum, and turning it anti-clockwise will increase the vacuum.
9 Once the correct vacuum reading is obtained, remove the instruments and reconnect the vacuum hoses to their original positions.

Choke vacuum diaphragm

10 Remove the air cleaner as described in Section 2.
11 Start the engine and check that the vacuum diaphragm plunger is

Chapter 4 Part A: Carburettor engines

13.4 Throttle positioner diaphragm unit (A) and adjustment screw (B) – Hitachi carburettor

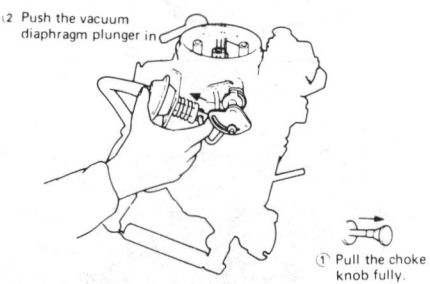

Fig. 4.18 Checking the choke vacuum diaphragm adjustment – Hitachi carburettor (Sec 13)

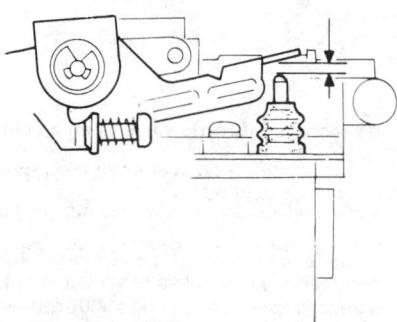

Fig. 4.19 Checking the secondary fuel cut-off valve plunger clearance – Hitachi carburettor (Sec 13)

pulled fully in with the engine idling. If not, check the condition of the vacuum hoses. If vacuum is present at the hose to diaphragm connection, then the diaphragm unit is faulty and must be renewed. Adjustment is as follows.
12 With the engine switched off, pull the choke knob fully out.
13 Push the vacuum diaphragm plunger in with your finger and, using feeler gauges, check the clearance between the choke flap and the carburettor bore (Fig. 4.18). If the clearance is not as given in the Specifications, carefully bend the connecting rod until the correct dimension is obtained.
14 Refit the air cleaner assembly as described in Section 2.

13.17 Secondary fuel cut-off system adjustment screw – Hitachi carburettor

Secondary fuel cut-off system

15 Remove the air cleaner as described in Section 2.
16 With the throttle linkage at rest, measure the clearance between the operating arm and the fuel cut-off valve plunger (Fig. 4.19) using feeler gauges.
17 If the clearance is not as given in the Specifications, turn the adjusting screw as necessary until the specified clearance is obtained (photo).
18 Refit the air cleaner assembly as described in Section 2.

Aisan carburettor

Throttle positioner system

19 A throttle positioner system is fitted to the carburettors of all B5 and E series engined models which are fitted with a manual gearbox. The system prevents engine stalling and reduces harmful exhaust emissions by reducing afterburning. The system consists of a vacuum diaphragm unit, a three-way solenoid valve and a control unit. Below approximately 1400 rpm, the solenoid valve supplies the vacuum from the inlet manifold to the throttle positioner diaphragm unit, which then opens the primary throttle valve by a small amount. When the engine speed increases, the control unit actuates the three-way solenoid valve which then cuts the vacuum supply to the throttle positioner diaphragm and returns the throttle valve to its original position.
20 On models equipped with power steering, an idle up system is also incorporated into the throttle positioner system. This system prevents the engine stalling when the power steering pump is operating. Every time the pump operates, a pressure switch actuates the throttle positioner three-way solenoid valve via a relay.
21 The throttle positioner system is adjusted as described in Section 3 of Chapter 1.

14 Shutter valve control system (B3 and B5 models) – general information and testing

General information

1 On B3 and B5 engined models a shutter valve control system is fitted to improve combustion efficiency and decrease engine noise. The system consists of the shutter valve body assembly, which is positioned between the inlet manifold and cylinder head, shutter valve vacuum diaphragm unit, two three-way solenoid valves, distributor vacuum diaphragm unit and an electrical control unit.
2 When the engine speed reaches 2500 rpm, the control unit actuates the three-way solenoid valves for the shutter valve diaphragm and distributor diaphragm unit. With this operation, vacuum to the shutter valve diaphragm is cut and vacuum is applied to the distributor vacuum

Chapter 4 Part A: Carburettor engines

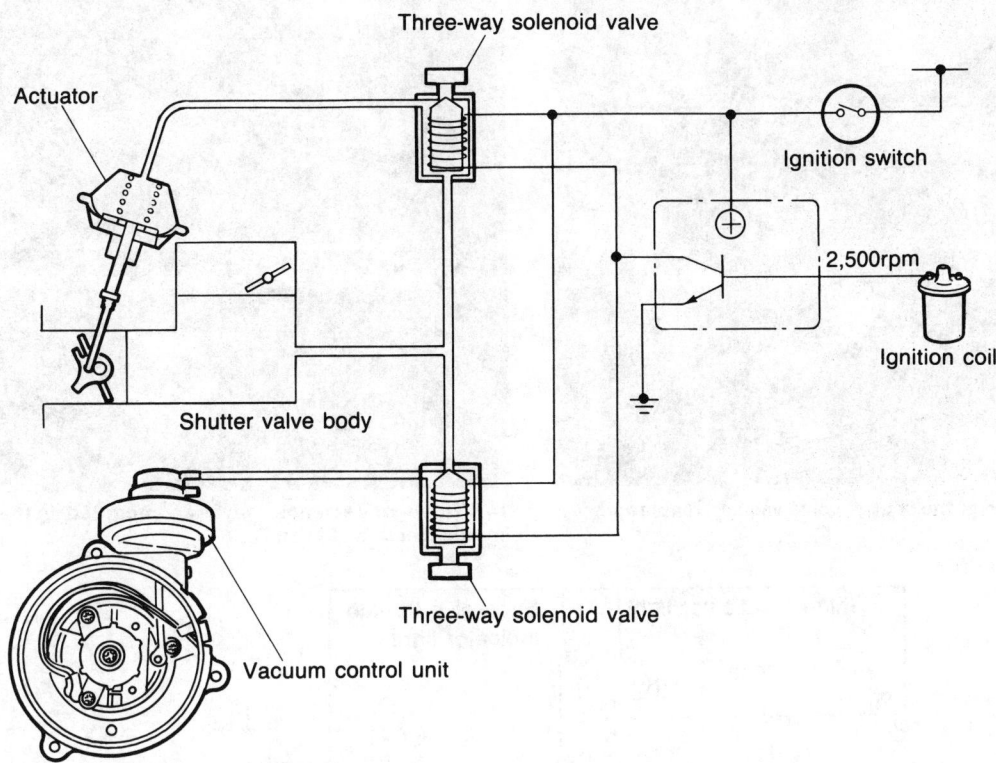

Fig. 4.20 Shutter valve control system components – B3 and B5 engined models (Sec 14)

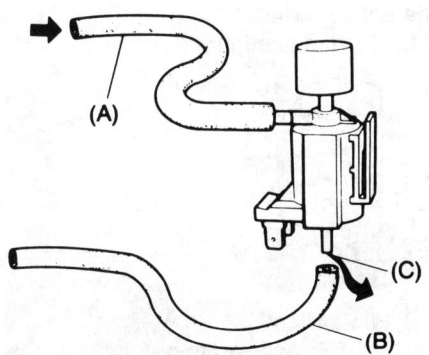

Fig. 4.21 Shutter valve diaphragm three-way solenoid valve operation (Sec 14)

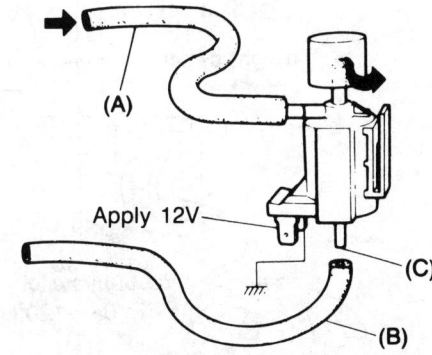

Fig. 4.22 Shutter valve diaphragm three-way solenoid valve operation with battery voltage applied (Sec 14)

unit. This then opens the shutter valve assembly and advances the ignition timing.

Testing

3 Start the engine and check that the shutter valve vacuum diaphragm pullrod is pulled fully in with the engine idling (photo). Increase the engine speed to over 2500 rpm and check that the pullrod is released and extends fully. If not, check the condition of the vacuum hoses for signs of cracking or splitting. Disconnect the vacuum hose from the shutter valve diaphragm and check that a vacuum is present in the hose. Increase the engine speed to over 2500 rpm and check that the vacuum disappears. If vacuum is present at the hose to diaphragm connection, then disappears at 2500 rpm, the diaphragm unit is faulty and must be renewed. **Note:** *On rare occasions the fault might be due to a sticky shutter valve mechanism, if this is the case, remove the assembly for inspection as described in Section 16.* If there is a fault in the vacuum supply check the control system components as follows.

Shutter valve diaphragm three-way solenoid valve

4 Using a voltmeter connect the meter positive (+) lead to the

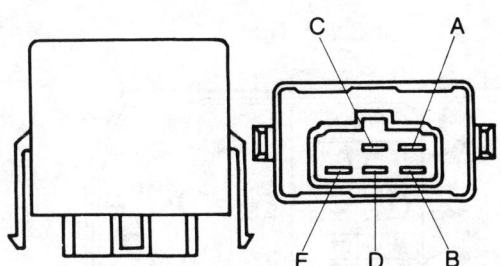

Fig. 4.23 Checking the control unit (Sec 14) Terminal Voltage

A Below 1.5V at below 1500 rpm Approx. 12V at above 1500 rpm
B 12 to 14V
C 0V
D Below 1.5V at below 2500 rpm Approx. 12V at above 2500 rpm
F 12 to 14V

144 Chapter 4 Part A: Carburettor engines

14.3 Check the operation of the shutter valve vacuum diaphragm pullrod (arrowed)

14.5 Three-way solenoid valves are mounted on the engine compartment bulkhead

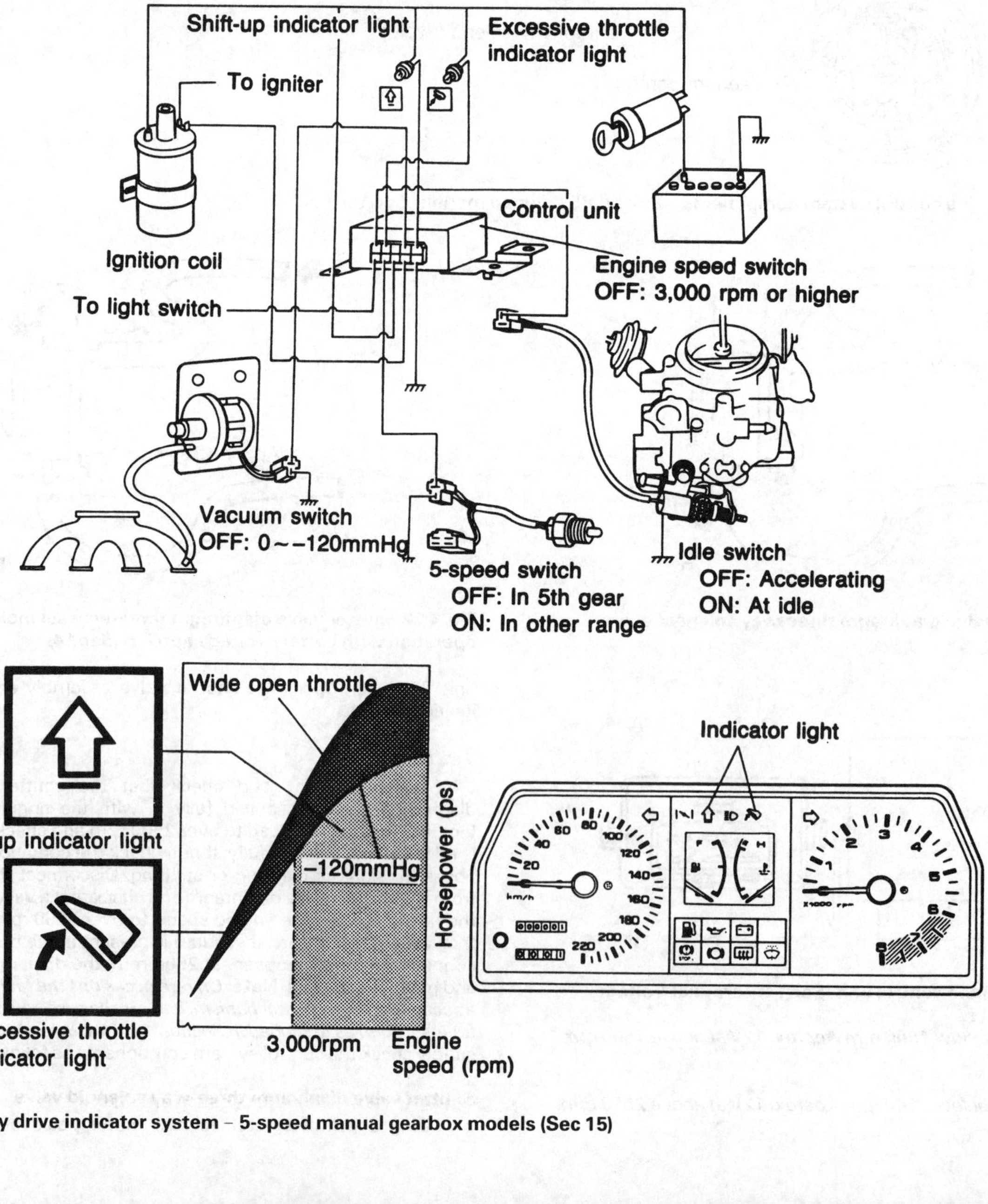

Fig. 4.24 Economy drive indicator system – 5-speed manual gearbox models (Sec 15)

Chapter 4 Part A: Carburettor engines 145

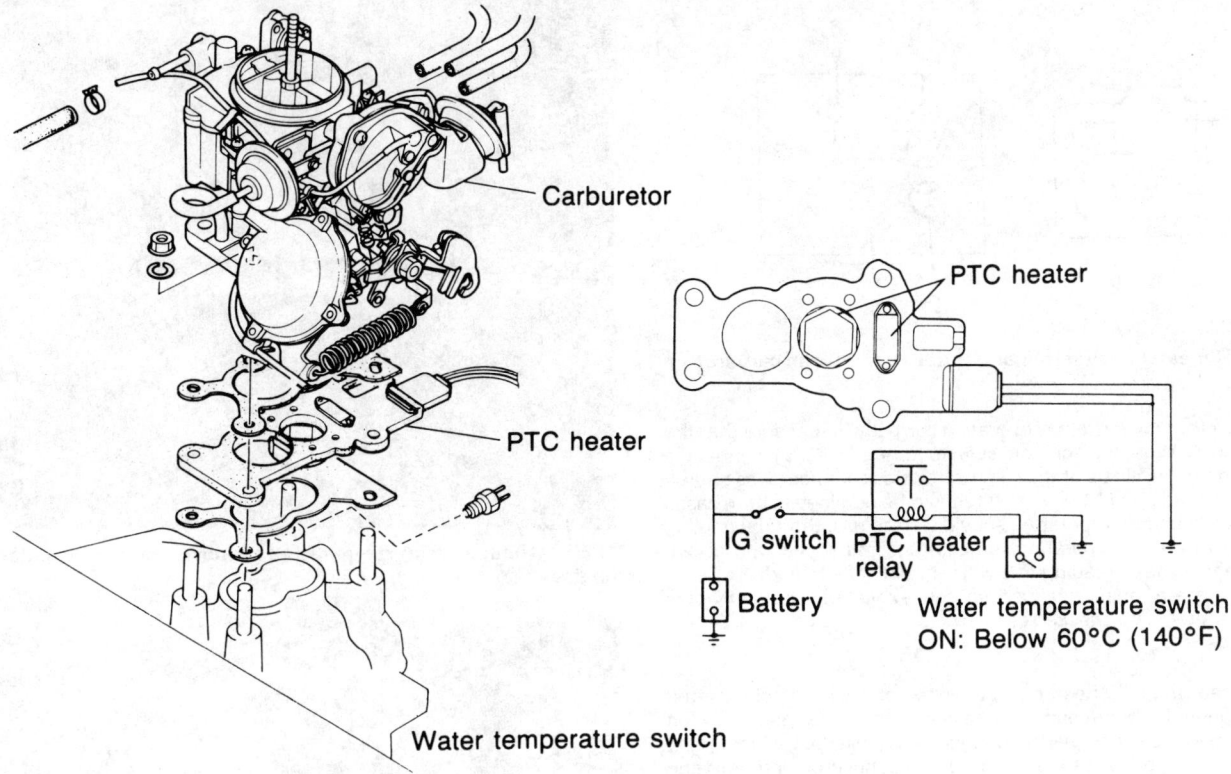

Fig. 4.25 PTC heater system – Aisan carburettor (Sec 16)

black/white terminal of the solenoid and the negative (–) lead to earth. Start the engine and allow it to idle then slowly increase the engine speed to above 2500 rpm, noting the readings obtained. Below 2500 rpm a reading of approximately 1.5 volts should be obtained, and above 2500 rpm a reading of 12 volts. If the specified readings are not obtained, first check the condition of all relevant wiring and connectors, then check the operation of the control unit as described below. If the correct voltages are obtained check the operation of the solenoid valve as follows.

5 Disconnect the hose from the shutter valve vacuum diaphragm unit and trace it back to the vacuum solenoid which is mounted on the engine compartment bulkhead (photo). Disconnect the other hose from the end of the solenoid, then disconnect the electrical connections and remove the solenoid from the car. Blow through the hose which was disconnected from the vacuum unit (A in Fig. 4.21) and check that the air comes out of the other solenoid hose union (C). Then, using a 12 volt battery and two auxiliary wires, connect the battery across the solenoid terminals and blow through the hose (A) again. The air should now be expelled through the solenoid exhaust port (Fig. 4.22). If not the solenoid is faulty and must be renewed.

Control unit

6 The control unit is situated inside the car where it is mounted on the left-hand side of the heater matrix assembly. To gain access to the unit remove the lower left-hand facia panel and peel back the carpet.
7 Warm the engine up to normal operating temperature. With the engine running check the voltages between the various control unit terminals and earth referring to Fig. 4.23 for terminal identification. If the readings obtained do not match those in the accompanying table the control unit is faulty and must be renewed.

15 Economy drive indicator system – general information

1 This system is fitted to September 1985 onward five-speed manual gearbox models. The system is designed to improve the fuel consumption by warning the driver when he is using excessive throttle,

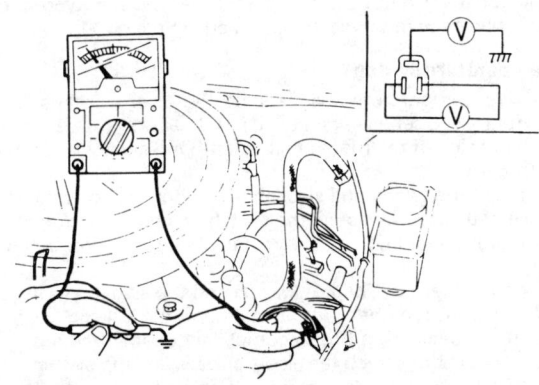

Fig. 4.26 Testing the PTC heater system – Aisan carburettor (Sec 16)

and when the car is being driven in an inappropriate gear via two warning lamps in the instrument panel. Fig. 4.24 shows the system components and layout.
2 As the throttle is pressed, the vacuum in the inlet manifold is reduced and when it drops to 120 mm Hg (4.72 in Hg), the warning lamp comes on.
3 A 'shift-up' indicator lamp is also fitted and this illuminates when the engine speed exceeds 3000 rpm in each gear (except fifth). When the engine is decelerating the idle switch is activated and the 'shift-up' lamp goes out.
4 If the system malfunctions the vehicle must be taken to a Mazda dealer for fault diagnosis and repair.

16 PTC heater system (Aisan carburettor) – testing

1 The system comprises of the PTC heater (situated between the carburettor and inlet manifold), a relay and a water temperature switch which is situated on the inlet manifold.

146 Chapter 4 Part A: Carburettor engines

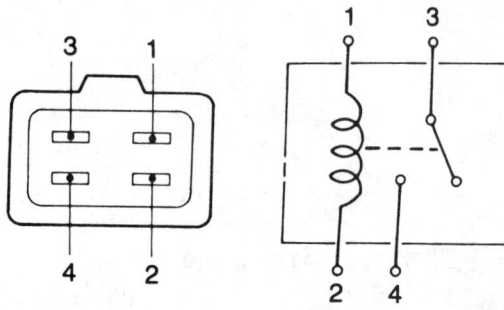

Fig. 4.27 PTC heater relay terminal identification – Aisan carburettor (Sec 16)

2 To check that the system is operating correctly first ensure that the engine is cold. With the ignition switch turned to 'ON', connect a voltmeter to the terminals of the PTC heater block connector as shown in Fig. 4.26. A reading of 11 to 12 volts should be obtained in both cases and the throttle body temperature should start to rise. Start the engine, warm it up to normal operating temperature and repeat the above checks. In both cases a reading of 0 volts should be obtained.
3 If the system does not perform as expected, the individual components should be checked as follows.

PTC heater
4 Disconnect the PTC heater block connector. Using an ohmmeter, check for continuity between the three heater terminals, and between the heater terminals and earth. Continuity should be present only between the two parallel (+) terminals of the connector, and between the single (-) terminal and earth. If this is not the case the PTC heater is faulty and must be renewed. The heater can be renewed once the carburettor has been removed as described in Section 11.

Water temperature switch
5 The water temperature switch is located on the underside of the inlet manifold. On E series engines the PTC heater switch is the one on the right-hand side of the manifold (the left-hand switch is for the FICB system) (photo).
6 Disconnect the wiring and slacken the switch until its retained only by a couple of threads. Unscrew the switch, then quickly remove it and plug the inlet manifold hole to prevent coolant loss. Mop up any spilt coolant.
7 Referring to Fig. 3.8 (Chapter 3), suspend the switch in a saucepan, or suitable vessel, together with a thermometer. Fill the vessel with water so that the switch probe is completely submerged. Connect an ohmmeter across the switch terminals or between the switch terminal and the switch body (as appropriate). Continuity should be present indicating that the switch is on.
8 Slowly heat the water whilst noting the ohmmeter reading. When the water temperature reaches 60°C an open circuit should be present indicating that the switch is off. Carry on heating the water until the temperature is well above 60°C then allow the water to cool. Ensure that the switch turns on again when the temperature falls to 60°C. If the switch does not perform as expected, it is faulty and must be renewed.
9 Prior to fitting, wrap a suitable sealing tape around the switch threads.
10 Remove the plug from the inlet manifold and quickly screw in the switch. Tighten the switch securely and connect the switch wiring.
11 Check, and if necessary top up, the coolant level as described in Chapter 1.

PTC heater relay
12 The PTC heater relay is located in the front left-hand corner of the engine compartment, directly below the horn and cooling fan relays (photo).
13 Referring to Fig. 4.27 for terminal identification, connect the positive (+) terminal of a 12 volt battery to the number 1 relay terminal and the negative (-) terminal to the number 2 relay terminal. Using an ohmmeter, check for continuity between terminals 3 and 4 of the relay. With the voltage applied continuity should be present, and when the voltage is disconnected there should be an open circuit. If this is not the case the relay is faulty and must be renewed.

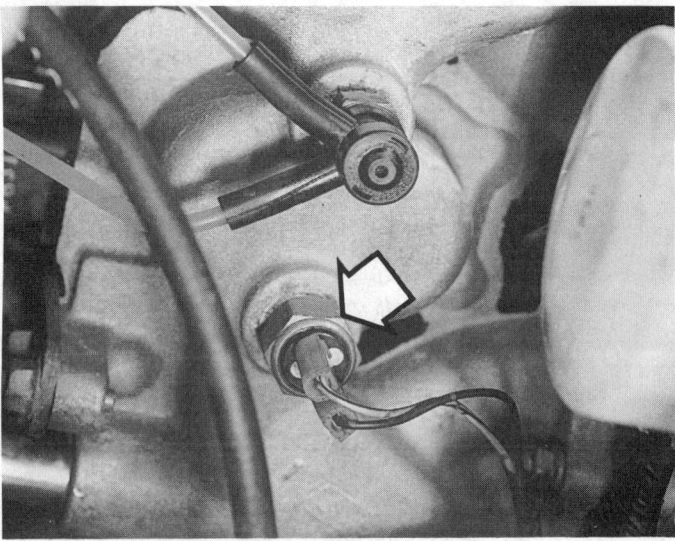

16.5 PTC heater water temperature switch (arrowed) – B series engine

16.12 PTC heater relay location (arrowed)

17 Inlet manifold – removal and refitting

Removal
1 Drain the cooling system as described in Chapter 1.
2 Remove the carburettor as described in Section 11.
3 Disconnect the brake servo vacuum hose and the water hose(s) from the inlet manifold. Make a note of the position of any remaining vacuum hoses which will impede inlet manifold removal and disconnect them from the manifold.
4 On September 1985 models onward disconnect the wiring from the switch(es) situated on the underside of the manifold.
5 Undo the nuts and bolts securing the inlet manifold to the cylinder head and withdraw the manifold and gasket off the studs.
6 On B3 and B5 engines also remove the shutter valve body assembly and gasket from the cylinder head.

Refitting
7 Remove all traces of old gasket from the cylinder head, manifold and shutter valve body sealing faces (as applicable).
8 On B3 and B5 engines fit a new gasket over the studs and refit the shutter valve body assembly (photos).
9 Fit a new inlet manifold gasket and install the inlet manifold (photo).

Chapter 4 Part A: Carburettor engines

17.8A On B3 and B5 engined models fit a new gasket over the inlet manifold studs...

17.8B ...then refit the shutter valve assembly...

17.9 ...followed by a new inlet manifold gasket

17.10 Reconnect the brake servo vacuum hose (arrowed) and secure in position with its retaining clip

Refit the manifold retaining nuts and bolts and tighten them evenly and progressively to the specified torque.
10   Reconnect the vacuum hoses, water hose(s) and the brake servo vacuum hose (as applicable) to their original positions on the inlet manifold. Ensure that all hoses are securely fastened by their retaining clips (photo).
11   Refit the carburettor as described in Section 11.
12   Refill the cooling system as described in Chapter 1.

18 Exhaust manifold – removal and refitting

Removal

1   Firmly apply the handbrake, then jack up the front of the car and support it on axle stands.
2   From underneath the car, undo the nuts securing the exhaust front pipe to the manifold and the bolt securing the front pipe to the support bracket (photos). Carefully separate the flange joint and recover the gasket.
3   From within the engine compartment, remove the duct joining the air cleaner assembly to the exhaust manifold stove.
4   Undo the four bolts and remove the stove from the manifold (photo).
5   Undo the nuts and bolts securing the manifold to the cylinder head. Note the location of the bypass pipe support bracket on the exhaust manifold stud and carefully ease the bracket off the stud (photo).
6   Withdraw the manifold and gasket from the cylinder head (photos).

Refitting

7   Refitting is the reverse of the removal sequence. Ensure that the manifold and cylinder head sealing faces are clean and always use new gaskets. Tighten the manifold nuts and bolts evenly and progressively to the specified torque.

19 Exhaust system – general information, removal and refitting

General information

1   On all pre-September 1985 models except 1500 GT models, the exhaust system consists of five sections, the front pipe, the front

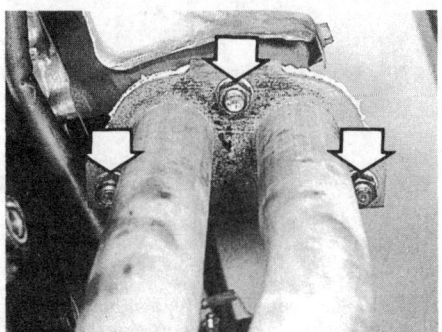

18.2A Undo the exhaust front pipe to manifold retaining nuts (arrowed)...

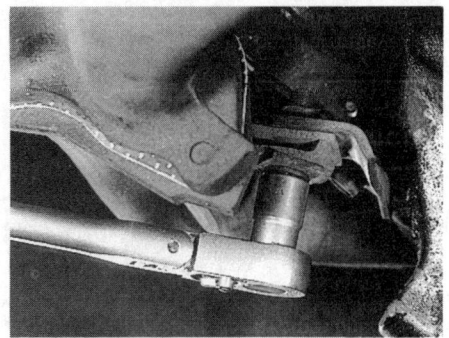

18.2B ...and the front pipe to mounting bracket bolt

18.4 Removing the exhaust manifold stove

Chapter 4 Part A: Carburettor engines

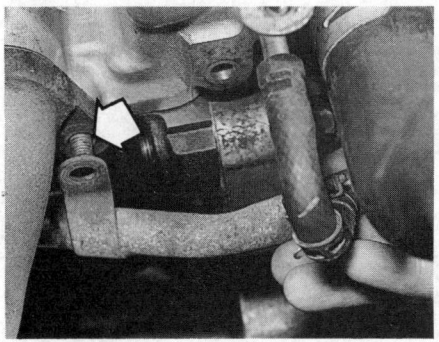

18.5 Remove the manifold nut and ease the bypass pipe support bracket off the stud (arrowed)

18.6A Undo the nuts and bolts and remove the exhaust manifold...

18.6B ...and gasket

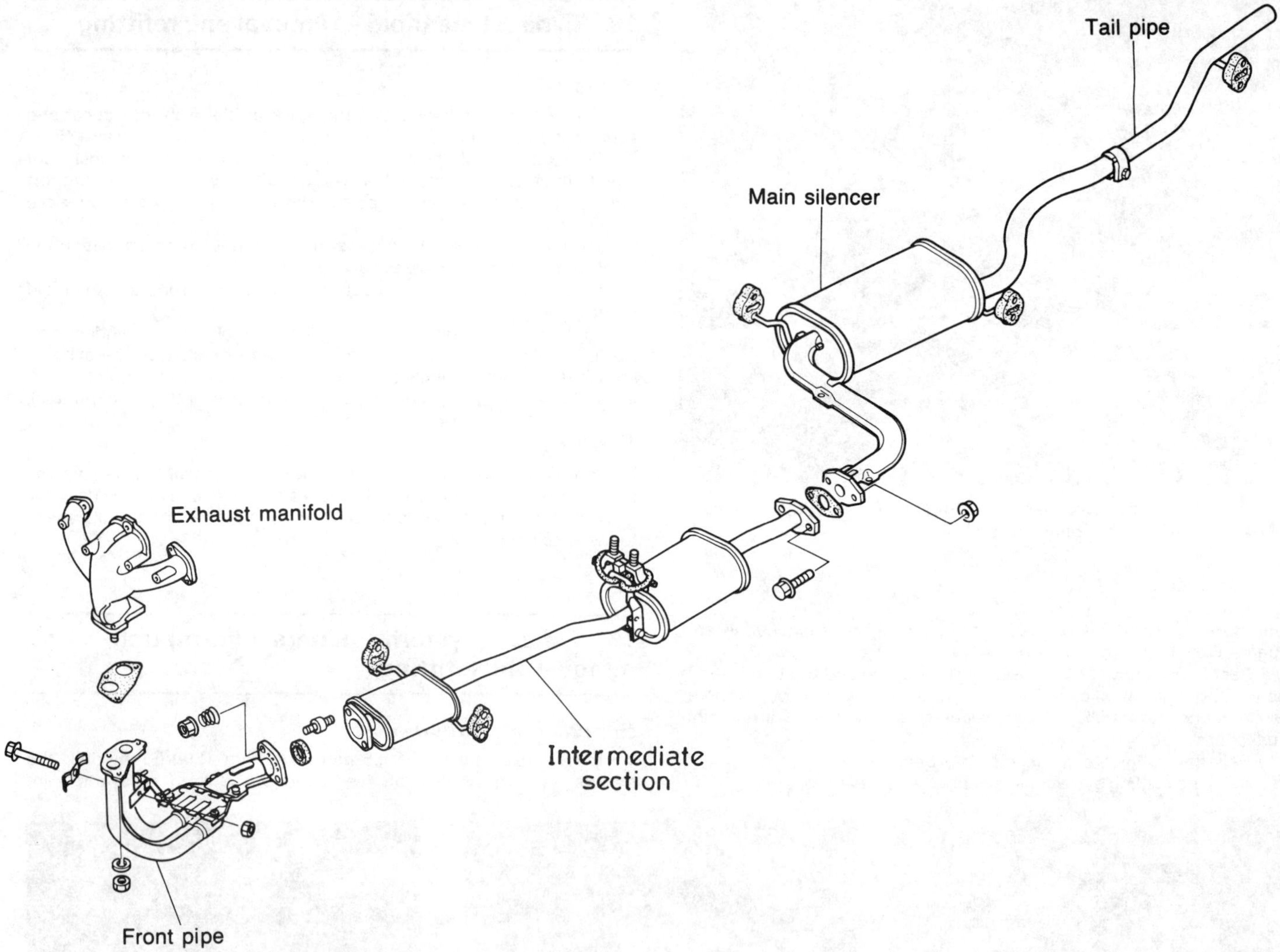

Fig. 4.28 Exhaust system components – later models shown (Sec 19)

silencer, an intermediate pipe, the main silencer and tailpipe section. On 1500 GT models the system is similar, but consists of four sections, the front silencer and intermediate pipe being one single component. On September 1985 models onward, the exhaust system also consists of four sections. These are the front pipe, an intermediate section which contains two silencer boxes, the main silencer section and the tailpipe.

On all models each exhaust section can be removed individually, with the exception of the main silencer which is removed complete with the tailpipe.

Removal

2 To remove the complete system or part of the system, jack up the

Chapter 4 Part A: Carburettor engines

front and/or rear of the car and support it securely on axle stands. Alternatively drive the front or rear wheels up on ramps or position the car over an inspection pit.

Front pipe

3 Undo the nuts securing the front pipe to the manifold and the bolt securing the front pipe to the support bracket. Carefully separate the flange joint and recover the gasket. Undo the flexible joint nuts and remove the springs securing the front pipe to the intermediate section/front silencer. Withdraw the front pipe and recover the sealing ring.

Front silencer/intermediate section

4 To remove the front silencer separately on pre-September 1985 models, excluding 1500 GT, first separate the flexible joint as described above. Undo the two nuts securing the silencer to the intermediate section flange, then separate the flange and remove the gasket. Free the silencer support bracket from both its mounting rubbers and remove it from under the car. To remove the intermediate section, undo the nuts securing the section to the front silencer and the bolts securing it to the main silencer. Carefully separate both flanges and recover the gaskets. Free the mounting hook from the rubber rings and remove the intermediate section.
5 On 1500 GT models, undo the nuts securing the front pipe to the intermediate section and remove the springs. Carefully separate the flexible joint and recover the sealing ring. Undo the bolts securing the intermediate section to the main silencer section and free the mounting hook from the rubber rings. Carefully separate the rear flange, recover the gasket and remove the intermediate section from under the car.
6 On September 1985 models onward, undo the nuts securing the front pipe to the intermediate section and remove the springs. Carefully separate the flexible joint and recover the sealing ring. Undo the bolts securing it to the main silencer section then carefully separate the flange and recover the gasket. Free the front silencer support bracket from its mounting rubbers and the rear silencer mounting hook from the rubber rings and remove the intermediate section.

Main silencer section

7 Remove the bolts securing the intermediate section to the main silencer section, then carefully separate the flange and recover the gasket. Free the main silencer and tailpipe support brackets/mounting hooks from the mounting rubbers/rubber rings (as appropriate) and manoeuvre the silencer and tailpipe assembly out of position.

19.8 Tailpipe retaining clamp

Tailpipe

8 Slacken the tailpipe retaining clamp bolt and free the support bracket from its mounting rubber (photo). The tailpipe can then be removed from the main silencer section using a twisting motion. If the tailpipe is corroded in place, apply liberal amounts of penetrating oil to the joint and allow it time to soak in before attempting to remove it.

Refitting

9 Each section is refitted by a reverse of the removal sequence. Ensure that all traces of corrosion have been removed from the flanges and renew all gaskets. Prior to tightening the flange nuts and bolts, ensure all exhaust system rubber mountings/rings are correctly seated and that there is adequate clearance between the exhaust system and vehicle underbody. When reassembling the flexible front pipe joint always renew the sealing ring. Tighten the flexible joint nuts so that the springs are well compressed, but not coil bound (photos).

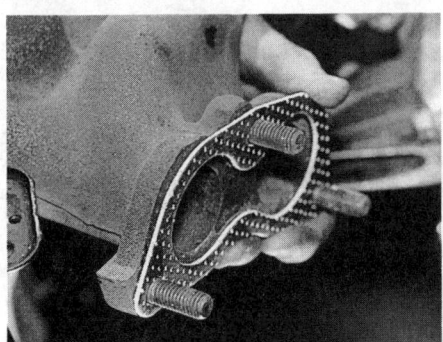

19.9A Always use new gaskets on refitting...

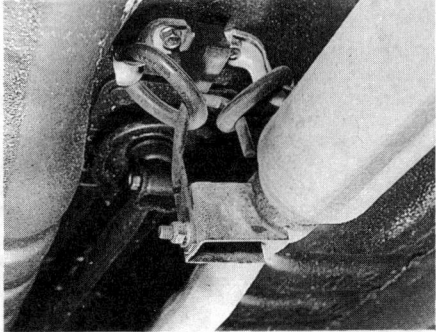

19.9B ...and ensure that the exhaust system mountings are in good condition

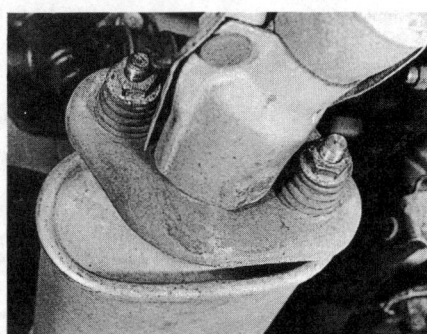

19.9C Tighten flexible joint nuts so springs are well compressed but are not coil bound

Part B: Fuel injected engines

20 General information and precautions

The fuel system consists of a centrally mounted fuel tank, an electric fuel pump and the various fuel injection components which are described further in Section 25.

Fuel is supplied from the tank by an electric pump, which is located inside the tank, via a pressure regulator, to the fuel rail. The fuel rail acts as a reservoir for the four fuel injectors, which inject fuel into the cylinder inlet tracts.

A fuel filter is incorporated in the fuel supply line to ensure that the fuel supplied to the injectors is clean.

The exhaust system consists of four sections secured by flanges, joints or push fit with clamps, and a cast iron exhaust manifold. A spring loaded semi ball and socket joint is used to connect the exhaust front pipe to the intermediate pipe thus catering for engine and exhaust system movement. The system is suspended throughout its length on rubber ring or block type mountings.

Warning: *Many of the procedures in this Chapter require the removal of fuel lines and connections. Residual pressure will remain in the fuel lines long after the car has last been used, therefore extra care must be taken when disconnecting a fuel line hose. Loosen any fuel hose slowly to avoid a sudden release of pressure which may cause fuel spray. As an added precaution place a rag over each union as it is disconnected to catch any fuel which is forcibly expelled. Before carrying out any operation on the fuel system refer to the precautions given in Safety First! at the beginning of this Manual and follow them implicitly. Petrol is a highly dangerous and volatile liquid and the precautions necessary when handling it cannot be overstressed.*

21 Air cleaner housing assembly – removal and refitting

Removal
1 Slacken the retaining clamp and disconnect the inlet duct from the airflow meter.
2 Disconnect the HT lead and LT wiring connector from the ignition coil.
3 Disconnect the wiring connector from the airflow meter.
4 Slacken and remove the nuts and washers securing the air cleaner housing assembly to the vehicle, noting the earth wire which is fitted beneath one of the rear retaining nuts, and lift the assembly out of position (photo).

Refitting
5 Refitting is the reverse of the removal sequence.

22 Accelerator cable – removal, refitting and adjustment

Removal
1 Working in the engine compartment, open the throttle linkage fully by hand and disconnect the cable end from the linkage lever.
2 Slacken the locknuts securing the cable to the bracket on the cylinder head cover and remove the cable from the bracket.
3 From inside the car, remove the cover under the facia.
4 Slip the cable end out of the slot on the accelerator pedal then withdraw the cable through the bulkhead grommet into the engine compartment.

Refitting
5 Refitting is a reversal of the removal sequence. Adjust the cable by means of the locknuts on the cylinder head cover bracket, so that there is a small amount of slack in the cable with the throttle linkage lever on its stop (photo). Have an assistant fully depress the accelerator pedal and check that the throttle linkage on the carburettor opens fully. If not, working from inside the car, slacken the accelerator pedal stop bolt locknut and screw the stop bolt in or out as necessary until the throttle linkage opens fully. Hold the stop bolt in this position and tighten the locknut securely.

23 Accelerator pedal – removal and refitting

Refer to Part A: Section 7.

21.4 Air cleaner housing assembly mounting nuts (arrowed). Note earth lead fitted below right-hand nut

22.5 Adjust the accelerator cable by slackening and repositioning the locknuts

Chapter 4 Part B: Fuel injected engines 151

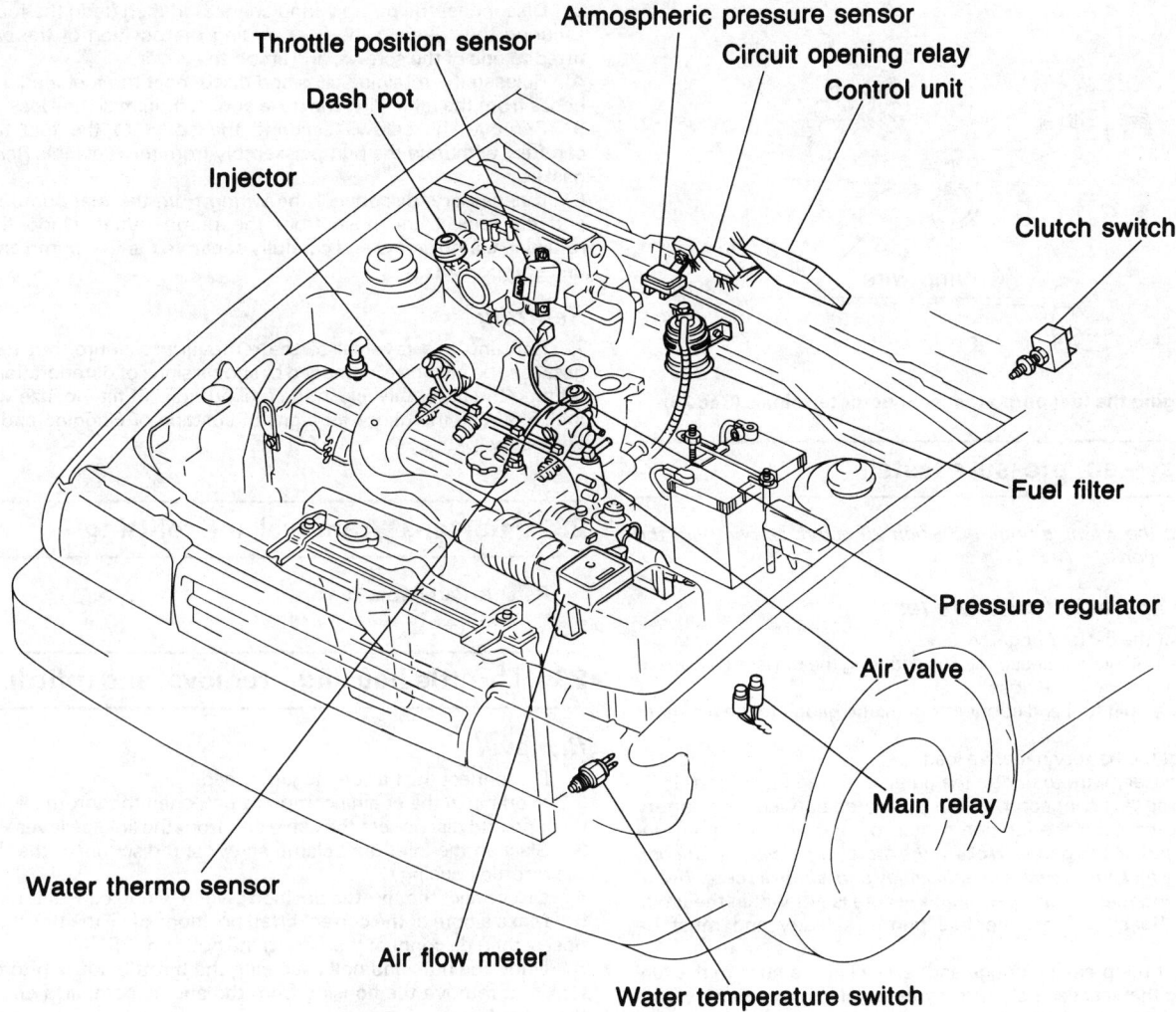

Fig. 4.29 Fuel injection system component locations (Sec 25)

24 Unleaded petrol – general information and usage

The fuel injected model covered in this Section can run on either leaded or unleaded fuel providing it meets the minimum octane rating requirement of 96 RON. No adjustments are necessary to the ignition timing.

25 Fuel injection system – general information and fault diagnosis

The fuel injection system is of the Mazda 'Electronic Gasoline Injection' (EGI) type.

Fuel is supplied from the centrally mounted tank by an electric pump, which is situated inside the tank, via a fuel filter and pressure regulator, to the fuel rail. The fuel rail acts as a reservoir for the four fuel injectors, which inject fuel into the cylinder head inlet tracts, upstream of the inlet valves. The fuel injectors receive a pulse once per crankshaft revolution. The duration of the electrical pulse determines the quantity of fuel injected, and is computed by the EGI control unit on the basis of information received from the various sensors.

Inducted air passes from the air cleaner through a vane type airflow meter before passing to the cylinder head inlet tracts via the throttle valve and surge tank. A flap in the vane airflow meter is deflected in proportion to the airflow; this deflection is then converted into an electrical signal and passed to the EGI control unit. An adjustable air bypass channel provides the means of mixture adjustment.

A throttle position switch enables the EGI control unit to compute not only the throttle position, but also its rate of change. Extra fuel can thus be provided for acceleration when the throttle is opened suddenly. Information from the throttle position switch is also used to cut off fuel on the overrun, thus improving fuel economy and reducing exhaust gas emissions.

Additional sensors inform the EGI control unit of air temperature, air pressure and engine coolant temperature so that the pulse to the injectors can be adjusted accordingly. Neutral and clutch switches are also fitted to inform the control unit if the engine is under load or not, so that the idle speed can be adjusted accordingly. On models equipped with power steering, an idle up system is incorporated to increase the idle speed when the power steering pump is operating. A switch on the pump informs the EGI control unit, which then opens an electrically operated solenoid valve. This allows air to bypass the throttle housing and so increase the idle speed.

Due to the complexity of the fuel injection system it is recommended that the car should be taken to a Mazda dealer for fault diagnosis, should any problems occur. They will have access to the digital code checker which will track down the injection fault quickly without the need to test all the injection system components individually. However, most components can be tested individually by the home mechanic as described in Section 31.

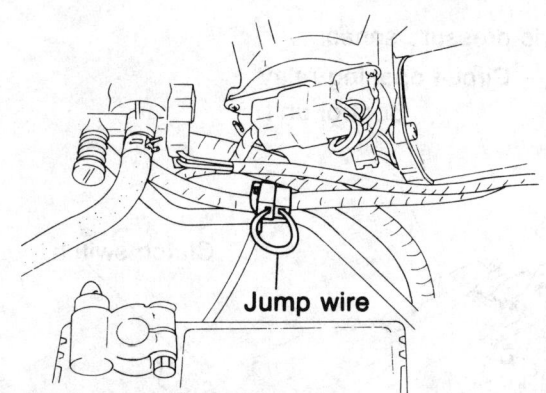

Fig. 4.30 Bridging the fuel pump test connector terminals (Sec 26)

26 Fuel system pressure tests

Note: Refer to the warning note in Section 20 before carrying out the following operations.

Fuel pump operating pressure test

1 Disconnect the battery negative lead.
2 Using a pair of pliers, release the clip securing the upper fuel hose to the filter and disconnect the hose.
3 Mop up any spilt fuel and connect a pressure gauge to the fuel filter outlet union.
4 Reconnect the battery negative lead.
5 Use an auxiliary wire to bridge the green/white and black terminals of the fuel pump test connector, which is situated between the battery and wiper motor. Turn the ignition switch on and note the pressure reading obtained on the gauge. **Note:** *If the fuel pump does not function at all, check the fuel pump circuit opening relay and wiring as described in Section 31.* If the fuel pump operating pressure is not within the limits given in the Specifications, the fuel pump is faulty and must be renewed.
6 Disconnect the pressure gauge and reconnect the hose to the fuel filter, ensuring that it is securely held by the retaining clip. Remove the wire from the fuel pump test connector.

Fuel pressure regulator test

7 Check the fuel pump operating pressure as described above.
8 Disconnect the battery negative lead.
9 Using a pair of pliers, release the clip securing the upper fuel hose to the filter and disconnect the hose.
10 Using a T-piece and spare fuel hose, connect a pressure gauge into the fuel line between the fuel filter and upper hose.
11 Reconnect the battery negative lead.
12 Start the engine and note the pressure gauge reading obtained when the engine is idling. Disconnect the vacuum hose, linking the pressure regulator to the surge tank, from the regulator and block the hose end with a finger. With the engine idling at the specified speed, note the fuel line pressure reading at the gauge. If both the readings are not within the limits given in the Specifications the fuel pressure regulator is faulty and must be renewed.
13 Reconnect the fuel hose to the filter, ensuring it is held securely by the retaining clip, and the vacuum hose to the regulator.

27 Fuel pump and fuel gauge sender unit assembly – removal and refitting

Note: Refer to the warning note in Section 20 before carrying out the following operation.

Removal

1 Disconnect the battery negative lead.
2 Remove the rear seat cushion as described in Chapter 11.
3 Disconnect the pump wiring connector, then undo the four screws securing the cover to the floor, noting the position of the earth wire fitted to one of the screws, and lift off the cover.
4 Release the retaining clips and disconnect the fuel feed and return hoses from the tank. Plug the hose ends to minimise fuel loss.
5 Remove the screws securing the cover to the fuel tank then carefully withdraw the pump assembly from the fuel tank. Remove the gasket.
6 If necessary, disconnect the wiring from the fuel pump terminals and disconnect the hose from the pump outlet. Undo the pump retaining clamp screw and carefully separate the fuel pump and sender unit components.

Refitting

7 Refitting is a reversal of the removal procedure, but use a new gasket if the old one is damaged or shows signs of deterioration. Prior to refitting the assembly, clean the fuel pump inlet filter gauze with clean fuel. Examine the gauze for signs of splitting or clogging and renew if necessary.

28 Fuel tank – removal and refitting

Refer to Part A: Section 5.

29 Throttle housing – removal and refitting

Removal

1 Disconnect the battery negative lead.
2 Working in the engine compartment, open the throttle linkage fully by hand and disconnect the cable end from the linkage lever.
3 Slacken the inlet duct clamp screw and disconnect the duct from the throttle housing.
4 Disconnect the throttle position switch wiring connector.
5 Make a note of the correct fitted positions of all the throttle housing hoses, then disconnect them from the housing.
6 Undo the nuts and bolts securing the throttle housing to the surge tank and remove the housing from the engine compartment. Recover the gasket.

Refitting

7 Refitting is a reversal of the removal sequence, bearing in mind the following points.

   (a) *Ensure that all traces of old gasket are removed from the throttle housing and surge tank sealing faces and fit a new gasket.*
   (b) *Adjust the accelerator cable as described in Section 22.*
   (c) *Refill the cooling system as described in Chapter 1.*
   (d) *On completion adjust the idle speed, CO content and throttle position switch as described in Chapter 1.*

30 Surge tank – removal and refitting

Removal

1 Disconnect the battery negative lead.
2 Drain the cooling system as described in Chapter 1.
3 Working in the engine compartment, open the throttle linkage fully by hand and disconnect the cable end from the linkage lever.
4 Slacken the inlet duct clamp screw and disconnect the duct from the throttle housing.
5 Disconnect the throttle position switch wiring connector.
6 Make a note of the correct fitted positions of all the coolant and air hoses which are connected to the throttle housing and surge tank, to use as a guide on refitting, then disconnect all the hoses.
7 Undo the nuts and bolts securing the surge tank to the inlet manifold and mounting brackets, then lift the surge tank upwards off the inlet manifold. Recover the gasket.

Chapter 4 Part B: Fuel injected engines

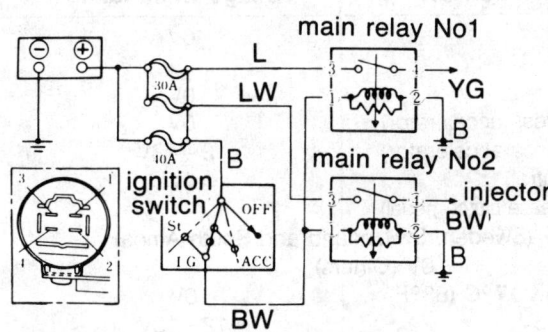

Fig. 4.31 EGI main relay terminal identification for twin circular relays (Sec 31)

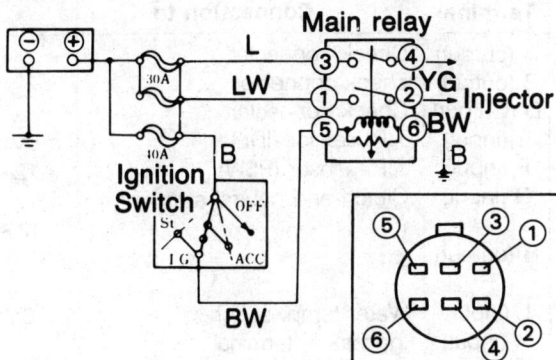

Fig. 4.32 EGI main relay terminal identification for rectangular relay (Sec 31)

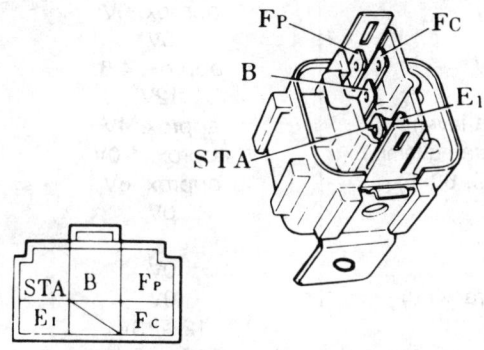

Fig. 4.33 Circuit opening relay terminal identification (Sec 31)

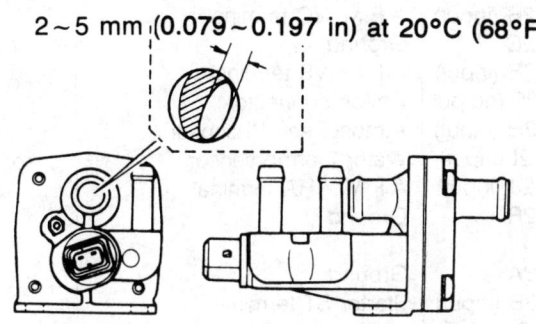

Fig. 4.34 Checking the auxiliary air valve operation (Sec 31)

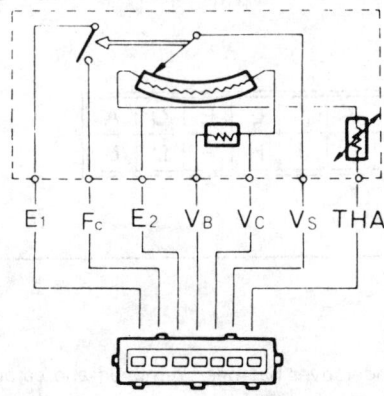

Fig. 4.35 Airflow meter terminal identification (Sec 31)

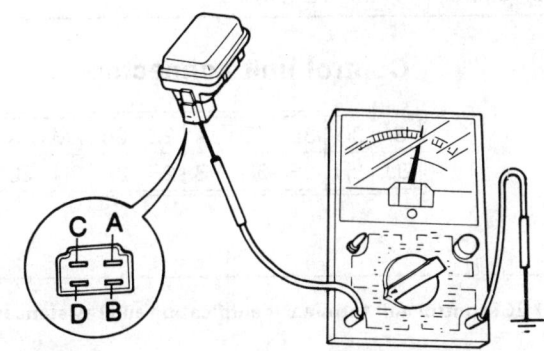

Fig. 4.36 Testing the atmospheric pressure sensor (Sec 31)

Refitting

8 Refitting is a reversal of the removal sequence, bearing in mind the following points.

  (a) Ensure that all traces of old gasket are removed from the inlet manifold and surge tank sealing faces and fit a new gasket.
  (b) Adjust the accelerator cable as described in Section 22.
  (c) Refill the cooling system as described in Chapter 1.
  (d) On completion adjust the idle speed, CO content and throttle position switch as described in Chapter 1.

31 Fuel injection system components – testing

1 Before checking any of the fuel injection components, ensure that the battery is fully charged and that the EGI system main fuse, which is situated in the main fuse box on the left-hand side of the engine compartment, is intact.

EGI main relay

2 Check that power is reaching the EGI main relay, situated in the front left-hand corner of the engine compartment (below the horn and cooling fan relay), by having an assistant turn the ignition switch on and off whilst checking that the relay clicks (photo). If a clicking sound is not emitted from the relay, disconnect the relay wiring connector(s) and use a voltmeter to check that 12 volts is present between the connector(s) black/white terminal(s) and earth. If this is not the case, the fault lies in the either the wiring between the ignition switch and relay or in the switch itself. If 12 volts is present, test the EGI main relay.

3 Disconnect the wiring connector(s), undo the retaining screw and remove the relay from the engine compartment. Using a 12 volt battery, two auxiliary wires and an ohmmeter, test the relay as follows.

Pre-July 1987 models (twin circular relays)

4 Referring to Fig. 4.31, connect the battery positive (+) terminal to the number 1 terminal of number 1 relay and the battery negative (–) terminal to the number 2 terminal. With the voltage applied, check for continuity between number 3 and 4 terminals. Disconnect the voltage and check again for continuity between terminal 3 and 4 of number 1 relay. Repeat the above test on number 2 relay. If all is well, there should

Terminal	Connection to	Voltage with ignition ON	Voltage when idling
A (output)	Check connector	0V	0V
B (output)	check connector	0V	0V
D (Output)	Check connector	0V	0V
E (input)	T.P. sensor (IDL)	0V (12~13V: depress accelerator)	0V
F (input)	T.P. sensor (PSW)	12~13V (0V: depress accelerator)	12~13V
G (input)	Clutch and neutral switch	0v (neutral) 12~13V (in-gear, release clutch pedal)	0V
H (output)		12~13V (Sweden, Switzerland and South Africa) 0V (Others)	
L (input)	Water temp. switch	0V (12~13V: below 17°C (63°F))	0V
M (input)	Ignition — terminal	12~13V	12~13V
N (input)	Distributor	0 or 12V	0.5V
2A (output)	Atmospheric P. sensor	approx. 5V	approx. 5V
2B (input)	A.F.M. VC terminal	approx. 8V	approx. 9V
2C	Ground	0V	0V
2E (input)	A.F.M. VS terminal	approx. 1.5V	approx. 4.5
2F (output	Check connector	12~13V	12V
2H (input)	Atmospheric P. sensor	approx. 4V (at sea level)	approx. 4V
2I (input)	Water thermo sensor	approx. 1.0V (normal operating temp.)	approx. 1.0V
2J (input)	A.F.M. THA terminal	approx. 6V (at 20°C, 68°F)	approx. 6V
2R	Ground	0V	0V
3A	Ground	0V	0V
3B (input)	Starter ST terminal	0V (approx. 10V: cranking)	0V
3C (output)	Injector No. 2, No. 4	12~13V	12~13V
3E (output)	Injector No. 1, No. 3	12~13V	12~13V
3G	Ground	0V	0V
3I (output)	A.F.M. VB terminal	12~13V	12~13V

Control unit connector

3I	3G	3E	3C	3A	2D	2o	2M	2K	2I	2G	2E	2C	2A	M	K	I	G	E	C	A
3J	3H	3F	3D	3B	2R	2P	2N	2L	2J	2H	2F	2D	2B	N	L	J	H	F	D	B

Fig. 4.37 EGI control unit terminal identification and resistances (Sec 31)

be continuity between terminals 3 and 4 with the voltage applied, and an open circuit with the voltage disconnected. If this is not the case for either number 1 or 2 relay, the EGI main relay must be renewed.

July 1987 models onward (rectangular relay)

5 Referring to Fig. 4.32, connect the battery positive (+) terminal to terminal 5 of the relay, and the battery negative (-) terminal to the relay terminal 6. With the voltage applied check for continuity between terminals 3 and 4, and terminals 1 and 2. Disconnect the voltage and repeat the check. If the relay is functioning correctly there should be continuity between the terminals with the voltage applied, and an open circuit with the voltage disconnected. If not the relay is faulty and must be renewed.

Fuel pump and circuit opening relay

6 The condition of the fuel pump can be determined by checking the fuel pump operating pressure as described in Section 26. If the pump fails to function when the test connector terminals are bridged, check the circuit opening relay and wiring as follows.
7 The circuit opening relay is situated inside the car on the right-hand side of the heater unit, where it is mounted on the EGI control unit. Access to the relay can be gained from the drivers side once the right-hand facia undercover has been removed and carpet has been released.
8 Check that power is reaching the circuit opening relay by turning the ignition switch on and off whilst checking that the relay clicks. If a clicking sound is not emitted from the relay, remove the relay and check the resistances between the relay terminals (Fig. 4.33). If the results do not agree with those in the table below, the relay is faulty and must be renewed.

Terminals	Resistance
STA to E1	15 to 30 ohms
B to FC	80 to 150 ohms
B to FP	Open circuit

9 If the circuit opening relay is in good condition, refit the relay to the car.
10 Remove the rear seat cushion, referring to Chapter 11 if necessary, to gain access to the fuel pump wiring connector.
11 Bridge the fuel pump test connector terminals with an auxiliary wire and turn the ignition switch on. Using a voltmeter, connect the meter positive (+) lead to the fuel pump wiring connector green/red terminal and the meter negative (-) lead to earth and note the reading obtained. If a reading of 12 volts is obtained, the fuel pump is faulty and must be renewed. If no reading is obtained, the fault lies in the wiring.

Chapter 4 Part B: Fuel injected engines

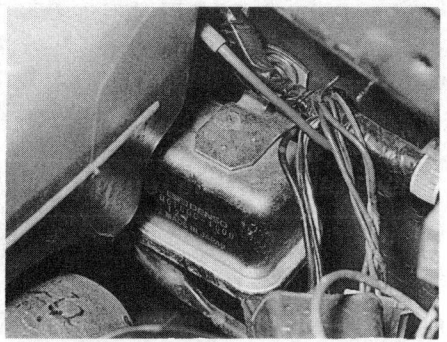

31.2 EGI main relay location (horn relay removed for clarity)

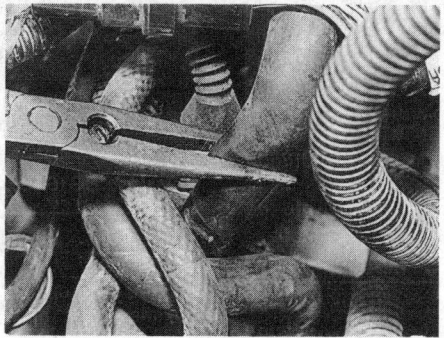

31.14 Pinch the auxiliary air valve to surge tank hose and note the effect on idle speed

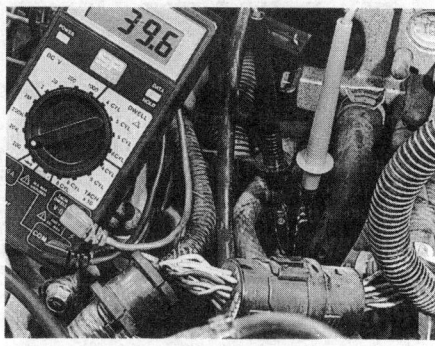

31.15 Checking the auxiliary air valve resistance

Fuel injectors

12 Disconnect the wiring connector from the top of each injector and measure the resistance between the injector terminals. Any injector on which the resistance reading is found not to be within the specified limits should be renewed.

Auxiliary air valve

13 Start the engine and allow it to idle at the specified speed.
14 Using a pair of pliers or grips, pinch the large hose which links the auxiliary air valve to the surge tank, whilst noting the effect on the engine idle speed (photo). When the engine is cold and the hose is pinched the idle speed should drop significantly, and when the engine is at operating temperature the idle speed should drop no more than 200 rpm. If this is not the case, proceed as follows.
15 Disconnect the wiring connector from the air valve and check the resistance between the air valve terminals (photo). If the resistance is not within the limits given in the Specifications, the valve can be considered faulty.
16 Disconnect the air hose from the top of the auxiliary air valve. Look down the valve bore, using a torch if necessary, and check the valve operation (photo). When the engine is cold, below 20°C, the valve should be open, and above 20°C the valve should be closed. Referring to Fig. 4.34, Mazda specify that at 20°C the valve opening should be 2 to 5 mm. If the valve operation is suspect it must be renewed.

Fuel pressure regulator

17 The fuel pressure regulator can be tested by performing the pressure tests described in Section 26.

Airflow meter and air temperature sensor

18 Remove the airflow meter as described in Section 32.
19 Refer to Fig. 4.35 and, using an ohmmeter, check the resistances between the various airflow meter terminals and compare them with those in the table below (photo).

Terminal	Resistance
E2 to VS	20 to 400 ohms
E2 to VC	100 to 300 ohms
E2 to VB	200 to 400 ohms
E1 to FC	Open circuit

20 Check the resistances between the terminals with the sensing flap in both the fully closed and fully open positions, and compare the readings with the values given in the following table.

Terminals	Sensing flap fully closed	Sensing flap fully open
E1 to FC	Open circuit	0 ohms
E2 to VS	20 to 400 ohms	20 to 1000 ohms

21 The air temperature sensor is an internal component of the airflow meter and can be checked by measuring the resistance between terminals E2 and THA of the airflow meter. Compare the readings obtained at various temperatures with those in the following table.

Temperature	Resistance
−20°C	10 k ohms to 20 k ohms
0°C	4 k ohms to 7 k ohms
20°C	2 k ohms to 3 k ohms
40°C	900 to 1300 ohms
60°C	Open circuit

22 If the measured air temperature and airflow meter resistances differ greatly from the those given in the above tables, it is likely that the airflow meter assembly is faulty. However, have your results confirmed by a Mazda dealer or auto electrical specialist before obtaining a replacement unit or having the original repaired.

Throttle position switch

23 The throttle position switch can be tested as described in Section 3 of Chapter 1.

Atmospheric pressure sensor

24 The atmospheric pressure sensor is mounted on the engine compartment bulkhead, just to the right of the fuel filter (photo).
25 Referring to Fig. 4.36, connect the positive (+) lead of a voltmeter to the sensor D terminal, and the negative (−) lead to earth. Turn the ignition switch on and note the voltage reading obtained. If the sensor is functioning correctly a reading of approximately 4 ± 0.5 volts should be obtained at sea level (a slightly lower voltage reading will be obtained at higher altitude). If this is not the case renew the atmospheric pressure sensor.

Water temperature thermo sensor

26 Remove the water temperature thermo sensor as described in Section 32.
27 Referring to Fig. 3.8 (Chapter 3), suspend the sensor in a saucepan or suitable vessel together with a thermometer. Fill the vessel with cold water so that the sensor probe is completely submerged. Connect an ohmmeter between the sensor terminals.
28 Gently heat the water whilst noting the resistance readings obtained at the temperatures given in the accompanying table. If the sensor does not perform as expected, it must be renewed.

Water temperature	Resistance
20°C	2450 ± 240 ohms
80°C	322 ± 32 ohms

Water temperature switch

29 Remove the water temperature switch as described in Section 32.
30 Referring to Fig. 3.8 (Chapter 3), suspend the switch in a saucepan or suitable vessel together with a thermometer. Fill the vessel with cold water so that the switch probe is completely submerged. Connect an ohmmeter between the switch terminals. There should be continuity between the switch terminals.
31 Gently heat the water and note the temperature at which the switch contacts open. Turn the source of heat off, then allow the water to cool and note the temperature at which the switch contacts close and continuity between the terminals is regained. If the switch is operating

Chapter 4 Part B: Fuel injected engines

31.16 Disconnect the hose from the top of the auxiliary air valve and check valve operation

31.19 Checking airflow meter terminal resistances

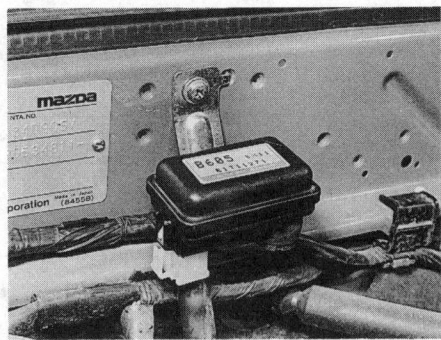

31.24 Atmospheric pressure sensor location

31.35 Checking the neutral switch operation

31.36 EGI control unit location (arrowed)

correctly there should be continuity between the terminals below 17°C, and an open circuit above 17°C. If this is not the case renew the switch.

Clutch switch

32 Remove the right-hand facia undercover to gain access to the clutch switch.
33 Disconnect the wires from the switch and connect an ohmmeter to the switch terminals. There should be continuity between the switch terminals when the clutch pedal is depressed, and an open circuit when the pedal is released. If this is not the case, attempt to adjust the switch by slackening and adjusting the locknuts as necessary, until it operates correctly. If it proves impossible to adjust the switch, it must be renewed.

Neutral switch

34 Disconnect the neutral switch wiring at the connector which is situated just below the battery tray.
35 Connect an ohmmeter to the terminals of the switch side of the wiring connector and check the operation of the neutral switch (photo). If the switch is in good condition there should be continuity between the terminals when the gearbox is in neutral, and an open circuit when the gearbox is in gear. If this is not the case the neutral switch must be renewed.

EGI control unit

36 The EGI control unit is situated inside the car where it is located behind the heater unit (photo). Remove the left-hand facia undercover and release the carpet to gain access to the control unit wiring connector.
37 Warm the engine up to normal operating temperature and check the voltages present between the various control unit terminals and earth, both with the ignition switched on and the engine idling. Compare the results with those given in Fig. 4.37. If the results differ greatly from those in the table it is likely that the control unit is faulty. However, have your findings confirmed by a Mazda dealer or auto electrical specialist before consigning the control unit to the scrap bin and buying a replacement unit.

32 Fuel injection system components – removal and refitting

Fuel injectors

Note: *Refer to the warning note in Section 20 before carrying out the following operation.*

Removal

1 Remove the surge tank as described in Section 30.

32.7 Auxiliary air valve wiring connector (A), upper hose (B) and coolant hoses (C)

Chapter 4 Part B: Fuel injected engines

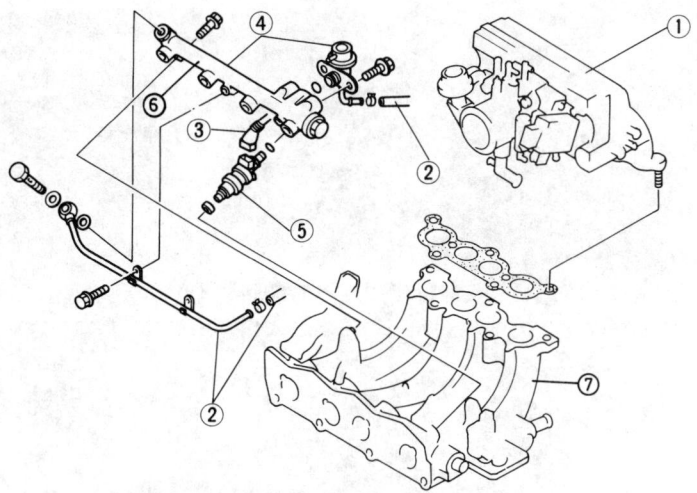

Fig. 4.38 Fuel injectors, pressure regulator and related components (Sec 32)

1 Surge tank assembly
2 Fuel return pipe
3 Injector wiring connector
4 Fuel pressure regulator
5 Injector
6 Fuel distribution rail
7 Inlet manifold

2 Using a pair of pliers, release the retaining clips and disconnect the fuel and vacuum hoses from the pressure regulator, and the fuel hose from the fuel pipe.
3 Disconnect the wiring connectors from the injectors.
4 Undo the bolts securing the fuel distribution rail to the inlet manifold and carefully remove the rail.
5 Withdraw the injectors and remove the injector insulator seals from the inlet manifold.

Refitting
6 Refitting is a reverse of the removal sequence, noting that the O-rings which are fitted to the upper end of the injectors must be renewed. Lubricate the O-rings with fuel to ease installation.

Auxiliary air valve
Removal
7 Disconnect the wiring connector from the top of the valve (photo).
8 Using a pair of pliers, release the retaining clips and disconnect both the upper and lower air hoses.
9 Disconnect the coolant hoses from the side of the valve and plug the hoses to minimise the loss of coolant. Mop up any spilt coolant.
10 Undo the two bolts which secure the valve to the left-hand side of the inlet manifold and remove it from the engine compartment.

Refitting
11 Refitting is a reversal of the removal procedure. If necessary replenish lost coolant as described in Chapter 1.

Fuel pressure regulator
Note: *Refer to the warning note in Section 20 before carrying out the following operation.*

Removal
12 Remove the surge tank as described in Section 30.
13 Using a pair of pliers, release the retaining clips and disconnect the fuel and vacuum hoses from the pressure regulator.
14 Undo the two fuel pressure regulator retaining bolts, then remove the regulator from the fuel distribution rail, along with its O-ring.

Refitting
15 Refitting is a reverse of the removal sequence using a new O-ring.

Airflow meter
Removal
16 Disconnect the wiring connector from the airflow meter, and the HT lead and LT wiring from the ignition coil. Undo the nut securing the

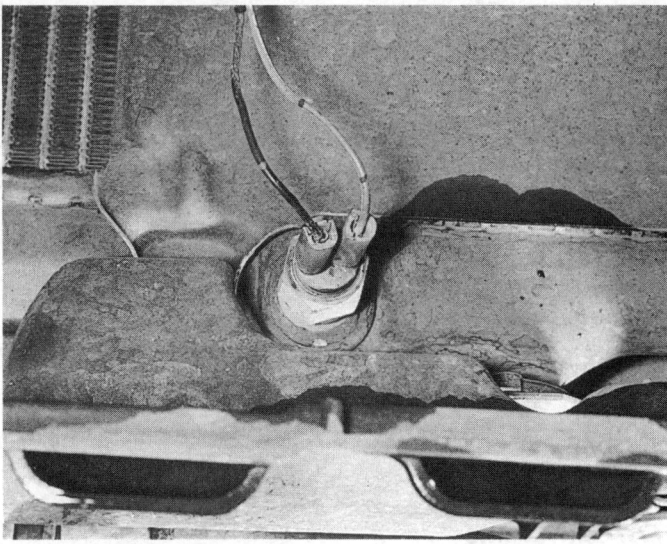

32.30 Water temperature switch location

earth wire to the air cleaner housing and disconnect the wire.
17 Slacken the retaining clamp and disconnect the inlet duct from the air cleaner housing.
18 Undo the bolts securing the air cleaner cover assembly to the housing and remove the assembly from the engine compartment.
19 From inside the cover, undo the four nuts which retain the airflow meter. Separate the meter and cover and remove the gasket.

Refitting
20 Refitting is a reverse of the removal procedure using a new gasket.

Throttle position switch
Removal
21 Disconnect the switch wiring connector.
22 Undo the two switch retaining screws and remove the switch from the throttle housing.

Refitting
23 Locate the throttle switch on the throttle valve spindle and refit the retaining screws.
24 Reconnect the wiring connector and adjust the switch as described in Section 3 of Chapter 1.

Water temperature thermo sensor
Removal
25 Drain the cooling system as described in Chapter 1.
26 Remove the throttle housing as described in Section 29.
27 Disconnect the wiring connector and unscrew the thermo sensor from the inlet manifold.

Refitting
28 Refitting is the reverse of the removal sequence, referring to Chapter 1 to refill the cooling system.

Water temperature switch
Removal
29 Drain the cooling system as described in Chapter 1.
30 Disconnect the wiring and unscrew the temperature switch from the bottom of the radiator (photo).

Refitting
31 Refitting is the reverse of the removal sequence, referring to Chapter 1 to refill the cooling system.

Clutch switch
Removal
32 Remove the right-hand facia undercover to gain access to the clutch switch.

158   Chapter 4   Part B: Fuel injected engines

33 Disconnect the switch wiring then slacken the locknuts and remove the switch from the car.

Refitting

34 Refit the switch and tighten the locknuts finger tight only. Connect an ohmmeter to the switch terminals and adjust the switch so that there is continuity between the terminals when the pedal is depressed, and an open circuit when the pedal is released. Once the switch is correctly adjusted tighten the locknuts securely.

Neutral switch

Removal

35 Remove the gearbox as described in Chapter 7.
36 Unscrew the neutral switch from the gearbox housing and remove it along with its sealing washer.

Refitting

37 Renew the sealing washer, then refit the switch to the gearbox and tighten it securely.
38 Refit the gearbox as described in Chapter 7.

EGI control unit

Removal

39 Remove both the facia undercovers and release the carpet to gain access to both sides of the control unit.
40 Undo the screws securing the control unit to the floor and remove it from behind the heater unit.

Refitting

41 Refitting is the reverse of the removal procedure.

33 Idle up system – general information and adjustment

1 On models equipped with power steering, an idle up system is fitted to prevent the engine stalling whilst idling when the power steering pump is operating. The system consists of the power steering pump pressure switch and an electrically operated solenoid. When the power steering pump is operating, the pump switch opens up the solenoid and allows air to bypass the throttle valve and flow into the surge tank, thus increasing the idle speed. The system can be adjusted as follows.
2 Check and, if necessary, adjust the idle speed as described in Chapter 1.
3 With the engine idling at the specified speed, disconnect the wire from the power steering pump pressure switch and earth it using an auxiliary wire (photo).
4 With the wire earthed the engine speed should rise to 1150 to 1250 rpm. If this is not the case, turn the adjusting screw on the power steering idle-up solenoid, situated in the right-hand rear corner of the engine compartment, until the engine speed is within the specified range (photo). Reconnect the wire to the power steering pump switch.

34 Economy drive indicator system – general information

Refer to Part A: Section 15.

35 Inlet manifold – removal and refitting

Removal

1 Remove the surge tank as described in Section 30.
2 Disconnect the brake servo vacuum hose from the inlet manifold, and the vacuum and fuel hoses from the fuel pressure regulator. Disconnect the wiring connectors from the injectors.
3 Undo the nuts and bolts securing the inlet manifold to the cylinder head then carefully remove the manifold and gasket.

33.3 Disconnect the wire from the power steering pump pressure switch (arrowed) and connect it to earth

33.4 Adjusting the power steering idle-up solenoid adjusting screw

Refitting

4 Ensure that all traces of old gasket are removed and that the inlet manifold and cylinder head sealing faces are clean.
5 Fit a new inlet manifold gasket and install the manifold. Tighten the manifold retaining nuts and bolts evenly and progressively to the specified torque.
6 Reconnect the vacuum and fuel hoses, and the brake servo vacuum hose ensuring that they are securely fastened by their retaining clips.
7 Refit the surge tank as described in Section 30.

36 Exhaust manifold – removal and refitting

Refer to Part A: Section 18.

37 Exhaust system – general information, removal and refitting

Refer to Part A: Section 19.

Part C: Emission control systems

38 General information

Fuel evaporative emission control

The function of this system is to reduce the amount of fuel vapour released into the atmosphere. The system is controlled by a check valve which is linked to the fuel tank.

Crankcase emission control

The function of the Positive Crankcase Ventilation (PCV) system is to draw blow-by gases from the crankcase and cylinder head cover chamber and direct them into the inlet manifold. From there the gases are drawn into the combustion chambers with the fuel/air mixture and burnt in the combustion process.

Exhaust emission control

The following components are fitted to reduce the amount of harmful hydrocarbons and carbon monoxide emitted in the exhaust.

Carburettor models

On pre-July 1987 models which are fitted with a manual gearbox, the system is controlled by the throttle positioner system (See Part A: Section 13). July 1987 onward B5 engined models with a manual gearbox are equipped with an anti-afterburn valve (AAV), which supplies fresh air into the inlet manifold during deceleration, as well as the throttle positioner system.

B3 engined models and all models equipped with automatic transmission are fitted with the anti-afterburn valve only.

Fuel injected models

On fuel injected models, the exhaust emissions are controlled by the dashpot and an anti-afterburn valve. The dashpot ensures that the throttle valve closes slowly during deceleration whilst the anti-afterburn valve supplies fresh air to the inlet manifold.

39 Emission control system components – testing and renewal

Fuel evaporative emission control

1 The check valve is situated just to the rear of the fuel tank, where it is mounted onto the vehicle underbody.
2 Firmly chock the front wheels then jack up the rear of the vehicle and support it on axle stands.
3 Using a pair of pliers, release the retaining clips and disconnect the three hoses from the check valve. Undo the retaining bolt and remove the valve from underneath the vehicle.
4 Referring to Fig. 4.39, blow through the check valve port A and check that the air comes out of port B. Then block port B and confirm that the air comes out of port C. Next block port B and suck through port A, air should come through from port C. If the check valve does not perform as described it is faulty and must be renewed.
5 Fit the valve to the underside of the vehicle and tighten its retaining bolt securely. Reconnect the hoses to their original positions and secure them in place with the retaining clips. Lower the vehicle to the ground.

Crankcase emission control

6 Warm up the engine to normal operating temperature and allow it to idle at the speed specified in Chapter 1.
7 Remove the PCV valve from the right-hand corner of the cylinder head cover (photo).
8 With the engine idling, cover the end of the PCV valve with a finger and check that a vacuum is present in the hose. If not, renew the PCV valve.

Exhaust emission control

Throttle positioner system

9 Refer to Part A: Section 13 for information on testing the throttle positioner system.

Anti-afterburn valve

10 Start the engine and warm it up to normal operating temperature.
11 With the engine idling, block the intake port on the underside of the valve making sure that the engine speed does not change. Increase the engine speed then release the throttle linkage quickly so that the engine speed decreases quickly. Check that air is sucked into the AAV intake port for approximately 1 to 2 seconds after the throttle is released. If this is not the case the anti-afterburn valve is faulty and must be renewed (photo).

Dashpot

12 Refer to Chapter 1, Section 3 for information on testing and adjustment of the dashpot.

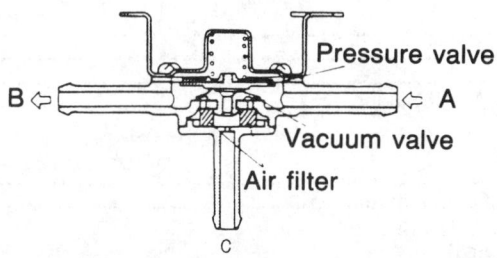

Fig. 4.39 Testing the fuel tank check valve (Sec 39)

39.7 Removing the PCV valve

39.11 Location of the anti-afterburn valve (arrowed)

Chapter 5 Ignition system

Contents

Part A: Contact breaker ignition system
Contact breaker points and condenser –
check, adjustment and renewal .... See Chapter 1
Distributor – removal, overhaul and refitting 3
General information 1
HT leads, distributor cap and rotor arm –
check and renewal .... See Chapter 1
Ignition coil – removal, testing and refitting 4
Ignition system – testing 2
Ignition timing – check and adjustment .... See Chapter 1
Spark plug check and renewal .... See Chapter 1

Part B: Electronic ignition system
Distributor – removal, overhaul and refitting 7
General information and precautions 5
HT leads, distributor cap and rotor arm –
check and renewal .... See Chapter 1
Ignition coil – removal, testing and refitting 8
Ignition system – testing 6
Ignition timing – check and adjustment .... See Chapter 1
Shutter valve ignition advance mechanism (B3 and B5 engine
models) – testing 9
Spark plug check and renewal .... See Chapter 1

Specifications

Part A: Contact breaker ignition system

General
System .... Conventional contact breaker and coil ignition system
Application .... Models manufactured before December 1981
Firing order 1-3-4-2
Location of No 1 cylinder .... Crankshaft pulley end

Distributor
Type .... Conventional with contact breaker points and condenser
Direction of rotor arm rotation .... Anti-clockwise (viewed from cap)

Ignition coil
Type .... Conventional, mounted on the engine compartment left-hand inner valance
Primary resistance .... 3.1 ohms
Secondary resistance .... 10 to 30 k ohms
Ballast resistance .... 1.6 ohms

Part B: Electronic ignition system

General
System type .... Electronic breakerless High Energy Ignition (H.E.I.) system
Application .... Models manufactured after December 1981
Firing order 1-3-4-2
Location of No 1 cylinder .... Crankshaft pulley end

Distributor
Type .... Conventional, containing signal generator and igniter unit assemblies
Direction of rotor arm rotation .... Anti-clockwise (viewed from cap)

Ignition coil
Type .... Conventional, mounted on the airflow meter on fuel injected models and the engine compartment left-hand inner valance on carburettor models
Secondary resistance .... 10 to 30 k ohms

Part A: Contact breaker ignition system

1 General information

Models of the Mazda 323 manufactured before December 1981 utilize a conventional contact breaker point ignition system.

In order that the engine may run correctly, it is necessary for an electrical spark to ignite the fuel/air mixture in the combustion chamber at exactly the right moment in relation to engine speed and load. The ignition system is based on feeding low tension voltage from the battery to the coil, where it is converted to high tension voltage. The high tension voltage is powerful enough to jump the spark plug gap in the cylinder many times a second under high compression pressure, provided that the ignition system is in good working order and that all adjustments are correct.

The ignition system consists of two individual circuits known as the low tension (LT), or primary circuit, and the high tension (HT), or secondary circuit.

The low tension circuit consists of the battery, a lead to the ignition switch, a lead to the low tension or primary coil windings and the lead from the low tension coil windings to the contact breaker points and condenser in the distributor.

The high tension circuit consists of the high tension or secondary coil winding, the heavily insulated lead from the centre of the coil to the centre of the distributor cap, the rotor arm, the spark plug leads and the spark plugs.

The complete ignition system operation is as follows. Low tension voltage from the battery is changed within the ignition coil to high tension voltage by the opening and closing of the contact breaker points in the low tension circuit. High tension voltage is then fed, via a contact in the centre of the distributor cap, to the rotor arm of the distributor. The rotor arm revolves inside the distributor cap, and each time it comes in line with one of the four metal segments in the cap, the opening and closing of the contact breaker points causes the high tension voltage to build up, jump the gap from the rotor arm to the appropriate metal segment and so, via the spark plug lead, to the spark plug where it finally jumps the gap between the two spark plug electrodes, one being earthed.

The ignition timing is advanced and retarded automatically to ensure the spark occurs at just the right instant for the particular load at the prevailing engine speed.

The ignition advance is controlled both mechanically and by a vacuum-operated system. The mechanical governor mechanism consists of two weights which move out under centrifugal force from the central distributor shaft as the engine speed rises. As they move outwards they rotate the cam relative to the distributor shaft, and so advance the spark. The weights are held in position by two light springs, and it is the tension of these springs which is largely responsible for correct spark advancement.

The vacuum control consists of a diaphragm, one side of which is connected via a small bore tube, to the carburettor and the other side to the contact breaker plate. Depression in the induction manifold and carburettor, which varies with engine speed and throttle opening, causes the diaphragm to move, so rotating the contact breaker plate and advancing or retarding the spark.

A ballast resistor is incorporated in the low tension circuit between the ignition switch and the coil primary windings. During starting this resistor is bypassed allowing all available battery voltage to be fed to the coil which is of a low voltage type. This ensures that during starting when there is a heavy drain on the battery, sufficient voltage is still available at the coil to produce a powerful spark. During normal running, battery voltage is directed through the ballast resistor before reaching the coil.

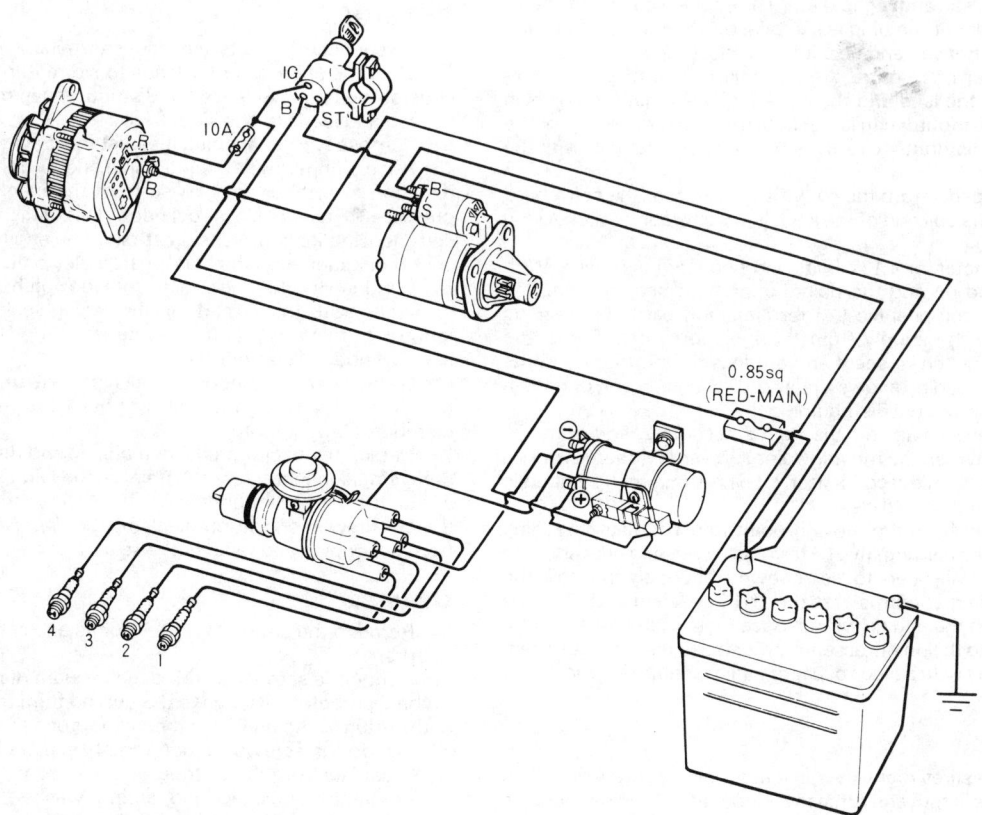

Fig. 5.1 Circuit layout of the conventional ignition system (Sec 1)

2 Ignition system – testing

1 There are two main symptoms indicating faults in the ignition system. Either the engine will not start or fire, or the engine is difficult to start and misfires. If it is a regular misfire, (ie the engine is running on only two or three cylinders), the fault is almost sure to be in the secondary or high tension circuit. If the misfiring is intermittent, the fault could be in either the high or low tension circuits. If the car stops suddenly, or will not start at all, it is likely that the fault is in the low tension circuit. Loss of power and overheating, apart from faulty carburation settings, are normally due to faults in the distributor or to incorrect ignition timing.

Engine fails to start

2 If the engine fails to start, and the engine was running normally when it was last used, first check that there is fuel in the fuel tank. If the engine turns over normally on the starter motor and the battery is evidently well charged, then the fault may be in either the high or low tension circuits. First check the HT circuit. If the battery is known to be fully charged, the ignition lights come on, but the starter motor fails to turn the engine, check the tightness of the leads on the battery terminals and how secure the earth lead connection is to the body. It is quite common for the leads to have worked loose, even if they look and feel secure. If one of the battery terminal posts gets very hot when trying to work the starter motor, this is a sure indication of a faulty connection to that terminal.
3 One of the most common reasons for bad starting is wet or damp spark plug leads and distributor. Remove the distributor cap. If condensation is visible internally, dry the cap with a rag and also wipe over the leads. Refit the cap.
4 If the engine still fails to start, check that the current is reaching the plugs, by disconnecting each plug lead in turn at the spark plug end, and holding the end of the lead, with insulated pliers to avoid electric shocks, about 5 mm away from the cylinder block. Spin the engine on the starter motor.
5 Sparking between the end of the lead and the block should be fairly strong with a good, regular blue spark. If current is reaching the plugs, then remove them, clean and regap them. The engine should now start.
6 If there is no spark at the plug leads, take off the HT lead from the centre of the distributor cap and hold it to the block as before. Spin the engine on the starter once more. A rapid succession of blue sparks between the end of the lead and the block indicates that the coil is in order and that the distributor cap is cracked, the rotor arm faulty, or the carbon brush in the distributor cap is not making good contact with the rotor arm.
7 If there are no sparks from the end of the lead from the coil, check the connections at the coil end of the lead. If it is in order start checking the low tension circuit.
8 Use a 12V voltmeter, or a 12V bulb, and two lengths of wire. With the ignition switched on and the points open, test between the low tension wire to the coil positive (+) terminal and earth. No reading indicates a break in the supply from the ignition switch. Check the connections at the switch to see if any are loose. Refit them and the engine should run. A reading shows a faulty coil or condenser, or broken lead between the coil and the distributor.
9 Take the condenser wire off the points assembly, and with the points open, test between the moving point and earth. If there is now a reading then the fault is in the condenser. Fit a new one, as described in Chapter 1, and the fault should clear.
10 With no reading from the moving point to earth, take a reading between earth and the coil negative (–) terminal. A reading here shows a broken wire which will need to be renewed between the coil and distributor. No reading confirms that the coil has failed and must be renewed, after which the engine will run once more. Remember to refit the condenser wire to the points assembly. For these tests it is sufficient to separate the points with a piece of paper while testing with the points open.

Engine misfires

11 If the engine misfires regularly, run it at a fast idling speed. Pull off each of the plug caps in turn and listen to the note of the engine. Hold the plug cap in a dry cloth or with a rubber glove as additional protection against a shock from the HT supply.
12 No difference in engine running will be noticed when the lead from the defective circuit is removed. Removing the lead from one of the good cylinders will accentuate the misfire.
13 Remove the plug lead from the end of the defective plug and hold it about 5 mm away from the block. Restart the engine. If the sparking is fairly strong and regular, the fault must lie in the spark plug.
14 The plug may be loose, the insulation may be cracked, or the electrodes may have burnt away, giving too wide a gap for the spark to jump. Worse still, one of the electrodes may have broken off.
15 If there is no spark at the end of the plug lead, or if it is too weak and intermittent, check the ignition lead from the distributor to the plug. If the insulation is cracked or perished, renew the lead. Check the connections at the distributor cap.
16 If there is still no spark, examine the distributor cap carefully for tracking. This can be recognised by a very thin black line running between two or more electrodes, or between an electrode and some other part of the distributor. These lines are paths which now conduct electricity across the cap, thus letting it run to earth. The only answer is a new distributor cap.
17 Apart from the ignition timing being incorrect, other causes of misfiring have already been dealt with under the Section dealing with the failure of the engine to start. To recap, these are that:

(a) The coil may be faulty giving an intermittent misfire.
(b) There may be a damaged wire or loose connection in the low tension circuit.
(c) The condenser may be short circuiting.
(d) There may be a mechanical fault in the distributor (broken driving spindle or contact breaker/advance mechanism spring).

18 If the ignition is too far retarded, it should be noted that the engine will tend to overheat, and there will be quite a noticeable drop in power. If the engine is overheating and the power is down, and the ignition timing is correct, then the carburettor should be checked, as it is likely that this is where the fault lies.

3 Distributor – removal, overhaul and refitting

Removal

1 Disconnect the battery negative terminal.
2 Mark the spark plug HT leads to aid refitting and pull them off the ends of the plugs. Release the distributor cap retaining clips and place the cap and leads to one side.
3 Remove No 1 spark plug (nearest the crankshaft pulley).
4 Place a finger over the plug hole and turn the engine in the normal direction of rotation (clockwise from the crankshaft pulley end) until pressure is felt in No 1 cylinder. This indicates that the piston is commencing its compression stroke. The engine can be turned with a socket or spanner on the crankshaft pulley bolt.
5 Continue turning the engine until the notch on the crankshaft pulley is aligned with the T mark on the timing scale just above the pulley (photo). In this position the engine is at Top Dead Centre (TDC) with No 1 cylinder on compression.
6 Using a dab of paint or a small file, make reference marks between the distributor base and the cylinder head, and the rotor arm and distributor body (photo).
7 Detach the vacuum advance pipe(s) and disconnect the LT lead at the wiring connector (photo). Release the wiring loom from the support clip on the distributor body.
8 Unscrew the distributor clamp bolt(s). Withdraw the distributor from the engine and recover the seal.

Overhaul

9 Remove the contact breaker points and condenser as described in Chapter 1.
10 Undo the screws securing the vacuum diaphragm unit to the side of the distributor. Disengage the pullrod from the peg on the baseplate and withdraw the unit from the distributor.
11 Undo the screws securing the baseplate in position and remove the baseplate from the distributor.
12 Extract the circlip and spring washer from the base of the distributor shaft then withdraw the shaft, complete with centrifugal governor, from the distributor. Recover the thrustwashers from the shaft.

Chapter 5 Part A: Contact breaker ignition system

Fig. 5.2 Exploded view of the conventional ignition distributor (Sec 3)

1. Distributor cap
2. Rotor arm
3. Contact breaker points
4. Vacuum diaphragm unit
5. Baseplate assembly
6. Distributor shaft circlip
7. Distributor shaft

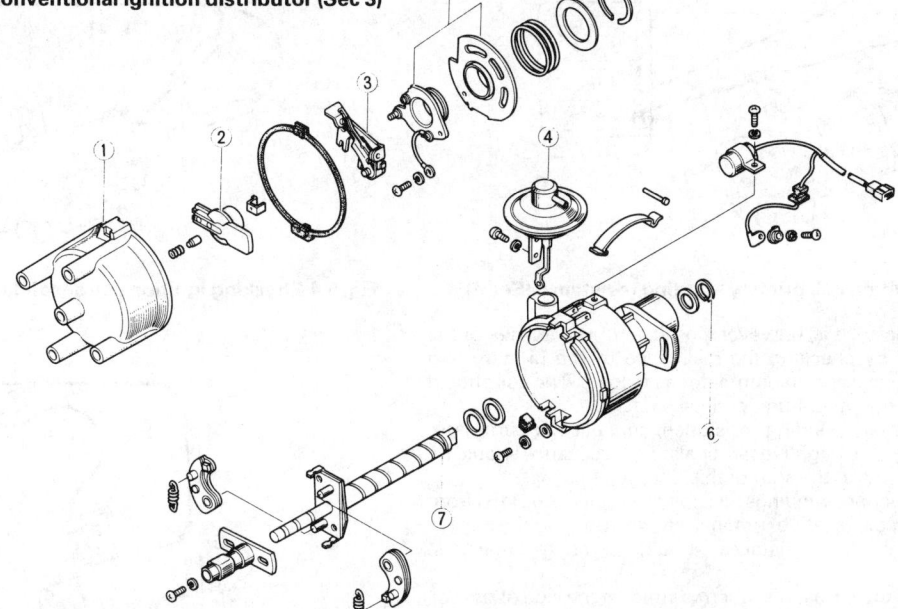

3.5 Crankshaft pulley notch aligned with TDC mark on timing scale (E series engine shown)

3.6 Reference marks (arrowed) made on distributor body and cylinder head

3.7 Disconnect the vacuum pipes from the vacuum diaphragm unit

13 With the distributor dismantled, renew any parts which show signs of wear or damage, and any that are known to be faulty. Pay particular attention to the centrifugal governor, checking for loose or broken springs, wear in the bob pivots and play in the distributor shaft. Apply suction to the vacuum diaphragm unit and check that the pullrod moves in as the suction is applied, and returns under spring pressure when the suction is released. If this is not the case renew the vacuum unit.

14 The distributor is reassembled by a reversal of the dismantling sequence, noting that the shaft should be lubricated with engine oil before inserting it into the body.

Refitting

15 To refit the distributor, first check that the engine is still at the TDC position with No 1 cylinder on compression. If the engine has been turned while the distributor was removed, return it to the correct position as previously described. Also make sure that the O-ring is in position on the base of the distributor and smear it with engine oil to aid refitting.

16 Align the previously made reference marks on the distributor body and rotor arm. If a new distributor is being fitted, position the rotor arm so that it is pointing towards the No 1 spark plug lead position in the cap.

17 With the vacuum diaphragm unit uppermost, insert the distributor into its location and turn the rotor arm slightly until the distributor shaft positively engages with the camshaft.

18 Turn the distributor until the alignment marks on the cylinder head and distributor base, made during removal, are aligned. If a new distributor is being fitted turn the unit until the rotor arm is pointing towards the No.1 spark plug lead segment in the cap.

19 Refit the distributor clamp bolt(s) and tighten securely.

20 Reconnect the LT lead at the connector, refit the vacuum advance pipe(s) and secure the wiring loom in the support clip.

21 Refit No 1 spark plug, the distributor cap and the spark plug HT leads.

22 Reconnect the battery then check and, if necessary, adjust the ignition timing as described in Chapter 1.

4 Ignition coil – removal, testing and refitting

Removal

1 The ignition coil is bolted to the inner valance on the left-hand side of the engine compartment.

2 To remove the coil disconnect the LT leads, at the coil positive and negative terminals, and the HT lead from the centre terminal of the coil.

3 Undo the mounting bracket retaining bolts and remove the coil.

Testing

4 Accurate checking of the coil output requires the use of specialist test equipment and should be left to a Mazda dealer or suitably

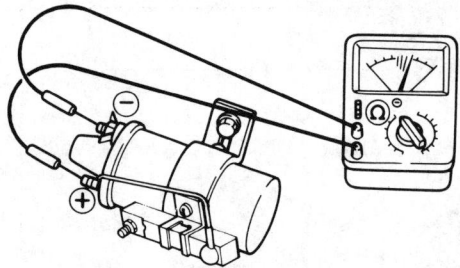

Fig. 5.3 Checking ignition coil primary winding resistance (Sec 4)

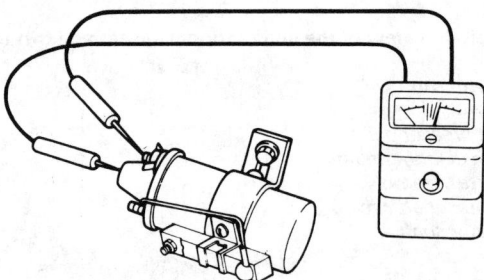

Fig. 5.4 Checking ignition coil secondary winding resistance (Sec 4)

equipped auto electrician. It is, however, possible to gain an idea of the condition of the coil by checking the resistance of the primary and secondary coil windings using an ohmmeter as follows. The coil should be at normal operating temperature for these checks.

5 To check the primary winding resistance, connect the ohmmeter across the coil positive and negative terminals. The resistance should be as given in Specifications at the start of this Chapter.

6 To check the secondary windings resistance, connect one lead from the ohmmeter to the coil negative terminal, and the other to the coil HT centre terminal. Again the resistance should be as given in the Specifications.

7 The resistance of the ballast resistor mounted on the side of the coil can also be checked by connecting the meter leads to the two resistor terminals.

Refitting

8 Refitting is a reversal of removal, but if necessary wipe clean the top of the coil to prevent any tracking of the HT current.

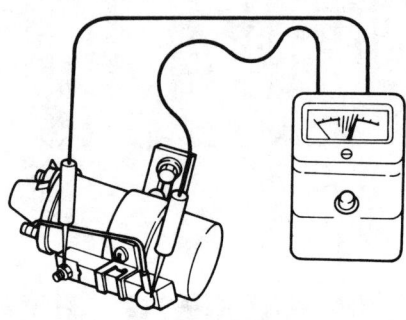

Fig. 5.5 Checking ignition coil ballast resistance (Sec 4)

Part B: Electronic ignition system

5 General information and precautions

Warning: *The voltages produced by the electronic ignition system are considerably higher than those produced by conventional systems. Extreme care must be taken when working on the system with the ignition switched on. Persons with surgically-implanted cardiac pacemaker devices should keep well clear of the ignition circuits, components and test equipment.*

Models of the Mazda 323 manufactured after December 1981 are equipped with an electronic High Energy Ignition (H.E.I.) system. The system is very similar in operation to that which is described in Section 1 with the exception of the distributor which operates as follows.

When the system is in operation, low tension voltage is changed in the coil into high tension voltage by the action of the igniter in conjunction with the signal rotor, permanent magnet and pick-up coil of the signal generator assembly. As each of the signal rotor teeth pass through the magnetic field around the pick-up coil, an electrical signal is sent to the igniter unit which triggers the coil in the same way as the contact breaker points on the earlier models. Otherwise the system is the same as that described in Section 1 noting that there is no ballast resistor incorporated in the low tension circuit.

Due to the sophisticated nature of the electronic ignition system the following precautions must be observed to prevent damage to the components and reduce risk of personal injury:

(a) *Ensure that the ignition is switched off before disconnecting any of the ignition wiring.*
(b) *Ensure that the ignition is switched off before connecting or disconnecting any ignition test equipment such as a timing light.*
(c) *Do not connect a suppression condenser or test lamp to the ignition coil negative terminal.*
(d) *Do not connect any test appliance or stroboscopic timing light requiring a 12 volt supply to the ignition coil positive terminal.*
(e) *Do not allow an HT lead to short out or spark against the computer control unit body.*
(f) *Do not earth the coil primary or secondary circuits.*

6 Ignition system – testing

The electronic ignition system can be tested using the information given for the contact breaker system in Section 2, ignoring the remarks made about testing the Low Tension circuit in paragraphs 8 to 10. The LT circuit can be checked as follows.

Using a 12V voltmeter, check for voltage between the low tension positive (+) terminal of the ignition coil and earth. With the ignition switch in the 'IG' position a reading of 12 volts should be obtained. No reading indicates a break in the supply from the ignition switch, this could be due to a blown fuse, or a fault in either the ignition switch itself or the relevant wiring. Check the connections at the switch to see if any are loose. Refit them and the engine should run.

If there is no still no reading, check for voltage between the low tension negative (–) terminal of the ignition coil and earth. With the ignition switch in the START position a reading of approximately 6 volts should be obtained. If this is not the case, it is likely that the fault lies in the HT ignition coil which should be tested as described in Section 8. If a reading of 6 volts is obtained the fault lies in one of the distributor components. Due to the complex nature of the distributor it is recommended that it should be taken to a Mazda dealer for examination since further testing is not possible without specialist equipment.

Chapter 5 Part B: Electronic ignition system

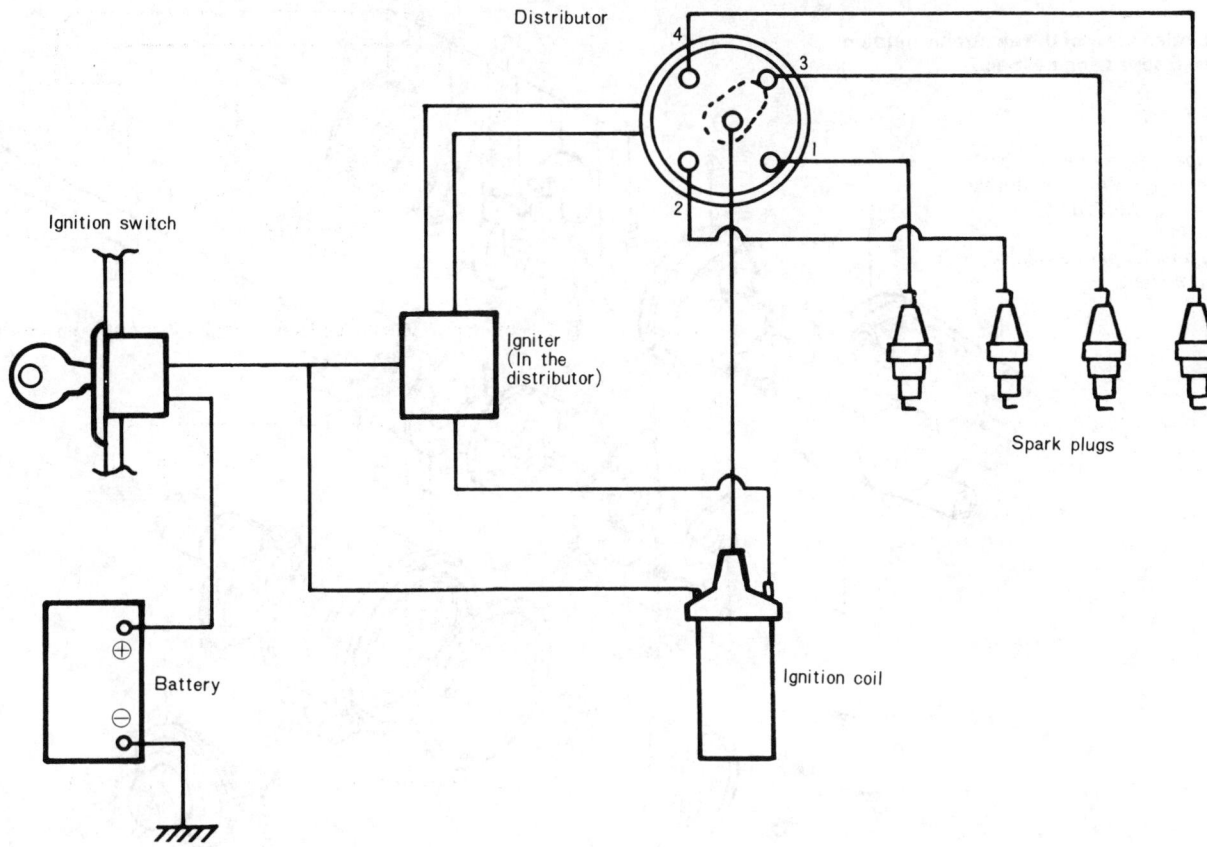

Fig. 5.6 Circuit layout of the electronic ignition system (Sec 5)

7 Distributor – removal overhaul and refitting

Removal

1 The distributor can be removed as described in Section 3 noting that the distributor cap is retained by two screws (photos).

Overhaul

E series engine

2 Undo the two screws and lift off the rotor (photo).
3 Undo the retaining bolt and lift the centrifugal governor and signal rotor assembly off the distributor shaft (photos).
4 Undo the two screws and remove the igniter and signal generator assembly together with sealing ring (photos).
5 Undo the screws securing the vacuum diaphragm unit to the distributor body. Disengage the unit pullrod from the signal plate peg and withdraw the unit from the distributor body (photo).
6 Undo the two screws and remove the signal plate assembly (photos).
7 Undo the two screws securing the bearing retainer plate assembly to the distributor and withdraw the plate (photos).
8 Drive out the roll pin from the base of the distributor shaft and remove the drive key (photo). Remove the circlip, where fitted, from the distributor shaft and press the shaft and bearing out of the distributor body.

B series engine

9 Remove the rotor, cover and sealing ring from the distributor assembly.
10 On B6 engined models, undo the screw from the end of the distributor shaft, and remove the cylinder identification rotor. Undo the two screws securing the cylinder identification signal generator baseplate in position, and remove the base plate and generator from the

7.1A Distributor cap is retained by two screws (arrowed)

7.1B Undo the distributor clamp bolts...

7.1C ...and remove the distributor from the cylinder head (B series engine shown)

Fig. 5.7 Exploded view of the electronic ignition distributor – E series engine (Sec 7)

1 Distributor cap
2 Rotor arm
3 Centrifugal governor assembly
4 Igniter and signal rotor assembly
5 Vacuum diaphragm unit
6 Signal plate
7 Bearing retainer plate assembly
8 Distributor shaft

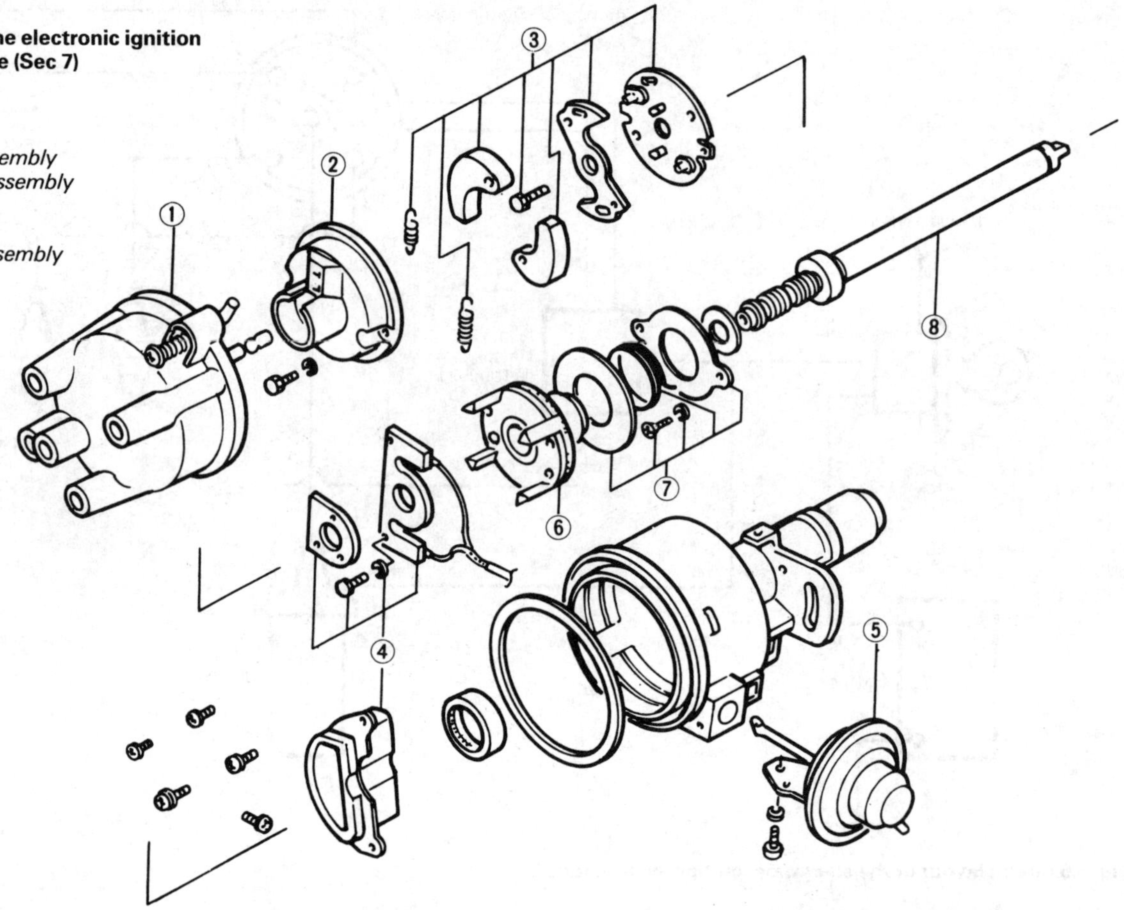

7.2 Undo the two screws (arrowed) and lift off the rotor

7.3A Undo the retaining bolt...

7.3B ...and withdraw the centrifugal governor and signal rotor assembly

distributor. Extract the circlip from the distributor shaft and slide off the signal rotor.

11 On B3 and B5 engined models withdraw the signal rotor and retaining pin noting that, if necessary, the rotor can be drawn off using a universal puller. Remove the screw from the centre of the rotor shaft (photos).

12 Undo the two screws securing the igniter and signal generator assembly to the mounting plate and remove the assembly. Undo the two screws securing the mounting plate to the distributor body and remove the plate from the distributor (photos).

13 Undo the screw(s) securing the vacuum diaphragm unit to the distributor body and remove the spring clip securing the pullrod to the signal plate (photo). Disengage the unit pullrod from the signal plate peg and withdraw the vacuum unit.

14 Undo the two screws and remove the signal plate assembly. Note that on reassembly it is possible to locate the signal plate slots with the centrifugal governor pins so that the plate assembly is 180° out. Mark the centrifugal governor weight pins and signal plate slots in some way to use as a guide on refitting (photo).

15 Drive out the roll pin from the base of the distributor shaft and remove the drive key and washer.

16 Withdraw the centrifugal governor and shaft assembly from the distributor body. Undo the two screws securing the bearing retainer plate (where fitted) to the distributor and remove the oil seal, washer and bearing from the body.

All models

17 With the distributor dismantled, renew any parts that show signs of wear or damage and any that are known to be faulty. Pay close attention to the centrifugal governor mechanism checking for loose or broken springs, wear in the weight pivots and play in the distributor shaft (photo). Apply suction to the vacuum diaphragm unit, and check that the pullrod moves in as the suction is applied, and returns under spring pressure when the suction is released. If this is not the case

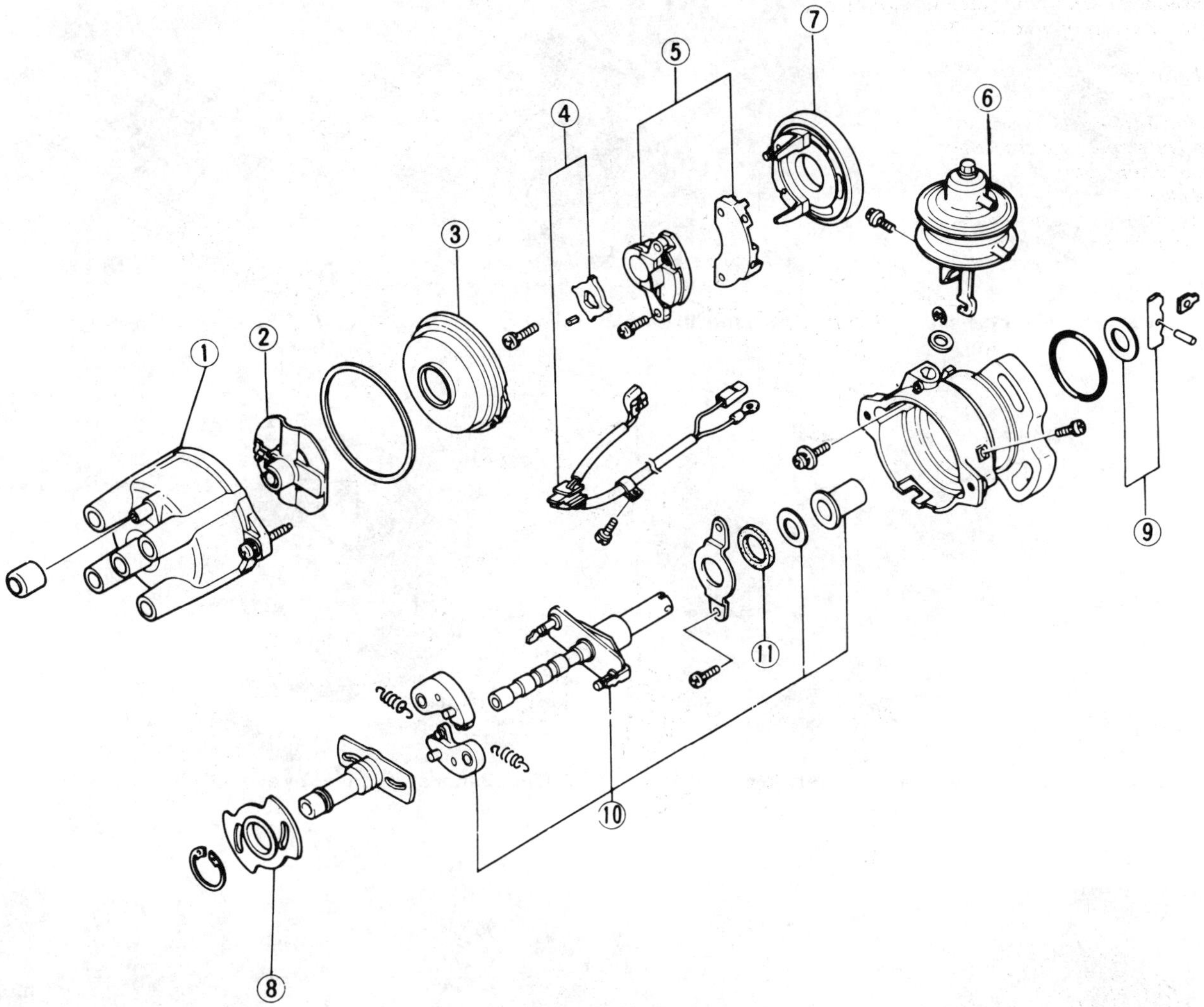

Chapter 5 Part B: Electronic ignition system

Fig. 5.8 Exploded view of the distributor – B3 and B5 engines (Sec 7)

1. Distributor cap
2. Rotor arm
3. Cover
4. Signal rotor and wiring
5. Igniter unit and signal generator assembly
6. Vacuum diaphragm unit
7. Signal plate
8. Plate
9. Distributor shaft drive key and washer
10. Centrifugal governor assembly and bush
11. Oil seal

7.4A Undo the igniter and signal generator retaining screws...

7.4B ...then remove the assembly along with the sealing ring

7.5 Disengage the pullrod and remove the vacuum diaphragm unit

Chapter 5   Part B: Electronic ignition system

7.6A Undo the signal plate retaining screws...

7.6B ...and lift out the assembly

7.7A Remove the bearing retainer plate screws...

7.7B and withdraw the retainer plate

7.8 Distributor shaft drive key and roll pin

7.11A If necessary, rotor and retaining pin (arrowed)...

7.11B ...can be withdrawn using a suitable puller...

7.11C ...then remove the screw from the centre of the rotor shaft

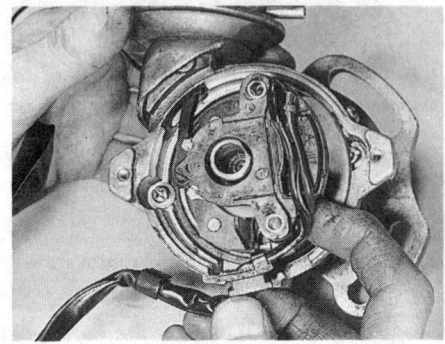
7.12A Remove the igniter and signal generator assembly...

7.12B ...and undo the mounting plate retaining screws

7.13 Undo the vacuum diaphragm retaining screws and remove spring clip (arrowed)

Chapter 5 Part B: Electronic ignition system

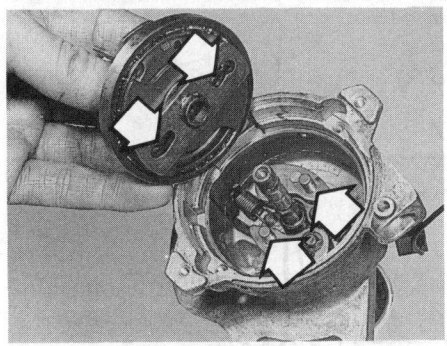

7.14 Mark the relative positions of the slots and pins to avoid confusion on reassembly

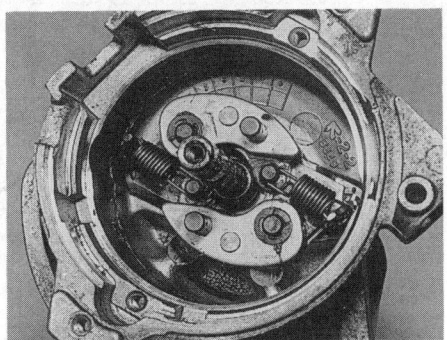

7.17 Closely examine the centrifugal governor assembly weight and springs for wear or damage

7.19 On B series engines, align cutaway on rotor drive key with mark on distributor body (arrowed)

renew the vacuum unit.
18 The distributor is reassembled by a reversal of the dismantling sequence, noting that the shaft should be lubricated with engine oil before inserting it into the body.

Refitting
19 The distributor can be refitted as described in Section 3, noting that when installing a distributor on B series engined models the rotor arm can be correctly positioned by aligning the cutaway on the end of the distributor drive key with the mark on the base of the distributor (photo).

8 Ignition coil – removal, testing and refitting

Removal
1 The ignition coil is located in the engine compartment where it is bolted onto the airflow meter housing on fuel injected models, and onto the left-hand inner valance on all carburettor models (photos).
2 To remove the coil, disconnect the LT leads at the coil positive and negative terminals, and the HT lead from the centre terminal of the coil.
3 Undo the mounting bracket retaining bolts and remove the coil.

Testing
4 Accurate checking of the coil output requires the use of specialist test equipment and should be left to a Mazda dealer or suitably equipped auto electrician. It is, however, possible to gain an idea of the condition of the coil by checking the resistance of the primary and secondary coil windings using an ohmmeter as follows. The coil should be at normal operating temperature for these checks.
5 To check the condition of the primary windings, connect the ohmmeter across the coil positive and negative terminals. Continuity should be present between the terminals.
6 To check the secondary windings resistance connect one lead from the ohmmeter to the coil negative terminal and the other to the coil HT centre terminal. The resistance should be as given in the Specifications at the start of this Chapter.
7 Also check the condition of the insulation between the primary windings and the body of the coil by connecting the meter leads to one of the coil low tension terminals and the other to the coil body. A reading of at least 10 M ohms should be obtained.

Refitting
8 Refitting is a reversal of removal, but if necessary wipe clean the top of the coil to prevent any tracking of the HT current.

9 Shutter valve ignition advance mechanism (B3 and B5 engine models) – testing

1 If when checking the ignition timing on B3 and B5 engined models, as described in Chapter 1, it is found that the timing is not advancing correctly, the shutter valve advance mechanism should be tested as follows.
2 Connect a timing light to the engine in accordance with the manufacturer's instructions (usually between No 1 spark plug and its HT lead).

8.1A Ignition coil location on fuel injected models

8.1B Ignition coil location on carburettor models

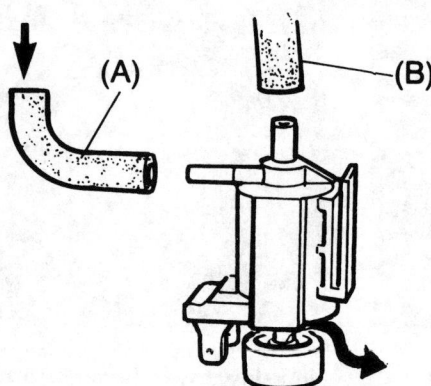

Fig. 5.9 Distributor three-way solenoid valve operation (Sec 9)

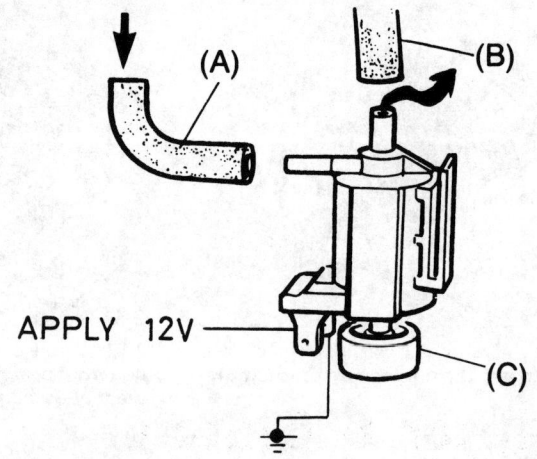

Fig. 5.10 Distributor three-way solenoid valve operation with battery voltage applied (Sec 9)

3 Start the engine and, with the timing light pointing at the timing marks, slowly increase the engine speed. When the engine reaches approximately 2500 rpm the ignition timing should suddenly advance. If this is not the case carefully check the condition of the relevant wiring and connectors before proceeding as follows.

4 Identify the hose which joins the vacuum diaphragm unit mounted on the distributor to the three-way solenoid valve mounted on the bulkhead, then disconnect it from the vacuum diaphragm. Start the engine and place a finger over the end of the hose. Increase the engine speed and check that a vacuum is present in the hose when the engine exceeds 2500 rpm. If this is the case the fault lies in the distributor. Check the operation of the vacuum diaphragm unit and distributor components as described in Section 7.

5 If there is no vacuum present, reconnect the hose to the vacuum diaphragm unit and trace the hose back to the three-way solenoid valve. Using a 12 V voltmeter, connect the meter positive (+) lead to the black and white terminal of the solenoid, and the negative lead (-) to earth. Start the engine and allow it to idle then slowly increase the engine speed to above 2500 rpm, noting the readings obtained. Below 2500 rpm a reading of approximately 1.5 volts should be obtained, and above 2500 rpm a reading of 12 volts. If the readings are not obtained, check the operation of the shutter valve control unit as described in Section 14 of Chapter 4. If all is well check the operation of the solenoid as follows.

6 Disconnect the hose from the distributor vacuum unit and trace it back to the three-way solenoid valve. Disconnect the other hose from the end of the solenoid then disconnect the electrical connections and remove the solenoid from the car. Blow through the hose which was disconnected from the vacuum unit (A in Fig. 5.9) and check that the air comes out of the solenoid valve exhaust port. Then, using a 12 volt battery and two auxiliary wires, connect the battery across the solenoid terminals and blow through the hose (A) again. The air should now be expelled through the vacuum hose union (B) on the end of the solenoid valve (Fig. 5.10). If not the three-way solenoid valve is faulty and must be renewed.

Chapter 6 Clutch

Contents

Clutch assembly – removal, inspection and refitting	7	Clutch pedal – adjustment	See Chapter 1
Clutch cable – adjustment	See Chapter 1	Clutch pedal – removal and refitting	3
Clutch cable – removal and refitting	2	Clutch release bearing – removal, inspection and refitting	8
Clutch hydraulic system – bleeding	6	Clutch slave cylinder – removal, overhaul and refitting	5
Clutch master cylinder – removal, overhaul and refitting	4	General information	1

Specifications

Type
Estate models from October 1989 onward .... Hydraulically operated single dry plate with diaphragm spring
All other models .... Cable operated single dry plate with diaphragm spring

Driven plate
Diameter:
   1100 cc and 1300 cc models .... 184 mm
   1500 cc and 1600 cc models .... 190 mm
Minimum lining thickness .... 0.3 mm above the rivet heads
Maximum permissible run-out:
   Pre-September 1985 models (lateral and vertical) .... 1.0 mm
   September 1985 models onward:
      Lateral .... 0.7 mm
      Vertical .... 1.0 mm

Torque wrench settings
	Nm	lbf ft
Clutch cable upper mounting bracket nuts	16 to 23	12 to 17
Clutch cover to flywheel	18 to 27	13 to 20
Release fork to lever retaining bolt	18 to 27	13 to 20

1 General information

All manual gearbox models are equipped with a single dry plate diaphragm spring clutch assembly. The unit consists of a steel cover which is dowelled and bolted to the rear face of the flywheel, and contains the pressure plate and diaphragm spring.

The clutch driven plate is free to slide along the splined gearbox input shaft, and is held in position between the flywheel and the pressure plate by the pressure of the diaphragm spring. Friction lining material is riveted to the clutch driven plate which has a spring cushioned hub to absorb transmission shocks and help ensure a smooth take-up of the drive.

The driven plate is located between the flywheel and the clutch pressure plate and slides on the splines of the gearbox input shaft. When the clutch is engaged, the diaphragm spring forces the pressure plate to grip the driven plate against the flywheel and drive is transmitted from the crankshaft, through the driven plate to the gearbox primary shaft. On disengaging, the clutch pressure plate is lifted to release the driven plate with the result that the drive to the gearbox is disconnected.

The clutch is operated by a foot pedal suspended under the facia. On Estate models from October 1989 onward, the pedal is hydraulically linked to the clutch release lever on the gearbox housing. The pedal operates the clutch master cylinder, which is linked to a slave cylinder mounted on the gearbox housing by means of a pipe and flexible hose. On all other models the clutch pedal is linked directly to the clutch release lever by means of a cable.

Depressing the clutch pedal actuates the release arm, pushing the release bearing against the diaphragm fingers, so moving the centre of the diaphragm spring inwards. As the centre of the spring is pushed in, the outside of the spring pivots out, so moving the pressure plate backwards and disengaging its grip on the driven plate.

When the pedal is released, the diaphragm spring forces the

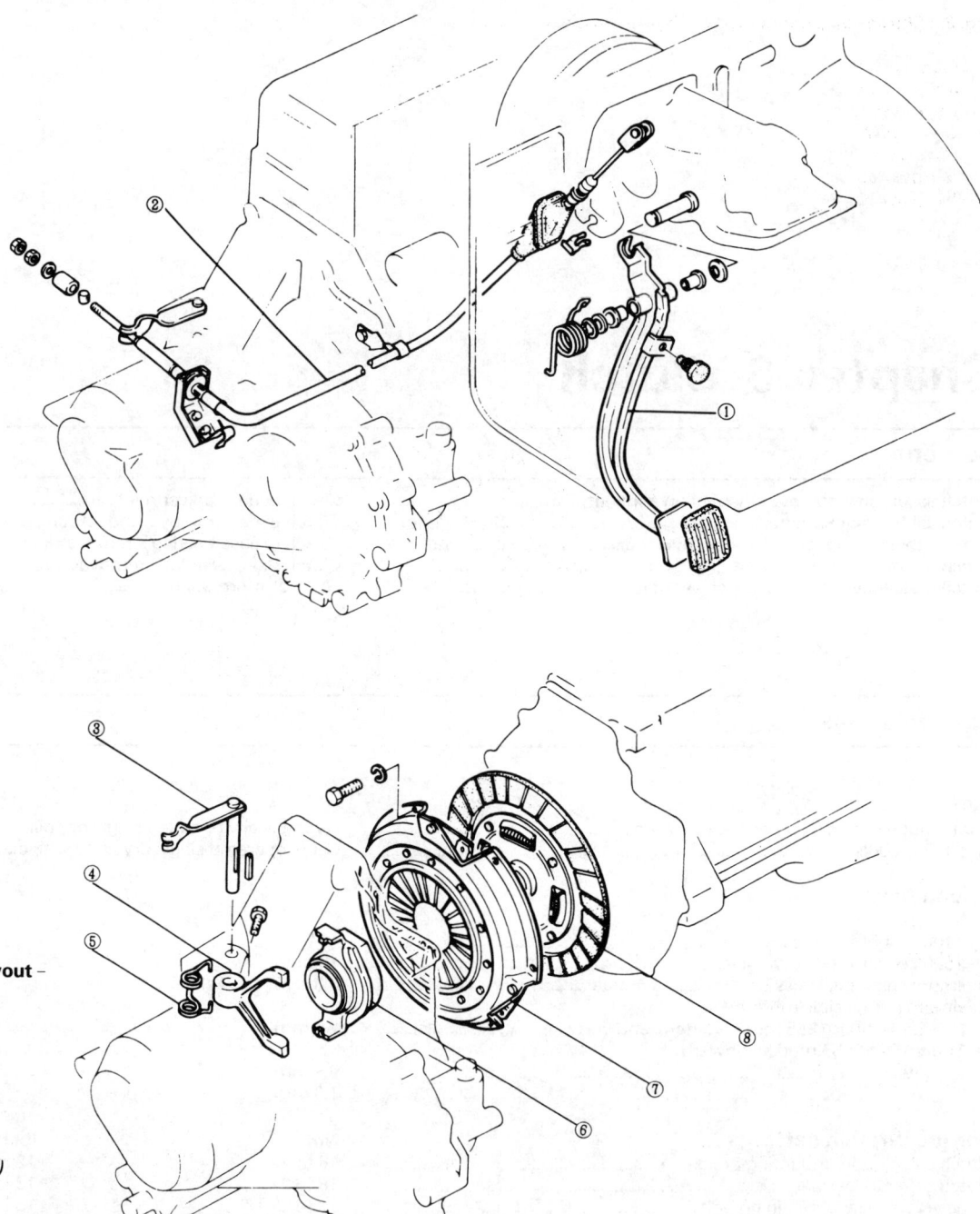

Fig. 6.1 Clutch component layout – cable operated clutch (pre-September 1985 models shown) (Sec 1)

1 Clutch pedal
2 Clutch cable
3 Release lever
4 Release fork
5 Return spring (where fitted)
6 Release bearing
7 Cover assembly
8 Driven plate

pressure plate into contact with the friction linings on the driven plate. The driven plate is now firmly sandwiched between the pressure plate and the flywheel, thus transmitting engine power to the gearbox.

2 Clutch cable – removal and refitting

Removal

1 Working in the engine compartment, using a suitable pair of pliers, disconnect the return spring from the clutch release lever (where fitted).
2 Unscrew the locknut (if fitted) then remove the adjuster nut and slide off the damper and roller (as applicable) from the end of the clutch cable.
3 Free the lower end of the cable from the release arm, mounting bracket and any relevant clips or guides.
4 From inside the car remove the cover from under the facia and disconnect the cable end from the hook on the pedal.
5 On pre-September 1985 models, extract the retaining clip and withdraw the upper end of the cable from the master cylinder bracket. Release the rubber boot and pull the cable through into the engine compartment.
6 On September 1985 models onward working from inside the engine compartment, undo the two nuts which secure the clutch cable upper mounting bracket to the engine compartment bulkhead (photo).
7 Release the cable from any remaining clips or guides and remove it from the engine compartment.
8 Inspect the cable for signs of fraying and for smoothness of operation. On pre-September 1985 models also examine the rubber mounting boot for damage or deterioration. Renew the cable if any of the above areas are unsatisfactory.

Chapter 6 Clutch

Fig. 6.2 Clutch component layout – hydraulically operated clutch (Sec 1)

1. Clutch pedal
2. Master cylinder
3. Hydraulic hose
4. Driven plate
5. Cover assembly
6. Release bearing
7. Slave cylinder

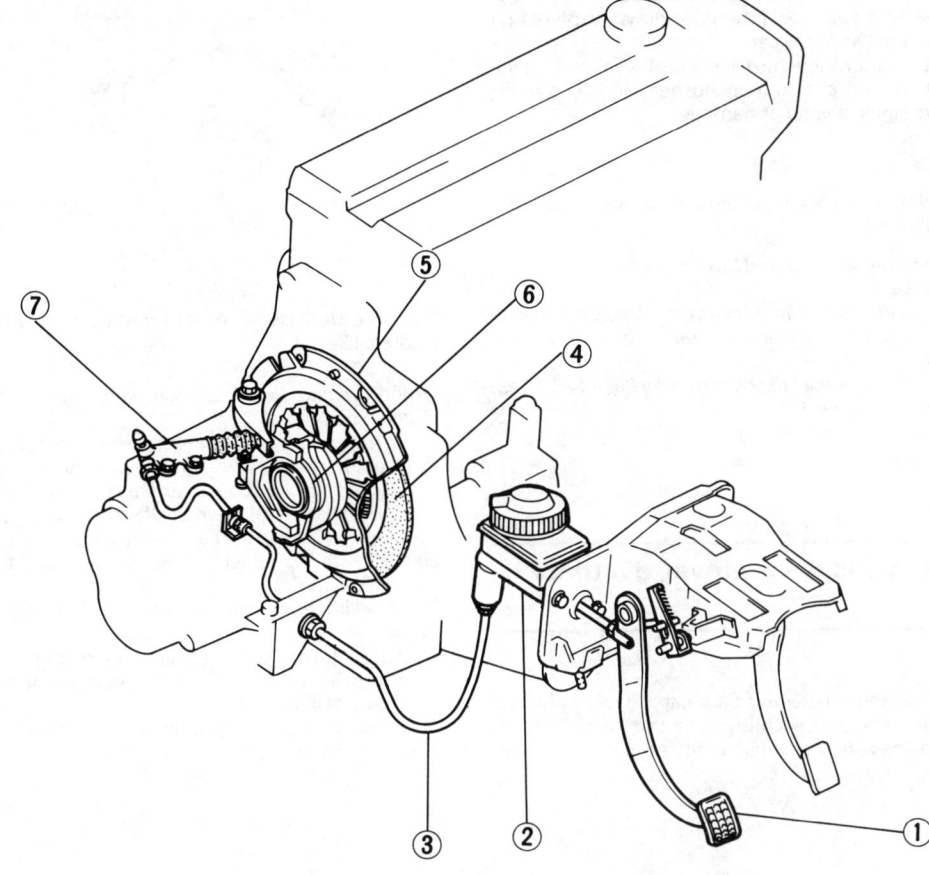

Refitting

9 Refitting is the reverse of the removal sequence, bearing in mind the following points.

(a) Lubricate the clutch pedal hook and the cable roller with multi-purpose grease.
(b) On pre-September 1985 models, apply a suitable sealant to the groove of the mounting rubber before refitting it to the bulkhead.
(c) On September 1985 models onward, tighten the cable upper mounting bracket nuts to the specified torque.
(d) With the cable installed, adjust the clutch pedal height and free play as described in Chapter 1.

3 Clutch pedal – removal and refitting

Removal

Pre-September 1985 models

1 Remove the cover from underneath the facia.
2 Release the pedal return spring, then extract the retaining clip from the end of the pedal pivot pin.
3 Withdraw the pivot pin and recover the washers and return spring.
4 Withdraw the pedal from its location and disconnect it from the clutch cable.
5 Examine all clutch pedal components for signs of wear or damage, paying particular attention to the mounting bushes, and renew any component which shows signs of wear or damage.

September 1985 models onward

6 Remove the cover underneath the facia.
7 On E1 and E3 engined models using a pair of suitable pliers, unhook the pedal return spring and remove it from the car.
8 On B3, B5, B6 and E5 engined models, remove the retaining clip

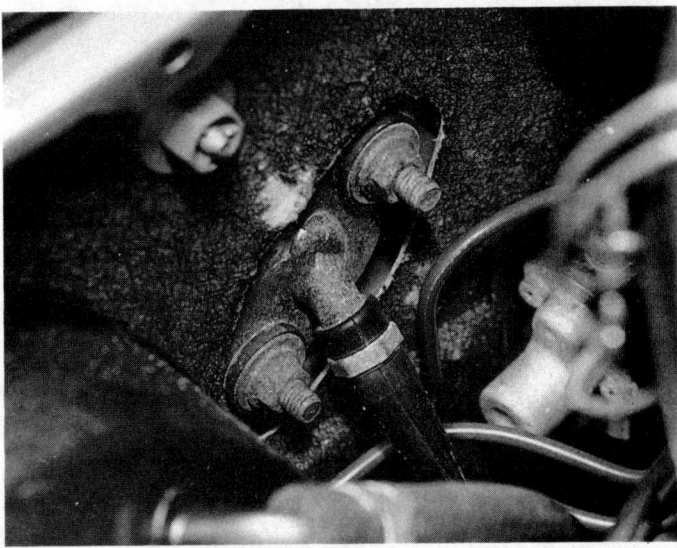

2.6 Clutch cable to bulkhead mounting nuts – September 1985 models onward

securing the return spring assembly to the master cylinder bracket and remove the return spring and lever assembly complete with bushes.
9   On October 1989 Estate models, remove the circlip and withdraw the clevis pin securing the master cylinder pushrod to the clutch pedal.
10 On all models undo the retaining nut from the pivot bolt, remove the cover, second nut, washers and spacer then withdraw the pivot bolt.
11 Remove the clutch pedal from the car.
12 Examine all clutch pedal components for signs of wear or damage, paying particular attention to the mounting bushes, and renew any component which shows signs of wear or damage.

Refitting

13 The pedal is refitted by a reverse of the removal sequence, bearing in mind the following points.

(a) Lubricate the pedal pivot points and bushes with a multi-purpose grease.
(b) Ensure that the pedal return spring is correctly located noting that on E1 and E3 models the short hook must be on the pedal end of the spring.
(c) With the pedal installed, adjust the clutch pedal height and free play as described in Chapter 1.

Fig. 6.3 Clutch pedal attachment details – pre-September 1985 models (Sec 3)

4 Clutch master cylinder – removal, overhaul and refitting

Removal

1   Unscrew the master cylinder reservoir filler cap. Place a piece of polythene over the filler neck and securely refit the cap. This will minimise hydraulic fluid loss during subsequent operations. As an added precaution place absorbent rags beneath the master cylinder clutch pipe union.
2   Wipe clean the area around the clutch pipe union on the bottom of the master cylinder, then unscrew the union nut and carefully withdraw the pipe. Plug or tape over the pipe and master cylinder orifice to minimise the loss of hydraulic fluid, and to prevent the entry of dirt into the system. Take great care not to allow any hydraulic fluid to come into contact with the vehicle paintwork. Wash off any spilt fluid immediately with cold water.
3   Working from inside the car, remove the cover from underneath the facia.
4   Undo the nut which secures the master cylinder to the bulkhead.
5   From inside the engine compartment undo the second master cylinder retaining nut.
6   Remove the master cylinder from the engine compartment along with its mounting spacer and gaskets.

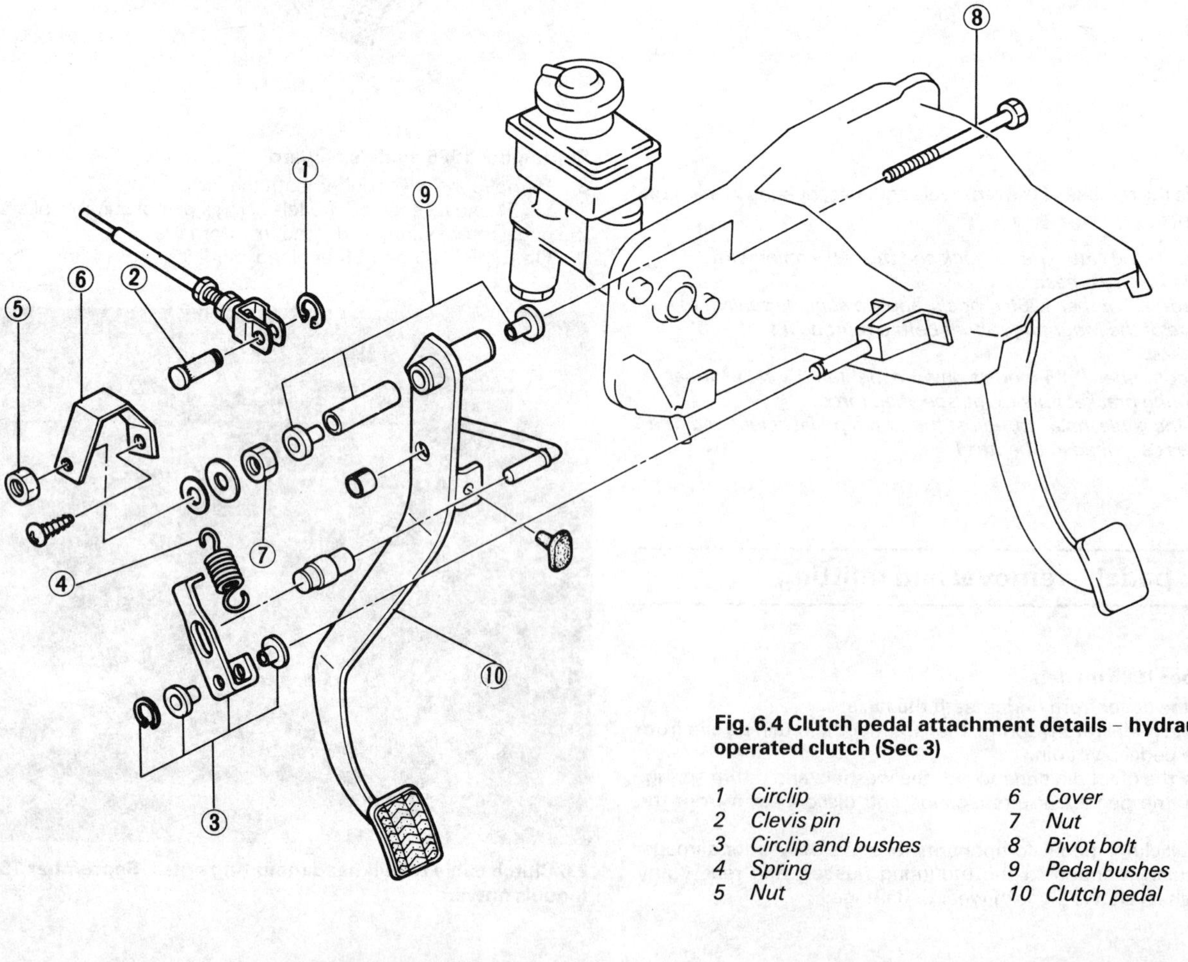

Fig. 6.4 Clutch pedal attachment details – hydraulically operated clutch (Sec 3)

1 Circlip
2 Clevis pin
3 Circlip and bushes
4 Spring
5 Nut
6 Cover
7 Nut
8 Pivot bolt
9 Pedal bushes
10 Clutch pedal

Chapter 6 Clutch

Fig. 6.5 Exploded view of the clutch master cylinder (Sec 4)

1 Circlip
2 Piston and secondary cup assembly
3 Spring seat
4 Primary cup
5 Spring
6 One-way valve bolt
7 Sealing washer
8 One-way valve piston
9 Spring
10 Cap
11 Reservoir
12 Mounting bush seal
13 Master cylinder body

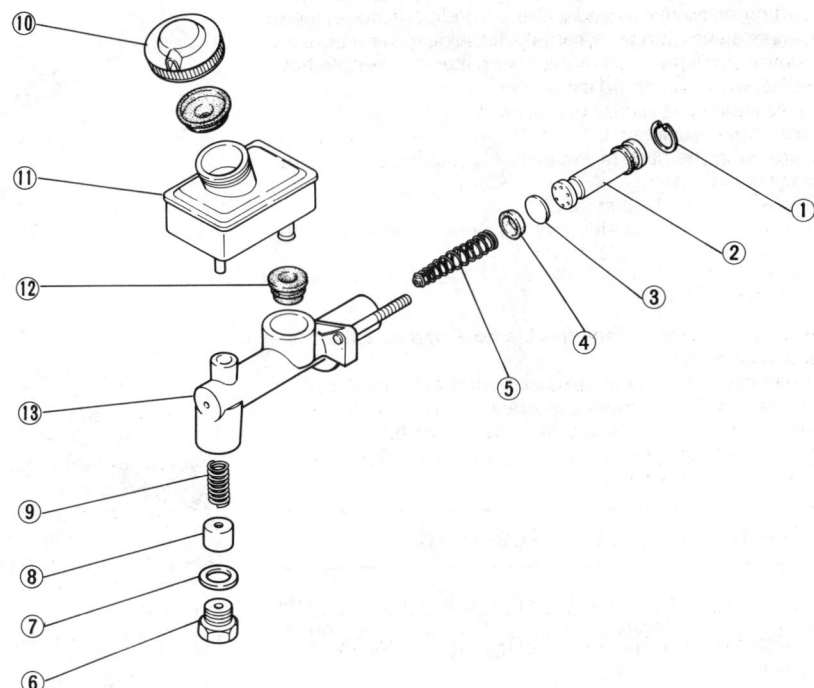

Overhaul

7 Remove the reservoir filler cap and pour out the hydraulic fluid into a suitable container. Prepare a clean uncluttered work surface on which to dismantle the unit.
8 Using a wooden dowel or similar, push the piston into the master cylinder bore then extract the piston retaining circlip using a suitable pair of circlip pliers.
9 Withdraw the piston assembly followed by the spring seat and spring. If necessary, tap gently on the master cylinder body with a soft faced mallet to release the piston. If this fails to release it, the piston can be pushed out by applying compressed air to the union bolt hole. Only low pressure compressed air should be required such as is generated by a foot pump.
10 Securely clamp the master cylinder body in a vice equipped with soft jaws, then slacken and remove the one-way valve bolt and sealing washer from the bottom of the master cylinder. Withdraw the valve piston, noting which way around it is fitted, and return spring. If necessary the piston can be removed using one of the methods described above.
11 Using a flat bladed screwdriver carefully prise the reservoir off the master cylinder and remove the mounting bush seal from the cylinder body.
12 With the master cylinder completely dismantled, clean all the components in methylated spirit, or clean hydraulic fluid, and dry with a lint-free rag.
13 Carefully examine the cylinder bore and pistons for signs of wear, scoring or corrosion. If damage is evident renew the complete assembly.
14 If the components are in a satisfactory condition, obtain a repair kit consisting of a new piston assembly, springs, one-way valve bolt washer and piston retaining circlip. Examine the master cylinder reservoir bush seal for damage and renew if necessary.
15 Lubricate the master cylinder bore, pistons, and seals thoroughly in clean hydraulic fluid and assemble them as follows.
16 Insert the piston return spring, ensuring that its tapered end is innermost, and spring seat.
17 Insert the piston into the cylinder bore using a twisting motion whilst taking great care not to distort the lips of the new seals as they enter the cylinder.
18 Secure the piston in position with the circlip, ensuring that it is correctly located in the groove in the master cylinder bore.
19 Insert the one-way valve return spring and piston ensuring that the piston is fitted the correct way around.
20 Refit the one-way valve bolt along with a new washer and tighten it securely.
21 Fit the reservoir mounting bush seal to the cylinder inlet port and push the reservoir firmly into place.

Refitting

22 Position a new gasket on each side of the master cylinder mounting spacer, and refit the spacer and master cylinder.
23 Refit both the master cylinder mounting nuts and tighten them securely.
24 Refit the cover to the underside of the facia.
25 Refit the clutch pipe to the master cylinder and tighten its union nut securely.
26 Refill the master cylinder reservoir and bleed the system as described in Section 6.

5 Clutch slave cylinder – removal, overhaul and refitting

Removal

1 Slacken the flexible hose to slave cylinder union.
2 Fit a brake hose clamp to the flexible hose leading to the slave cylinder. This will minimise hydraulic fluid loss during subsequent operations.
3 Undo both the slave cylinder mounting bolts.
4 Unscrew the slave cylinder from the end of the flexible hose and remove it from the engine compartment.

Overhaul

5 Prise the rubber gaiter off the end of the cylinder and separate the gaiter and pushrod.
6 Withdraw the piston and spring from the cylinder bore. If necessary, tap gently on the slave cylinder body with a soft faced mallet to release the piston. If this fails to release it, the piston can be pushed out by applying compressed air to the hose union hole. Only low pressure compressed air should be required such as is generated by a foot pump.
7 Remove the slave cylinder bleed screw and tip out the steel ball.

8 Clean all parts in methylated spirit, or clean hydraulic fluid, and wipe dry using a lint-free cloth. Inspect the piston and cylinder bore for signs of damage, scuffing or corrosion and if these conditions are evident, renew the complete slave cylinder assembly. If the piston and bore are satisfactory, obtain a new piston seal, cylinder gaiter and flexible hose sealing washer. Renew the pushrod if it is bent.
9 Lubricate the piston and cylinder bore with clean hydraulic fluid and refit the steel ball and bleed screw.
10 Refit the spring to the back of the piston and carefully insert the piston into the cylinder bore.
11 Insert the pushrod into the piston.
12 Install the rubber gaiter ensuring that it is correctly located in the grooves in both the pushrod and slave cylinder.

Refitting

13 Fit a new sealing washer to the flexible hose and refit the slave cylinder onto the hose end.
14 Refit the slave cylinder to the gearbox housing and tighten both the mounting bolts and the hose union securely.
15 Bleed the hydraulic system as described in Section 6.

6 Clutch hydraulic system – bleeding

1 The clutch hydraulic system must be bled whenever any part of the system is disconnected. The system is bled via the bleed screw on the slave cylinder using a similar procedure to that given for the braking system in Section 5 of Chapter 9 (Fig. 6.7).

7 Clutch assembly – removal, inspection and refitting

Warning: *Dust created by clutch wear and deposited on the clutch components may contain asbestos which is a health hazard. DO NOT blow it out with compressed air or inhale any of it. DO NOT use petrol or petroleum based solvents to clean off the dust. Brake system cleaner or methylated spirit should be used to flush the dust into a suitable receptacle. After the clutch components are wiped clean with rags, dispose of the contaminated rags and cleaner in a sealed, marked container.*

Removal

1 Access to the clutch may be gained in one of two ways. Either the engine/gearbox unit can be removed, as described in Chapter 2, and the gearbox separated from the engine, or the engine may be left in the car and the gearbox unit removed independently, as described in Chapter 7.
2 Having separated the gearbox from the engine, unscrew and remove the clutch cover retaining bolts, working in a diagonal sequence and slackening the bolts only a few turns at a time. If necessary hold the crankshaft pulley bolt with a spanner to prevent the flywheel rotating.
3 Ease the clutch cover off its locating dowels and be prepared to catch the clutch driven plate which will drop out as the cover is removed. Note which way round the driven plate is fitted.

Inspection

4 With the clutch assembly removed, clean off all traces of asbestos dust using a dry cloth. This is best done outside or in a well ventilated area.
5 Examine the linings of the clutch driven plate for wear and loose rivets, and the driven plate rim for distortion, cracks, broken torsion springs and worn splines. The surface of the friction linings may be highly glazed, but, as long as the friction material pattern can be clearly seen this is satisfactory. If there is any sign of oil contamination, indicated by a continuous, or patchy, shiny black discolouration, the driven plate must be renewed and the source of the contamination traced and rectified. This will be either a leaking crankshaft oil seal or gearbox input shaft oil seal or both. The renewal procedure for the former is given in Chapter 2, however renewal of the gearbox input shaft oil seal should be entrusted to a Mazda dealer as it involves

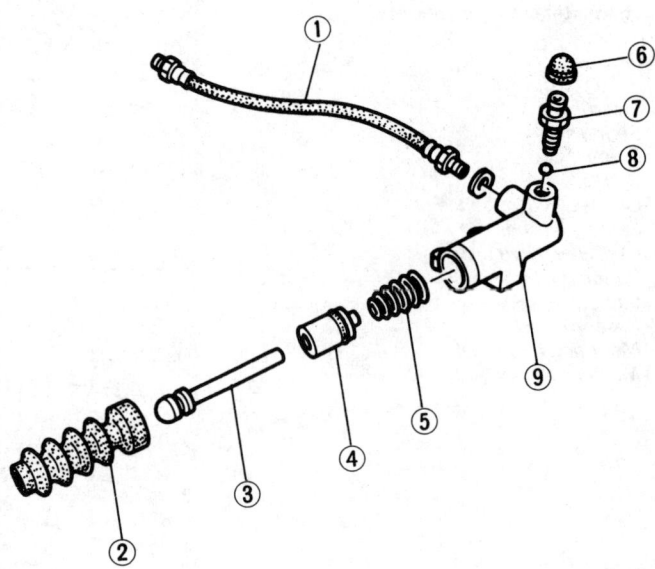

Fig. 6.6 Exploded view of the clutch slave cylinder (Sec 5)

1 Hydraulic hose   6 Dust cap
2 Gaiter   7 Bleed screw
3 Pushrod   8 Steel ball
4 Piston and seal   9 Slave cylinder body
5 Spring

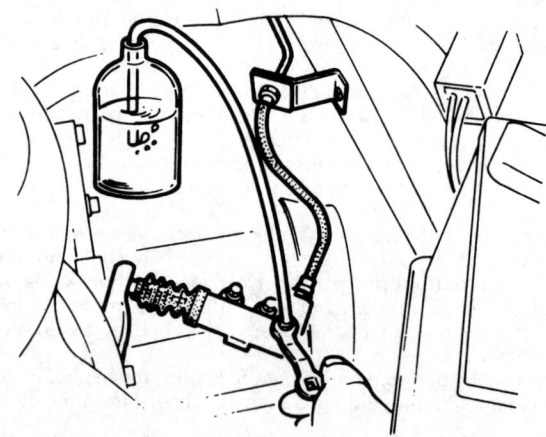

Fig. 6.7 Bleeding the hydraulic clutch system (Sec 6)

dismantling the gearbox. The driven plate must also be renewed if the friction lining thickness has worn down to less than 0.3 mm above the level of the rivet heads (photo).
6 Check the machined faces of the flywheel and pressure plate. If either is grooved, or heavily scored, renewal is necessary. The pressure plate must also be renewed if any cracks are apparent, or if the diaphragm spring is damaged or its pressure suspect.
7 With the gearbox removed it is advisable to check the condition of the release bearing, as described in Section 8.

Refitting

8 It is important that no oil or grease is allowed to come into contact with the friction material of the clutch driven plate or the pressure plate and flywheel faces. It is advisable to refit the clutch assembly with clean hands and to wipe down the pressure plate and flywheel faces with a clean dry rag before assembly begins.
9 Begin reassembly by placing the clutch driven plate against the

Chapter 6 Clutch

7.5 Checking the driven plate friction lining thickness

7.9 Refit the driven plate with its largest hub offset (arrowed) facing away from the flywheel...

7.10 ...and refit the cover assembly

7.12 Centralise the driven plate using either a round bar...

7.15 ...or a suitable clutch aligning tool

7.16 Tighten the clutch cover retaining bolts to the specified torque

flywheel so that the side with the larger hub offset is facing away from the flywheel (photo).

10 Place the clutch cover over the dowels, refit the retaining bolts and tighten them finger tight so that the clutch driven plate is gripped, but can still be moved (photo).

11 The clutch driven plate must now be centralised so that, when the engine and transmission are mated, the splines of the gearbox input shaft will pass through the splines in the centre of the clutch driven plate hub.

12 Centralisation can be carried out quite easily by inserting a round bar through the hole in the centre of the clutch driven plate so that the end of the bar rests in the hole in the end of the crankshaft (photo).

13 Using the support bearing as a fulcrum, moving the bar sideways or up and down will move the clutch driven plate in whichever direction is necessary to achieve centralisation.

14 Centralisation is easily judged by removing the bar and viewing the clutch driven plate hub in relation to the support bearing. When the support bearing appears exactly in the centre of the clutch driven plate hub, all is correct.

15 An alternative and more accurate method of centralisation is to use a commercially available clutch aligning tool obtainable from most accessory shops (photo).

16 Once the clutch is centralised, progressively tighten the cover bolts in a diagonal sequence to the torque setting given in the Specifications (photo).

17 The transmission can now be refitted to the engine by referring to the relevant Chapter of this manual.

8 Clutch release bearing – removal, inspection and refitting

Removal

1 To gain access to the release bearing it is necessary to separate the engine and gearbox either by removing the gearbox individually, or by

8.3A Disengage the release bearing and slide it off the input shaft

8.3B Remove the release fork retaining bolt...

8.3C ...then withdraw the release lever upward and remove the Woodruff key (arrowed) and fork

removing both the engine and gearbox as an assembly and separating them after removal. Depending on the method chosen, the appropriate procedures will be found in Chapter 2 or Chapter 7.

2   On October 1989 Estate models the clutch release bearing can be slid straight off the gearbox input shaft. If necessary the release fork can then be freed from its pivot ball and rubber gaiter and removed from the gearbox.

3   On all other models unhook the clutch release arm return spring (where fitted), then twist the release arm and slide the bearing off the input shaft. If necessary the release lever and fork can then be removed after undoing the retaining bolt. Withdraw the release lever upwards then lift out the Woodruff key and slide the release fork and, where fitted, the return spring off the end of the lever shaft (photos).

Inspection

4   Check the bearing for smoothness of operation and renew it if there is any roughness or harshness as the bearing is spun.

5   Inspect the release fork components for any signs of cracks or distortion and check the return spring (where fitted) for signs of fatigue. Renew the components if necessary.

Refitting

6   Refitting is the reverse of the removal sequence, bearing in mind the following points.

(a) *Lubricate the inner circumference of the release bearing and the release fork contact area with molybdenum disulphide grease.*

(b) *If a return spring is fitted, ensure that it is correctly located on the release fork.*

(c) *On October 1989 onwards Estate models, ensure that the release fork spring retainer is correctly located behind the flat shoulder of the pivot ball.*

Chapter 7
Manual gearbox and automatic transmission

Contents

Part A: Manual gearbox
Driveshaft oil seals – renewal	4
Gearchange linkage/mechanism – removal, overhaul and refitting	2
General information	1
Manual gearbox – removal and refitting	6
Manual gearbox oil level check	See Chapter 1
Manual gearbox overhaul – general information	7
Reversing lamp switch – removal and refitting	5
Speedometer drive – removal and refitting	3

Part B: Automatic transmission
Automatic transmission – removal and refitting	17
Automatic transmission fluid level check	See Chapter 1
Automatic transmission overhaul – general information	18
Driveshaft oil seals – renewal	16
General information	8
Kickdown solenoid – testing	12
Kickdown switch – adjustment	11
Selector cable/mechanism – adjustment	9
Selector cable/mechanism – removal and refitting	10
Speedometer drive – removal and refitting	15
Starter inhibitor switch – testing	14
Vacuum diaphragm – testing	13

Specifications

Part A: Manual gearbox

Type .... Four or five forward speeds (all synchromesh) and reverse. Final drive differential integral with main gearbox

Gear ratios

	B6 engine	All other engines
1st	3.153:1	3.416:1
2nd	1.842:1	1.842:1
3rd	1.290:1	1.290:1
4th	1.028:1	0.972:1
5th	0.820:1	0.775:1

Final drive ratios

E1 engine	4.388:1
E3 and B6 engines	4.105:1
E5, B3 and B5 engines	3.850:1

Torque wrench settings

	Nm	lbf ft
Remote control housing extension bar nut	32 to 47	24 to 35
Gearchange rod bolts	16 to 22	12 to 17
Gearbox mounting bolts:		
12 mm bolt	65 to 95	48 to 70
14 mm bolt	90 to 120	66 to 88
Crossmember mounting nuts and bolts	65 to 91	48 to 67
Engine/transmission mounting to crossmember nuts	32 to 47	23 to 35
Front mounting to gearbox housing bolts (pre-September 85)	19 to 26	14 to 19
Front engine mounting bolt (September 1985 on)	37 to 52	27 to 38

Chapter 7 Manual gearbox and automatic transmission

Part B: Automatic transmission

Type .... Three forward speeds and reverse, final drive differential integral with transmission

Ratios
1st	2.841:1
2nd	1.541:1
3rd	1.000:1
Reverse	2.400:1
Final drive	3.631:1

Torque wrench settings

	Nm	lbf ft
Gear selector lever mounting bracket bolt	12 to 17	9 to 13
Gear selector knob locknut	15 to 20	11 to 15
Selector cable locknuts	8 to 11	6 to 8
Torque converter to drive plate	35 to 50	26 to 37
Transmission to engine:		
12 mm bolt	65 to 95	48 to 70
14 mm bolt	90 to 120	66 to 88
Crossmember mounting nuts and bolts	61 to 87	45 to 64
Engine/transmission mounting to crossmember bolts	32 to 47	23 to 35
Front mounting to transmission housing bolts (pre-September 85)	19 to 26	14 to 19
Front engine mounting bolt (September 1985 on)	37 to 52	27 to 38

Part A: Manual gearbox

1 General information

The manual gearbox is either of four or five-speed type with one reverse gear. Baulk ring synchromesh gear engagement is used on all the forward gears. The final drive (differential) unit is integral with the main gearbox and is located between the clutch housing and the gearbox case. The gearbox and differential both share the same lubricating oil.

Gear selection is by means of a floor-mounted lever connected by a remote control housing and gearchange rod to the gearbox shift rod.

2 Gearchange linkage/mechanism – removal, overhaul and refitting

Removal

1 Apply the handbrake, chock the rear wheels then jack up the front of the car and support it on axle stands.
2 From under the car, undo the nuts and remove the bolts securing the gearchange rod to the gear lever and shift rod (photos). Remove the gearchange rod from under the car.
3 On pre-September 1985 models, working from inside the car, carefully prise the centre console upper panel out of its locating catches using a screwdriver.
4 On September 1985 models onward, undo the centre console retaining screws and remove the console. Remove both the left and right-hand front side wall panels, each being retained by a single screw.
5 Unscrew the gear lever knob and slide off the gear lever rubber gaiter. On September 1985 models onward also remove the mounting rubber.
6 Prise the lever retaining spring out of position using a screwdriver and lift the gear lever assembly out of the remote control housing.
7 Slide the retaining spring and upper ball seat off the upper end of the lever, and if necessary, remove the gaiter, lower ball seat holder and lower ball seat from the lower end of the gear lever.
8 If necessary, the remote control housing can be removed by first undoing the four nuts which secure the housing to the floor then, from underneath the car, removing the nut and washers which secure the remote control housing extension rod to the gearbox housing (photo). The housing can then be removed from under the car noting the fitted position of the rubber seal.

Overhaul

9 Examine the upper and lower seat components for signs of wear or damage and renew as necessary.
10 Check that the gearchange rod is completely straight and free from any signs of wear or damage. Check the pivot bushes in the gear lever and gearbox shift rod for signs of wear and renew if necessary.
11 If removed, examine the remote control housing components for signs of wear or damage, paying particular attention to the extension bar rubber bush and the rubber seal. Renew components as necessary.

Refitting

12 Refitting is the reverse of the removal sequence bearing in mind the following points.

(a) Tighten all nuts and bolts to the specified torque.
(b) Lubricate the gear lever seat components with multi-purpose grease.
(c) Ensure that the hooked part of the gear lever retaining spring is correctly located in the housing groove.

3 Speedometer drive – removal and refitting

Removal

1 The speedometer drive is located in the top of the gearbox housing.

Chapter 7 Part A: Manual gearbox

Fig. 7.1 Cross-sectional view of the five-speed manual gearbox (Sec 1)

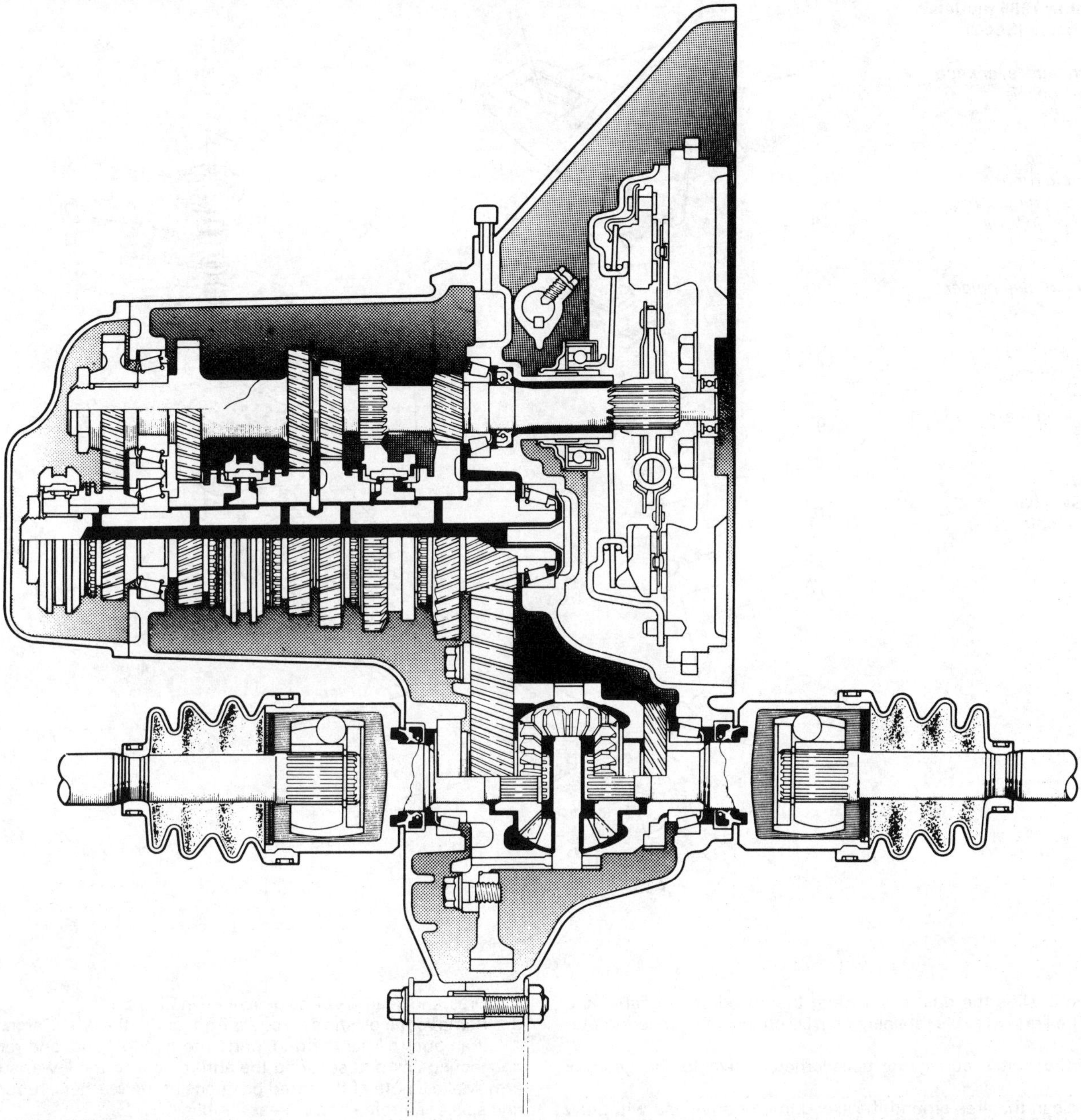

2.2A Remove the gearchange rod to gearbox shift rod bolt...

2.2B ...and the gearchange rod to gear lever retaining bolt

2.8 Remove the remote control housing extension bar from the gearbox housing stud

Chapter 7 Part A: Manual gearbox

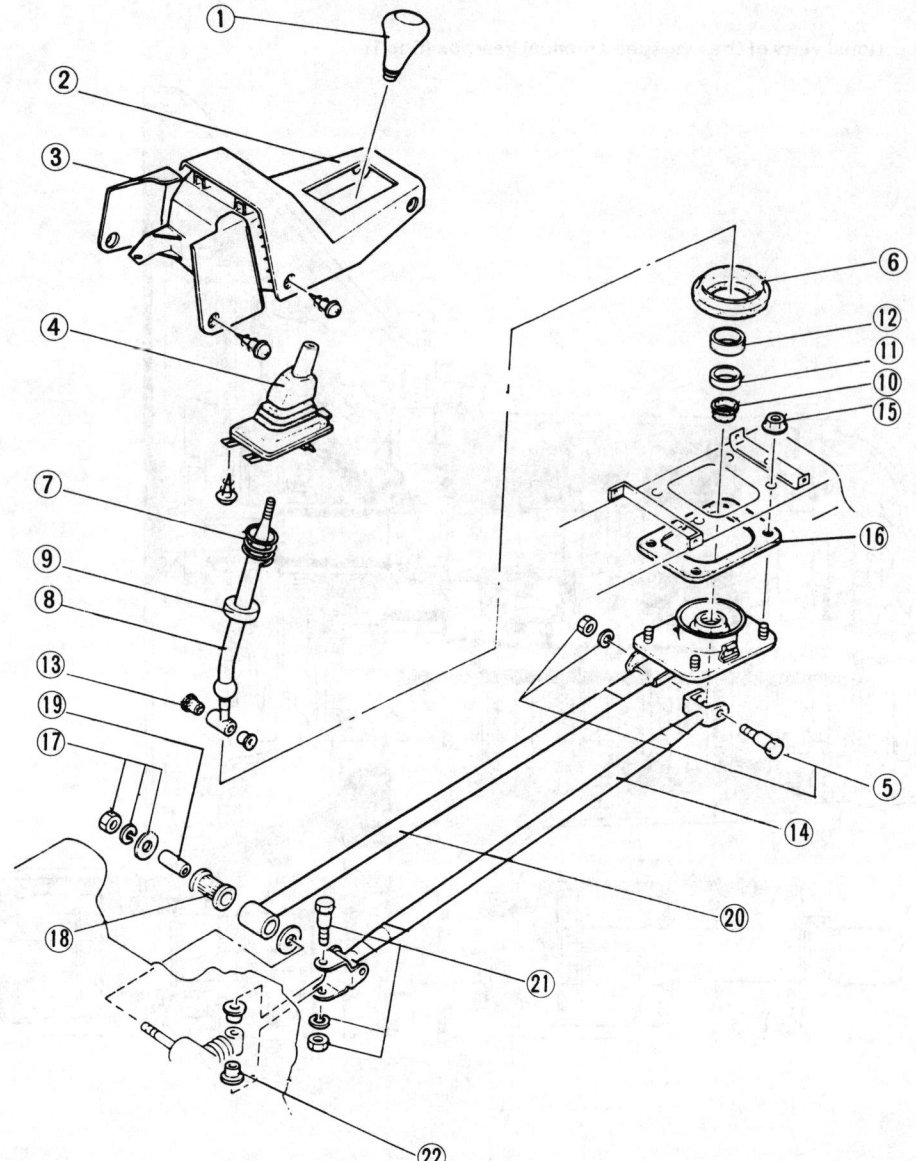

Fig. 7.2 Exploded view of the gearchange linkage components – September 1985 models onward shown (Sec 2)

1 Gearchange lever knob
2 Centre console
3 Side wall
4 Gaiter
5 Nut and bolt
6 Mounting rubber
7 Lever retaining spring
8 Gearchange lever
9 Upper ball seat
10 Gaiter
11 Lower ball seat holder
12 Lower ball seat
13 Bush
14 Gearchange rod
15 Self-locking nut
16 Rubber seal
17 Nut, spring washer and washer
18 Bush
19 Spacer
20 Extension bar
21 Nut and bolt
22 Bush

To remove it, slide the dust cover along the speedometer cable and unscrew the knurled cable retaining ring. Disconnect the speedometer cable.
2 Undo the bolt securing the speedometer drive to the gearbox housing.
3 Wipe clean the area around the speedometer drive and withdraw the speedometer drive from the gearbox housing.

Refitting

4 Prior to refitting, check the speedometer drive O-ring for signs of damage and renew if necessary.
5 Apply a smear of oil to the O-ring and push the speedometer drive into position in the gearbox housing.
6 Refit the speedometer drive retaining bolt and tighten it securely.
7 Reconnect the speedometer cable to the drive, tightening its retaining ring securely, and slide the speedometer cable dust cover back into position.

4 Driveshaft oil seals – renewal

1 Apply the handbrake, chock the rear wheels then jack up the front of the car and support it on axle stands. Remove the appropriate front roadwheel.

2 Drain the gearbox oil as described in Chapter 1.
3 Remove the plastic side cover from under the wheel arch.
4 If an anti-roll bar is fitted, undo the two locknuts and remove the connecting link bolt securing the anti-roll bar to the lower suspension arm. Make a note of the fitted positions of the washers, rubber bushes and spacer for reference on reassembly.
5 Undo the nut and remove the pinch-bolt securing the lower suspension arm balljoint to the swivel hub.
6 Using a long stout bar, carefully lever the lower suspension arm down to release the balljoint from the swivel hub, whilst taking great care not to damage the balljoint rubber gaiter.
7 The inner constant velocity joint can be released from the gearbox by pulling the swivel hub firmly outwards. If this fails to release the inner joint insert a suitable bar between the inner joint and the gearbox housing and carefully lever the joint out of position.
8 Lever the gearbox oil seal out of position using a suitable flat bladed screwdriver (photo).
9 Apply a smear of oil to the outer edge of the new seal, and tap it into position using a hammer and suitable tubular drift which bears only on the hard outer edge of the seal. Tap the seal in until it contacts the housing lip (photo).
10 Renew the driveshaft retaining circlip and apply a smear of grease to the oil seal lip and driveshaft joint splines.
11 Engage the inner driveshaft joint splines with those of the differential sun gear and carefully insert the inner joint fully into position by pushing the swivel hub inwards (photo). Once fully home, check that

Chapter 7 Part A: Manual gearbox

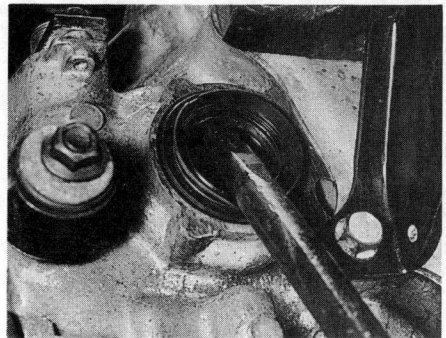

4.8 Lever the driveshaft oil seal out of position...

4.9 ...and fit the new seal using a suitable tubular drift

4.11 Take care not to damage the seal lip when refitting the driveshaft

the inner driveshaft is held firmly in position by gently pulling the swivel hub outwards.
12   Insert the lower suspension arm balljoint into the swivel hub and refit the pinch-bolt and nut. Tighten the pinch-bolt to the specified torque (See Chapter 10).
13   Refit the anti-roll bar (where fitted) with reference to Section 6, Chapter 10.
14   Refit the cover to the wheel arch.
15   If necessary, repeat the procedure for the remaining oil seal.
16   Refit the roadwheel(s), then lower the car to the ground and tighten the wheel nuts to the specified torque.
17   Refill the gearbox with the correct type and quantity of lubricant as described in Chapter 1.

5 Reversing lamp switch – removal and refitting

Removal

1   Apply the handbrake, chock the rear wheels then jack up the front of the car and support it on axle stands.
2   Drain the gearbox oil as described in Chapter 1.
3   Disconnect the reversing lamp switch wiring at the harness wiring connector.
4   If necessary, remove the undertray to gain access to the reversing lamp switch (photo).
5   Unscrew the reversing lamp switch from the underside of the gearbox and remove the washer.

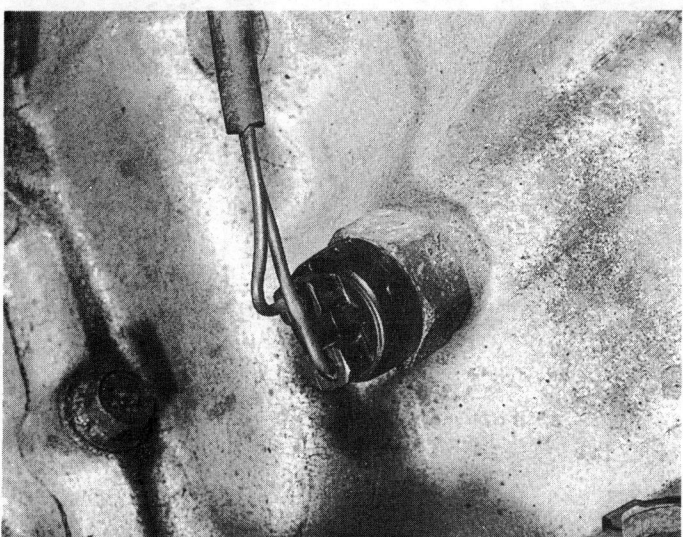

5.4 Reversing lamp switch is located on the underside of the gearbox

Refitting

6   Clean the gearbox housing and reversing lamp switch threads.
7   Insert the switch together with a new washer and tighten it securely.
8   Pass the switch wiring through the cable guide on the side of the gearbox housing and connect it to the main wiring harness.
9   Refit the undertray if removed.
10   Lower the car to the ground and refill the gearbox with the correct quantity and type of oil as described in Chapter 1.

6 Manual gearbox – removal and refitting

Removal

1   Remove the starter motor as described in Chapter 12.
2   Disconnect the speedometer cable by unscrewing its knurled retaining ring.
3   On late Estate models, undo the two bolts securing the clutch operating cylinder in position and remove it from the gearbox. Tie the operating cylinder to its mounting bracket. Remove the bolt which secures the earth strap to the gearbox.
4   On all other models fitted with a cable operated clutch, unscrew the locknut (if fitted) and remove the adjuster nut at the end of the cable, then withdraw the washer and roller (as applicable). Remove the two clutch cable mounting bracket bolts, noting the cable guide and earth strap which are retained by the bolts (photos). Release the cable from the operating arm and place it to one side. Refit the clutch cable adjuster components to the cable for safekeeping.
5   Slacken the upper gearbox housing to engine mounting bolts, then remove the bolts along with the cooling system bypass pipe support bracket.
6   Release all the relevant wiring harnesses and cables that are likely to impede gearbox removal from their respective cable clips.
7   Disconnect the reversing light switch at the wiring harness connector.
8   Ensure the handbrake is applied, then jack up the front of the car and support it securely on axle stands. Remove the front roadwheels.
9   Drain the gearbox oil as described in Chapter 1.
10   Remove the undertray and side covers to gain full access to the underside of the gearbox.
11   Where fitted remove the front anti-roll bar as described in Chapter 10.
12   Slacken and remove the pinch-bolt securing the lower arm balljoint to the swivel hub.
13   Using a suitable bar, lever the suspension arm down to release the balljoint shank from the swivel hub.
14   The inner constant velocity joint can be released from the gearbox by pulling the swivel hub firmly outwards. If this fails to release the inner joint, insert a suitable bar between the inner joint and the gearbox housing and carefully lever the joint out of position. Support the inner constant velocity joint as it is removed to avoid damaging the driveshaft oil seal (photo).
15   Undo the nut and bolt securing the gearchange rod to the gearbox shift rod.
16   Slacken and remove the nut which secures the remote control

6.4A Unscrew the clutch cable locknut and adjusting nut (arrowed)

6.4B Clutch cable mounting bracket retaining bolts, earth strap and cable guide

6.14 Take care not to damage driveshaft oil seals when removing inner constant velocity joints

6.17 Remove the cover plate bolts (arrowed) and the plate

6.18 Supporting the engine with an engine support bar

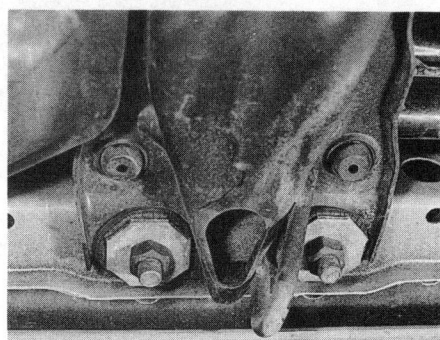
6.21A Undo the crossmember front...

6.21B ...and rear mounting nuts and bolts (arrowed) (pre-September 1985 models shown)

6.25 Removing the gearbox unit

6.26 Ensure the locating dowels are in position before refitting the gearbox

housing extension bar to the gearbox. Remove the outer washers, slide the bar off the stud and recover the inner washers.
17 Undo the bolts securing the flywheel cover plate to the clutch housing and remove the plate (photo).
18 Place a jack with an interposed block of wood to take the weight of the engine. Alternatively fit a hoist or support bar to the engine lifting eyes to take the weight of the engine (photo).
19 Place a jack and block of wood beneath the gearbox.
20 Undo all the nuts securing the front and rear engine/gearbox mounting rubbers to the crossmember.
21 Slacken and remove the crossmember retaining nuts and bolts (as applicable) and lower the crossmember away from the car (photos).
22 On Pre-September 1985 models slacken and remove the bolts securing the front mounting bracket to the left-hand end of the gearbox.
23 On later models undo the nut and remove the bolt from the front engine mounting rubber.
24 Remove all the remaining bolts that secure the gearbox clutch housing to the engine. Make a final check that all necessary components have been disconnected.

25 Lower the engine and gearbox slightly and withdraw the gearbox from the engine. It may initially be tight owing to the locating dowels. Once the gearbox is free, lower the jack and remove the unit out from under the car (photo).

Refitting

26 The gearbox is refitted by a reversal of the removal procedure bearing in mind the following points (photo).

(a) Make sure the dowels are correctly positioned prior to installation.

(b) Apply a little molybdenum disulphide grease to the splines of the gearbox input shaft. Do not apply too much otherwise there is a possibility of the grease contaminating the clutch friction disc.

(c) On completion refill the gearbox with the specified type and quantity of lubricant as described in Chapter 1.

(d) On models with a cable operated clutch, adjust the cable as described in Chapter 1.

(e) Refit the anti-roll bar with reference to Section 6, Chapter 10 (if applicable).
(f) Tighten all nuts and bolts to the specified torque.

7 Manual gearbox overhaul – general information

Overhauling a manual gearbox is a difficult and involved job for the DIY home mechanic. In addition to dismantling and reassembling many small parts, clearances must be precisely measured and, if necessary, changed by selecting shims and spacers. Gearbox internal components are also often difficult to obtain and in many instances, extremely expensive. Because of this, if the gearbox develops a fault or becomes noisy, the best course of action is to have the unit overhauled by a repair specialist or to obtain an exchange reconditioned unit.

Nevertheless, it is not impossible for the more experienced mechanic to overhaul a gearbox if the special tools are available and the job is done in a deliberate step-by-step manner so that nothing is overlooked.

The tools necessary for an overhaul include internal and external circlip pliers, bearing pullers, a slide hammer, a set of pin punches, a dial test indicator and possibly a hydraulic press. In addition, a large, sturdy workbench and a vice will be required.

During dismantling of the gearbox, make careful notes of how each component is fitted to make reassembly easier and accurate.

Before dismantling the gearbox, it will help if you have some idea what area is malfunctioning. Certain problems can be closely related to specific areas in the gearbox which can make component examination and replacement easier. Refer to the Fault diagnosis Section at the beginning of this manual for more information.

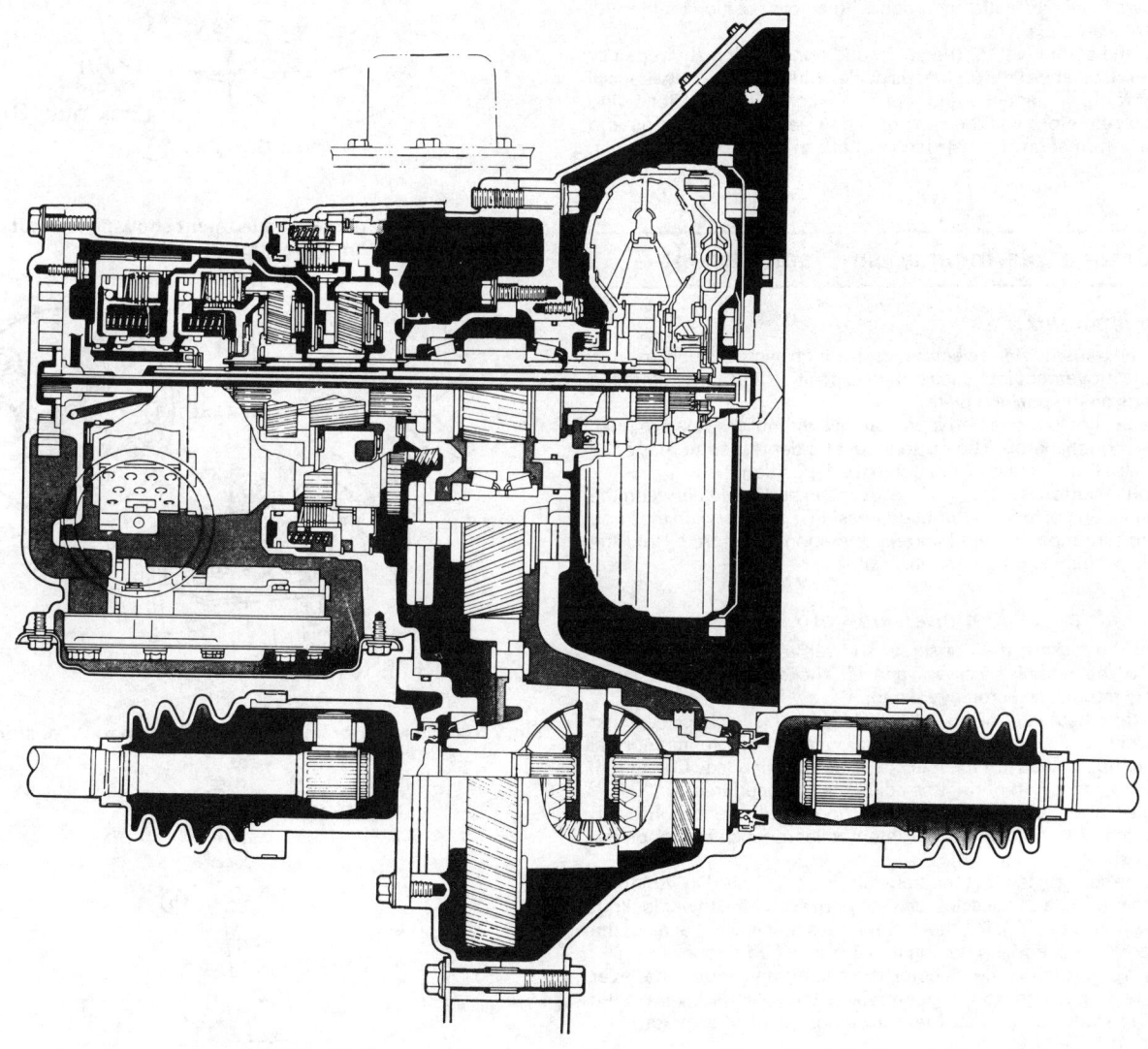

Fig. 7.3 Cross-sectional view of automatic transmission (Sec 8)

Part B: Automatic transmission

8 General information

Mazda 1500 Automatic models are equipped with a three-speed fully automatic transmission consisting of a torque converter, an epicyclic geartrain, hydraulically-operated clutches and brakes, and an electronic control unit.

The torque converter provides a fluid coupling between engine and transmission which acts as an automatic clutch, and also provides a degree of torque multiplication when accelerating.

The epicyclic geartrain provides either of the three forward or one reverse gear ratios according to which of its component parts are held stationary or allowed to turn. The components of the geartrain are held or released by brakes and clutches which are activated by a hydraulic control unit. An oil pump within the transmission provides the necessary hydraulic pressure to operate the brakes and clutches.

Driver control of the transmission is by a six position selector lever which allows fully automatic operation with a hold position on the first and second gear ratios.

Due to the complexity of the automatic transmission any repair or overhaul work must be left to a Mazda dealer with the necessary special equipment for fault diagnosis and repair. The contents of the following Sections are therefore confined to supplying general information and any service information and instructions that can be used by the owner.

9 Selector cable/mechanism – adjustment

Selector lever knob

1 Move the transmission selector lever through each position, making sure that the movement into each detent is positive and corresponds to the markings on the position plate.
2 The lever should move between D and N without the need to push in the button on the knob. The button must be depressed to move the lever between D and R and in and out of the P position.
3 If the push button is loose, or if the lever can be moved between the above mentioned positions without depressing it, then slacken the knob locknut and turn the knob until correct operation is restored. Hold the knob in this position and tighten the locknut.

Selector cable – 1982 models onward

4 Loosen the locknut then unscrew and remove the selector lever knob. Undo the retaining screws and lift the centre console clear. Temporarily refit the selector lever knob.
5 Check that the handbrake is fully applied and place chocks against the roadwheels. The engine should be warmed up to its normal operating temperature with the idle speed as specified (see Chapter 1), but switch the engine off during the adjustment procedures.
6 At the lever end of the selector cable, loosen off the locknuts and then move the selector lever to 'N', (detent roller must be fully engaged in the 'N' position - see Fig. 7.4).
7 Move the selector lever at the transmission to 'N' (see Fig. 7.5) then, reverting to the cable at the selector lever end, screw the lower locknut up until it contacts the 'T' joint, then screw the upper locknut against the 'T' joint from the top end, tightening it to the specified torque.
8 Press the button on the selector lever knob and move the lever towards the 'R' position, to the point where the selector lever on the transmission starts to operate. Measure the amount of selector lever movement ('a' in Fig. 7.6).
9 Now pull the selector lever back to the 'D' position and measure the amount of movement at 'b'. The movements measured for 'a' and 'b' should be equal. If they are not, loosen off the upper cable locknuts at the selector lever end and adjust then to suit. When movements 'a' and 'b' are equal, tighten the locknuts, then check the linkage for satisfactory operation and engagement throughout the range.
10 If the selector lever operation is not satisfactory, set it in the 'P' position, loosen off the detent roller mounting screws and move the detent roller to suit. Retighten the detent roller mounting screw and recheck the selector cable adjustment as described above.
11 Remove the selector lever knob and refit the centre console. Screw the adjuster locknut fully onto the selector lever, then screw the knob fully down onto the locknut. From this position unscrew the knob

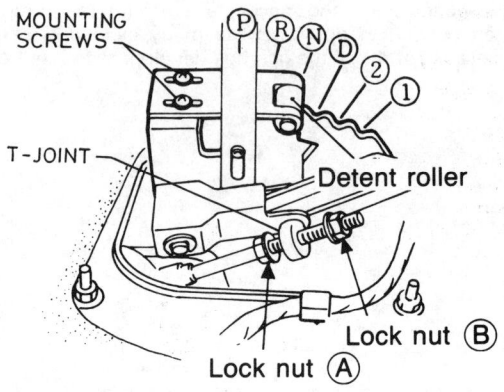

Fig. 7.4 Shift control cable adjustment showing locknuts and detent roller engaged in the 'N' position (Sec 9)

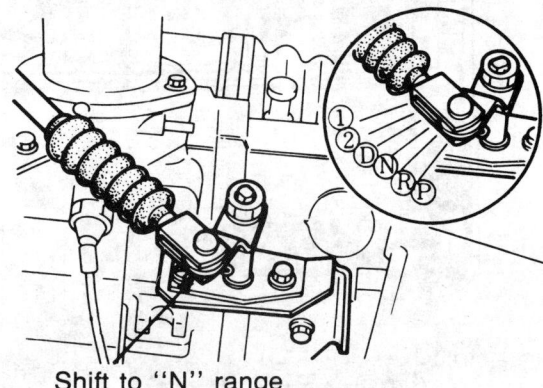

Fig. 7.5 Move selector lever on transmission to 'N' position (Sec 9)

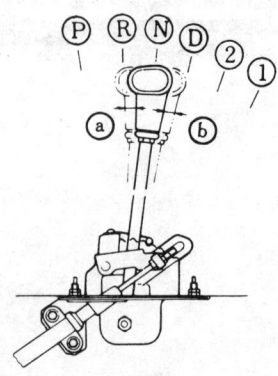

Fig. 7.6 Selector lever movements between 'a' and 'b' must be equal (Sec 9)

Chapter 7 Part B: Automatic transmission

Fig. 7.7 Exploded view of the selector rod linkage – pre-1982 models (Sec 10)

1 Selector lever knob
2 Locknut
3 Pushrod
4 Clevis pin
5 Bolt
6 Bolt and nut
7 Selector lever bracket
8 Spring
9 Detent pin
10 Selector lever
11 Clevis pin
12 Rear selector rod
13 Rear counter lever
14 Rear counter bracket
15 Clevis pin
16 Front selector rod
17 Counter rod
18 Front counter lever
19 Front counter bracket

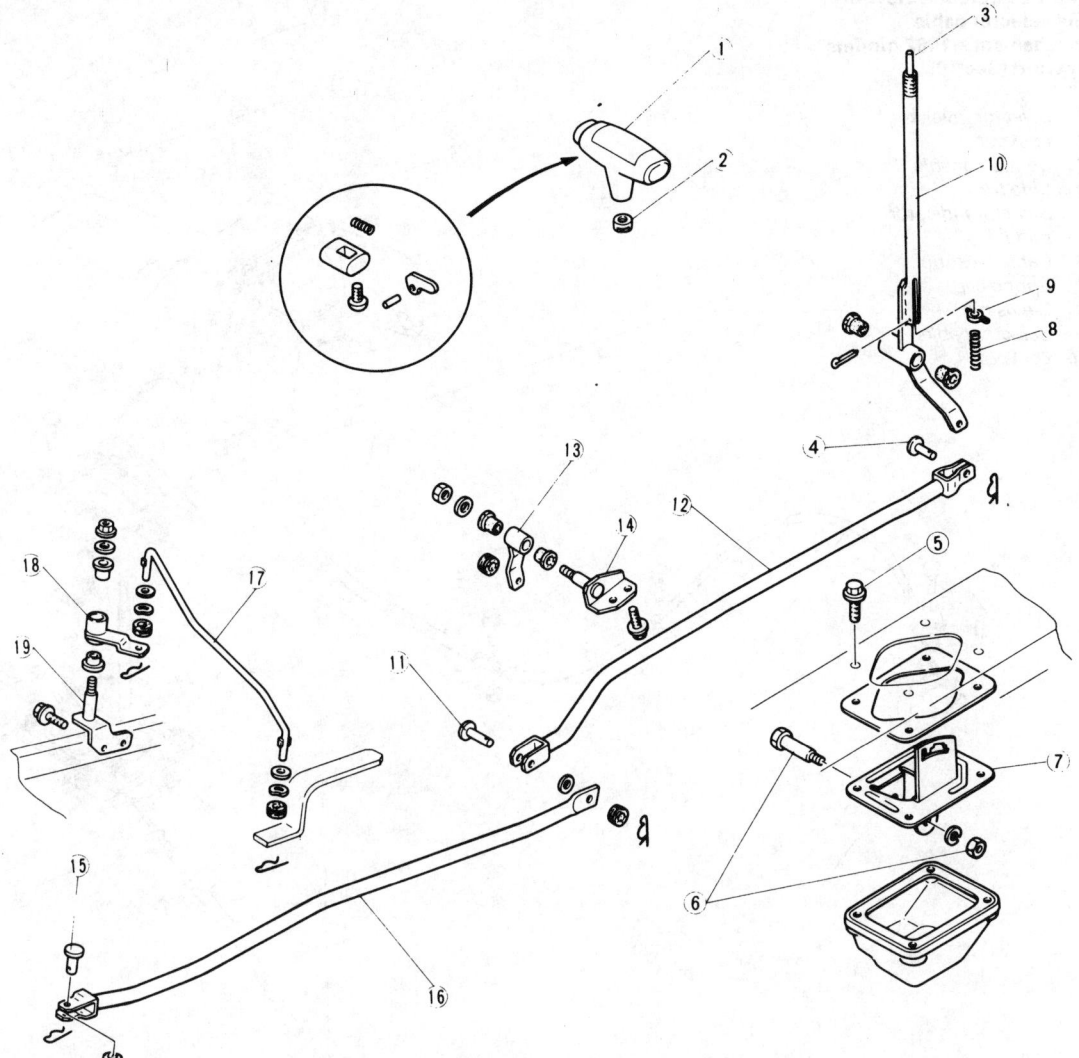

no more than one full turn to position it correctly, and tighten the locknut securely. Check that the lever movement and engagement is satisfactory throughout its range.
12   Remove the wheel chocks and road test the vehicle.

10 Selector cable/mechanism – removal and refitting

Removal

1   Apply the handbrake, chock the rear wheels then jack up the front of the car and support it on axle stands.

Pre-1982 models (rod mechanism)

2   Working from under the car, extract the spring clip and withdraw the clevis pin securing the rear gear selector rod to the base of the lever. Remove the spring clip and clevis pin which retains the front end of the rear selector rod then remove the rod from under the car.
3   Extract the spring clip, and withdraw the clevis pin securing the front selector rod to the transmission selector lever, then remove the rod from under the car.
4   If necessary the front and rear counter lever assemblies can also be removed once their retaining bolts and spring clips have been removed.
5   From inside the car, slacken the selector lever locknut then unscrew the selector lever knob and locknut from the top of the lever.
6   Withdraw the pushrod from the centre of the selector lever.
7   Remove all the centre console retaining screws and lift the console off the lever.
8   Slacken and remove the four bolts securing the selector lever assembly to the floor then remove the lever assembly, rubber seal and dust cover from underneath the car.

1982 models onward (cable mechanism)

9   Loosen the selector knob locknut then unscrew both the knob and locknut from the end of the lever.
10   Withdraw the pushrod from the centre of the lever.
11   Remove the four screws which retain the selector lever position indicator and lift it off the lever.
12   Loosen the upper selector cable locknut and remove it from the cable end.
13   Working from underneath the car, unscrew the two bolts which secure the rear selector cable mounting bracket and release the cable from the selector lever.
14   Extract the spring clip and withdraw the clevis pin which secures the front end of the cable to the transmission, then remove the two front cable mounting bracket retaining bolts and remove the cable from underneath the car.
15   From inside the car, slacken and remove the four nuts securing the selector lever assembly to the floor then lower the assembly and rubber seal out of position and remove it from underneath the car.

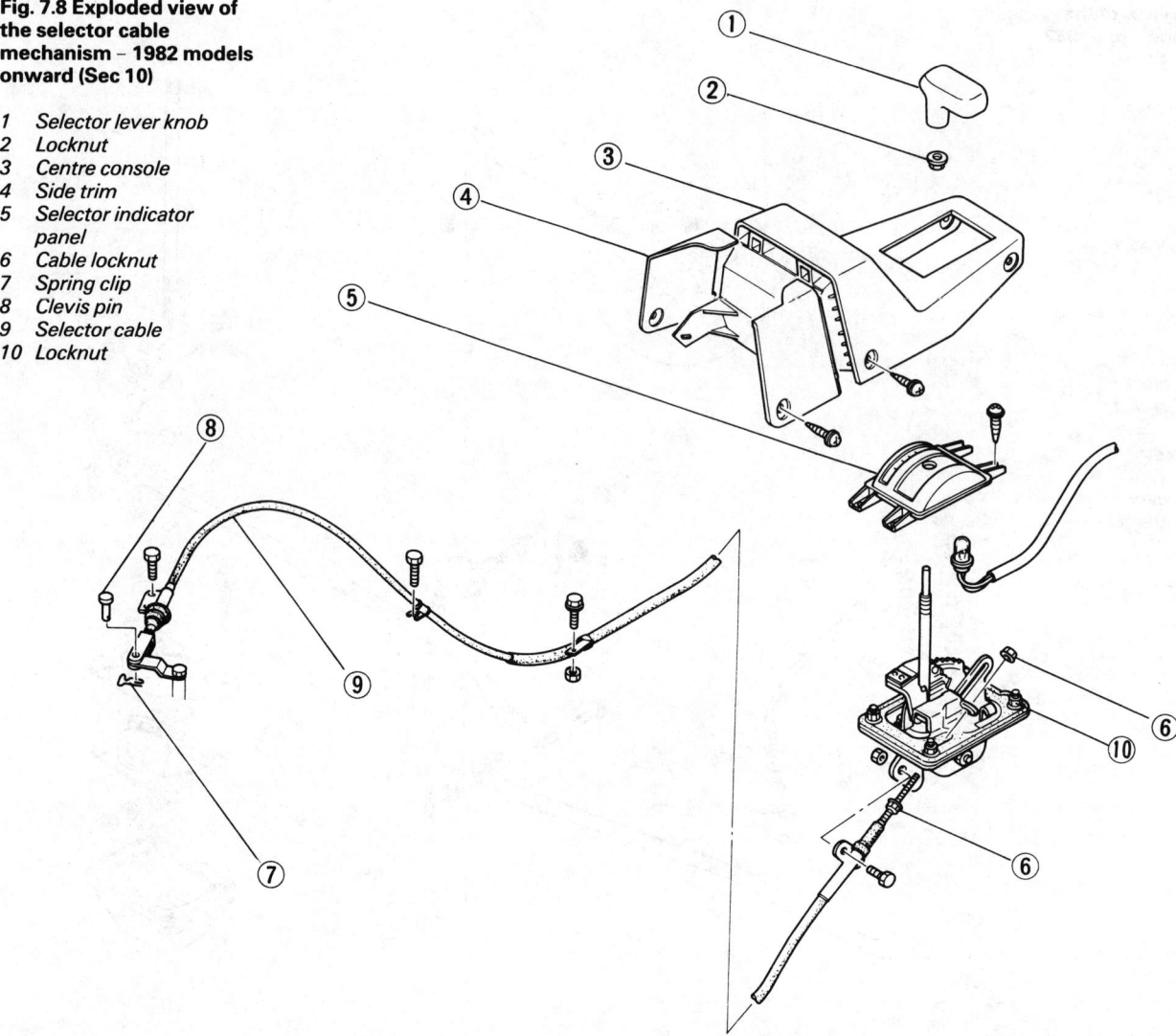

Fig. 7.8 Exploded view of the selector cable mechanism – 1982 models onward (Sec 10)

1 Selector lever knob
2 Locknut
3 Centre console
4 Side trim
5 Selector indicator panel
6 Cable locknut
7 Spring clip
8 Clevis pin
9 Selector cable
10 Locknut

Refitting

16 Refitting is the reverse of the removal procedure noting the following points.

(a) Tighten all nuts and bolts securely.
(b) Apply grease to the selector lever and rod or cable pivots.
(c) On cable mechanism models adjust the cable as described in Section 9.

11 Kickdown switch – adjustment

1 The kickdown switch should be adjusted if the transmission does not downshift under full throttle application. The switch is fitted to the accelerator pedal mounting bracket.
2 To adjust the switch, a multimeter or ohmmeter will be needed.
3 From inside the car remove the cover from under the facia.
4 Set the multimeter to the ohms x1 scale and connect its probes across the switch terminals, an open circuit should be present. Slowly depress the accelerator pedal whilst noting the reading on the meter. When the pedal is depressed approximately $\frac{7}{8}$ of its stroke the meter reading should suddenly change to 0 ohms, indicating continuity through the switch. If not, slacken the switch locknut and rotate the switch body until the kickdown switch operation is correct. Tighten the switch locknut securely and recheck the adjustment.

5 If correct adjustment proves impossible, the switch is faulty and must be renewed. If the switch adjustment was found to be correct, test the kickdown solenoid as described in Section 12.
6 Refit the cover to the underside of the facia.

12 Kickdown solenoid – testing

1 A fault in the kickdown solenoid is indicated if the transmission does not downshift under full application, or if the downshift is hesitant or rough. The solenoid location is shown on Fig. 7.9, and can be tested as follows.
2 Apply the handbrake, chock the rear wheels then jack up the front of the car and support it on axle stands.
3 Remove the undertray to gain access to the transmission unit.
4 Drain the transmission fluid as described in Chapter 1.
5 Disconnect the solenoid wiring at the harness connector, then unscrew the kickdown solenoid from the transmission and remove it from under the car.
6 To test the switch a fully charged 12 volt battery and two auxiliary wires will be required. Connect the positive terminal of the battery to the solenoid terminal using one wire, and connect the negative battery terminal to the solenoid body (Fig. 7.10). When battery voltage is applied to the solenoid, the plunger should operate quickly and smoothly and return smoothly as the battery voltage is disconnected. If

Chapter 7 Part B: Automatic transmission

Fig. 7.9 Layout of transmission external components (Secs 12 to 14)

1 Kickdown switch
2 Vacuum solenoid
3 Starter inhibitor switch
4 Fluid level/filler tube
5 Oil pan
6 Housing mating faces
7 Side oil seal
8 Fluid pipes
9 Speedometer driven gear
10 Servo retainer
11 Drain plug
12 Fluid pressure test connection

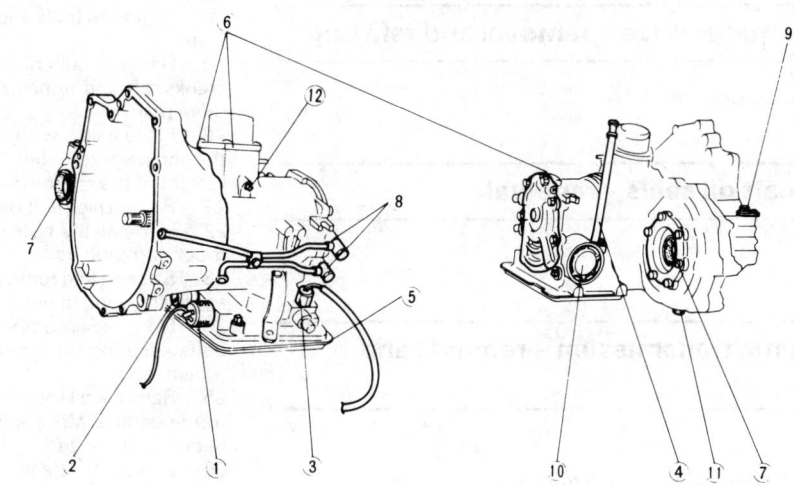

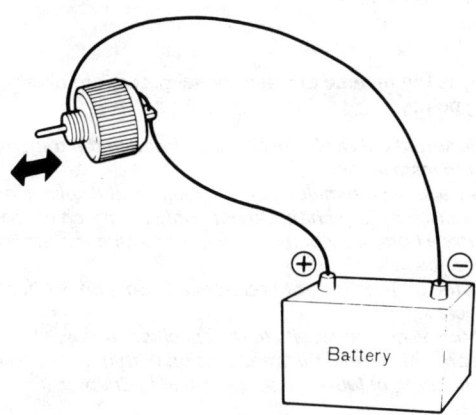

Fig. 7.10 Testing the kickdown solenoid (Sec 12)

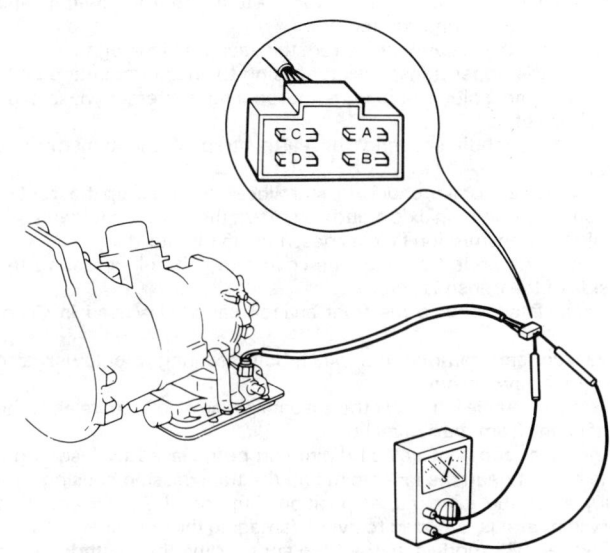

Fig. 7.11 Testing the starter inhibitor switch (Sec 14)

not the solenoid is faulty and must be renewed.
7 Prior to fitting, check the solenoid O-ring for signs of damage and renew if necessary.
8 Apply a smear of oil to the O-ring and refit the kickdown solenoid to the transmission, tightening it securely.
9 Refit the undertray.
10 Lower the car to the ground and fill the transmission with the specified type and quantity of fluid as described in Chapter 1.

13 Vacuum diaphragm – testing

1 Malfunction of the vacuum diaphragm can cause harsh gear changes and gear changes occurring at higher than normal engine speeds. The vacuum diaphragm location is shown on Fig. 7.9, and can be tested as follows.
2 Apply the handbrake, chock the rear wheels then jack up the front of the car and support it on axle stands.
3 Remove the undertray to gain access to the transmission unit.
4 Drain the transmission fluid as described in Chapter 1.
5 Disconnect the vacuum hose and unscrew the diaphragm from the transmission.
6 Apply a suction to the diaphragm outlet port and check that the rod moves as suction is applied. If not, the unit is faulty and must be renewed. When obtaining a new unit, make sure that the new part is the same as the original because a number of different units are available, each having a different length operating rod.

7 Examine the diaphragm vacuum hose for signs of damage or deterioration and renew if necessary.
8 Fit the vacuum diaphragm to the transmission and tighten it securely. Refit the vacuum hose.
9 Lower the car to the ground and refill the transmission with the specified type and quantity of fluid as described in Chapter 1.

14 Starter inhibitor switch – testing

1 If the starter inhibitor switch is operating correctly the engine should only start when the transmission selector lever is in the P or N position and the reversing lamps should be illuminated when the selector lever is in the R position. The location of the switch is shown on Fig. 7.9.
2 If a fault is suspected, separate the switch wiring connector and check for continuity using a multimeter. Referring to Fig. 7.11 for terminal identification, there should be continuity at connector terminals A and B with the transmission in the P and N positions, and at terminals C and D with the transmission in the R position. If this is not the case, renew the starter inhibitor switch.
3 The switch can be removed and refitted using the information given for the reversing lamp switch in Section 5.

15 Speedometer drive – removal and refitting

Refer to Section 3.

16 Driveshaft oil seals – renewal

Refer to Section 4.

17 Automatic transmission – removal and refitting

Removal
1 Disconnect the battery negative terminal.
2 Unscrew the speedometer cable retaining ring and disconnect the cable from the transmission housing.
3 Disconnect the starter inhibitor switch, neutral switch and kickdown solenoid wiring connectors.
4 Disconnect the vacuum hose from the vacuum diaphragm.
5 Slacken the upper transmission housing to engine mounting bolts and remove the bolts along with the cooling system bypass pipe support bracket.
6 Remove the bolt securing the earth strap to the transmission housing.
7 Apply the handbrake, chock the rear wheels then jack up the front of the car and support it on axle stands. Remove the front roadwheels.
8 Drain the transmission fluid as described in Chapter 1.
9 Remove the undertray and side covers to gain full access to the underside of the transmission.
10 Where fitted remove the front anti-roll bar as described in Chapter 10.
11 Slacken and remove the pinch-bolt securing the lower arm balljoint to the swivel hub.
12 Using a suitable bar, lever the suspension arm down to release the balljoint shank from the swivel hub.
13 The inner constant velocity joint can be released by inserting a suitable bar between the inner joint and the transmission housing, and carefully lever the joint out of position. Support the inner constant velocity joint as it is removed to avoid damaging the driveshaft oil seal.
14 On pre-1982 models, extract the spring clips then withdraw the clevis pins, then disconnect both the front selector rod and counter rod from the transmission.
15 On 1982 models onward, remove the spring clip then extract the clevis pin and free the selector cable from the transmission. Remove both the cable front mounting bracket retaining bolts and place the cable to one side.
16 Loosen the hose clamps and disconnect the transmission feed and return hoses from the oil cooler (where fitted). Plug the hoses to prevent the ingress of dirt or excessive fluid spillage.
17 Refer to Chapter 12 and remove the starter motor.
18 Undo the bolts securing the flywheel cover plate, and remove the plate.
19 Using a suitable spanner on the crankshaft pulley, rotate the crankshaft and remove the bolts securing the torque converter to the drive plate.
20 Place a jack with interposed block of wood to take the weight of the engine. Alternatively fit a hoist to the engine lifting eyes and take the weight of the engine on the hoist.
21 Place a jack and block of wood beneath the transmission.
22 Undo all the nuts securing the front and rear engine/transmission mounting rubbers to the crossmember.
23 Slacken and remove the crossmember retaining nuts and bolts (as applicable) and lower the crossmember away from the car.
24 On pre-September 1985 models slacken and remove the three bolts securing the front mounting bracket to the left-hand end of the transmission.
25 Remove all the remaining bolts securing the transmission housing to the engine. Make a final check that all necessary components have been disconnected.
26 Lower the transmission slightly and withdraw the transmission from the engine. It may initially be tight owing to the locating dowels. Once the transmission is free, lower the jack and remove the unit out from under the car, ensuring that the torque converter stays in position on the transmission shaft.

Refitting
27 Refitting is the reverse of the removal procedure bearing in mind the following points.

  (a) Make sure the dowels are correctly fitted to the transmission prior to installation.
  (b) Apply a little high melting point grease to the splines of the transmission input shaft. Do not apply too much otherwise there is a possibility of the grease contaminating the torque convertor.
  (c) Refit the anti-roll bar with reference to Section 6, Chapter 10 (if applicable).
  (d) Tighten all nuts and bolts to the specified torque.
  (e) On completion refill the transmission with the specified type and quantity of lubricant as described in Chapter 1.

18 Automatic transmission overhaul – general information

In the event of a fault occurring on the transmission, it is first necessary to determine whether it is of an electrical, mechanical or hydraulic nature and to do this special test equipment is required. It is therefore essential to have the work carried out by a Mazda dealer if a transmission fault is suspected.

Do not remove the transmission from the car for possible repair before professional fault diagnosis has been carried out, since most tests require the transmission to be in the vehicle.

Chapter 8 Driveshafts

Contents

Driveshafts – removal and refitting 2
Driveshaft rubber gaiter and constant velocity joint
check ....See Chapter 1
General information 1
Inner constant velocity joint rubber gaiter – renewal 4
Outer constant velocity joint rubber gaiter – renewal 3
Vibration damper – removal and refitting 5

Specifications

Type .... Unequal length solid steel, splined to inner and outer constant velocity joints

Driveshaft length	**Manual transmission**	**Automatic transmission**
E1 and E3 engine:		
Pre-September 1985 models:		
Right-hand side	659.0 mm	–
Left-hand side	376.0 mm	–
September 1985 models onward:		
Right-hand side	660.6 mm	–
Left-hand side	384.1 mm	–
E5 engine:		
Pre-September 1985 models:		
Right-hand side	659.0 mm	659.0 mm
Left-hand side	376.0 mm	376.0 mm
September 1985 models onward:		
Right-hand side	657.0 mm	652.7 mm
Left-hand side	380.5 mm	373.7 mm
B3 and B5 engine:		
Right-hand side	907.5 mm	907.5 mm
Left-hand side	628.5 mm	628.5 mm
B6 engine:		
Pre-July 1987 models:		
Right-hand side	657.0 mm	–
Left-hand side	380.5 mm	–
July 1987 onwards models:		
Right-hand side	907.5 mm	–
Left-hand side	628.5 mm	–

Driveshaft diameter
Pre-September 1985 models .... 25 mm
September 1985 models onward .... 22 mm

Torque wrench setting	**Nm**	**lbf ft**
Driveshaft retaining nut	160 to 240	118 to 177

1 General information

Drive is transmitted from the differential to the front wheels by means of two unequal length, solid steel driveshafts.

On all manual gearbox and early automatic transmission models, both driveshafts are fitted with ball and cage constant velocity joints at each end. Later automatic transmission models have a ball and cage constant velocity joint on the outer end of each shaft but employ a tripod type constant velocity joint on the inner end of each driveshaft. On all models the outer joints are splined to accept the driveshaft and wheel hub flange, while the inner joints are splined to accept the driveshaft and differential sun gears. Both inner and outer constant velocity joints are packed with special grease during manufacture and sealed with a rubber gaiter.

Where necessary, to eliminate driveshaft induced harmonic vibrations and resonance, a rubber mounted steel damper is attached to the longer right-hand driveshaft.

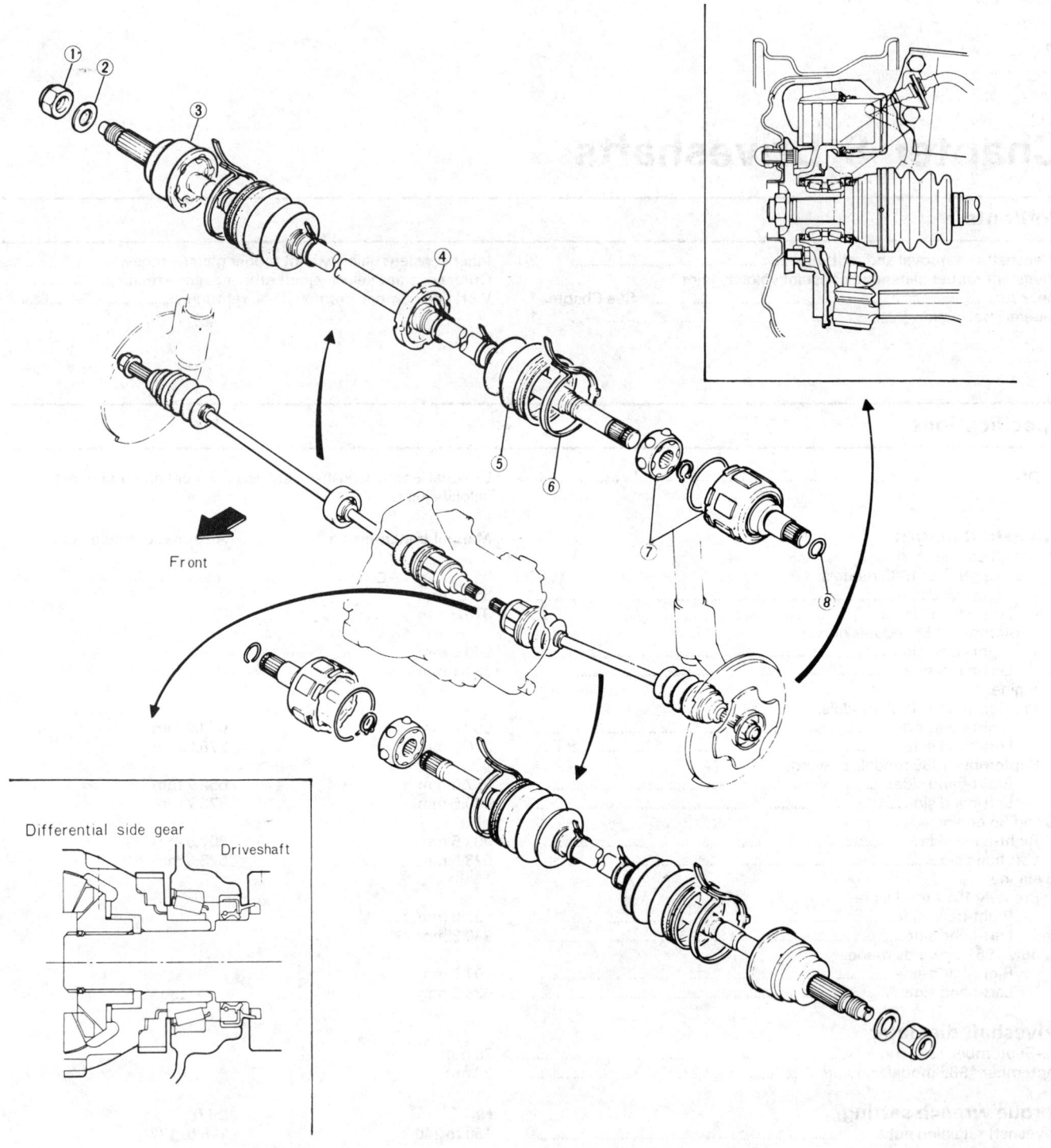

Fig. 8.1 Layout of the driveshaft assemblies and related components – manual transmission and early automatic transmission models (Sec 1)

1 Driveshaft retaining nut
2 Thrustwasher
3 Outer constant velocity joint
4 Vibration damper (where fitted)
5 Gaiter
6 Retaining clip
7 Inner constant velocity joint
8 Circlip

Chapter 8 Driveshafts

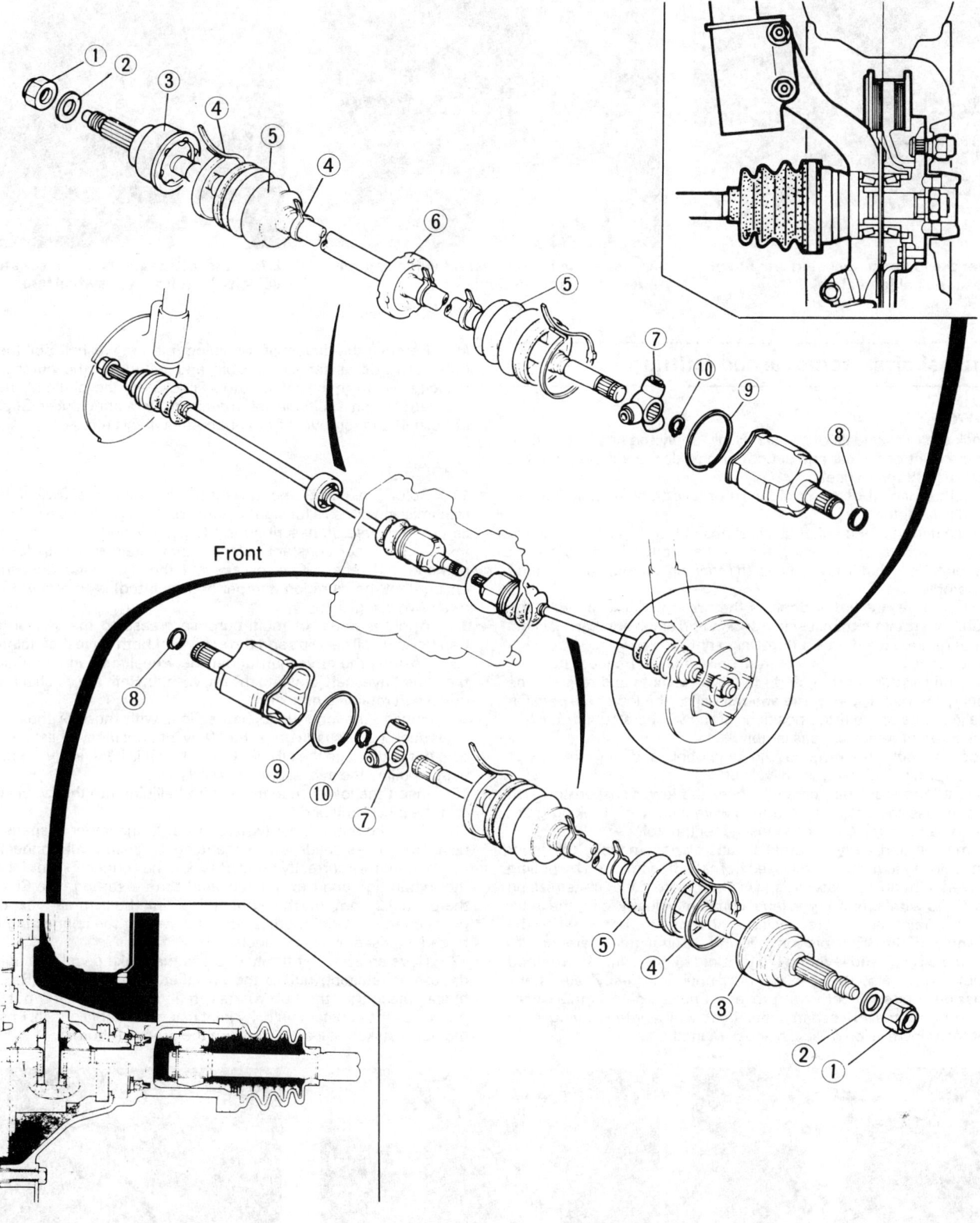

Fig. 8.2 Layout of the driveshaft assemblies and related components – later automatic transmission models (Sec 1)

1 Driveshaft retaining nut
2 Thrustwasher
3 Outer constant velocity joint
4 Retaining clip
5 Gaiter
6 Vibration damper (where fitted)
7 Tripod joint
8 Circlip
9 Circlip
10 Circlip

Chapter 8 Driveshafts

2.8 Lever the lower suspension arm down to release the balljoint shank

2.9 Lever the inner constant velocity joint out of the transmission...

2.10 ...and withdraw the outer constant velocity joint from the swivel hub

2 Driveshafts – removal and refitting

Removal

1 Chock the rear wheels of the car, firmly apply the handbrake then jack up the front of the car and support it on axle stands. Remove the appropriate front roadwheel.
2 Drain the manual gearbox oil or automatic transmission fluid as described in Chapter 1.
3 Using a hammer and suitable chisel nosed tool, tap up the staking securing the driveshaft retaining nut to the groove in the constant velocity joint. Note that a new driveshaft retaining nut must be obtained for reassembly.
4 Have an assistant firmly depress the brake pedal to prevent the front hub from rotating then, using a socket and extension bar, slacken, but do not remove the driveshaft retaining nut.
5 Remove the plastic side cover from under the inner wheel arch.
6 If an anti-roll bar is fitted, undo the two locknuts and remove the connecting link bolt securing the anti-roll bar to the lower suspension arm. Make a note of the fitted positions of the washers, rubber bushes and spacer for reference on reassembly.
7 Undo the nut and remove the pinch-bolt securing the lower suspension arm balljoint to the swivel hub.
8 Using a long stout bar, carefully lever the lower suspension arm down to release the balljoint from the swivel hub, whilst taking great care not to damage the balljoint rubber gaiter (photo).
9 On manual and early automatic transmission models, the inner constant velocity joint can be released from the transmission by pulling the swivel hub firmly outwards. On later automatic transmission models, fitted with a tripod type inner constant velocity joint, the inner joint must be released by carefully inserting a suitable bar between the inner joint and the transmission housing, taking great care not to damage the oil seal, and levering the joint out of position. This method of removal can also be used on manual and early automatic transmission models if removal proves troublesome (photo). On all models support the driveshaft inner joint whilst withdrawing it to prevent the transmission oil seal being damaged.

10 Remove the driveshaft retaining nut and washer, pull the swivel hub outwards as far as possible and withdraw the outer constant velocity joint from the hub (photo). If necessary the joint and hub can be separated using a suitable soft metal drift or bearing puller. Support the driveshaft and remove it from under the wheel arch.

Refitting

11 Before installing the driveshaft examine the swivel hub and transmission oil seals for signs of wear or damage. If necessary the seals can be renewed as described in Chapters 7 andor 10 (as applicable). Inspect the inner constant velocity joint retaining circlip for signs of damage and renew if necessary. On the right-hand driveshaft also ensure that the vibration damper (where fitted) is correctly positioned as shown in Fig. 8.7.
12 Apply a smear of multi-purpose grease to the swivel hub and transmission oil seal lips and the splines of both driveshaft joints.
13 Position the driveshaft under the wheel arch and carefully insert the outer driveshaft joint into the swivel hub. Refit the washer and a new driveshaft retaining nut.
14 Engage the inner driveshaft splines with those of the differential sunwheel and push the driveshaft fully into position, whilst taking great care not to damage the oil seal. Check that it is held firmly in position by gently pulling the swivel hub outwards.
15 Insert the lower suspension arm balljoint into the swivel hub and refit the pinch-bolt and nut.
16 Refit the anti-roll bar (where fitted) to the lower suspension arm, using the notes made on dismantling to ensure all connecting link components are correctly fitted. Position the connecting link locknuts so that when tightened to the specified torque setting (see Chapter 10) there is 6.2 mm of the connecting link bolt thread exposed on pre-September 1985 models, and 10.8 mm of the connecting link bolt thread exposed on later models.
17 Have an assistant firmly depress the brake pedal and tighten the driveshaft retaining nut to the specified torque setting. Release the brake, check that the hub rotates freely, then stake the nut into the groove on the constant velocity joint using a suitable punch. Ensure that the nut is staked at least 4 mm into the groove (photos).

2.17A Tighten the driveshaft retaining nut to the specified torque...

2.17B ...then use a suitable punch...

2.17C ...to stake the nut firmly into the driveshaft groove

Chapter 8 Driveshafts

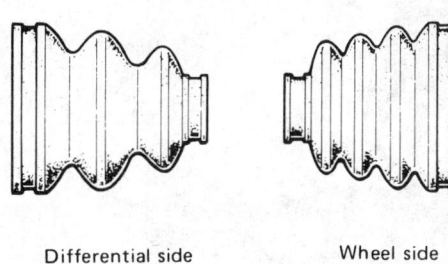

Fig. 8.3 Constant velocity joint rubber gaiter identification (Secs 3 and 4)

18 Tighten the lower suspension arm balljoint to swivel hub pinch-bolt to the specified torque setting (Chapter 10).
19 Refit the roadwheel then lower the car to the ground and tighten the wheel nuts to the specified torque.
20 Refill the gearbox or transmission with the correct type and quantity of lubricant as described in Chapter 1.

3 Outer constant velocity joint rubber gaiter – renewal

1 Remove the driveshaft from the car as described in Section 2.
2 Secure the driveshaft in a vice equipped with soft jaws, and release the two rubber gaiter retaining clips by raising the locking tangs with a screwdriver and then raising the end of the clip with pliers.
3 Slide the rubber gaiter down the shaft to expose the outer constant velocity joint.
4 Using a soft faced mallet, sharply strike the outer edge of the joint to drive it off the end of the shaft. The outer joint is retained on the driveshaft by a circular section circlip and striking the joint in this manner forces the circlip fully into its groove, so allowing the joint to slide off.
5 Withdraw the rubber gaiter from the driveshaft.
6 With the constant velocity joint removed from the driveshaft, thoroughly clean the joint using paraffin, or a suitable solvent, and dry it thoroughly. Carry out a visual inspection of the joint.
7 Move the inner splined driving member from side to side to expose each ball in turn at the top of its track. Examine the balls for cracks, flat spots or signs of surface pitting.
8 Inspect the ball tracks on the inner and outer members. If the tracks have widened, the balls will no longer be a tight fit. At the same time check the ball cage windows for wear or cracking between the windows.
9 If on inspection any of the constant velocity joint components are found to be worn or damaged, it will be necessary to renew the complete joint assembly, since no components are available separately. If the joint is in satisfactory condition, obtain a repair kit consisting of a new gaiter, retaining clips and the correct type and quantity of grease. Also inspect the joint retaining circlip for signs of damage and renew if necessary.
10 Tape over the splines on the end of the driveshaft, then fit the small retaining clip onto the gaiter and carefully slide the gaiter onto the shaft (photo).
11 Remove the tape and ensure that the constant velocity joint retaining circlip is correctly located in its groove on the end of the driveshaft (photo). Engage the help of an assistant for the following operations.
12 Position the constant velocity joint over the splines on the driveshaft until it abuts the circlip (photo).
13 Using two small screwdrivers placed either side of the circlip, compress the clip and at the same time have your assistant firmly strike the end of the joint with a soft faced mallet (photos). This should not require an undue amount of force. If the joint does not spring into place, remove it, reposition the circlip and try again. Do not force the joint, otherwise the circlip will be damaged.
14 Check that the joint is securely retained on the driveshaft end, then

3.10 Fit the outer constant velocity joint gaiter and clip to the driveshaft

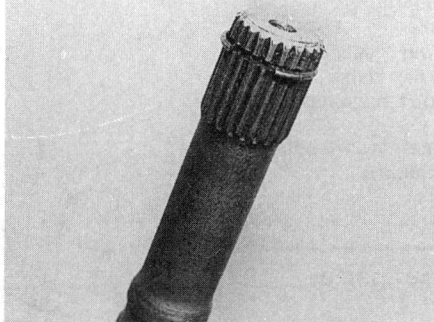

3.11 Ensure that the circlip is correctly located in the driveshaft groove

3.12 Position the joint over the splines so that it abuts the circlip

3.13A With the circlip compressed using two screwdrivers...

3.13B ...strike the joint firmly into position on the driveshaft

3.14 Pack the joint with the grease supplied in the repair kit

3.15 Position the large retaining clip over the joint...

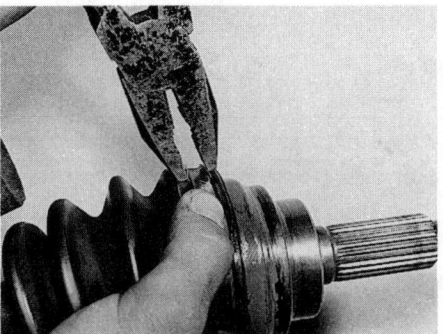

3.16 ...and secure it in position using pliers

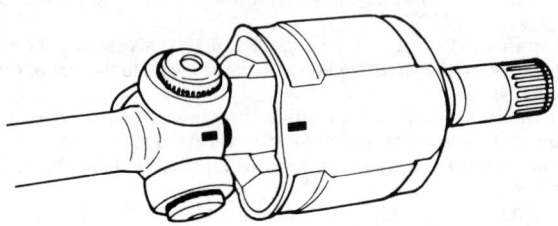

Fig. 8.4 Mark the tripod joint and outer ring for alignment prior to dismantling (Sec 4)

pack it thoroughly with the grease supplied. Work the grease well into the ball tracks whilst twisting the joint, and fill the rubber gaiter with any excess (photo).

15 Ease the gaiter over the joint and place the large retaining clip in position. Ensure that the gaiter is correctly located in the grooves on both the driveshaft and constant velocity joint (photo).

16 Using pliers, pull the large retaining clip and fold it over until the end locates between the two raised tangs. Hold the clip in this position whilst bending the tangs over to lock the clip (photo). Secure the small retaining clip using the same procedure.

17 Check that the constant velocity joint moves freely in all directions then refit the driveshaft to the car as described in Section 2.

4 Inner constant velocity joint rubber gaiter – renewal

1 Remove the driveshaft from the car as described in Section 2.
2 Secure the driveshaft in a vice equipped with soft jaws, and release the two rubber gaiter retaining clips by raising the locking tangs with a screwdriver, and then raising the end of the clip with pliers.
3 Slide the rubber gaiter down the shaft to expose the outer constant velocity joint.

Ball and cage type joint

4 Mark the relative position of the constant velocity joint outer member and driveshaft.
5 Using a screwdriver, carefully prise out the large circlip from inside the outer member and slide the outer member off the shaft.
6 Mark the relative position of the inner driving member assembly and driveshaft then, using circlip pliers, extract the circlip securing the inner member to the driveshaft. Withdraw the inner member and ball cage assembly as a unit.
7 Remove the rubber gaiter from the driveshaft.
8 Carry out a careful visual inspection of the joint as described in Section 3, paragraphs 6 to 9 inclusive. The large circlip which retains the constant velocity joint outer member should be renewed regardless of its apparent condition.

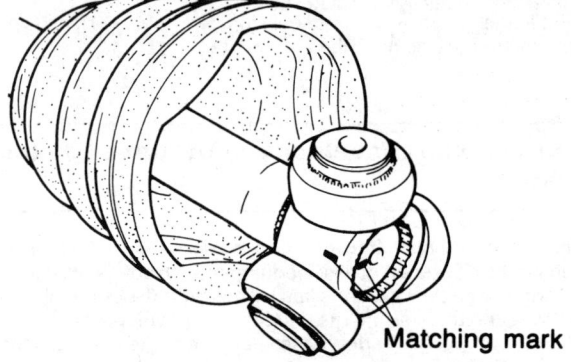

Fig. 8.5 Mark the tripod joint and shaft for alignment prior to removal (Sec 4)

Fig. 8.6 Removing the tripod joint (Sec 4)

9 Tape over the splines on the end of the driveshaft, then fit the small retaining clip onto the gaiter and carefully slide the gaiter onto the shaft. Once the gaiter is in position remove the tape from the driveshaft.
10 Refit the inner driving member and ball cage assembly, using the marks made on dismantling to ensure it is correctly positioned on the driveshaft splines, and secure with the circlip. Ensure that the member is fitted with the balls offset towards the end of the driveshaft.
11 Pack the inner member, ball cage and outer member thoroughly with the grease supplied in the repair kit.
12 Align the marks made on dismantling and slide the outer member over the balls, then secure it in position with the large circlip.

Tripod type joint

13 Using circlip pliers, remove the large circlip from inside the joint outer member. Mark the relative position of the outer member and

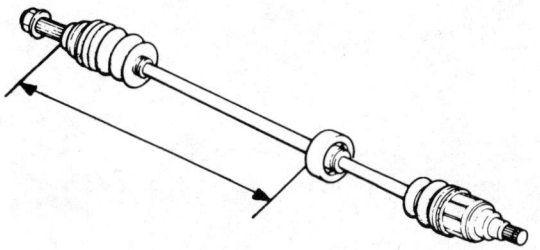

Fig. 8.7 Vibration damper positioning dimension (Sec 5)

Pre-September 1985 models – 367 ± 2 mm
September 1985 models onward – 405 ± 3 mm

tripod joint, then slide the outer member off the driveshaft.
14 Remove the circlip from the end of the driveshaft and mark the relative position of the tripod joint and driveshaft.
15 Remove the tripod joint from the shaft using a hammer and a suitable soft metal drift, taking great care not to allow the drift to contact the tripod joint rollers or the driveshaft splines.
16 Remove the rubber gaiter from the driveshaft.
17 Clean the tripod joint and outer member with a suitable solvent, dry them completely, and examine them as follows.
18 Check the tripod joint rollers and joint outer member for signs of wear, pitting or scuffing on their bearing surfaces. Also ensure that all rollers rotate smoothly and easily with no traces of roughness.
19 If on inspection the tripod joint or outer member reveals signs of wear or damage, it will be necessary to renew the complete joint assembly, since no components are available separately. If the joint components are in satisfactory condition, obtain a repair kit consisting of a new gaiter, retaining clips and the correct type and quantity of grease.
20 Tape over the splines on the end of the driveshaft then fit the small retaining clip onto the gaiter and carefully slide the gaiter onto the shaft. Once the gaiter is in position remove the tape from the driveshaft.
21 Align the marks made on dismantling and engage the tripod joint with the driveshaft splines. Use a soft metal drift to tap the joint onto the shaft, again taking great care not to contact the rollers or driveshaft splines, and secure it in position with the circlip.
22 Liberally apply the grease supplied with the kit to the tripod joint rollers and pack the excess into the outer member.
23 Slide the outer member over the tripod joint, using the marks made on dismantling to ensure it is correctly positioned, and secure it in position with the large circlip.

All joints

24 Ease the gaiter over the joint and place the large retaining clip in position. Ensure that the gaiter is correctly located in the grooves on both the driveshaft and constant velocity joint.
25 Using pliers pull the large retaining clip and fold it over until the end locates between the two raised tangs. Hold the clip in this position whilst bending the tangs over to lock the clip. Secure the small retaining clip using the same procedure.
26 Check that the constant velocity joint moves freely in all directions then refit the driveshaft to the car as described in Section 2.

5 Vibration damper – removal and refitting

Removal

1 Remove the right-hand driveshaft from the car as described in Section 2, then remove the inner constant velocity joint as described in Section 4.
2 Release the vibration damper retaining clip and slide the damper off the end of the driveshaft.
3 Visually inspect the damper and renew it if it is obviously damaged, or if there is any sign of deterioration of the rubber.

Refitting

4 Slide the damper onto the driveshaft ensuring that its retaining clip side is facing the inner constant velocity joint, and position it as shown in Fig. 8.7.
5 Secure the damper in position with the retaining clip and refit the inner constant velocity joint as described in Section 4.

Chapter 9 Braking system

Contents

Brake pedal – removal and refitting	2
Dual proportioning valve – testing, removal and refitting	11
Front brake caliper – removal, overhaul and refitting	9
Front brake disc – inspection, removal and refitting	10
Front brake pads – renewal	8
Front brake pad wear check	See Chapter 1
General information	1
Handbrake cable(s) – removal and refitting	18
Handbrake check and adjustment	See Chapter 1
Handbrake lever – removal and refitting	17
Hydraulic fluid level check	See Chapter 1
Hydraulic fluid renewal	See Chapter 1
Hydraulic pipes and hoses – inspection, removal and refitting	6
Hydraulic system – bleeding	5
Master cylinder – removal, overhaul and refitting	7
Rear brake caliper – removal, overhaul and refitting	15
Rear brake disc – inspection, removal and refitting	16
Rear brake pads – renewal	14
Rear brake pad wear check	See Chapter 1
Rear brake shoe wear check	See Chapter 1
Rear brake shoes – renewal	12
Rear wheel cylinder (drum brakes) – removal, overhaul and refitting	13
Stop lamp switch – removal, refitting and adjustment	19
Vacuum servo unit – testing, removal and refitting	3
Vacuum servo unit check valve – removal, testing and refitting	4

Specifications

System type .... Dual hydraulic circuit split diagonally with dual proportioning valve in rear hydraulic circuit. Disc front brakes with ventilated discs on models from September 1985 on, drum rear brakes, except on 1.6i models which have rear disc brakes. Vacuum servo assistance on all. Cable-operated handbrake on rear brakes

Front brakes
Type .... Disc with single piston sliding calipers
Disc diameter:
    Pre-September 1985 models .... 227 mm
    September 1985 models onward .... 260 mm
Disc thickness:
    Pre-September 1985 models:
        New .... 11.0 mm
        Minimum .... 10.0 mm
    September 1985 models onward:
        New .... 18.0 mm
        Minimum .... 16.0 mm
Maximum disc run-out .... 0.1 mm
Disc pad thickness:
    Pre-September 1985 models:
        New .... 10.0 mm
        Minimum .... 1.0 mm
    September 1985 models onward:
        New .... 8.0 mm
        Minimum .... 1.0 mm

Rear drum brakes
Type .... Single leading shoe drum, self-adjusting
Drum internal diameter:
    Pre-September 1985 models:
        New .... 180 mm
        Maximum .... 181 mm
    September 1985 onward Saloon and Hatchback models:
        New .... 200 mm
        Maximum .... 201 mm

Chapter 9 Braking system

Rear drum brakes (continued)
Drum internal diameter:
  May 1986 onward Estate models:
    New .... 228.6 mm
    Maximum .... 229.6 mm
Brake lining thickness:
  Pre-September 1985 models:
    New .... 4.0 mm
    Minimum .... 1.0 mm
  September 1985 models onward:
    New .... 5.0 mm
    Minimum .... 1.0 mm
Wheel cylinder bore diameter .... 17.46 mm

Rear disc brakes
Type .... Disc with single piston sliding calipers
Disc diameter .... 222 mm
Disc thickness:
  New .... 10.0 mm
  Minimum .... 8.0 mm
Maximum disc run-out .... 0.1 mm
Disc pad thickness:
  New .... 8.0 mm
  Minimum .... 1.0 mm

Torque wrench settings

	Nm	lbf ft
Brake pedal pivot bolt	20 to 35	15 to 26
Brake caliper guide pin bolts (pre-September 1985)	45 to 55	33 to 41
Brake caliper to mounting bracket bolts (September 1985 on):		
Upper bolt	16 to 25	12 to 18
Lower bolt	20 to 29	15 to 22
Brake caliper mounting bracket bolts:		
Pre-September 1985 models	56 to 66	41 to 49
September 1985 models onward:		
Front caliper	40 to 50	29 to 36
Rear caliper	50 to 70	36 to 51
Brake disc to hub flange	44 to 55	33 to 41
Brake hose union bolts	22 to 30	16 to 22
Brake pipe union nuts	13 to 22	10 to 16
Wheel cylinder to backplate bolts:		
Pre-September 1985 models	13 to 16	10 to 12
September 1985 models onward	10 to 13	8 to 10
Backplate to stub axle bolts	46 to 68	34 to 50
Vacuum servo to bulkhead nuts	13 to 16	10 to 12
Master cylinder to servo nuts	13 to 16	10 to 12
Driveshaft retaining nut	160 to 240	118 to 177
Roadwheel nuts	90 to 110	65 to 80

1 General information

The braking system is of the servo-assisted, dual circuit hydraulic type. The 1.6i model is fitted with both front and rear disc brakes, whereas all other models are equipped with disc brakes at the front, and drum brakes at the rear. The arrangement of the hydraulic system is such that each circuit operates one front and one rear brake from a tandem master cylinder. Under normal circumstances both circuits operate in unison. However, in the event of hydraulic failure in one circuit, full braking force will still be available at two wheels. A dual proportioning valve is also incorporated in the hydraulic circuit to regulate the pressure applied to the rear brakes and reduce the possibility of the rear wheels locking under heavy braking.

The disc brakes are actuated by single piston sliding type calipers. The rear drum brakes (where fitted) leading and trailing shoes are actuated by twin piston wheel cylinders which are self-adjusting by footbrake application.

The handbrake provides an independent mechanical means of rear brake application.

Warning: *Dust created by the braking system may contain asbestos, which is a health hazard. Never blow it out with compressed air and don't inhale any of it. An approved filtering mask should be worn when working on the brakes. DO NOT use petroleum based solvents to clean brake parts. Use brake cleaner or methylated spirit only.*

2 Brake pedal – removal and refitting

Removal
1 Working inside the car remove the cover from under the facia panel.
2 Extract the split pin, washer and clevis pin securing the servo unit pushrod to the brake pedal (photo).
3 Using pliers, carefully unhook the return spring from the brake pedal and pedal mounting bracket.
4 Slacken and remove the nut, spring washer and flat washer from the brake pedal pivot bolt, then withdraw the pivot bolt and remove the brake pedal.
5 Examine all brake pedal components for signs of wear, paying particular attention to the pedal bushes and pivot bolt, and renew as necessary. Renew the split pin as a matter of course.

Refitting
6 Refitting is a reverse of the removal procedure, but lubricate the bushes, pivot bolt and clevis pin with a multi-purpose grease.

Fig. 9.1 Exploded view of the brake pedal components (Sec 2)

1. Split pin
2. Clevis pin
3. Nut
4. Spring washer
5. Flat washer
6. Pivot bolt
7. Brake pedal
8. Return spring
9. Bushes
10. Return stop
11. Pedal pad

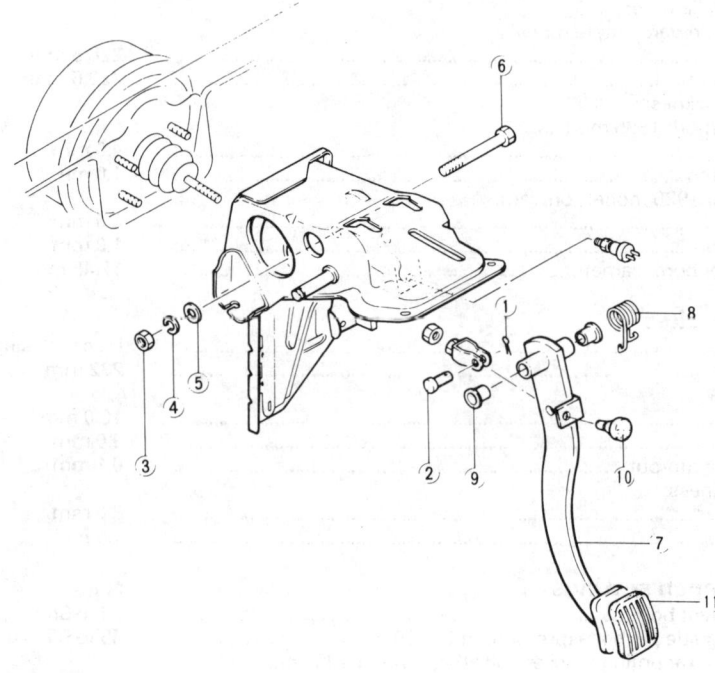

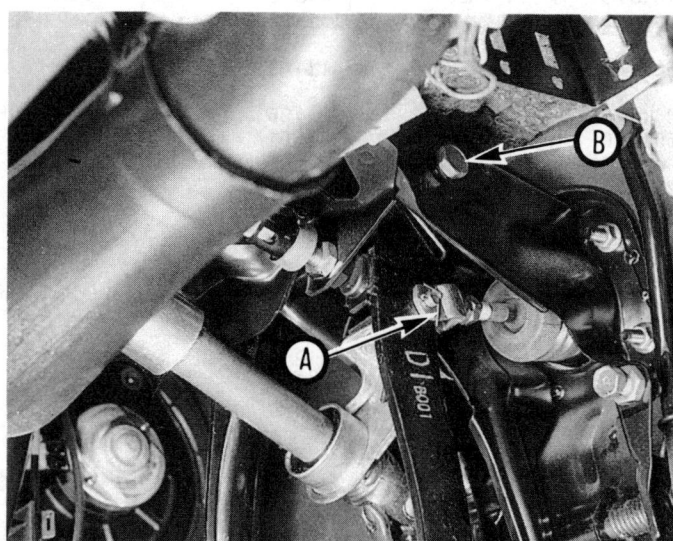

2.2 Brake pedal servo unit retaining split pin and clevis pin (A) and pivot bolt (B)

3 Vacuum servo unit – testing, removal and refitting

Testing

1 To test the operation of the servo unit depress the footbrake several times to exhaust the vacuum, then start the engine whilst keeping the pedal firmly depressed. As the engine starts there should be a noticeable 'give' in the brake pedal as the vacuum builds up. Allow the engine to run for at least two minutes then switch it off. If the brake pedal is now depressed it should feel normal, but further applications should result in the pedal feeling firmer, with the pedal stroke decreasing with each application.

2 If the servo does not operate as described, inspect the servo unit check valve as described in Section 4.

3 If the servo unit still fails to operate satisfactorily the fault lies within the unit itself. Repairs to the unit are possible, but special tools are required and the work should be entrusted to a suitably equipped Mazda dealer.

Removal

4 Remove the master cylinder as described in Section 7.
5 Disconnect the vacuum hose at the elbow connection on the servo unit.
6 Working from inside the car, remove the cover from under the facia panel.
7 Extract the split pin, washer and clevis pin securing the servo unit pushrod to the brake pedal (photo).
8 Slacken the four nuts securing the servo unit to the bulkhead and remove the unit from the engine compartment.

Refitting

9 Before refitting, it is necessary to check the servo unit output rod

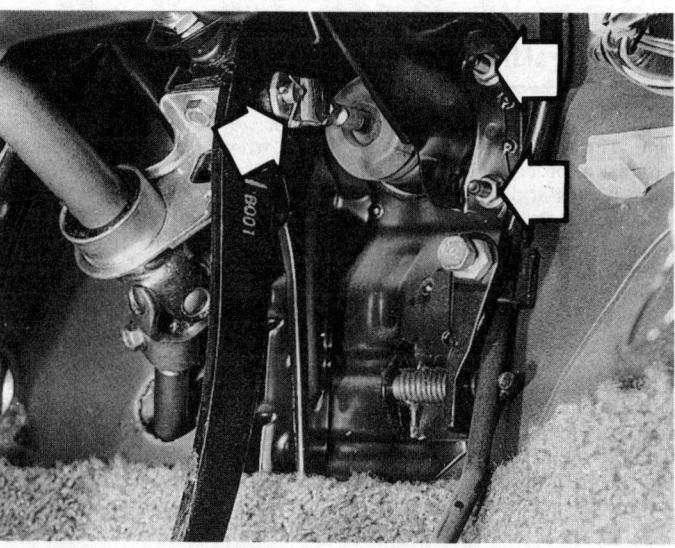

3.7 Vacuum servo unit mounting nuts and pushrod split pin and clevis pin

Chapter 9 Braking system

Fig. 9.2 Vacuum servo unit attachments – pre-September 1985 models shown (Sec 3)

1. Split pin
2. Clevis pin
3. Servo pushrod yoke
4. Master cylinder
5. Vacuum hose
6. Vacuum servo unit

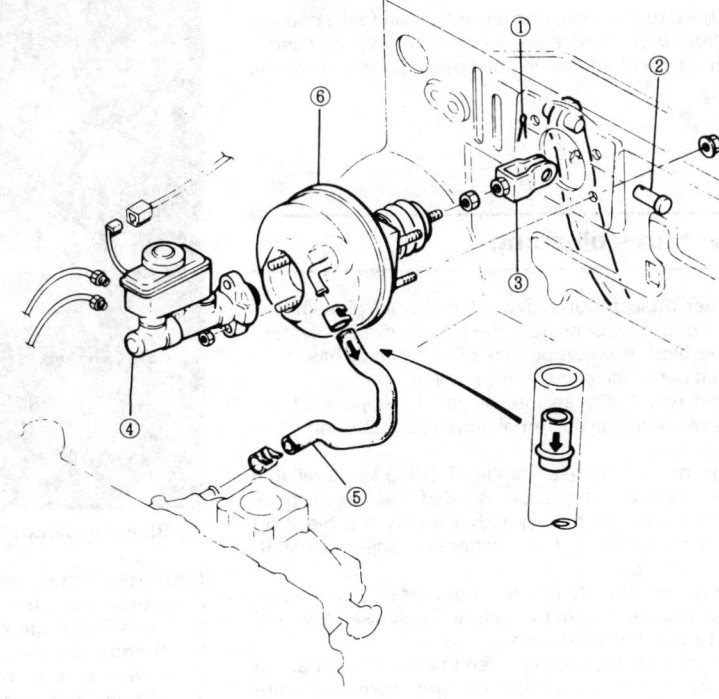

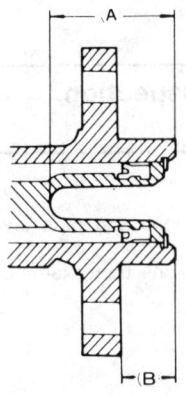 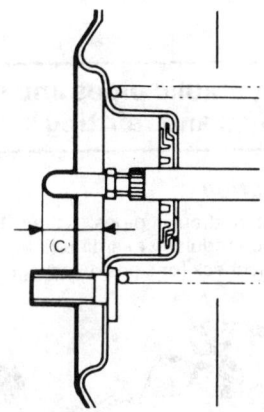 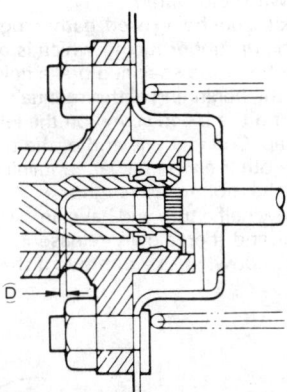

Fig. 9.3 Servo unit output rod adjustment (Sec 3)

A Master cylinder edge to base of primary piston bore
B Master cylinder edge to mounting flange
C Master cylinder mating face on servo to end of output rod

Fig. 9.4 Servo unit output rod clearance (Sec 3)

D Specified clearance

clearance using the following procedure.

10 Using a depth gauge, measure the distance from the end of the master cylinder to the bottom of the primary piston bore (Fig. 9.3). Call this figure A.
11 Now measure the distance from the end of the master cylinder to the edge of the mounting flange (Fig. 9.3). Call this figure B.
12 On the servo unit measure the distance from the master cylinder mating face to the end of the output rod (Fig 9.3). Call this figure C. Calculate the output rod clearance D using the following formula:

$D = (A - B) - C$

The output rod clearance D should be 0.1 to 0.5 mm. If adjustment is necessary, slacken the locknut and turn the end of the output rod until its length is correct to achieve the specified clearance. Tighten the locknut once the specified clearance is obtained.
13 The servo unit can then be fitted using the reverse of the removal sequence. Tighten the servo unit mounting nuts to the specified torque and secure the servo pushrod clevis in position with a new split pin.
14 Refit the master cylinder as described in Section 7.

4 Vacuum servo unit check valve – removal, testing and refitting

1 The servo unit check valve is situated within the vacuum hose which connects the servo unit to the inlet manifold.

Removal

2 Release the hose clips and disconnect the vacuum hose from the servo unit and inlet manifold and remove it from the car.

Testing

3 Examine the hose for damage, splits, cracks or general deterioration. Make sure that the check valve inside the hose is working correctly by blowing through the hose from the servo end. Air should flow in this direction, (direction of the arrow painted on the hose) but not when blown through from the engine end. Renew the hose and check valve assembly if it is at all suspect.

Chapter 9 Braking system

Refitting

4 Refit the vacuum hose to the servo unit and inlet manifold, ensuring that the arrow on the hose is situated at the servo end of the hose and is pointing towards the inlet manifold. Secure the hose in position with the clips.

5 Hydraulic system – bleeding

1 If the master cylinder, dual proportioning valve or brake pipes/hoses have been disconnected and reconnected, then the complete system (both circuits) must be bled. If a component of one circuit has been disturbed then only that particular circuit need be bled.
2 Bleed the left-hand rear brake and its diagonally opposite front brake, then repeat this sequence on the remaining circuit if the complete system is to be bled.
3 There are a variety of do-it-yourself brake bleeding kits available from motor accessory shops, and it is recommended that one of these kits is used wherever possible as they greatly simplify the bleeding operation. Follow the kit manufacturer's instructions in conjunction with the following procedure (photo).
4 During the bleeding operation do not allow the brake fluid level in the reservoir to drop below the minimum mark, and only use new fluid for topping-up. Never re-use fluid bled from the system.
5 Before starting, check that all rigid pipes and flexible hoses are in good condition and that all hydraulic unions are tight. Take great care not to allow hydraulic fluid to come into contact with the vehicle paintwork, otherwise the finish will be seriously damaged. Wash off any spilt fluid immediately with cold water.
6 If a brake bleeding kit is not being used, gather together a clean jar, a suitable length of plastic or rubber tubing which is a tight fit over the bleed screw and a new tin of the specified brake fluid (see 'Lubricants, fluids and capacities' at the beginning of this manual).
7 Clean the area around the bleed screw on the left-hand rear brake and remove the dust cap. Connect one end of the tubing to the bleed screw and immerse the other end in the jar containing sufficient brake fluid to keep the end of the rubber submerged.
8 Open the bleed screw half a turn and have an assistant depress the brake pedal to the floor and then slowly release it. Tighten the bleed screw at the end of each downstroke to prevent the expelled air/fluid

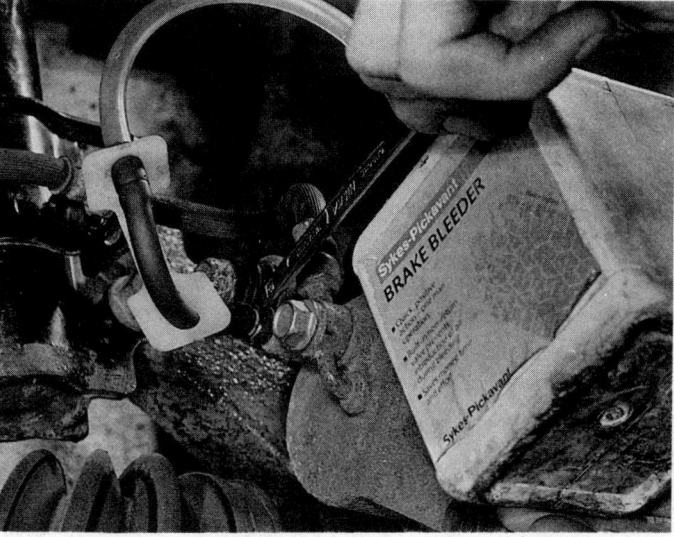

5.3 Bleeding a front brake caliper

from being drawn back into the system. Continue this procedure until clean brake fluid, free from air bubbles, can be seen flowing into the jar, and then finally tighten the bleed screw.
9 Remove the tube, refit the dust cap and repeat this procedure on the diagonally opposite front brake caliper.
10 Repeat the procedure on the remaining circuit.

6 Hydraulic pipes and hoses – inspection, removal and refitting

Inspection

1 The hydraulic pipes, hoses, hose connections and pipe unions should be regularly examined.
2 First check for signs of leakage at the pipe unions, then examine the

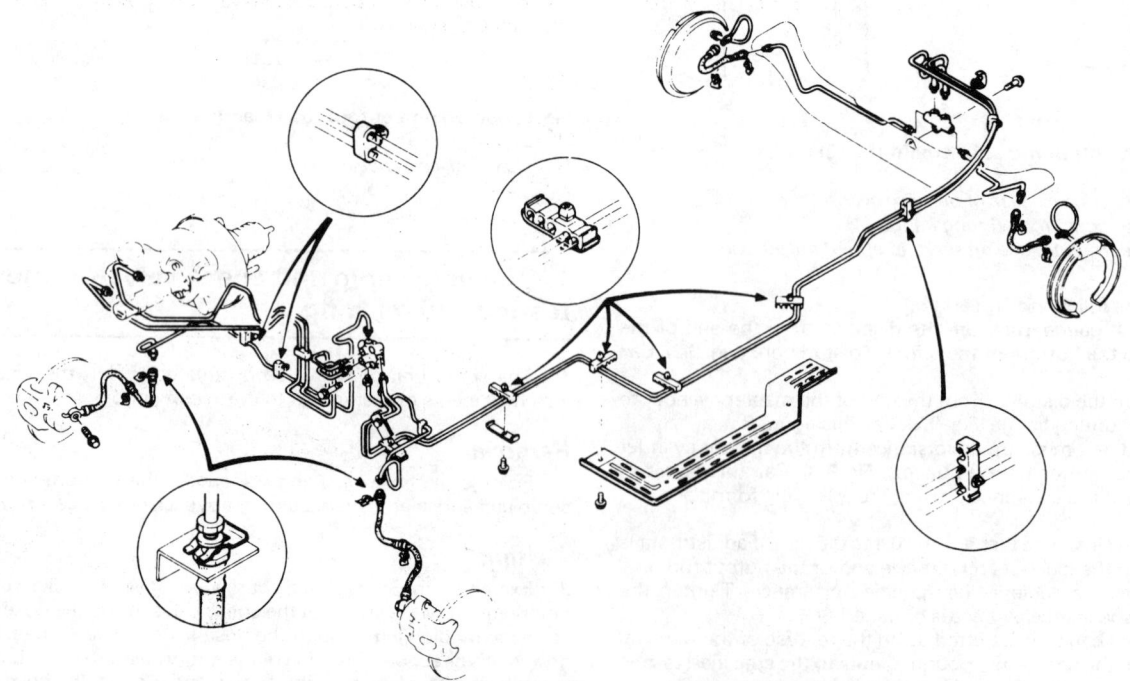

Fig. 9.5 Layout of the braking system hydraulic pipes and hoses – pre-September 1985 models (Sec 6)

Chapter 9 Braking system

6.2 Examine the flexible hoses for signs of cracking or chafing

flexible hoses for signs of cracking, chafing and fraying (photo).
3   The brake pipes should be examined carefully for signs of dents, corrosion or other damage. Corrosion should be scraped off, and if the depth of pitting is significant, the pipes renewed. This is particularly likely in those areas underneath the vehicle body where the pipes are exposed and unprotected.

Removal

4   If any section of pipe or hose is to be removed, the loss of fluid may be reduced by removing the hydraulic fluid reservoir filler cap, placing a piece of polythene over the filler neck, then refitting and tightening the filler cap. If a section of pipe is to be removed from the master cylinder, the reservoir should be emptied by syphoning out the fluid or drawing out the fluid with a pipette.
5   To remove a section of pipe, unscrew the union nuts at each end of the pipe and release it from the clips attaching it to the body. Where the union nuts are exposed to the full force of the weather, they can sometimes be quite tight. If an open-ended spanner is used, burring of the flats on the nuts is not uncommon, and for this reason it is preferable to use a split ring spanner which will engage all the flats. If such a spanner is not available, self-locking grips may be used although this is not recommended.

6   To remove a flexible hose first clean the ends of the hose and the surrounding area, then unscrew the union nut(s) from the hose end(s). Recover the spring clip and withdraw the hose from the serrated mounting in the support bracket. Where applicable, unscrew the hose from the caliper.
7   Brake pipes with flared ends and union nuts in place can be obtained individually or in sets from Mazda dealers or accessory shops. The pipe is then bent to shape, using the old pipe as a guide, and is ready for fitting to the car.

Refitting

8   Refitting the pipes and hoses is a reversal of removal. Make sure that brake pipes are securely supported in their clips and ensure that the hoses are not kinked. Check also that the hoses are clear of all suspension components and underbody fittings and will remain clear during movement of the suspension and steering. After refitting, remove the polythene from the reservoir and bleed the brake hydraulic system as described in Section 5.

7 Master cylinder – removal, overhaul and refitting

Removal

1   Unscrew the hydraulic fluid reservoir filler cap, place a piece of polythene over the filler neck and refit the cap securely. This will minimise brake fluid loss during subsequent operations. As an added precaution place absorbent rags beneath the master cylinder brake pipe unions.
2   Trace the wiring back from the fluid reservoir level switch and disconnect it at the wiring connector.
3   Wipe clean the area around the brake pipe unions on the side of the master cylinder. Unscrew the two union nuts and carefully withdraw the pipes (photo). Plug, or tape over, the pipe ends and master cylinder orifices to minimise the loss of brake fluid and to prevent the entry of dirt into the system. Take great care not to allow any brake fluid to come into contact with the vehicle paintwork. Wash off any spilt fluid immediately with cold water.
4   Slacken the two nuts securing the master cylinder to the vacuum servo unit and withdraw the unit from the engine compartment.

Overhaul

5   Remove the reservoir filler cap and pour out the brake fluid into a suitable container. Prepare a clean uncluttered work surface on which to dismantle the unit.

Fig. 9.6 Exploded view of the brake master cylinder (Sec 7)

1   Reservoir filler cap assembly
2   Float
3   Reservoir
4   Mounting bush seal
5   Stopper bolt
6   Circlip
7   Primary piston assembly
8   Secondary piston assembly

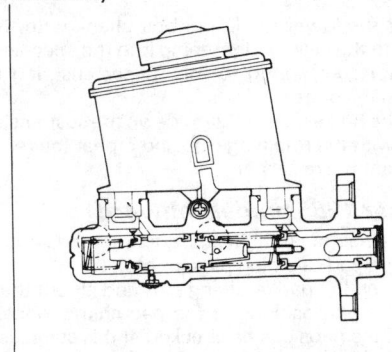

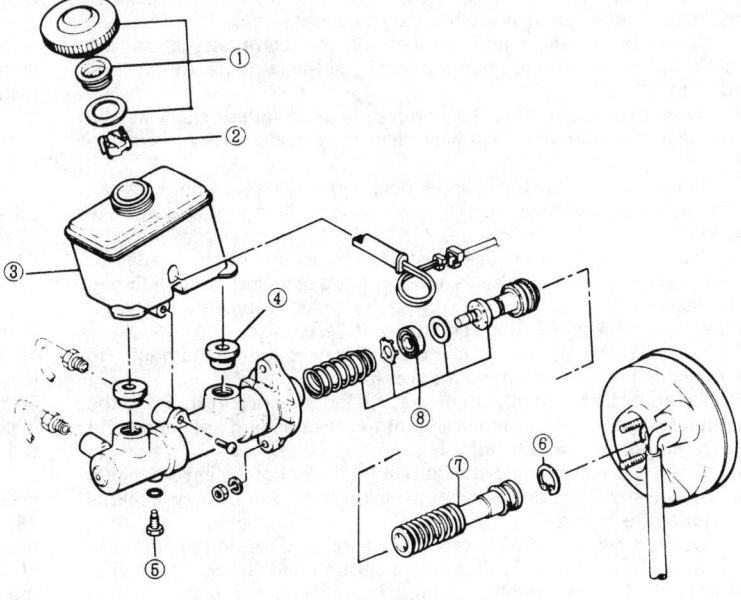

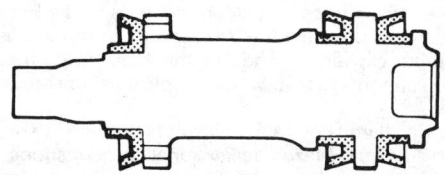

Fig. 9.7 Secondary piston seal arrangement (Sec 7)

7.3 Using a brake pipe spanner to undo the master cylinder union nuts

6 Remove the single screw which secures the reservoir to the master cylinder body and ease the reservoir out of position.
7 Carefully extract the rubber bush seals from the master cylinder inlet ports.
8 Remove the secondary piston retaining bolt from the underside of the cylinder body noting its sealing O-ring.
9 Using circlip pliers, extract the primary piston retaining circlip from the end of the cylinder bore.
10 Withdraw the primary piston assembly followed by the secondary piston assembly. If necessary, tap gently on the master cylinder body with a soft faced mallet to release the pistons. If this fails to release them, the pistons can be pushed out by applying compressed air to the secondary piston union bolt hole. Only low pressure compressed air should be required such as is generated by a foot pump.
11 Note the location and position of the components on the secondary piston then remove the spring, spreader plate, rubber seal and shim.
12 With the master cylinder completely dismantled, clean all the components in methylated spirit, or clean brake fluid, and dry with a lint-free rag.
13 Carefully examine the cylinder bore and pistons for signs of wear, scoring or corrosion, if damage is evident renew the complete assembly.
14 If the components are in a satisfactory condition, obtain a repair kit consisting of new pistons, seals and springs, a new stopper bolt O-ring and piston retaining circlip. Examine the master cylinder reservoir rubber bush seals for damage and renew if necessary.
15 Lubricate the master cylinder bore, pistons and seals thoroughly in clean brake fluid and assemble them as follows.
16 Carefully fit the shim, rubber seal, spreader plate and spring to the secondary piston using the reverse of the dismantling sequence and with reference to Figs. 9.6 and 9.7.
17 Insert the secondary piston into the cylinder bore using a twisting motion whilst taking great care not to distort the lips of the new seals as they enter the cylinder.
18 Using a wooden dowel, push the secondary piston fully into the master cylinder bore and refit the stopper bolt and O-ring. Check that the stopper bolt is properly located by pushing the piston in and releasing it several times.
19 Insert the primary piston assembly into the cylinder bore, again taking great care not to distort the piston seals.
20 Secure the primary piston in position with its retaining circlip, ensuring that it is correctly located in the groove in the master cylinder bore.
21 Fit the reservoir rubber bush seals to the cylinder inlet ports and push the reservoir firmly into place. Refit the reservoir retaining screw.

Refitting

22 Before refitting the master cylinder, clean the mounting faces and check the servo unit output rod clearance as described in Section 3.
23 The master cylinder is refitted by a reverse of the removal sequence. Tighten the master cylinder retaining nuts and brake pipe unions to the specified torque settings. On completion refill the master cylinder reservoir and bleed the hydraulic system as described in Section 5.

8 Front brake pads – renewal

Warning: *Disc brake pads must be renewed on both front wheels at the same time – never renew the pads on only one wheel as uneven braking may result. Also, the dust created by wear of the pads may contain asbestos, which is a health hazard. Never blow it out with compressed air and don't inhale any of it. An approved filtering mask should be worn when working on the brakes. DO NOT use petroleum based solvents to clean brake parts. Use brake cleaner or methylated spirit only.*

1 Apply the handbrake, then jack up the front of the car and support it on axle stands. Remove the front roadwheels.

Pre-September 1985 models

2 Using a screwdriver, remove the retaining clip which secures the brake hose to the suspension strut then prise out the pad return spring from the top of the caliper, noting its correctly fitted position (photo).
3 Remove the lower caliper guide pin bolt and pivot the caliper away from the disc to gain access to the brake pads. Tie the caliper to the suspension strut using a piece of wire (photo).
4 Remove the inner shim which is fitted to the caliper piston (photo).
5 Remove the brake pads from the caliper mounting bracket whilst noting the correct position of the spring clips and shim (photo). If required, the thickness of the pads can be checked at this stage using a steel rule.
6 Before refitting the pads, check that the guide pin bolts are free to slide in the caliper bracket, and check that the rubber guide pin gaiters fitted to the bracket are undamaged. Brush the dust and dirt from the caliper and piston but *do not inhale it as it is injurious to health*. Inspect the dust seal around the piston for damage and the piston for evidence of fluid leaks, corrosion or damage. If attention to any of these components is necessary, refer to Section 9.
7 Renew the pad return spring, spring clips and shims as a matter of course. These are available as a service kit, which also contains a sachet of special grease with which to lubricate the springs and shims with on installation.
8 To refit the pads, first smear a small amount of the supplied grease onto the spring clips and shims. Install the spring clip and shim on the outer pad and fit the spring clip to the inner pad.
9 Install both pads in the caliper mounting bracket.
10 Push the piston fully into the caliper bore. This can be achieved using firm hand pressure or if necessary a G-clamp.
11 Fit the inner shim to the caliper piston and swing the caliper back down into position.
12 Smear the lower guide pin bolt shank with the supplied grease then refit it to the caliper, tightening it to the specified torque.
13 Fit the return spring to the caliper ensuring that it is correctly located with the pads.
14 Refit the brake hose to its guide on the suspension strut, securing it in position with the retaining clip, and repeat the renewal procedure on the other front brake caliper.

September 1985 models onward

15 Using a pair of suitable pliers, extract the spring retainer from the outer pad and pad retaining pins (photo).
16 Withdraw the pad retaining pins and lift out the brake pads noting the correct fitted positions of the pad shims (photos). If required the thickness of the pads can be checked at this stage using a steel rule.
17 Before refitting the pads check that the caliper is free to slide easily

Chapter 9 Braking system

Fig. 9.8 Front brake caliper and associated components – pre-September 1985 models (Secs 8 and 9)

1. Guide pin
2. Shims
3. Spring clips
4. Disc pads
5. Mounting bracket
6. Guide pin bush
7. Return spring
8. Caliper

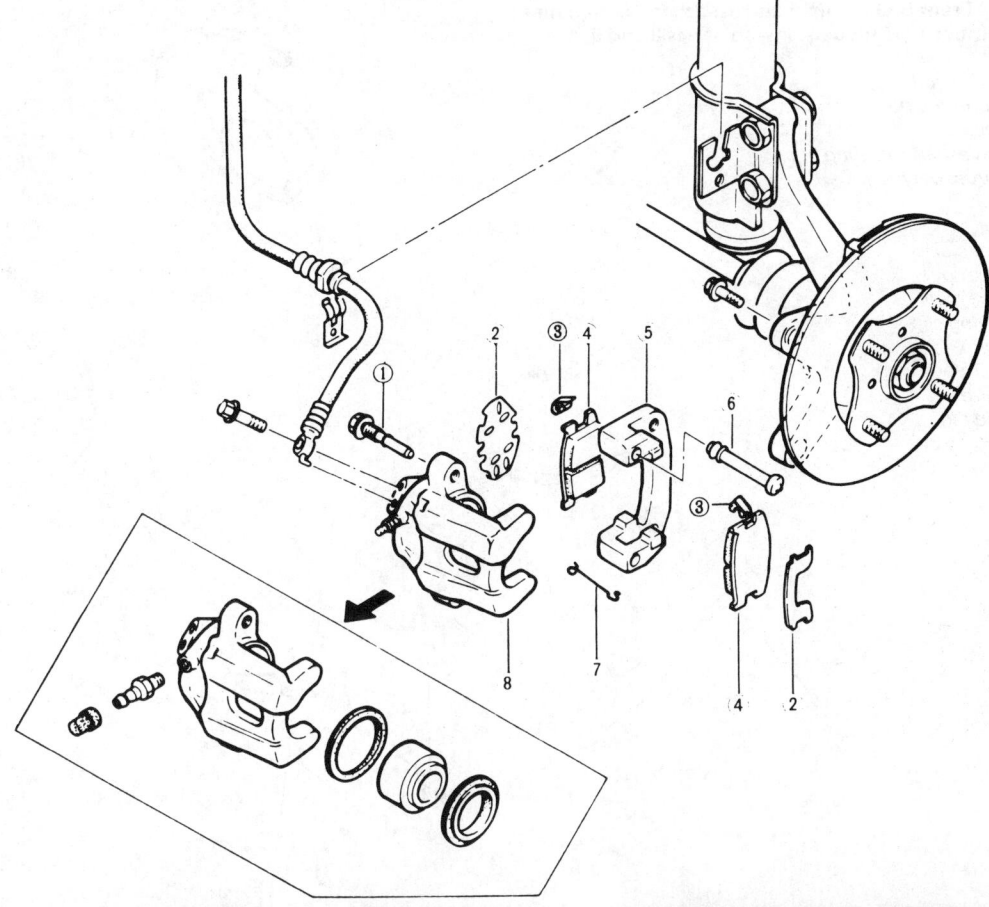

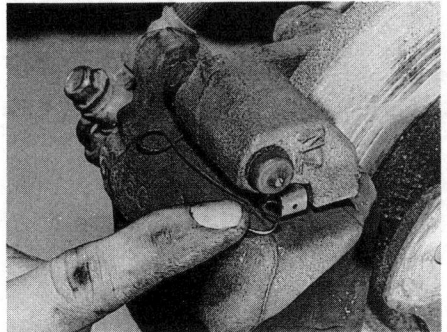

8.2 On pre-September 1985 models remove the pad return spring

8.3 Slacken and remove the lower guide pin bolt...

8.4 ...then pivot the caliper away from the disc and remove the shim from the piston

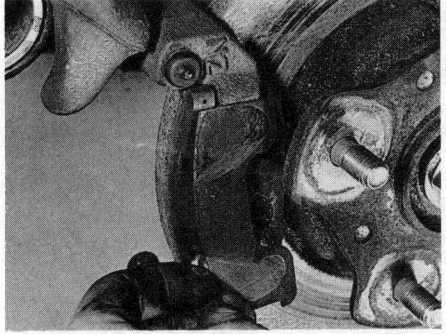

8.5 Pads and shims can then be removed from the mounting bracket

8.15 On September 1985 models onward remove the spring retainer...

Fig. 9.9 Front brake caliper and associated components – September 1985 models onward (Secs 8 and 9)

1 Union bolt
2 Mounting bolts
3 Caliper
4 Driveshaft retaining nut
5 Thrustwasher
6 Hub
7 Bolt
8 Brake disc
9 Disc cover
10 Disc pads
11 Spring clip
12 Shim
13 Shim
14 Shim
15 Pad pin

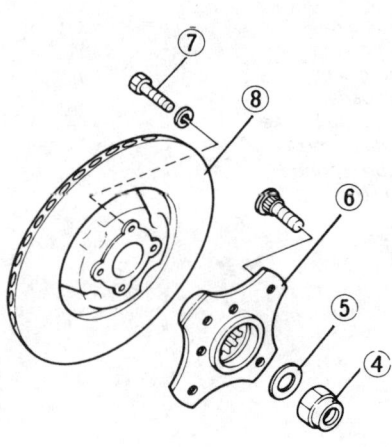
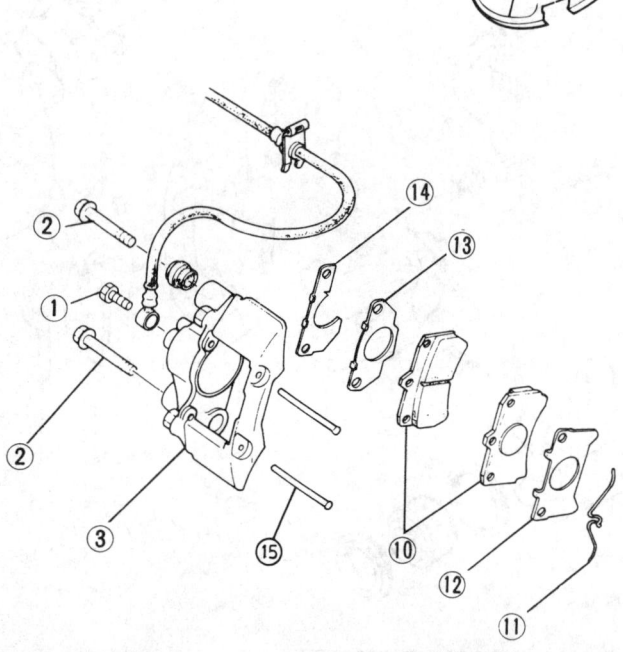

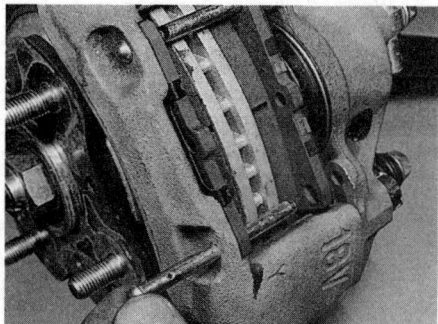

8.16A Withdraw the pad retaining pins...

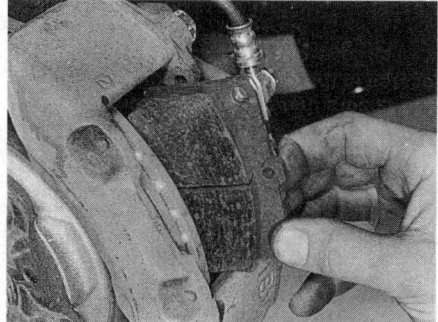

8.16B ...and remove the pads from the caliper

8.19A Locate the inner pad first shim...

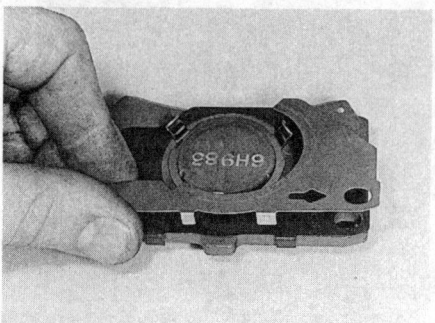

8.19B ...then fit the second shim so that the arrow points in the direction of normal disc rotation

8.22 Ensure the spring retainer is correctly located in the outer pad and pad retaining pins (arrowed)

Chapter 9 Braking system

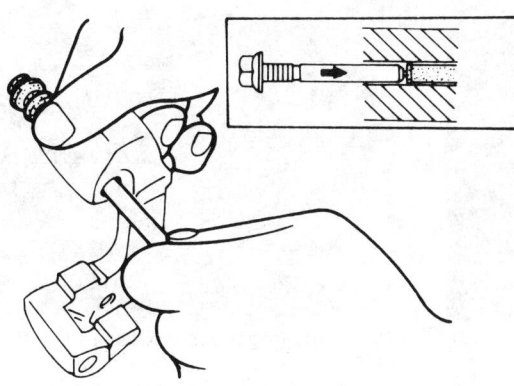

Fig. 9.10 Removing the guide pin bushes from the caliper – pre-September 1985 models (Sec 9)

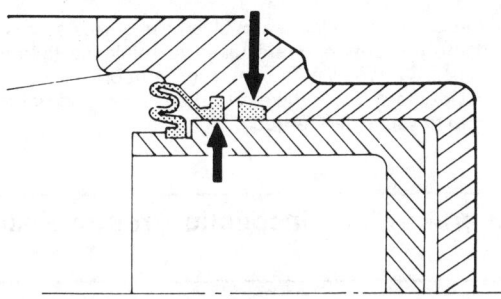

Fig. 9.12 Correct positioning of the piston and dust seals in the caliper – pre-September 1985 models (Sec 9)

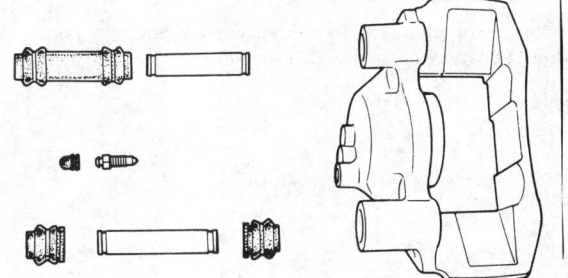

Fig. 9.11 Check the condition of the caliper mounting bushes – September 1985 models onward (Sec 9)

9.5 Brake hose clamp fitted to a front flexible brake hose

9.6 Brake hose union bolt (A) and front caliper mounting bolts (B) on September 1985 models onward

on its mountings and that the mounting bush rubber gaiters are undamaged. Brush the dust and dirt from the caliper and piston but *do not inhale it as it is injurious to health.* Inspect the dust seal around the piston for damage, and the piston for evidence of fluid leaks, corrosion or damage. If attention to any of these components is necessary, refer to Section 9.
18 Examine the shims, pad retaining pins and spring retainer for corrosion, wear or damage and renew as necessary.
19 To refit the pads, first fit the two shims to the inner brake pad and the single shim to the outer pad using the notes made on dismantling to ensure they are correctly positioned (photos).
20 Make sure that the caliper piston is fully retracted in its bore. If not, carefully push it in using a flat bar or screwdriver as a lever or preferably use a G-clamp.
21 Slide both the brake pads into position in the caliper and insert the pad retaining pins ensuring that the holes are closest to the outer edge of the caliper.
22 Insert the spring retainer into the holes in the pad retaining pins, then locate the centre of the retainer with the hole in the outer brake pad (photo). Repeat the renewal procedure on the other front brake caliper.

All models
23 Depress the footbrake two or three times to bring the pistons into contact with the pads, then refit the roadwheels and lower the car to the ground. Tighten the roadwheel nuts to the specified torque when the car is on its wheels, then check the fluid level in the master cylinder reservoir.

9 Front brake caliper – removal, overhaul and refitting

Removal
1 Apply the handbrake, then jack up the front of the car and support it on axle stands. Remove the appropriate front roadwheel.
2 On pre-September 1985 models proceed as described in paragraphs 2 to 4 of Section 8, noting that it is not necessary to tie the caliper to the suspension strut.
3 On September 1985 models onward, remove the brake pads as described in Section 8.
4 On all models very slowly depress the brake pedal until the piston has been ejected just over half way out of its bore.
5 Fit a brake hose clamp to the flexible brake hose leading to the front

10.2A Checking brake disc thickness with a micrometer

10.2B Checking the brake disc run-out with a dial test gauge

10.7 Removing the brake disc to hub flange bolts

brake caliper (photo). This will minimise brake fluid loss during subsequent operations.

6 Undo the brake hose union bolt and remove the hose from the caliper. Plug the end of the hose and the caliper orifice to prevent dirt entering the hydraulic system (photo).

7 Remove the caliper guide pin/mounting bolt(s) (as applicable) and lift the caliper away from the vehicle.

Overhaul

8 With the caliper on the bench, wipe away all traces of dust and dirt, but *avoid inhaling the dust as it is injurious to health*.

9 Withdraw the partially ejected piston from the caliper body and remove the dust seal. The piston can be withdrawn by hand, or if necessary pushed out by applying compressed air to the union bolt hole. Only low pressure should be required such as is generated by a foot pump.

10 Using a small screwdriver, extract the piston hydraulic seal whilst taking great care not to damage the caliper bore.

11 On pre-September 1985 models, remove the guide pin bushes from the mounting bracket by pushing them out from the inside using the guide pin bolts.

12 On September 1985 models onward, withdraw the mounting bushes from the caliper body and remove the bush rubber gaiters.

13 Clean all parts in methylated spirit, or clean brake fluid, and wipe dry using a lint-free cloth. Inspect the piston and caliper bore for signs of damage, scuffing or corrosion and if these conditions are evident, renew the complete caliper body assembly.

14 If the components are in a satisfactory condition, a repair kit consisting of new seals and special lubricant should be obtained. Note on pre-September 1985 models this kit also contains new guide pin bushes. On later models examine the mounting bushes and gaiters and renew if necessary.

15 Lubricate the new seals with the special lubricant in the repair kit and carefully fit the new piston seal into the caliper bore. Lubricate the caliper bore and piston with clean brake fluid.

16 On pre-September 1985 models, position the dust seal over the innermost end of the piston so that the caliper bore sealing lip protrudes beyond the base of the piston. Engage the sealing lip of the dust seal with the groove in the caliper, then carefully insert the piston into the caliper bore until the outer sealing lip of the dust seal can be engaged with the groove in the piston. Having done this push the piston fully into the bore. Ease the piston out again slightly and check that the dust seal lip remains correctly seated. Lubricate the guide pin bushes with the special lubricant provided and refit them to the caliper mounting bracket. Refit the shim to the caliper piston.

17 On September 1985 models onward, carefully insert the piston into the caliper bore whilst taking great care not to distort the piston seal. Ease the dust seal over the piston and press it into position in the caliper body. Lubricate the mounting bushes with the special lubricant and refit them to the caliper body. Ensure that each gaiter is correctly located in its grooves in the caliper and bush.

Refitting

18 Position the caliper over the disc, refit the guide pin/mounting bolts and tighten to the specified torque.

19 On pre-September 1985 models, refit the brake pad return spring and brake hose retaining clip, and on later models refit the brake pads. Refer to Section 8 for further information.

20 Refit the brake hose to the caliper and tighten the union bolt to the specified torque.

21 Remove the brake hose clamp and bleed the brakes as described in Section 5. If the precautions were taken to minimise fluid loss, it should only be necessary to bleed the relevant front brake.

22 Refit the roadwheel, lower the car to the ground and tighten the roadwheel nuts to the specified torque.

10 Front brake disc – inspection, removal and refitting

Inspection

1 Apply the handbrake, then jack up the front of the car and support it on axle stands. Remove the appropriate front roadwheel.

2 Rotate the disc by hand and examine it for deep scoring, grooving or cracks. Light scoring is normal and may be removed with emery tape, but, if excessive, the disc must be renewed. Any loose rust and scale around the outer edge of the disc can be removed by lightly tapping it with a small hammer while rotating the disc. Measure the disc thickness with a micrometer if available. Disc run-out can be checked using a dial test gauge, although a less accurate method is to use a feeler gauge together with a metal base block (photos).

Removal

3 Using a hammer and suitable chisel nosed tool, tap up the staking

10.11 Ensure the retaining nut is staked securely into the driveshaft groove

Chapter 9 Braking system

securing the driveshaft retaining nut to the groove in the constant velocity joint.
4   Have an assistant firmly depress the brake pedal, then using a socket and extension bar, slacken the nut and remove it along with the washer. Discard the nut as a new one must be used on refitting.
5   Slacken and remove the two bolts which secure the brake caliper to the swivel hub. Lift the caliper off the disc and tie it to the front suspension coil spring to avoid straining the brake hose.
6   The hub flange and disc assembly can then be pulled off the driveshaft splines using a suitable puller and removed from the vehicle.
7   Remove the four bolts from the rear of the disc and separate the disc and hub flange (photo).

Refitting

8   Reassemble the brake disc and hub flange and tighten the four retaining bolts to the specified torque.
9   Engage the disc and hub assembly with the driveshaft splines and drive the assembly into position using a hammer and suitable drift. Fit the washer and new driveshaft retaining nut.
10   Slide the brake caliper assembly over the disc and tighten both its mounting bolts to the specified torque.
11   Have an assistant firmly depress the brake pedal and tighten the driveshaft retaining nut to the specified torque setting. Release the brake, check that the hub rotates freely, then stake the nut into the groove on the constant velocity joint using a suitable punch. Ensure that the nut is staked at least 4 mm into the groove (photo).
12   Refit the roadwheel, lower the car to the ground and tighten the roadwheel nuts to the specified torque.

11   Dual proportioning valve – testing, removal and refitting

Testing

1   A dual proportioning valve is incorporated in the hydraulic braking circuit to regulate the pressure applied to the rear brakes, and reduce the risk of the rear wheels locking under heavy braking. The dual proportioning valve is mounted on the bulkhead in the engine compartment (photo).
2   Specialist equipment is required to check the performance of the valve, therefore if the valve is thought to be faulty the car should be taken to a suitably equipped Mazda dealer for testing. If the valve is found to be faulty it must be renewed as repairs are not possible.

Removal

3   Unscrew the master cylinder reservoir filler cap, place a piece of polythene over the filler neck and refit the cap securely. This will minimise brake fluid loss during subsequent operations. As an added precaution place absorbent rags beneath the dual proportioning valve brake pipe unions.
4   Wipe clean the area around the brake pipe unions on the dual proportioning valve, then make a note of how the pipes are arranged to use as a reference on refitting. Unscrew the union nuts and carefully

11.1 Dual proportioning valve location

withdraw the pipes. Plug, or tape over, the pipe ends and valve orifices to minimise the loss of brake fluid and to prevent the entry of dirt into the system. Take great care not to allow any brake fluid to come into contact with the vehicle paintwork. Wash off any spilt fluid immediately with cold water.
5   Slacken the two bolts which secure the valve to the bulkhead and remove it from the engine compartment.

Refitting

6   Refit the dual proportioning valve to the bulkhead and tighten its mounting bolts securely.
7   Refit the brake pipes to the valve, using the notes made on dismantling to ensure they are correctly positioned, and tighten the union nuts to the specified torque.
8   Remove the polythene from the master cylinder reservoir filler neck and bleed the complete hydraulic system as described in Section 5.

12   Rear brake shoes – renewal

Warning: *Drum brake shoes must be renewed on both rear wheels at the same time – never renew the shoes on only one wheel as uneven braking may result. Also, the dust created by wear of the shoes may contain asbestos, which is a health hazard. Never blow it out with compressed air and don't inhale any of it. An approved filtering mask should be worn when working on the brakes. DO NOT use petroleum based solvents to clean brake parts. Use brake cleaner or methylated spirit only.*

1   Chock the front wheels, then jack up the rear of the car and support

12.2A Tap up the rear hub retaining nut staking...

12.2B ...then remove the nut...

12.3 ...and withdraw the brake drum assembly from the stub axle

Fig. 9.13 Exploded view of the rear brake shoes and associated components – September 1985 models onward (Sec 12)

1 Brake drum
2 Holding pins and spring clips
3 Brake shoes
4 Return springs
5 Anti-rattle springs
6 Self-adjust mechanism
7 Wheel cylinder
8 Backplate

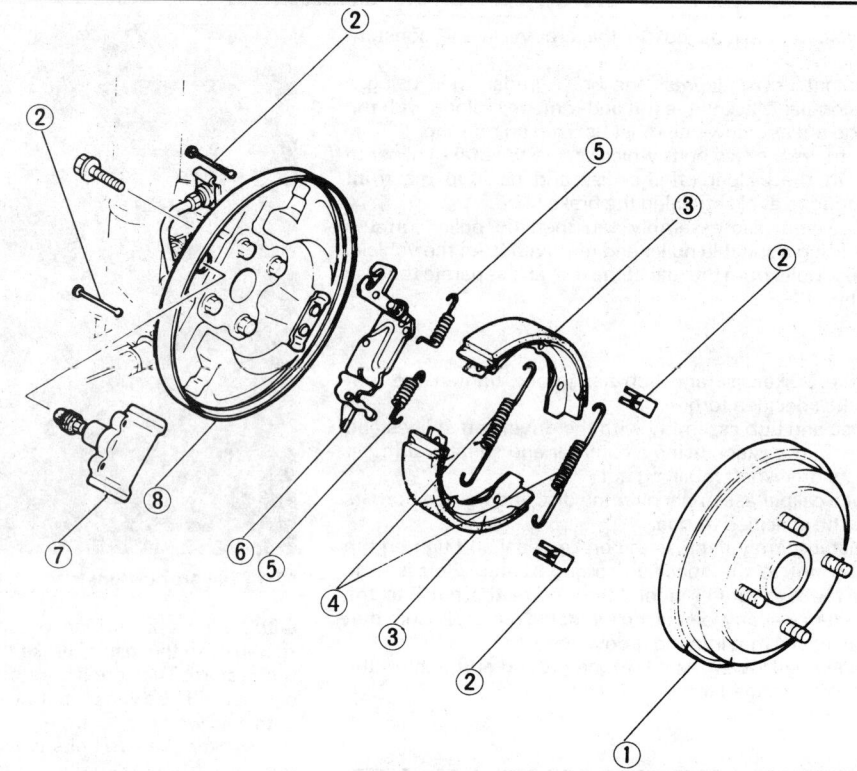

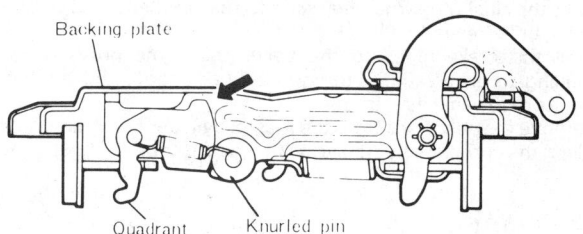

Fig. 9.14 Self-adjust mechanism quadrant (arrowed) set in the fully retracted position – pre-September 1985 models (Sec 12)

it on axle stands. Remove the appropriate rear roadwheel and release the handbrake. Carefully lever the hub cap out of position.
2   Using a suitable pointed tool, tap up the staking securing the rear hub retaining nut to the groove in the stub axle. Slacken and remove the hub nut and discard it (photos). A new nut must be obtained for refitting. **Note:** *The right-hand hub nut on later models has a left-hand thread and must therefore be turned clockwise to loosen it.* This nut is galvanised for easy identification.
3   Pull the brake drum and hub assembly squarely off the stub axle whilst holding the outer bearing and thrustwasher in position with your thumbs to prevent them dropping out (photo). If it's not possible to remove the assembly due to the brake shoes binding on the drum, it will be necessary to release the handbrake cable. On early models this can be achieved by extracting the clevis pin which secures the cable to the operating mechanism lever, and then moving the lever in until it touches the backplate. On later models remove the two bolts which secure the cable to the rear of the backplate, disconnect the cable from the self-adjusting mechanism and release it in a similar way. This will increase the clearance between brake shoes and drum and allow the drum to be removed.
4   With the brake drum assembly removed, brush or wipe the dust from the drum, brake shoes, wheel cylinder and backplate. *Take great care not to inhale the dust as it is injurious to health.*
5   Note the fitted positions of the various springs and linkages as an aid to refitting (photo).

12.5 Fitted positions of shoes and springs on September 1985 models onward

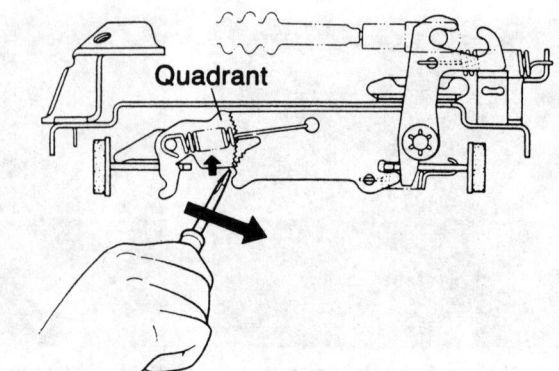

Fig. 9.15 Resetting the brake self-adjust mechanism quadrant – September 1985 models onward (Sec 12)

Chapter 9 Braking system

12.13 Set the self-adjusting mechanism quadrant in the fully retracted position (pre-September 1985 models shown)

12.12 Apply a trace of high melting-point grease to the areas arrowed

6 Depress the spring clip of the leading brake shoe holding pin whilst supporting the pin from the rear of the backplate. Turn the clip through 90° and lift it off, then withdraw the holding pin from the rear of the backplate.
7 Release the leading shoe from its lower pivot and disconnect the lower return spring. Remove the spring from both brake shoes.
8 Ease the upper end of the leading shoe out of its wheel cylinder location and detach it from the handbrake self-adjusting mechanism. Disconnect the upper return spring and anti-rattle spring (where fitted) and remove the spring(s) along with the leading shoe.
9 Remove the trailing shoe holding pin and spring clip as described in paragraph 6.

10 Disconnect the anti-rattle spring from the trailing shoe, release the shoe from its pivot locations and remove the shoe.
11 Disconnect the self-adjusting mechanism from the handbrake, if not already having done so, and remove it from the backplate.
12 Before fitting the new brake shoes clean off the brake backplate with a rag and apply a trace of high melting-point grease to the brake shoe contact areas, lower pivots and wheel cylinder pistons (photo).
13 Set the self-adjusting mechanism in the fully retracted position by easing the quadrant away from the knurled pin or ratchet (as applicable) and turning it so that it touches the backplate when the self-adjusting mechanism is fitted (photo).
14 Engage the handbrake lever on the self-adjusting mechanism with

12.15A Refit the trailing shoe anti-rattle spring...

12.15B ...and secure the shoe in position with the holding pin and spring clip (pre-September 1985 models shown)

12.16 Fit the upper and lower return springs to both shoes

12.17 Locate the leading shoe with the self-adjust mechanism...

12.18 ...and secure it in position with its holding pin and spring clip (pre-September 1985 models shown)

12.19 Ensure the self-adjust mechanism is in the fully retracted position (pre-September 1985 models shown)...

12.20A ...then refit the brake drum assembly...

12.20B ...and screw on a new hub retaining nut

12.21 Tighten the rear hub nut to the specified torque and turn the drum to settle the bearings

12.22 Setting the rear hub bearing preload

12.23A Once the bearing preload is correct use a suitable punch...

12.23B ...to stake the retaining nut firmly into the stub axle groove

the slot on the trailing brake shoe, and place the shoe in position on the backplate.
15 Refit the trailing shoe anti-rattle spring followed by the holding pin and spring clip (photos).
16 Fit the upper and lower return springs to the trailing shoe, then connect them with the leading shoe (photo).
17 Locate the leading shoe over the self-adjusting mechanism quadrant, then engage the shoe with its lower pivot and the wheel cylinder piston (photo).
18 Refit the leading shoe holding pin and spring clip (photo).
19 Tap the brake shoes up or down as necessary so that they are central on the backplate, and make sure that the self-adjusting mechanism is fully retracted as described in paragraph 13 (photo).
20 Refit the brake drum and hub assembly complete with thrustwasher then screw on the new hub retaining nut and set the bearing preload as follows (photos).
21 Tighten the hub nut to a torque of 25 to 30 Nm (18 to 22 lbf ft) then rotate the wheel hub a few times to settle the hub bearings. Slacken the hub nut one complete turn (photo).
22 Using a spring balance connected to one of the wheel studs on the brake drum, measure the torque at which the drum just starts to turn. This should be 0.4 to 1.0 kg on early models, and 0.26 to 0.87 kg on later models. Tighten the hub retaining nut as necessary until the correct starting torque is obtained (photo).
23 When the hub bearing preload is correctly set, stake the retaining nut into the axle groove using a suitable punch (photos). Ensure that the nut is staked at least 2 mm into the groove. Refit the hub cap.
24 Refit the handbrake cable to the self-adjusting mechanism lever. Secure the clevis pin in position with a new split pin on early models, and on later models refit the handbrake cable to the backplate tightening its retaining bolts securely.
25 Repeat the operation on the remaining rear brake.
26 Depress the footbrake two or three times and check that the brakes are operating correctly. Also check the operation of the handbrake mechanism then refit the roadwheel, lower the car to the ground and tighten the roadwheel nuts to the specified torque.

13 Rear wheel cylinder (drum brakes) – removal, overhaul and refitting

Removal
1 Remove the rear brake shoes as described in Section 12.
2 Fit a brake hose clamp to the flexible hose leading to the rear wheel cylinder. This will minimise the loss of fluid during the subsequent operation.
3 Wipe away all traces of dirt around the brake pipe union at the rear of the wheel cylinder.
4 Unscrew the union nut securing the brake pipe to the wheel cylinder. Carefully ease out the pipe and plug, or tape over, its end to prevent dirt entry (photo).
5 Unscrew the two bolts securing the wheel cylinder to the backplate, and remove the cylinder from the backplate noting the gasket fitted behind the cylinder.

Overhaul
6 With the cylinder on the bench, remove the bleed screw and the steel ball from the bleed screw orifice. It may be necessary to tap the cylinder on a block of wood to dislodge the ball.
7 Remove the dust covers and withdraw the pistons from each end of the cylinder.
8 Remove the fluid seals, filling blocks and spring by pushing them out from one end of the wheel cylinder bore.

Chapter 9 Braking system

Fig. 9.16 Exploded view of a rear wheel cylinder (Sec 13)

1. Dust covers
2. Pistons
3. Fluid seals
4. Filling blocks
5. Spring
6. Dust cap
7. Bleed screw
8. Steel ball

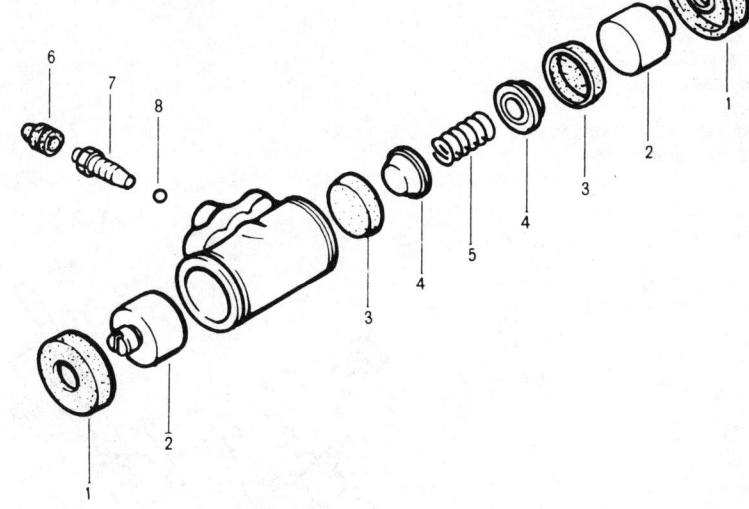

9 Thoroughly clean all components in methylated spirit or clean brake fluid and dry with a lint-free cloth.
10 Carefully examine the surfaces of the pistons and cylinder bore for wear, score marks or corrosion. If damage of this nature is evident then the complete wheel cylinder must be renewed. If the pistons and cylinder bore are in a satisfactory condition, obtain a repair kit consisting of new seals and dust covers.
11 Dip the new seals and pistons in clean brake fluid and assemble them wet as follows.
12 Refer to Fig. 9.16 and place the spring in the cylinder bore followed by the two filling blocks.
13 Insert a fluid seal into each end of the cylinder bore ensuring that the flat side of the seals are facing outwards.
14 Insert the pistons into the cylinder bore and fit the dust covers. Ensure that the dust cover lips are correctly located in the groove on the wheel cylinder body.
15 Place the steel ball in the bleed screw orifice and refit the bleed screw.

Refitting

16 Fit a new gasket to the rear of the wheel cylinder.
17 Place the cylinder in position on the backplate and engage the brake pipe and union nut. Screw the union nut in two or three turns to ensure the thread has started.
18 Refit the wheel cylinder retaining bolts and tighten them to the specified torque.
19 Tighten the brake pipe union nut to the specified torque and remove the clamp from the brake hose.
20 Refit the brake shoes as described in Section 12.
21 Bleed the hydraulic braking system as described in Section 5.

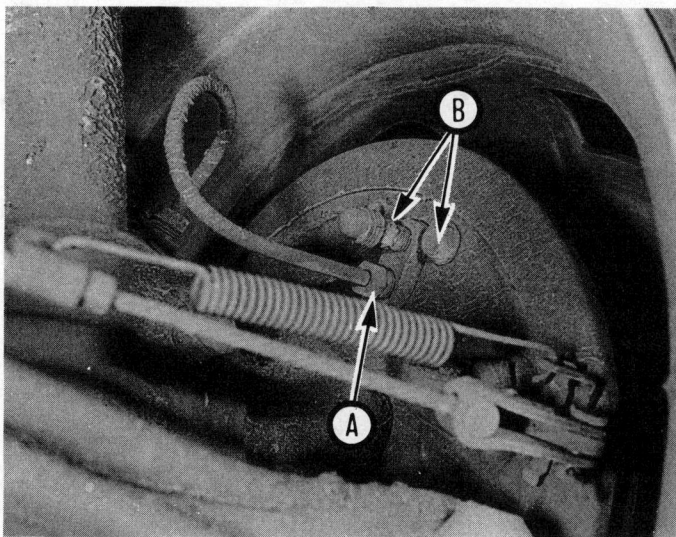

13.4 Brake pipe union nut (A) and rear wheel cylinder retaining bolts (B)

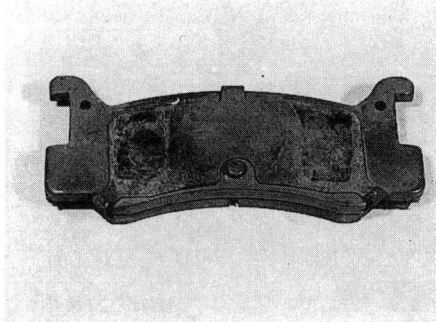

14.6A Fit the shims onto the pads...

14.6B ...and the pad springs onto the mounting bracket...

14.6C ...then install the pads...

Fig. 9.17 Rear brake caliper and associated components (Sec 14)

1 Bolt
2 Mounting bolt
3 Brake hose
4 Caliper
5 Pad shims and springs
6 Brake pads and shims

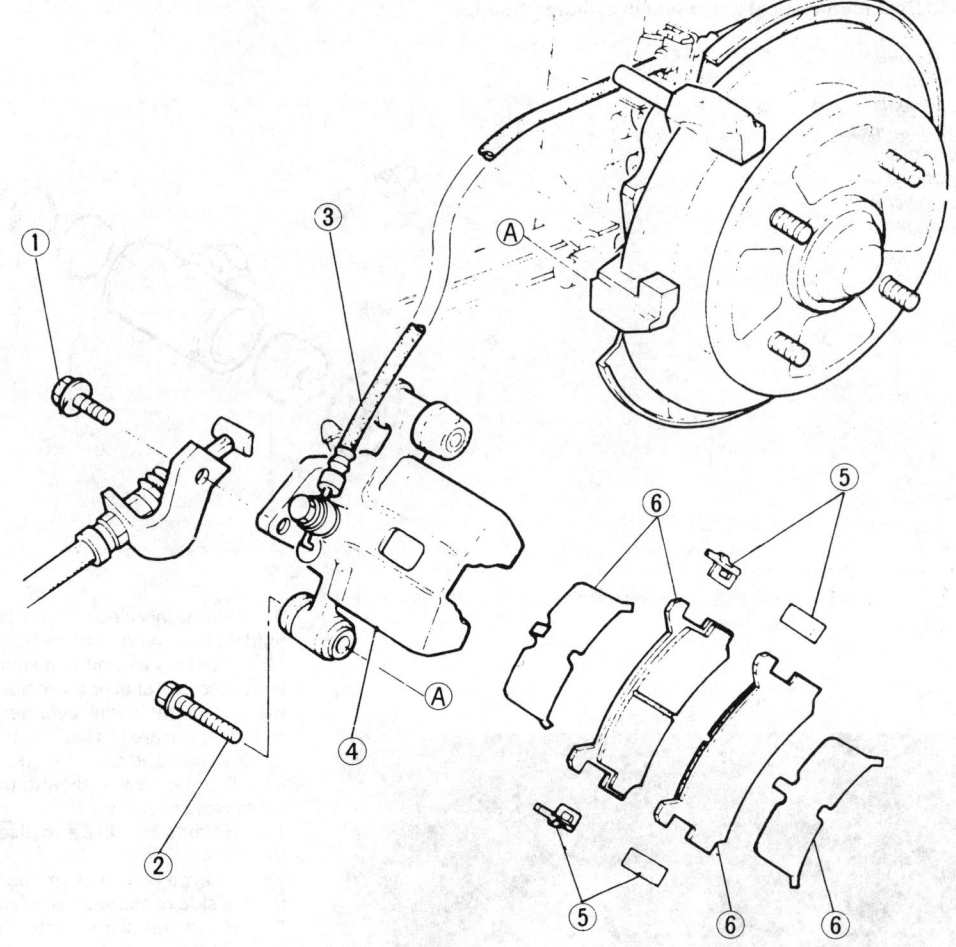

14 Rear brake pads – renewal

Warning: *Disc brake pads must be renewed on both rear wheels at the same time – never renew the pads on only one wheel as uneven braking may result. Also, the dust created by wear of the pads may contain asbestos, which is a health hazard. Never blow it out with compressed air and don't inhale any of it. An approved filtering mask should be worn when working on the brakes. DO NOT use petroleum based solvents to clean brake parts. Use brake cleaner or methylated spirit only.*

1 Chock the front wheels, then jack up the rear of the car and support it on axle stands. Remove the rear wheels and release the handbrake.
2 Slacken and remove the caliper mounting bolt. Pivot the caliper away from the disc and tie it to the rear suspension coil spring to avoid straining the brake hose.
3 Remove the upper pad spring (where fitted), and withdraw the pads from the caliper mounting bracket, noting the correct fitted positions of the pad springs and shims. If required the thickness of the pads can be checked at this stage using a steel rule.
4 Before refitting the pads check that the caliper is free to slide easily

noting that if precautions were taken to minimise fluid loss it should only be necessary to bleed the relevant rear brake. On completion check that both the footbrake and handbrake function correctly before taking the car on the road.

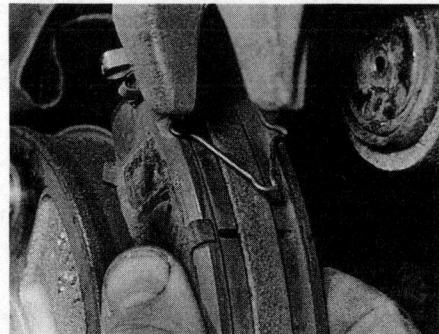

14.6D ...and fit the upper pad spring (where fitted)

14.7 Using a pair of circlip pliers to retract the caliper piston

14.8 Pivot the caliper down over the pads and refit the mounting bolt (arrowed)

on its mountings, and that the mounting bush rubber gaiters are undamaged. Brush the dust and dirt from the caliper and piston but *do not inhale it as it is injurious to health*. Inspect the dust seal around the piston for damage, and the piston for evidence of fluid leaks, corrosion or damage. If attention to any of these components is necessary, refer to Section 15.
5   Examine the shims and pad springs for signs of corrosion, wear or damage and renew as necessary.
6   To refit the pads, first fit the shims onto the pads and the pad springs onto the caliper mounting bracket. Install the pads in the caliper mounting bracket and refit the upper pad spring (where fitted) (photos).
7   Retract the piston fully into the caliper bore by rotating it in a clockwise direction. This can be achieved using a suitable pair of pliers as a peg spanner or by fabricating a peg spanner for the task (photo).
8   Pivot the caliper back down onto the disc and refit the caliper mounting bolt (photo). Tighten the mounting bolt to the specified torque.
9   Repeat the procedure on the remaining brake caliper.
10 Depress the footbrake two or three times to bring the pistons into contact with the pads, then refit the roadwheels and lower the car to the ground. Tighten the roadwheel nuts to the specified torque when the car is on its wheels, then check the fluid level in the master cylinder reservoir and top up if necessary.

15.5 Slide the caliper off the mounting bracket guide pin

15 Rear brake caliper – removal, overhaul and refitting

Removal

1   Chock the front wheels, then jack up the rear of the car and support it on axle stands. Remove the rear wheel and release the handbrake.
2   Remove the bolt which secures the handbrake cable bracket to the caliper, and disconnect the cable from its operating lever on the caliper.
3   Fit a brake hose clamp to the flexible brake hose leading to the rear brake caliper. This will minimise brake fluid loss during subsequent operations.
4   Undo the brake hose union bolt and remove the hose from the caliper. Plug the end of the hose and the caliper orifice to prevent dirt entering the hydraulic system.
5   Undo the caliper mounting bolt, pivot the caliper away from the disc and slide it off the mounting bracket guide pin (photo). Note that it is not necessary to disturb the brake pads.

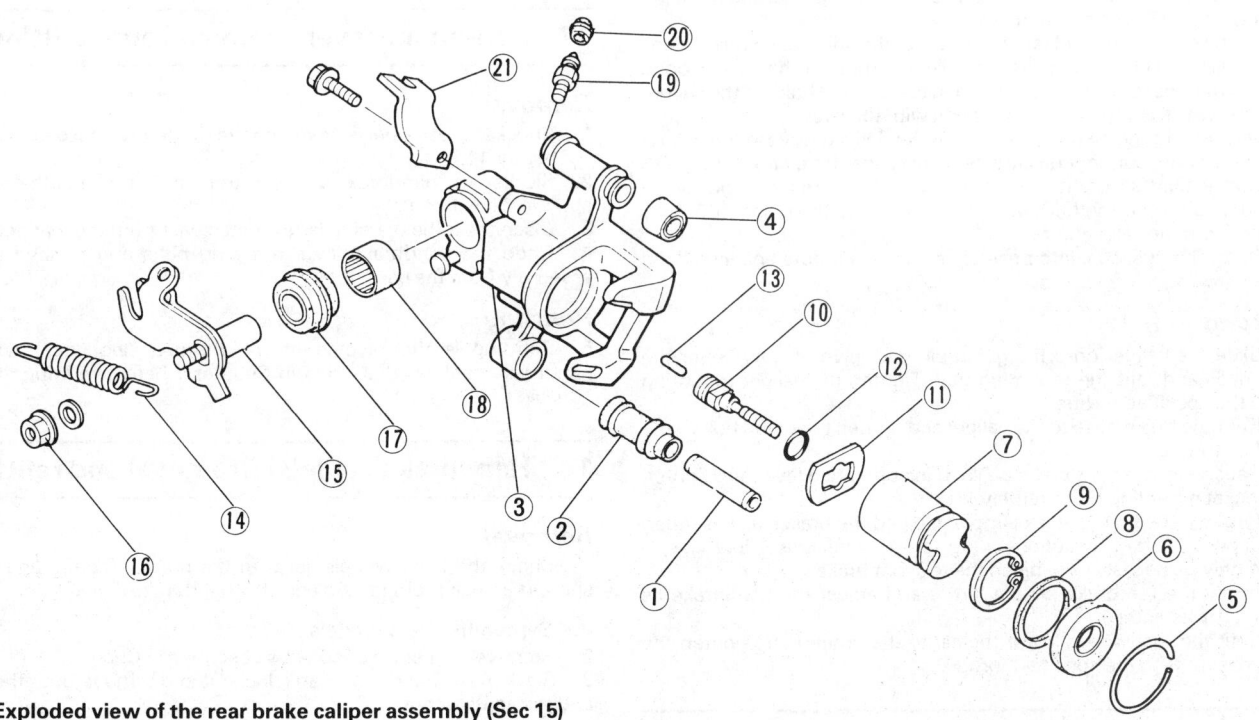

Fig. 9.18 Exploded view of the rear brake caliper assembly (Sec 15)

1 Mounting bush	6 Dust seal	11 Retaining plate
2 Seal	7 Piston	12 O-ring
3 Caliper body	8 Piston seal	13 Connecting link
4 Seal	9 Circlip	14 Return spring
5 Circlip	10 Self-adjusting spindle	15 Handbrake lever

16 Nut
17 Seal
18 Needle roller bearing
19 Bleed screw
20 Dust cap
21 Handbrake cable bracket

Overhaul

6 Remove all traces of dirt from the external surfaces of the caliper and grip the caliper securely in a vice equipped with soft jaws.
7 Using a small screwdriver, carefully prise out the circlip and remove the dust seal from the caliper bore.
8 Remove the piston from the caliper bore by rotating it in an anti-clockwise direction. This can be achieved using a suitable pair of pliers as a peg spanner or by fabricating a peg spanner for the task. Once the piston turns freely but does not come out any further the piston can be withdrawn by hand, or if necessary pushed out by applying compressed air to the union bolt hole. Only low pressure should be required such as is generated by a foot pump.
9 Remove the piston seal whilst taking great care not to scratch the caliper bore.
10 Extract the circlip from the bottom of the caliper bore using circlip pliers. Remove the handbrake retaining plate, self-adjusting spindle, O-ring and connecting link.
11 Unhook the handbrake operating lever return spring and withdraw the lever from the caliper. Prise out the dust seal.
12 If necessary unscrew and remove the bleed screw.
13 Clean all the components in methylated spirit, or clean brake fluid and dry them with a lint-free cloth.
14 Inspect the caliper bore and piston for signs of wear, corrosion or scoring. If damage of this nature is present renew the complete caliper assembly. Examine the handbrake operating lever, needle roller bearing and self-adjusting spindle components for wear or damage and renew if necessary. The needle roller bearing can be extracted from the caliper using a suitable puller and pressed, or tapped, into position using a suitable size tubular drift which bears only on the outer edge of the bearing. Check the caliper mounting bush, mounting bracket guide pin and gaiters for wear, corrosion or damage and renew as necessary.
15 If components are in a satisfactory condition obtain a repair kit consisting of seals, adjuster spindle O-ring and three special greases – a red, orange and white one. The red grease is for the piston fluid seal, the white grease for the adjuster spindle O-ring and the orange grease for the dust seal, needle roller bearing, handbrake spindle and caliper mounting bush and guide pin.
16 Lubricate all components as described above and then fit the handbrake spindle dust seal to the caliper, refit the spindle and hook the return spring back into position.
17 Fit the new O-ring and retaining plate to the adjuster spindle. Install the connecting link and adjuster spindle assembly in the caliper bore, ensuring that the retaining plate pins engage with the holes in the caliper body. Secure all components in position with the circlip.
18 Manoeuvre the piston seal into position in its groove in the caliper bore using only your fingers. Lubricate the caliper bore and piston with clean brake fluid and refit the piston. Turn the piston in a clockwise direction, using the method employed on dismantling, until it is fully retracted into the caliper bore.
19 Press the dust seal into position in the caliper bore and install the circlip.

Refitting

20 Slide the caliper onto the guide pin then pivot it down over the brake pads and refit the mounting bolt. Tighten the caliper mounting bolt to the specified torque.
21 Refit the brake hose to the caliper and tighten the union bolt to the specified torque.
22 Reconnect the handbrake cable to the operating lever and tighten the bracket mounting bolt securely.
23 Remove the brake hose clamp and bleed the brakes as described in Section 5. If the precautions were taken to minimise fluid loss, it should only be necessary to bleed the relevant brake.
24 Apply the footbrake several times and adjust the handbrake as described in Chapter 1.
25 Refit the roadwheel, lower the car to the ground and tighten the roadwheel nuts to the specified torque.

16 Rear brake disc – inspection, removal and refitting

Inspection

1 Chock the front wheels, then jack up the rear of the car and support it on axle stands. Remove the appropriate rear roadwheel and release the handbrake.
2 Rotate the disc by hand and examine it for deep scoring, grooving or cracks. Light scoring is normal, but, if excessive, the disc must be renewed. Any loose rust and scale around the outer edge of the disc can be removed by lightly tapping it with a small hammer while rotating the disc. Measure the disc thickness with a micrometer if available. Disc run-out can be checked using a dial test gauge, although a less accurate method is to use a feeler gauge together with a metal base block.

Removal

3 Prise out the hub cap from the centre of the disc.
4 Using a hammer and suitable chisel nosed tool, tap up the staking securing the retaining nut to the groove in the hub spindle.
5 Have an assistant depress firmly on the brake pedal then using a socket and extension bar, slacken the nut and remove it along with the thrustwasher. Discard the nut as a new one must be used on refitting.
Note: *The right-hand rear hub nut has a left-hand thread and must therefore be turned clockwise to remove it.*
6 Slacken and remove the two bolts which secure the brake caliper mounting bracket to the stub axle, and free the brake hose from the suspension strut by removing the retaining clip. Lift the caliper off the disc and tie it to the rear suspension coil spring to avoid straining the brake hose.
7 The disc assembly can then be pulled off the stub axle using a suitable puller and removed from the vehicle.

Refitting

8 Fit the disc assembly onto the stub axle and fit the thrustwasher and a new retaining nut.
9 Slide the brake caliper assembly onto the disc and tighten the caliper bracket mounting bolts to the specified torque. Refit the brake hose to its guide on the suspension strut and secure it in position with the retaining clip.
10 Set the bearing preload as described in Section 12, paragraphs 21 to 23 inclusive.
11 Refit the roadwheel, lower the car to the ground and tighten the roadwheel nuts to the specified torque.

17 Handbrake lever – removal and refitting

Removal

1 Chock the rear wheels, then remove the centre console as described in Chapter 11.
2 Slacken the handbrake cable locknut and remove both the locknut and the adjuster nut.
3 Disconnect the warning lamp switch wire from the rear of the lever.
4 Undo the handbrake lever retaining bolts and remove the lever assembly from the car.

Refitting

5 Refitting is the reverse of the removal sequence. Adjust the handbrake cable as described in Chapter 1 before refitting the centre console.

18 Handbrake cable(s) – removal and refitting

Removal

1 Chock the front wheels, jack up the rear of the car and securely support it on axle stands. Release the handbrake lever.

Pre-September 1985 models

2 Remove the centre console as described in Chapter 11.
3 Slacken the handbrake cable locknut and remove both the locknut and the adjuster nut.
4 From under the rear of the car, disconnect the return spring, extract the split pin and remove the clevis pin securing the cable ends to the levers at the rear of each brake backplate (photo).
5 Undo the two nuts and release the cable compensator assembly from the support bracket on the underbody (photo).
6 Extract the retaining clips and remove the cable guides from the rear crossmember.

Chapter 9 Braking system

Fig. 9.19 Handbrake lever and cable attachments – pre-September 1985 models (Secs 17 and 18)

1. Handbrake lever assembly
2. Cable adjusting nut
3. Handbrake warning light switch
4. Handbrake cable
5. Cable guide retaining clip
6. Split pin
7. Clevis pin
8. Return spring

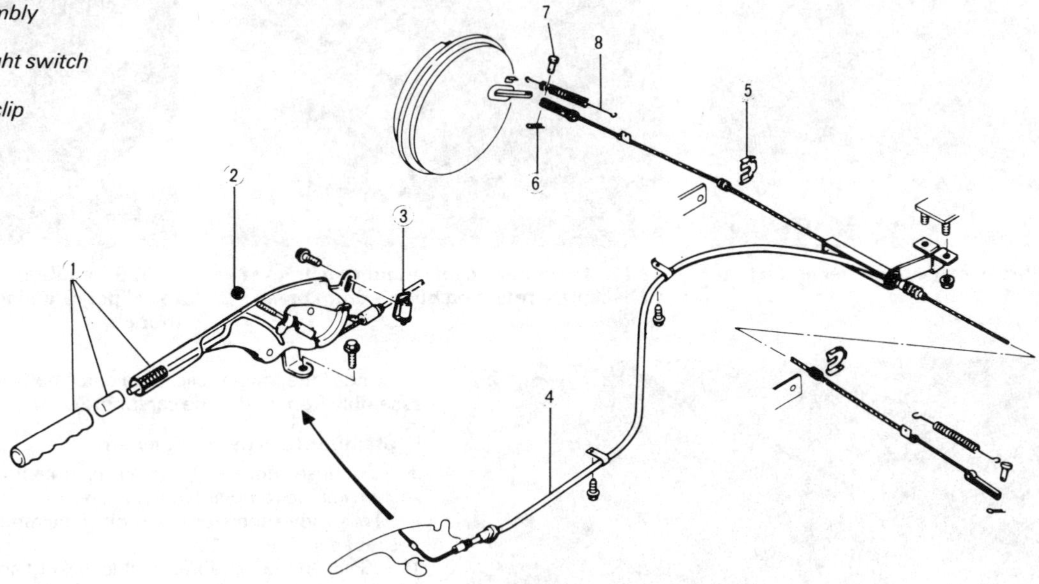

Fig. 9.20 Handbrake lever and cable attachments – September 1985 models onward (rear drum brake models shown) (Secs 17 and 18)

1. Adjuster nut
2. Equalizer
3. Return spring
4. Front cable section
5. Rear cable sections

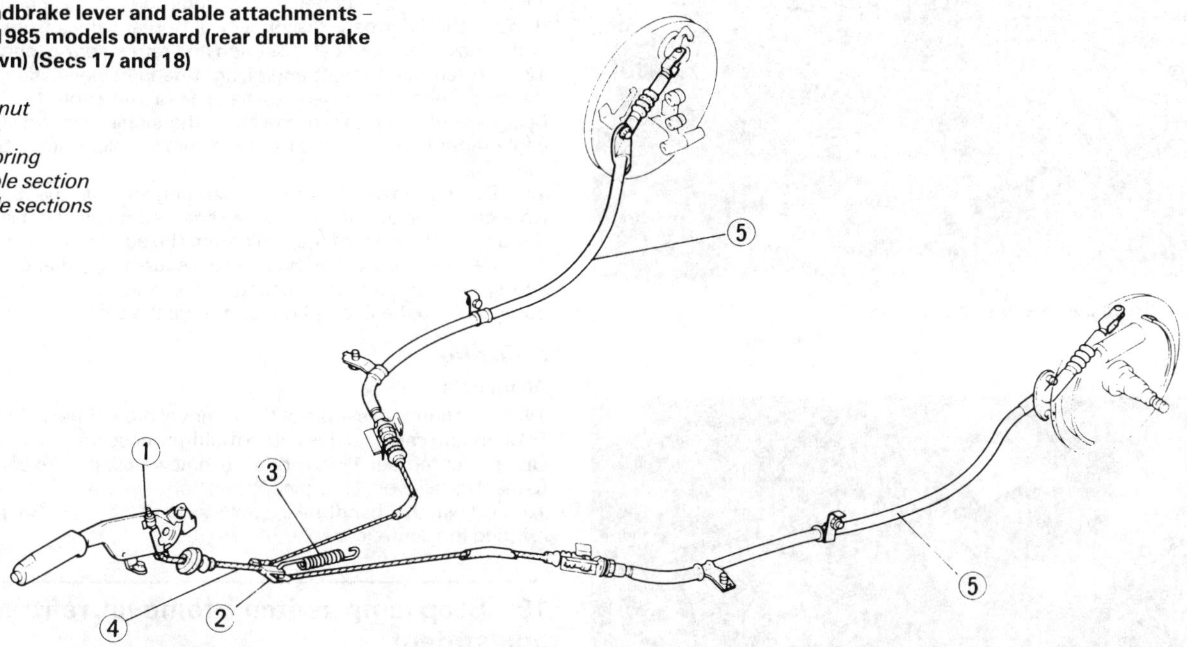

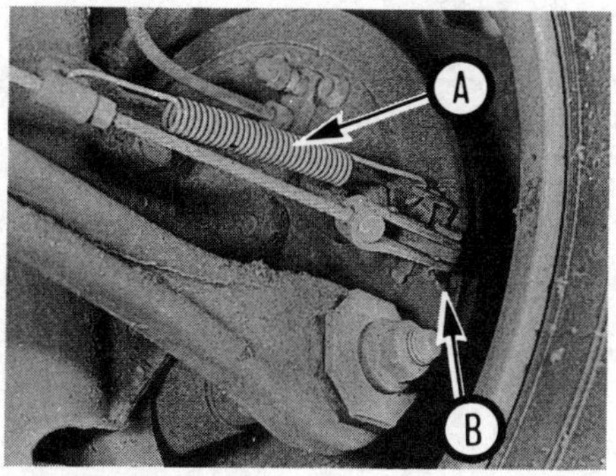

18.4 Handbrake cable return spring (A) and clevis pin (B)

18.5 Handbrake cable compensator retaining nuts

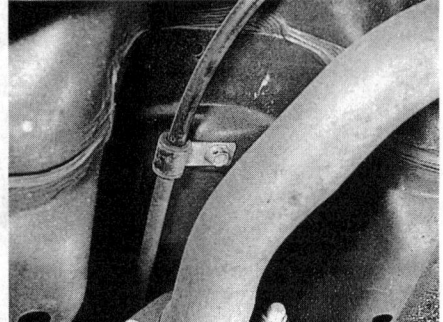

18.7 Handbrake cable clip retaining bolt

Chapter 9 Braking system

18.11 Handbrake cable equalizer and return spring

18.12A Handbrake cable mounting bracket to backplate retaining bolts – drum brake models

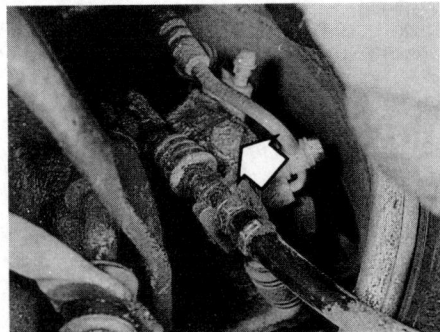
18.12B Handbrake cable mounting bracket to caliper retaining bolt – disc brake models

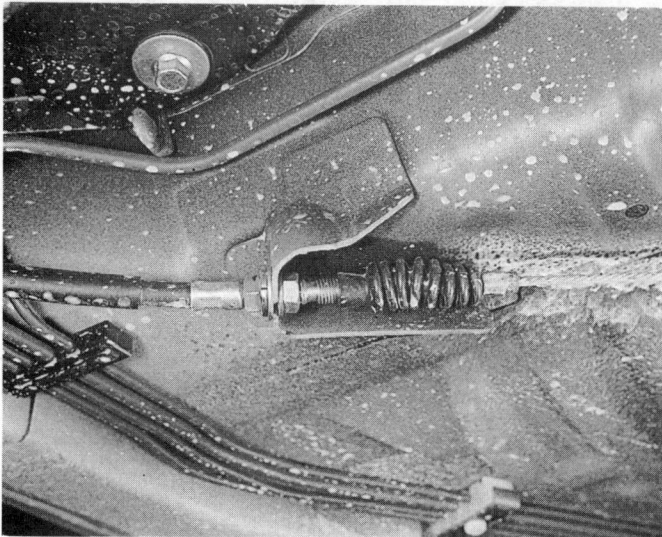
18.14 Handbrake cable adjuster locknuts

7 Undo the cable clip retaining bolts and withdraw the cable assembly from under the car (photo).

September 1985 models onward

8 On these models the handbrake cable is in three sections. Each section can be removed individually.
9 To remove the front section, first remove the centre console as described in Chapter 11.
10 Slacken the handbrake cable locknut and remove both the locknut and the adjuster nut.
11 Working from under the car, pull the front section of the cable through the floor pan. Disconnect both rear cables from the equalizer and remove the front cable section and return spring (photo).
12 To remove the left and/or right-hand cable section, first remove the two bolts which secure the rear of the cable to the rear brake backplate (drum brake models), or the single bolt which secures the cable mounting bracket to the rear brake caliper (disc brake models) (photos).
13 Detach the cable from its operating lever on the rear brake. If the front cable section has not been removed it will also be necessary to disconnect the front of the cable from the equalizer.
14 Remove the nut and bolt which secures the outer cable to the floor and slacken the adjuster locknuts which secure the front end of the outer cable to the floor (photo). Remove the cable from car.

Refitting

All models

15 Refitting is a reverse of the removal procedure. Lubricate exposed linkages and cable guides with a multi-purpose grease during assembly. On pre-September 1985 models use new split pins to secure the cable to rear brake lever clevis pins in position.
16 Adjust the handbrake cable as described in Chapter 1 before refitting the centre console.

19 Stop lamp switch – removal, refitting and adjustment

Removal

1 Remove the right-hand facia undercover.
2 Slacken the stop lamp switch locknut and unscrew the switch from its mounting bracket (photo).
3 Disconnect the wiring and remove the stop lamp switch from the car.

Refitting and adjustment

4 Refitting is a reverse of the removal procedure. Prior to tightening the stop lamp switch locknut, adjust the brake pedal height as described in Section 8 of Chapter 1.

19.2 Stop lamp switch locknut (arrowed)

Chapter 10 Suspension and steering

Contents

Front hub bearings – checking, removal and refitting	3
Front suspension and steering check	See Chapter 1
Front suspension anti-roll bar – removal and refitting	6
Front suspension balljoint – removal and refitting	8
Front suspension lower arm – removal and refitting	7
Front suspension strut – dismantling, inspection and reassembly	5
Front suspension strut – removal and refitting	4
Front swivel hub assembly – removal and refitting	2
General information	1
Ignition switch/steering lock – removal and refitting	18
Intermediate shaft – removal and refitting	19
Power steering drivebelt check, adjustment and renewal	See Chapter 1
Power steering fluid level check	See Chapter 1
Power steering pump – removal and refitting	22
Power steering system – bleeding	23
Rear hub bearings – checking, removal and refitting	9
Rear stub axle – removal and refitting	10
Rear suspension anti-roll bar – removal and refitting	13
Rear suspension lateral links – removal and refitting	14
Rear suspension strut – dismantling, inspection and reassembly	12
Rear suspension strut – removal and refitting	11
Rear suspension trailing arms – removal and refitting	15
Steering column – removal, checking and refitting	17
Steering gear – removal, overhaul and refitting	21
Steering gear rubber gaiter – renewal	20
Steering wheel – removal and refitting	16
Track rod – removal and refitting	25
Track rod outer balljoint – removal and refitting	24
Wheel alignment and steering angles – general information	26
Wheel and tyre maintenance and tyre pressure checks	See Chapter 1

Specifications

Front suspension

Type	Independent by MacPherson struts with coil springs and integral shock absorbers. Anti-roll bar on 1500 GT and September 1985 onward manual transmission models
Coil spring free length:	
Pre-September 1985 models	364.5 mm
September 1985 models onward:	
Standard suspension	365.5 mm
Hard suspension	380.5 mm
Spring wire diameter (maximum):	
Pre-September 1985 models	12.5 mm
September 1985 models onward:	
Standard suspension	12.6 mm
Hard suspension	12.8 mm
Spring coil diameter:	
Pre-September 1985 models	132.5 mm
September 1985 models onward:	
Standard suspension	137.2 mm
Hard suspension	136.6 mm
Number of spring coils:	
Pre-September 1985 models	5.8
September 1985 models onward:	
Standard suspension	4.9
Hard suspension	5.5

Rear suspension

Type	Independent by MacPherson struts with coil springs and integral shock absorbers located by lateral links. Anti-roll bar on all models
Coil spring free length:	
Pre-September 1985 models	357.7 mm
September 1985 onward Hatchback and Saloon models:	
Standard suspension	345.0 mm
Hard suspension	361.5 mm
September 1985 onward Estate models	359.0 mm

Rear suspension (continued)
Spring coil diameter .... 114 mm
Number of spring coils:
    Pre-September 1985 models 7.05
    September 1985 onward Hatchback and Saloon models:
        Standard suspension 4.6
        Hard suspension 5.2
    September 1985 onward Estate models 4.9
Rear wheel toe setting:
    Pre-September 1985 models .... Parallel
    September 1985 onward Hatchback and Saloon models .... −3 to +3 mm
    September 1985 to July 1987 Estate models .... −3 to +3 mm
    July 1987 onward Estate models .... −1 to +5 mm

Steering
Type:
    Manual steering .... Rack and pinion
    Power-assisted steering .... Rack and pinion with hydraulic assistance
Turns lock-to-lock:
    Pre-September 1985 models 3.2
    September 1985 models onward:
        Constant gear ratio type manual steering 3.6
        Variable gear ratio type manual steering 4.2
        Power-assisted steering 3.2
Steering wheel diameter .... 380 mm
Camber angle:
    Pre-September 1985 models .... 0° 55'
    September 1985 to July 1987 models .... 0° 48' ± 45'
    July 1987 models onward .... 0° 49' ± 30'
Castor angle:
    Pre-September 1985 models .... 1° 45'
    September 1985 to July 1987 models .... 1° 35' ± 35'
    July 1987 models onward .... 2° 09' ± 45'
Steering axis inclination:
    Pre-September 1985 models .... 12° 10'
    September 1985 models onward .... 12° 22'
Toe setting:
    Pre-September 1985 models .... −3 to +3 mm
    September 1985 models onward .... −1 to +5 mm

Roadwheels
Type .... Pressed steel or aluminium alloy depending on model
Wheel size .... 4½J x 13, 5J x 13 or 5½ JJ x 14 depending on model

Tyres
Tyre size .... 6.15-13-4PR, 155 SR 13, 175/75 SR 13, 175/70 HR 13 or 185/60 HR 14 depending on model

Tyre pressures (cold)
Front:
    Pre-September 1985 models .... 1.8 bar (26 lbf/in^2)
    September 1985 models onward .... 2.0 bar (29 lbf/in^2)
Rear (all models)* .... 1.8 bar (26 lbf/in^2)
*When fully loaded, increase the rear tyre pressure to 1.9 bar (28 lbf/in^2)

Torque wrench settings

Front suspension	Nm	lbf ft
Swivel hub to suspension strut	93 to 117	69 to 86
Lower arm balljoint pinch bolt	44 to 55	33 to 40
Balljoint to lower suspension arm	93 to 117	69 to 86
Suspension strut upper mounting to body	23 to 30	17 to 22
Suspension strut piston rod nut:		
Pre-September 1985 models	76 to 95	56 to 70
September 1985 models onward	56 to 69	40 to 51
Lower suspension arm rear mounting bracket to body	60 to 75	44 to 55
Lower suspension arm front mounting bracket	95 to 119	70 to 88
Lower suspension arm front pivot bolt	95 to 119	70 to 88
Lower suspension arm to mounting bracket nuts:		
Pre-September 1985 models	76 to 95	56 to 70
September 1985 models onward	44 to 55	33 to 40
Anti-roll bar connecting link locknut	12 to 18	9 to 13

Chapter 10 Suspension and steering

Torque wrench settings (continued)	Nm	lbf ft
Rear suspension		
Suspension strut upper mounting to body	23 to 30	17 to 22
Suspension strut piston rod nut	56 to 69	40 to 51
Trailing arm to suspension strut	55 to 69	40 to 51
Trailing arm to body	60 to 75	44 to 55
Lateral links to stub axle	64 to 76	47 to 56
Lateral links to crossmember:		
Pre-September 1985 models	64 to 76	47 to 56
September 1985 models onward	95 to 119	70 to 88
Lateral link locknuts	55 to 64	40 to 47
Stub axle to suspension strut	95 to 119	70 to 88
Backplate to stub axle bolts	46 to 68	34 to 50
Brake pipe union nuts	13 to 22	10 to 16
Disc cover to stub axle	46 to 68	34 to 50
Anti-roll bar to trailing arm	32 to 47	23 to 35
Anti-roll bar connecting link locknuts	12 to 18	9 to 13
Steering		
Steering wheel retaining nut	40 to 50	29 to 37
Steering column retaining bolts	16 to 23	12 to 17
Universal joint pinch bolts	18 to 27	13 to 20
Steering gear mounting brackets to bulkhead	32 to 47	23 to 35
Steering gear track rod to rack	85 to 95	63 to 70
Track rod outer balljoint retaining nut	30 to 45	22 to 33
Track rod outer balljoint locknut	35 to 40	26 to 29
Roadwheels		
Wheel nuts	90 to 110	65 to 80

1 General information

The independent front suspension is of the MacPherson strut type, incorporating coil springs and integral telescopic shock absorbers. Lateral and longitudinal location of each strut assembly is by pressed steel lower suspension arms utilizing rubber inner mounting bushes and incorporating a balljoint at their outer ends. On 1500 GT and all later manual transmission models, both lower suspension arms are connected by an anti-roll bar. The front swivel hubs, which carry the wheel bearings, brake calipers and the hub/disc assemblies, are bolted to the MacPherson struts and connected to the lower arms via the balljoints.

The fully independent rear suspension also utilizes MacPherson struts located by lateral and trailing links and interconnected by a rear anti-roll bar.

The steering gear is of the conventional rack and pinion type located behind the front wheels. On later models two types of manual steering were fitted; a constant gear ratio type similar to that which is fitted to earlier models, or a variable gear ratio type. The advantage of the variable gear ratio steering is that the gear ratio of the steering changes as the movement of the steering increases and this has the effect of reducing the amount of steering force required. Power-assisted steering was also available on some models. The main components being a rack and pinion steering gear unit, a hydraulic pump which is belt-driven off the crankshaft, and the hydraulic feed and return lines between the pump and steering gear.

Movement of the steering wheel is transmitted to the steering gear by an intermediate shaft with two universal joints. The front wheels are connected to the steering gear by track rods, each having an inner and outer balljoint.

2 Front swivel hub assembly – removal and refitting

Removal

1 Chock the rear wheels, firmly apply the handbrake, then jack up the front of the car and support it on axle stands. Remove the appropriate front roadwheel.
2 Using a hammer and suitable chisel nosed tool, tap up the staking securing the driveshaft retaining nut to the groove in the constant velocity joint.
3 Have an assistant firmly depress the footbrake then using a socket and extension bar, slacken and remove the driveshaft retaining nut and washer. Note that a new driveshaft retaining nut must be obtained for reassembly.
4 If an anti-roll bar is fitted, undo the two locknuts and remove the connecting link bolt securing the anti-roll bar to the lower suspension arm. Make a note of the correct fitted positions of the washers, rubber bushes and spacer to use as a reference on reassembly.
5 Remove the clip which secures the brake hose to its guide on the suspension strut, and undo the two bolts which secure the brake caliper mounting bracket to the swivel hub. Slide the caliper assembly off the disc and suspend it from suspension strut coil spring using string or wire.
6 Extract the split pin then unscrew the nut securing the track rod outer balljoint to the swivel hub. Release the balljoint from the swivel hub using a suitable balljoint separator whilst taking care not to damage the rubber gaiter.
7 Slacken the nut and remove the pinch-bolt securing the lower arm balljoint to the swivel hub.
8 Using a suitable bar, lever the lower suspension arm down to release the balljoint shank from the swivel hub, taking great care not to damage the balljoint gaiter.
9 Undo the two nuts securing the swivel hub to the suspension strut and remove the bolts whilst supporting the swivel hub assembly.
10 Release the swivel hub from the strut and pull it off of the driveshaft splines. If necessary the hub and driveshaft can be separated using a suitable puller. Remove the swivel hub assembly from the car.

Refitting

11 The swivel hub is refitted by a reversal of the removal procedure noting the following points.
    (a) Tighten all nuts and bolts to the specified torque, referring to Chapter 9 for the brake caliper mounting plate.
    (b) Ensure the anti-roll bar locknuts are positioned as described in Section 6, paragraph 9.
    (c) When fitting the new driveshaft retaining nut, tighten it to the specified torque (Chapter 8), then stake it into the groove in the constant velocity joint using a suitable punch. Ensure the nut is staked at least 4 mm into the groove.
    (d) Use a new split pin to secure the track rod balljoint retaining nut in position.

3 Front hub bearings – checking, removal and refitting

Note: *The front hub bearings should only be removed from the swivel hub if they are to be renewed. The removal procedure renders the*

Chapter 10 Suspension and steering

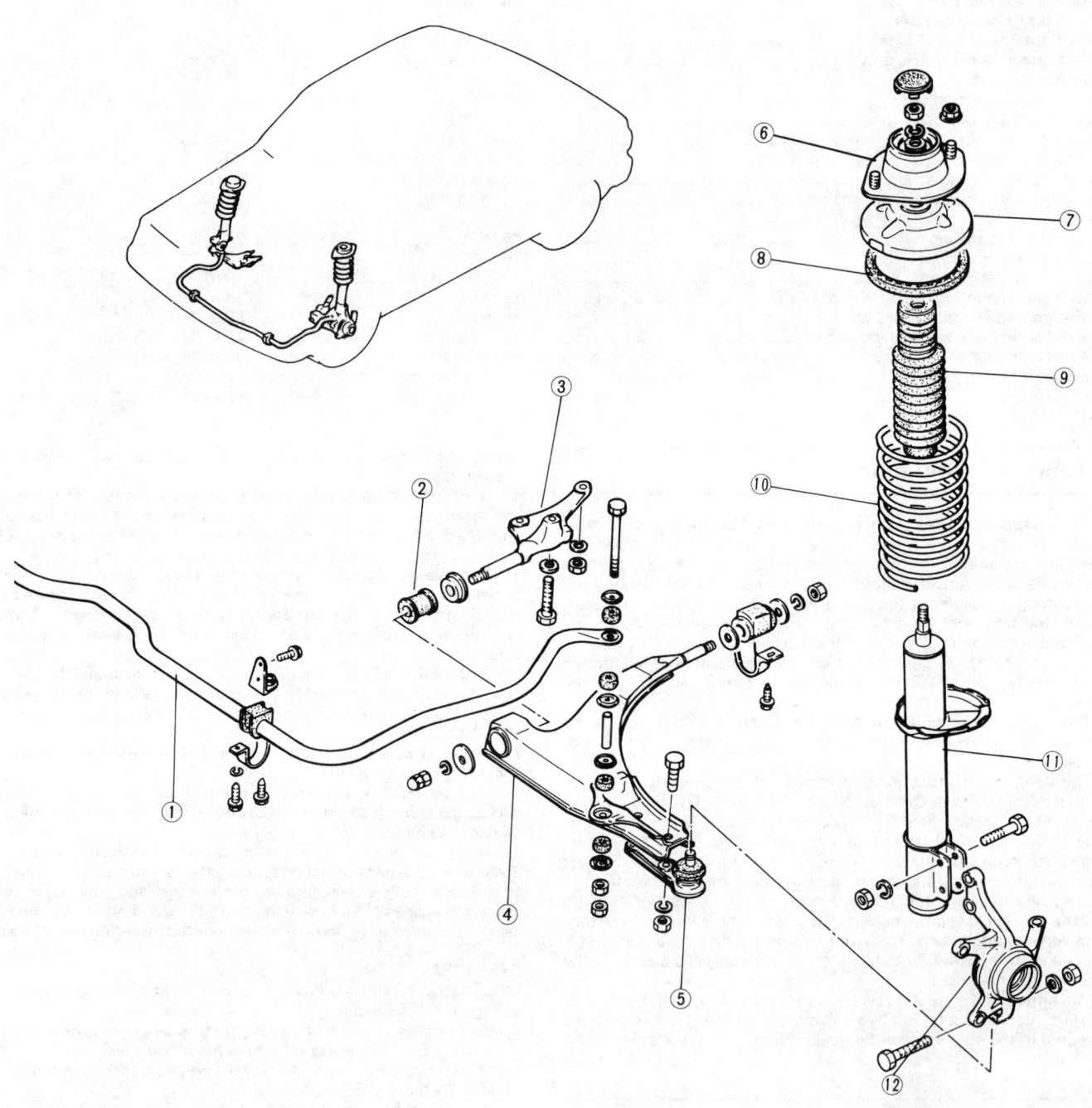

Fig. 10.1 Exploded view of the front suspension – pre-September 1985 models (Sec 1)

1 Anti-roll bar
2 Bush
3 Lower suspension arm front mounting bracket
4 Lower suspension arm
5 Lower suspension arm balljoint
6 Suspension strut upper mounting
7 Spring seat
8 Rubber ring
9 Strut piston dust cover
10 Coil spring
11 Suspension strut
12 Swivel hub

Chapter 10 Suspension and steering

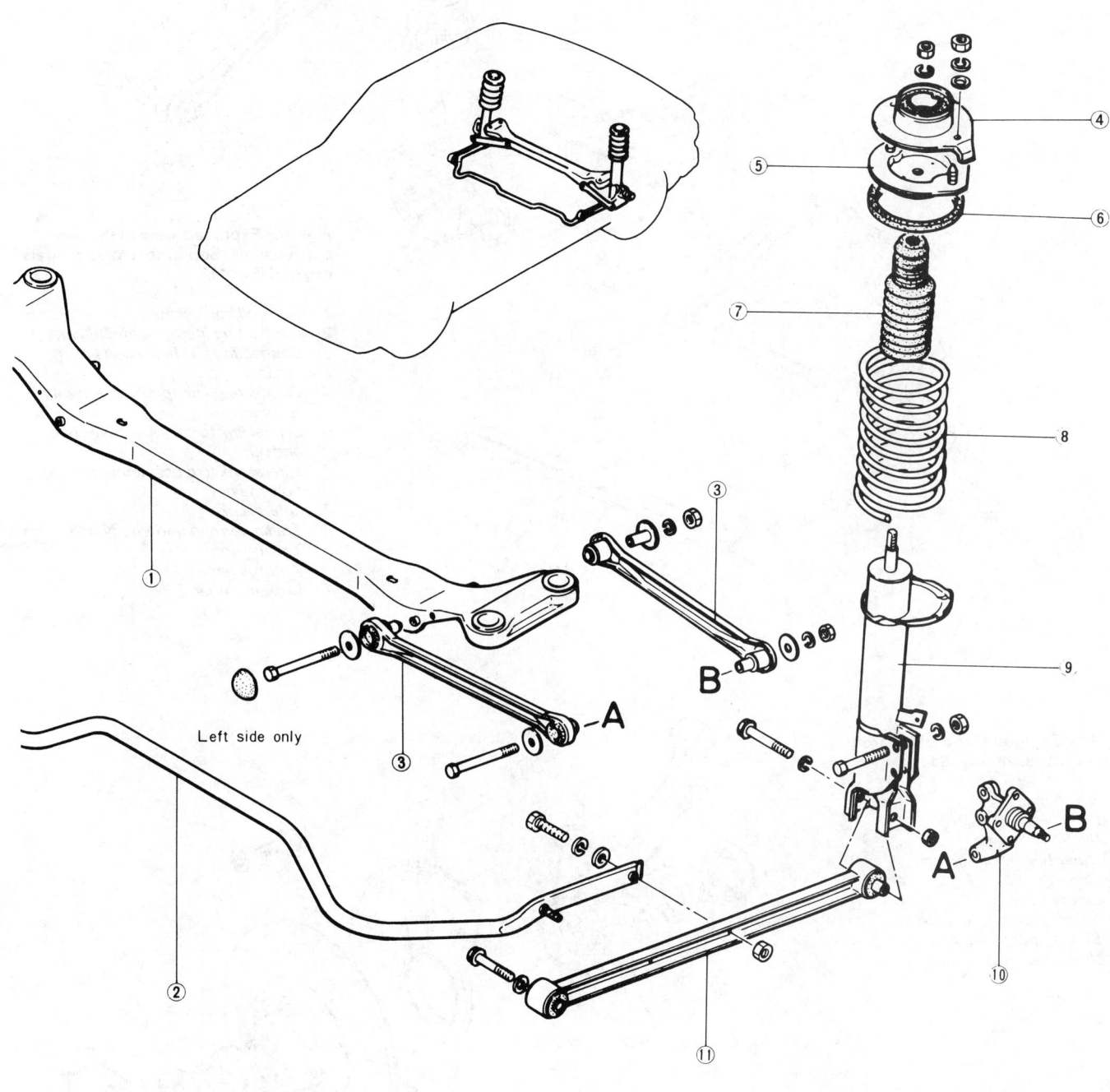

Fig. 10.2 Exploded view of the rear suspension – pre-September 1985 models (Sec 1)

1 Crossmember
2 Anti-roll bar
3 Lateral link
4 Suspension strut upper mounting
5 Spring seat
6 Rubber ring
7 Strut piston dust cover
8 Coil spring
9 Rear suspension strut
10 Rear stub axle
11 Trailing arm

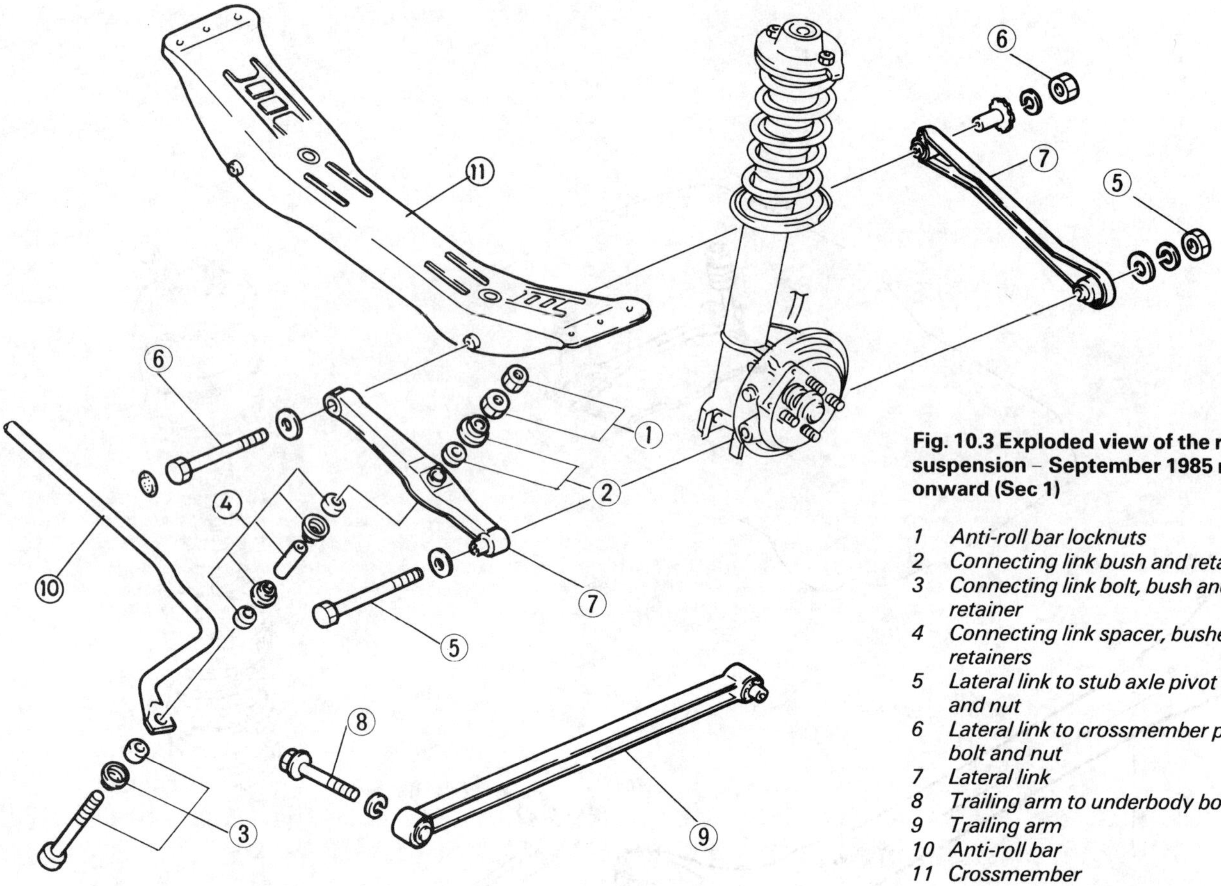

Chapter 10 Suspension and steering

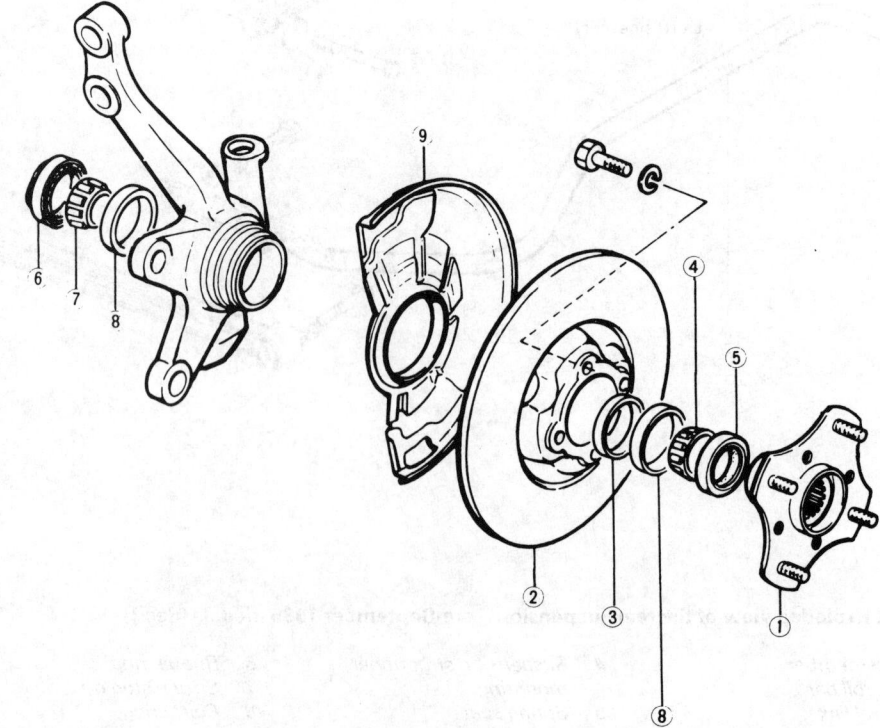

Fig. 10.3 Exploded view of the rear suspension – September 1985 models onward (Sec 1)

1. Anti-roll bar locknuts
2. Connecting link bush and retainer
3. Connecting link bolt, bush and retainer
4. Connecting link spacer, bushes and retainers
5. Lateral link to stub axle pivot bolt and nut
6. Lateral link to crossmember pivot bolt and nut
7. Lateral link
8. Trailing arm to underbody bolt
9. Trailing arm
10. Anti-roll bar
11. Crossmember

Fig. 10.4 Exploded view of the front swivel hub assembly (Sec 2)

1. Hub flange
2. Brake disc
3. Bearing spacer
4. Outer bearing inner race
5. Outer oil seal
6. Inner oil seal
7. Inner bearing inner race
8. Bearing outer races
9. Brake disc dust cover

Chapter 10 Suspension and steering

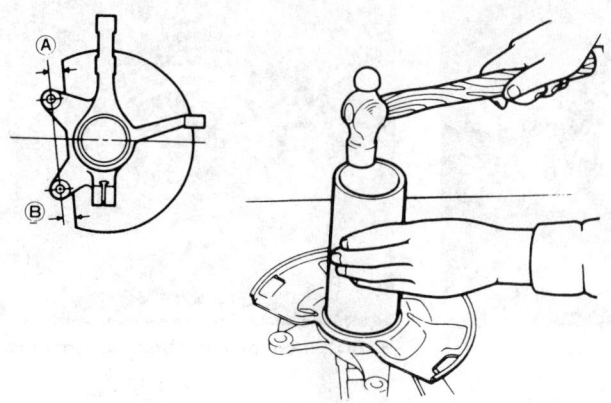

Fig. 10.5 Refitting the brake disc cover (Sec 3)

Dimensions A and B must be equal

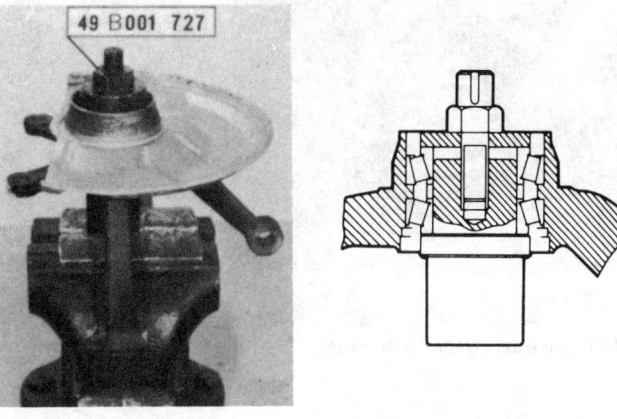

Fig. 10.6 Using the Mazda preload adjuster tool to check front hub bearing preload (Sec 3)

bearings unserviceable and they must not be re-used. Also note that special tool number 49 B001 727 will be required to set the bearing preload on reassembly. In the absence of this tool, front hub bearing renewal should be entrusted to a suitably equipped Mazda dealer.

Checking

1 Wear in the front hub bearings can be checked by measuring if there is any side-play present. To do this, chock the rear wheels, apply the handbrake, jack up the front of the car and support it on axle stands. Remove the relevant wheel. A dial test indicator should be fixed so that its probe is in contact with the disc face of the hub. With the indicator set in place, there should be no detectable movement of the disc when rocked top to bottom. If there is any trace of endplay, the bearings are worn excessively and should be renewed.

Removal

2 Remove the swivel hub assembly from the car as described in Section 2.
3 Securely support the swivel hub in a vice, and separate the hub flange and disc assembly by driving the hub out of the bearings using a hammer and tubular drift of suitable diameter.
4 The oil seal and outer bearing will remain on the hub flange as it is removed. The bearing can then be withdrawn using a suitable puller. With the bearing removed, slide off the oil seal.
5 Remove the spacer from the inner face of the bearing remaining in the swivel hub.
6 If the disc or hub flange require attention, undo the four bolts and separate the disc from the hub.
7 From the rear of the swivel hub extract the oil seal by prising it out with a flat-bladed screwdriver, then lift out the bearing inner race.
8 Using a hammer and suitable drift, drive out one of the bearing outer races from the centre of the swivel hub whilst taking care not to damage the hub bore. Remove the bearing carefully by tapping evenly around the outer race, whilst keeping the race as square as possible in the hub bore. After removal of the first race, turn the hub over and drive out the remaining outer race in a similar way.
9 If necessary the disc dust cover can be removed by prising it off the hub with a suitable screwdriver. Do not remove the disc unless it is damaged and requires renewal or is being transferred to a new swivel hub.
10 Renew both bearings as a pair along with both oil seals. Check for any signs of damage to the swivel hub bore and hub flange and renew these components if necessary. Minor burrs or score marks can be removed from the hub bore using a fine file or emery paper.

Refitting

11 Refit the disc cover to the swivel hub using a hammer and large bore tube to drive it into place. Make sure that the dust cover is positioned so that the distance from the flat edge to the caliper bracket mounting bolt holes in the swivel hub is the same at both the top and bottom (Fig. 10.5).

3.13 Fit the outer bearing outer race to the swivel hub

12 Support the swivel hub in a vice with the disc dust cover uppermost.
13 Place the outer bearing outer race in position in the swivel hub and carefully tap it into position in the hub using a hammer and a suitable tubular drift which bears only on the outer edge of the race (photo). Tap the race down until it contacts the shoulder in the centre of the swivel hub.
14 Turn the swivel hub over and repeat the above procedure for the inner bearing outer race.
15 To set the bearing preload pack both the bearing inner races with lithium based grease and place them in the swivel hub with the spacer positioned in between them (photos).
16 Fit the special tool 49 B001 727 to the centre of the swivel hub, then securely clamp the tool in a vice and tighten its retaining nut to 200 Nm (148 lbf ft). Tighten the nut in stages, increasing the torque by 50 Nm (37 lbf ft) each time and turning the hub through at least one revolution to settle the bearings in position before tightening further.
17 Using a spring balance connected to one of the brake caliper mounting bolt holes in the swivel hub, check the load required to start the swivel hub turning from rest. This should be 230 to 900 g.
18 If the load required to turn the swivel hub is excessive a thicker spacer is required, and if the load required is less than specified a thinner spacer is required. Spacers are available in 21 different thicknesses increasing in increments of 0.04 mm. Each spacer is stamped with a number (1 to 21) on one face. If the bearing preload is to be altered, try the next size up or down as necessary by removing the special tool and

226   Chapter 10   Suspension and steering

3.15A Fit the outer bearing inner race...

3.15B ...followed by the spacer...

3.15C ...then the inner bearing inner race

3.19A Fit a new inner...

3.19B ...and outer oil seal to the swivel hub

3.21 Fit the swivel hub assembly to the hub flange and disc

bearing race, changing the spacer then repeating the checking procedure.
19   When the correct bearing preload is obtained, remove the tool and pack the space between the hub and bearings with grease then fit the new inner and outer oil seals (photos). Use a hammer and tube of suitable diameter to tap the seals into the hub until their edges are flush with the edges of the swivel hub.
20   If removed, refit the disc to the hub flange and tighten the four retaining bolts to the specified torque (Chapter 9).

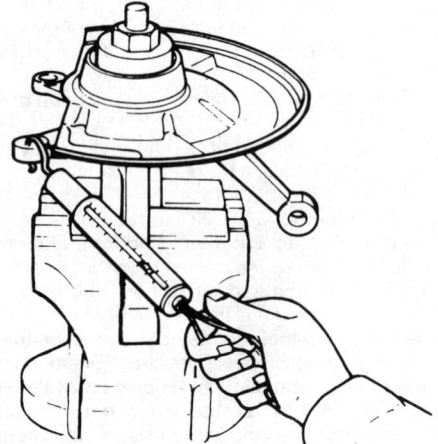
Fig. 10.7 Using a spring balance to check the load required to start the swivel hub turning (Sec 3)

21   Fit the assembled swivel hub assembly to the hub flange and disc then, using a hammer and suitable size tube which contacts the inner bearing race, drive the swivel hub fully home (photo). Alternatively the swivel hub can be pressed on to the disc.
22   Refit the swivel hub assembly to the car as described in Section 2.

4   Front suspension strut – removal and refitting

Removal

1   Chock the rear wheels, firmly apply the handbrake, then jack up the front of the car and support it on axle stands. Remove the appropriate front roadwheel.
2   Remove the clip which secures the brake hose to the suspension strut and lift the hose out of its guide (photos).
3   Slacken the nuts and remove the two bolts securing the strut to the swivel hub (photo).
4   Working in the engine compartment, on pre-September 1985 models prise off the plastic cap in the centre of the strut upper mounting, and check that there is a triangular arrow marked on the mounting rubber which is pointing towards the outside of the vehicle. On September 1985 models onward there should be a circular dot on the inner facing edge of the strut upper mounting (photo). If no mark can be seen, use a dab of white paint to mark the outer or inner (as appropriate) facing side of the mounting to use as a guide on refitting.
5   Undo the two nuts that secure the suspension strut upper mounting to the body. Separate the strut from the swivel hub and remove the unit from under the wheel arch (photos).

Chapter 10 Suspension and steering

4.2A Remove the brake hose retaining clip...

4.2B ...and free the brake hose from the suspension strut

4.3 Remove the suspension strut to swivel hub bolts

4.4 Circular dot (arrowed) on strut upper mounting on September 1985 models onward

4.5A Undo the strut upper mounting nuts...

4.5B ...and remove the strut from the car

Refitting

6 Refitting is by a reversal of the removal procedure noting the following points (photo).

(a) Ensure that the mark on the strut upper mounting is positioned as described in paragraph 4.
(b) Tighten all nuts and bolts to the specified torque.

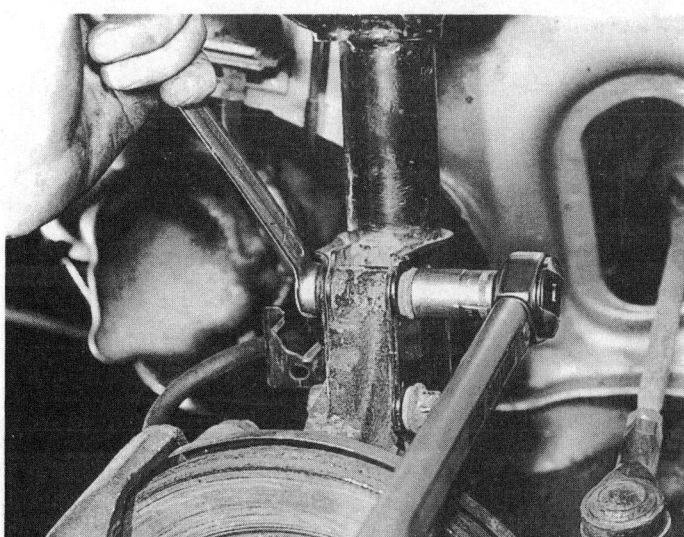

4.6 Tighten suspension strut to swivel hub mounting bolts to the specified torque

5 Front suspension strut – dismantling, inspection and reassembly

Note: *Before attempting to dismantle the front suspension strut, a suitable tool to hold the coil spring in compression must be obtained. Adjustable coil spring compressors are readily available and are recommended for this operation. Any attempt to dismantle the strut without such a tool is likely to result in damage or personal injury.*

Dismantling

1 With the strut removed from the car clean away all external dirt, then mount it upright in a vice.
2 Remove the plastic cap from the centre of the upper strut mounting and slacken **but do not remove** the strut piston retaining nut (photo).
3 Fit the spring compressor tool and compress the coil spring until all tension is relieved from the upper mounting (photo).
4 Remove the piston retaining nut and washer, then lift off the upper mounting block followed by the upper spring seat, rubber ring, strut piston dust seal and coil spring. On later models also lift off the lower spring seat.

Inspection

5 With the strut assembly now completely dismantled, examine all the components for wear, damage or deformation and check the bearing for smoothness of operation. Renew any of the components as necessary.
6 Examine the strut for signs of fluid leakage. Check the strut piston for signs of pitting along its entire length and check the strut body for signs of damage or elongation of the mounting bolt holes. Test the operation of the strut, while holding it in an upright position, by moving the piston through a full stroke and then through short strokes of 50 to

Chapter 10 Suspension and steering

5.2 Remove the plastic cap to gain access to the strut piston retaining nut

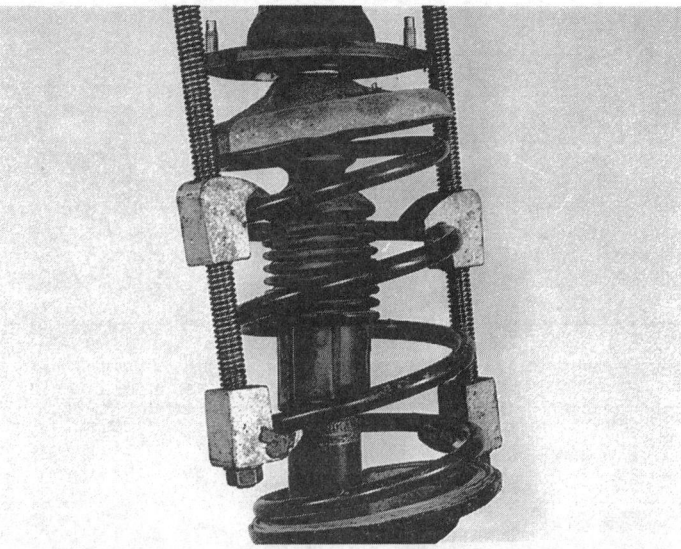

5.3 Using spring compressors to compress front suspension strut coil spring

100 mm. In both cases the resistance felt should be smooth and continuous. If the resistance is jerky, or uneven, or if there is any visible sign of wear or damage to the strut, renewal is necessary.

7 If any doubt exists about the condition of the coil spring, carefully remove the spring compressors and check the spring for distortion. Measure the free length of the spring and compare this with the figure given in the Specifications at the start of this Chapter. Renew the spring if it is distorted or outside the service length.

8 Inspect all other components for signs of damage or deterioration and renew any that are suspect.

Reassembly

9 Reassembly is a reversal of dismantling, however make sure that the spring ends are correctly located in the upper and lower seats and tighten the piston retaining nut to the specified torque.

6 Front suspension anti-roll bar – removal and refitting

Removal

1 Chock the rear wheels, firmly apply the handbrake then jack up the front of the car and support it on axle stands. Remove both front roadwheels.

2 If necessary, remove the engine undertray to gain access to the anti-roll bar mountings.

3 Undo the two locknuts and remove the connecting link bolt securing the ends of the anti-roll bar to the lower suspension arms. Make a note of how the connecting link washers, rubber bushes and spacers are arranged to use as a reference on refitting.

4 Undo the bolts securing the anti-roll bar mounting clamps to the chassis member, then remove the bar from underneath the vehicle.

5 Carefully examine the mountings, connecting links and anti-roll bar for signs of cracks, damage or deformation, paying particular attention to the rubber mounting bushes. Renew any worn component.

Refitting

6 Ensure that the anti-roll bar mounting bushes are positioned with their splits facing the front of the vehicle, then offer up the bar and install the mounting clamps. Tighten the clamp bolts finger tight only at this stage.

7 Using the notes made on dismantling, refit the connecting link components to their correct fitted positions and tighten the locknuts finger tight only (photo).

8 Tighten the anti-roll bar mounting clamp bolts and nuts to the specified torque (photo).

6.7 Ensure connecting link bolt components are correctly fitted...

6.8 ...then tighten the anti-roll bar mounting clamp nuts and bolts (arrowed) to the specified torque

Chapter 10 Suspension and steering

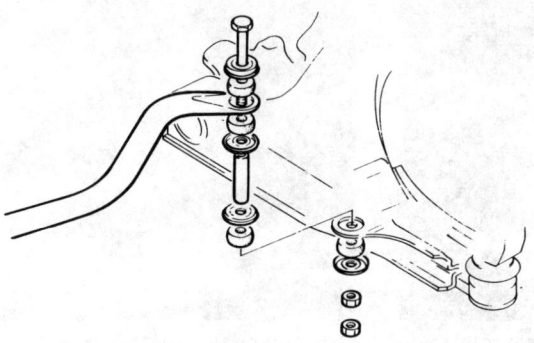

Fig. 10.8 Front anti-roll bar connecting link arrangement (Sec 6)

9 Position the connecting link locknuts so that when tightened to the specified torque setting there is 6.2 mm of the connecting link bolt thread exposed on pre-September 1985 models, and 10.8 mm of the connecting link bolt thread exposed on later models (Fig. 10.9).
10 Refit the undertray (if removed) and the roadwheels then lower the car to the ground. Tighten the roadwheel nuts to the specified torque.

7 Front suspension lower arm – removal and refitting

Removal

1 Chock the rear wheels, firmly apply the handbrake then jack up the front of the car and support it on axle stands.
2 If an anti-roll bar is fitted, undo the two locknuts and remove the connecting link bolt securing the anti-roll bar to the lower suspension arm. Make a note of the correct fitted positions of the rubber bushes, washers and spacer to use as a reference on refitting.
3 Undo the nut and remove the pinch-bolt securing the lower arm balljoint to the swivel hub.
4 Using a suitable bar, lever the lower suspension arm down to release the balljoint shank from the swivel hub.

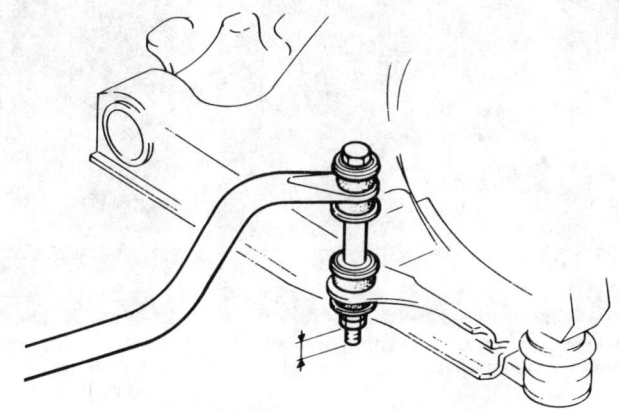

Fig. 10.9 Position the connecting link locknuts so that the specified amount of connecting link bolt thread is exposed (Sec 6)

5 On pre-September 1985 models undo the nuts and bolts securing the front and rear lower arm mounting brackets to the underbody. On September 1985 models onward, undo the two bolts which secure the rear lower arm mounting bracket to the underbody and remove the front pivot bolt (photos). Remove the arm from the car.
6 With the arm removed from the car, carefully examine the rubber mounting bushes for swelling or deterioration, the balljoint for slackness, and check for damage to the balljoint gaiter. Renewal of the balljoint is described in Section 8. Renewal of the suspension arm bushes may be carried out as follows.
7 To renew the rear bush, undo the retaining nut and withdraw the rear mounting and washers from the lower arm spindle. Renew the rear mounting as an assembly if the bushes are worn or damaged.
8 On early models, remove the front mounting bracket retaining nut and remove the bracket and washers from the arm.
9 To renew the front mounting bush, cut off the forward facing flange of the bush using a hacksaw then press the bush out using suitable mandrels.
10 Lubricate the new bush with soapy water then press it into position from the front of the arm.

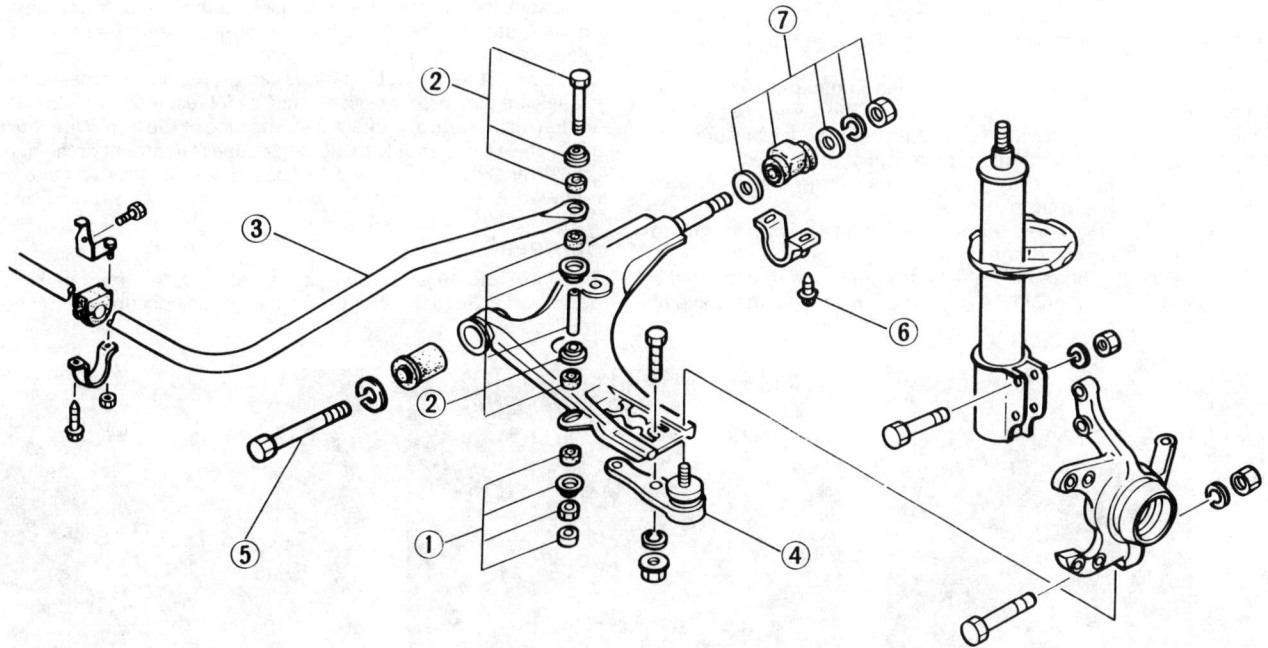

Fig. 10.10 Exploded view of front suspension lower arm and associated components – September 1985 models onward (Secs 6 and 7)

1 Locknuts, bush and retainer assembly
2 Connecting link bolt assembly
3 Anti-roll bar
4 Lower arm balljoint
5 Lower arm front mounting pivot bolt
6 Bolt
7 Lower arm rear mounting nut and bush assembly

Chapter 10 Suspension and steering

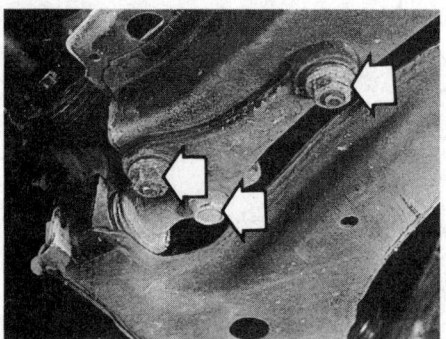

7.5A Lower arm front mounting bracket retaining nuts (arrowed)...

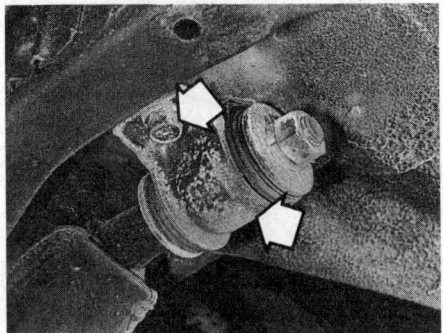

7.5B ...and rear mounting bracket retaining bolts (arrowed) on pre- September 1985 models

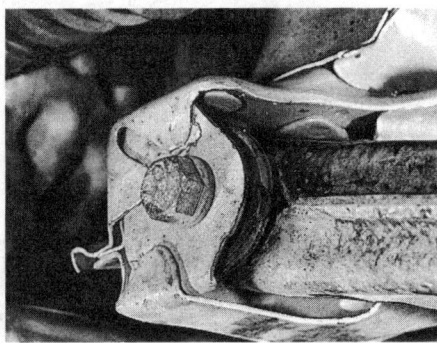

7.5C Lower arm front pivot bolt – September 1985 models onward

11 Refit the front and rear mounting brackets (as applicable), but do not tighten the retaining nuts fully at this stage.

Refitting
12 Refitting the suspension arm is the reverse sequence of removal bearing in mind the following points.

(a) Tighten all retaining nuts and bolts to the specified torque.
(b) Ensure the anti-roll bar connecting link locknuts are positioned as described in Section 6, paragraph 9.
(c) Do not fully tighten the lower arm to mounting bracket nut(s) or pivot bolt (as applicable) until the weight of the car is standing on its wheels.
(d) After refitting, check the front wheel toe setting as described in Section 26.

8 Front suspension balljoint – removal and refitting

Removal
1 Chock the rear wheels, firmly apply the handbrake, then jack up the front of the car and support it on axle stands. Remove the appropriate front roadwheel.
2 If an anti-roll bar is fitted, undo the two locknuts and remove the connecting link bolt securing the anti-roll bar to the lower suspension arm. Make a note of the correct fitted positions of the rubber bushes, washers and spacer to use a reference on refitting.
3 Undo the nut and remove the pinch-bolt securing the lower arm balljoint to the swivel hub (photo).
4 Using a suitable bar, lever the lower suspension arm down to release the balljoint shank from the swivel hub (photo).
5 Undo the two nuts and remove the bolts and washers securing the balljoint to the lower suspension arm (photo). Withdraw the balljoint from the end of the arm.

6 If the balljoint is excessively slack, or in anyway damaged, it must be renewed as an assembly. If, however, only the rubber gaiter is damaged a new gaiter may be obtained separately.

Refitting
7 The balljoint is refitted by a reverse of the removal sequence bearing in mind the following points.

(a) Tighten all nuts and bolts to the specified torque.
(b) Ensure the anti-roll bar locknuts are positioned as described in Section 6, paragraph 9.

9 Rear hub bearings – checking, removal and refitting

Note: *The rear hub bearings should only be removed from the swivel hub if they are to be renewed. The removal procedure renders the bearings unserviceable and they must not be re-used.*

Checking
1 Chock the front wheels, then jack up the rear of the car and support on axle stands. Remove the appropriate rear roadwheel and fully release the handbrake.
2 Wear in the rear hub bearings can be checked by measuring if there is any side play present. To do this, a dial test indicator should be fixed so that its probe is in contact with the face of the hub. With the indicator set in place there should be no detectable movement of the hub. If there is any trace of endplay, the bearings are worn excessively and should be renewed.

Removal
3 With reference to Chapter 9, remove the rear brake drum as described in Section 12 or the brake disc as described in Section 16 (as applicable).

8.3 Slacken and remove the balljoint pinch-bolt...

8.4 ...and lever the balljoint shank out of the swivel hub

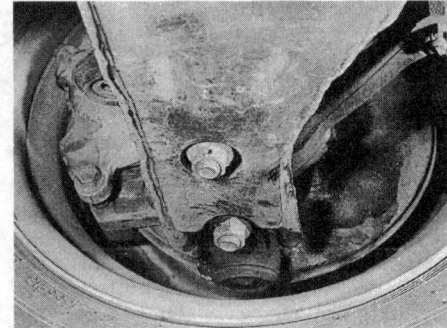

8.5 Lower arm balljoint retaining nuts

Chapter 10 Suspension and steering

9.5 Levering out the hub oil seal using a suitable screwdriver/tyre lever

9.8 Tap the bearing outer race into position in the drum using a suitable tubular drift

9.10 ...then fit the bearing inner race

4   Lift out the thrustwasher and outer bearing inner race.
5   Turn the hub over and, using a suitable screwdriver/tyre lever, carefully lever out the oil seal (photo). Lift out the inner bearing inner race.
6   Wipe away the surplus grease from the centre of the hub then, using a hammer and suitable drift, drive out one of the bearing outer races from the centre of the hub whilst taking care not to damage the hub bore. Remove the race carefully by tapping evenly around the outer race whilst keeping the race as square as possible in the hub bore. After removal of the first race, turn the hub over and drive out the remaining outer race in a similar way.then using a hammer and suitable drift carefully drive out one of the bearing inner races.
7   Renew both bearings as a pair along with the oil seal. Check for any signs of damage to the hub bore and renew if necessary. Minor burrs or score marks can be removed from the hub bore using a fine file or emery paper.

Refitting

8   Place the outer bearing outer race in position in the rear hub and carefully tap it into position in the hub using a hammer and a suitable tubular drift which bears only on the outer edge of the race (photo). Tap the race down until it contacts the shoulder in the centre of the swivel hub.
9   Turn the rear hub over and repeat the above procedure for the inner bearing outer race.
10 Pack the inner bearing inner race with lithium based grease and place it in position in the hub (photo).
11 Lubricate the lip of the oil seal with grease then tap it into position using a hammer and suitable sized tubular drift which bears only on the hard outer edge of the seal. Tap the seal in until it is flush with the edge of the hub.
12 Pack the outer bearing inner race with grease and also the area between the bearings in the hub bore.
13 Place the outer bearing inner race and thrustwasher in position and refit the brake drum or disc (as applicable) as described in Chapter 9.

10 Rear stub axle – removal and refitting

Removal

Pre-September 1985 models

1   Remove the rear brake drum as described in Section 12 of Chapter 9.
2   Using a brake hose clamp or similar tool, clamp the flexible brake hose. This will minimise the loss of fluid during the subsequent operations.
3   Wipe clean the area around the brake pipe union at the rear of the wheel cylinder. Undo the union nut and withdraw the brake pipe from the wheel cylinder. Plug the pipe and wheel cylinder orifice to prevent the entry of dirt into the hydraulic system.
4   Disconnect the handbrake cable return spring, then extract the split pin, remove the clevis pin and disconnect the cable from the lever.
5   Undo the four bolts securing the brake backplate to the stub axle and remove the brake backplate assembly from the vehicle.
6   Undo the nut and remove the pivot bolt and washers securing the lateral links to the stub axle.
7   Undo the two nuts and remove the bolts securing the stub axle to the rear suspension strut, then withdraw the stub axle from its location.
8   Examine the stub axle for any signs of cracks or damage, or for wear ridges on the oil seal flange. Surface corrosion or light scoring can be removed using fine emery cloth.

September 1985 onward (rear drum brake models)

9   Remove the brake drum as described in Section 12 of Chapter 9.
10 Remove the clip which secures the brake hose to the rear suspension strut and free the hose from its guide.
11 Undo the four bolts securing the brake backplate to the stub axle then carefully pull the backplate assembly outwards until it clears the stub axle spindle. Tie the assembly to the suspension strut coil spring to avoid straining the hydraulic hose.
12 Remove the stub axle as described in paragraphs 6 to 8.

10.16A Refit the stub axle retaining bolts...

10.16B ...and tighten them to the specified torque whilst applying an upward load on the stub axle

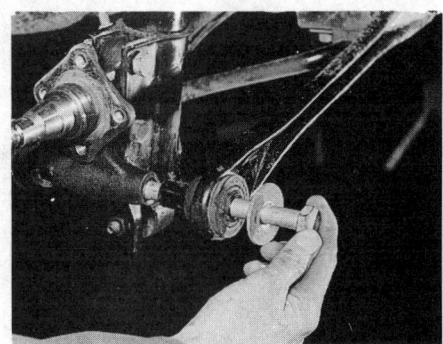

10.17 Refit the lateral link to stub axle bolt

10.18A Refit the brake backplate assembly...

10.18B ...and tighten its retaining bolts to the specified torque

September 1985 onward (rear disc brake models)
13 Remove the brake disc as described in Section 16 of Chapter 9.
14 Undo the four bolts which secure the dust cover to the stub axle and withdraw the cover.
15 Remove the stub axle as described in paragraphs 6 to 8.

Refitting
Pre-September 1985 models
16 Offer up the stub axle and refit the stub axle to suspension strut bolts, washers and nuts. Tighten the nuts to the specified torque whilst applying an upward load to the stub axle spindle (photos).
17 Refit the pivot bolt, washers and nut securing the lateral links to the stub axle (photo). Do not fully tighten the nut at this stage.
18 Refit the brake backplate assembly to the stub axle and tighten its retaining bolts to the specified torque (photos).
19 Reconnect the handbrake cable to the lever and refit the clevis pin. Secure the clevis pin in position using a new split pin and refit the handbrake cable return spring.
20 Refit the brake pipe to the wheel cylinder and tighten the union nut to the specified torque. Remove the clamp from the brake hose.
21 Refit the brake drum as described in Chapter 9, Section 12.
22 Bleed the hydraulic system as described in Chapter 9 then refit the roadwheel and lower the car to the ground. With the weight of the car standing on its wheels, tighten the lateral links to stub axle pivot bolt to the specified torque.

September 1985 onward (rear drum brake models)
23 Refit the stub axle and brake backplate as described in paragraphs 16 to 18.

24 Position the brake hose in its guide on the rear suspension strut and secure it in position with the retaining clip.
25 Install the brake drum as described in Chapter 9, Section 12.
26 Refit the roadwheel and lower the car to the ground. With the weight of the car standing on its wheels, tighten the lateral links to stub axle pivot bolt to the specified torque.

September 1985 onward (rear disc brake models)
27 Refit the stub axle as described in paragraphs 16 and 17.
28 Refit the rear disc dust cover to the stub axle and tighten its retaining bolts to the specified torque.
29 Install the brake disc as described in Chapter 9.
30 Refit the roadwheel and lower the car to the ground. With the weight of the car standing on its wheels, tighten the lateral links to stub axle pivot bolt to the specified torque.

11 Rear suspension strut – removal and refitting

Removal
1 Chock the front wheels, jack up the rear of the car and support it on axle stands. Remove the appropriate roadwheel.
2 On pre-September 1985 models disconnect the handbrake cable

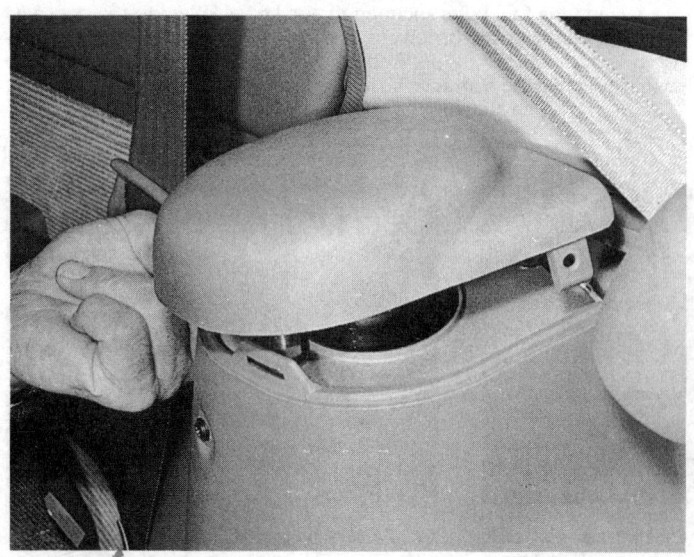
11.7A Remove the trim cap...

11.7B ...to gain access to the upper suspension strut mounting nuts (arrowed) (Estate model shown)

Chapter 10 Suspension and steering

return spring from the rear of the brake backplate. Extract the split pin and remove the clevis pin securing the handbrake cable end to the lever.
3 Undo the nut and remove the bolt and washer securing the trailing arm to the base of the suspension strut.
4 Undo the nut and remove the bolt and washers securing the lateral links to the stub axle assembly.
5 Remove the clip which secures the flexible brake hose to the rear suspension strut and lift the hose out of the guide.
6 Slacken the two nuts and remove the bolts and washers which secure the stub axle assembly to the suspension strut, then separate the stub axle assembly from the strut. Support the stub axle assembly in some way to avoid placing any strain on the hydraulic hose.
7 From inside the luggage compartment, remove the trim cap to gain access to the suspension strut upper mounting (photos).
8 Check that there is a dot of white paint on the inner edge of the suspension strut mounting block. If this is not the case, mark the mounting with a dab of white paint to use as a guide on refitting.
9 Undo the two nuts and remove the washers securing the strut upper mounting to the body. Ease the strut down, release the trailing and lateral links then remove the strut from the vehicle.

Refitting

10 Manoeuvre the suspension unit into position, ensuring that the white dot is facing towards the inside of the car, and refit the washers and strut upper mounting nuts. Tighten the mounting nuts to the specified torque.
11 Offer up the stub axle assembly and refit the stub axle to suspension strut bolts, washers and nuts. Tighten the nuts to the specified torque whilst applying an upward load to the stub axle.
12 Refit the pivot bolt, washers and nut securing the lateral links to the stub axle, and the pivot bolt, washers and nut which secures the trailing arm to the suspension strut. Do not fully tighten either nut at this stage.
13 Refit the brake hose to its guide on the suspension strut and secure it in position with the retaining clip.
14 On pre-September 1985 models reconnect the handbrake cable to the lever and refit the clevis pin. Secure the clevis pin in position using a new split pin and refit the handbrake cable return spring.
15 Refit the roadwheel and lower the car to the ground. With the weight of the car standing on its wheels, tighten the lateral links to stub axle pivot bolt and the trailing arm to suspension strut pivot bolt to the specified torque.

12 Rear suspension strut – dismantling, inspection and reassembly

Dismantling

1 With the strut removed from the car clean away all external dirt then mount it upright in a vice.
2 Slacken **but do not remove** the strut piston retaining nut from the centre of the upper mounting block.
3 Fit the spring compressor tool and compress the coil spring until all tension is relieved from the upper mounting.
4 Remove the piston retaining nut and washer then lift off the upper mounting block followed by the upper spring seat, rubber ring, strut piston dust seal and coil spring.

Inspection

5 Examine all rear suspension strut components using the information given for the front suspension strut in paragraphs 5 to 8 in Section 5.

Reassembly

6 Reassembly is a reversal of dismantling, however make sure that the spring ends are correctly located in the upper and lower seats and tighten the piston retaining nut to the specified torque.

13 Rear suspension anti-roll bar – removal and refitting

Removal

1 Chock the front wheels, jack up the rear of the car and support it on axle stands. Remove both rear roadwheels.

Pre-September 1985 models

2 Undo the two nuts and remove the bolts securing the anti-roll bar to each rear trailing arm (photo).
3 Withdraw the anti-roll bar from the trailing arms and manoeuvre it out from under the car.
4 Carefully examine the anti-roll bar components for signs of cracks, damage or deformation, paying particular attention to the rubber mounting bushes. Renew any worn component.

September 1985 models onward

5 Before removing the anti-roll bar, check that the white lines used to align the bar with its mounting bushes are clearly visible. If not, mark the bar using white paint.
6 Undo the two locknuts and remove the connecting link bolt securing the ends of the anti-roll bar to the lateral links. Make a note of the correct fitted positions of the connecting link washers, rubber bushes and spacers to use as a reference on refitting.
7 Undo the bolts securing the anti-roll bar mounting clamps to the crossmember, then remove the bar from underneath the vehicle (photo).
8 Carefully examine the mountings, connecting links and anti-roll bar for signs of cracks, damage or deformation, paying particular attention to the rubber mounting bushes. Renew any worn component.

Refitting

Pre-September 1985 models

9 Refitting is the reverse of the removal sequence noting the anti-roll bar mounting bolts should be tightened to the specified torque when the car is standing on its wheels.

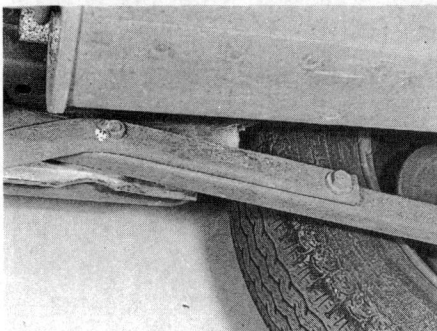

13.2 Rear anti-roll bar to trailing arm mounting bolts – pre-September 1985 models

13.7 Rear anti-roll bar mounting clamp bolt (A) and connecting link bolt (B) – September 1985 models onward

13.13 Position locknuts so that the specified amount of connecting link bolt thread is exposed when tightened to the specified torque

Chapter 10 Suspension and steering

September 1985 models onward

10 Ensure that the anti-roll bar mounting bushes are positioned with their splits facing the front of the vehicle then offer up the bar and install the mounting clamps. Align the white lines on the anti-roll bar with the mounting bushes and tighten the clamp bolts finger tight only at this stage.
11 Using the notes made on dismantling, refit the connecting link components to their correct fitted positions and tighten the locknuts finger tight only.
12 Tighten the anti-roll bar mounting clamp bolts to the specified torque.
13 Position the connecting link locknuts so that when tightened to the specified torque setting there is 18.0 mm of the connecting link bolt thread exposed (photo).
14 Refit the roadwheels and lower the car to the ground. Tighten the roadwheel nuts to the specified torque.

14 Rear suspension lateral links – removal and refitting

Removal

1 Chock the front wheels, jack up the rear of the car and support it on axle stands. Remove the appropriate roadwheel.
2 On September 1985 models onward, undo the locknuts and remove the connecting link bolt which secures the anti-roll bar to the front lateral link. Make a note of the correct fitted positions of the rubber bushes, washers and spacer to use as a reference on refitting.
3 On all models undo the nut and remove the pivot bolt and washers securing the lateral links to the stub axle.
4 On all except later Estate models, wipe clean the face of the notched eccentric spacer located behind the lateral to crossmember retaining nut (photo). On the face of the spacer will be found a positioning mark and a corresponding lug on the link which will be engaged with one of the spacer notches. Make a note of the number of notches, clockwise or anti-clockwise, the positioning mark is situated in relation to the lug. The position of the spacer determines the rear toe setting and it is important to refit it in the same place otherwise the setting will be lost.
5 Having noted the position of the eccentric spacer (if fitted), undo the retaining nut, remove the pivot bolt and washers and withdraw the lateral links from under the car. On pre-September 1985 models, if the forward facing link on the left-hand side is being removed, it will be necessary to slacken the rear crossmember mounting nuts and lower the crossmember to provide sufficient clearance for removal of the bolt.
6 Examine the lateral links for signs of damage or distortion or for any signs of deterioration of the mounting bushes. Renew the links if these conditions are found.

Refitting

7 Offer up the lateral links and insert the inner pivot bolt and washer. Refit the eccentric spacer (where fitted) to the position noted on removal and refit the washer and retaining nut. Do not fully tighten the bolt at this stage.
8 If loosened, tighten the rear crossmember mounting bolts to the specified torque.
9 Refit the pivot bolt, washers and nut securing the lateral links to the stub axle. Do not fully tighten the bolt at this stage.
10 On September 1985 models onward install the anti-roll bar connecting link components using the notes made on dismantling and refit the locknuts. Position the locknuts so that when tightened to the specified torque setting there is 18.0 mm of the connecting link bolt thread exposed. Tighten both the lateral link pivot bolts to the specified torque.
11 Refit the roadwheels, lower the car to the ground and tighten the roadwheel nuts to the specified torque. On pre September 1985 models then tighten the lateral link pivot bolts to the specified torque with the car standing on its wheels.
12 If new lateral links were fitted check the rear toe setting as described in Section 26.

15 Rear suspension trailing arms – removal and refitting

Removal

1 Chock the front wheels, jack up the rear of the car and support it on axle stands. Remove the appropriate rear roadwheel.
2 On pre-September 1985 models undo the two nuts and remove the bolts securing the anti-roll bar to the trailing arm.
3 On all models undo the nut and remove the bolt and washer securing the trailing arm to the rear suspension strut (photo). Separate the trailing arm from the strut.
4 Slacken and remove the bolt securing the trailing arm to the underbody then withdraw the arm and remove it from the car.
5 Examine the trailing arm for signs of damage or distortion or for signs of deterioration in the mounting bushes. Renew the trailing arm if these conditions are found.

Refitting

6 Refitting is the reverse of the removal sequence bearing in mind the following points.

14.4 Lateral link to crossmember retaining nut (A) and eccentric spacer (B) for toe setting adjustment

15.3 Trailing arm to suspension strut bolt (pre-September 1985 models shown)

Chapter 10 Suspension and steering

(a) On September 1985 models onward tighten the trailing arm front and rear pivot bolts to the specified torque whilst the car is supported on the axle stands.
(b) On pre-September 1985 models do not tighten the trailing arm pivot bolts and anti-roll bar mounting bolts to the specified torque until the car is standing on its wheels.

16 Steering wheel – removal and refitting

Removal

1 Set the front wheels to the straight-ahead position.
2 Remove the horn cap either by removing its retaining screws (if fitted) and/or gently prising it off the steering wheel.
3 Using a suitable socket, unscrew the nut securing the steering wheel to the column shaft (photo).
4 Using a dab of paint, mark the position of the steering wheel in relation to the shaft to use as a guide on refitting.
5 Using a suitable puller secured to the threaded holes on either side of the shaft, draw the steering wheel off the steering column shaft.

Refitting

6 Refitting is the reverse of the removal sequence. Align the marks made during removal and tighten the retaining nut to the specified torque.

16.3 Steering wheel retaining nut (arrowed)

17 Steering column – removal, checking and refitting

Removal

1 Disconnect the battery negative terminal.
2 Refer to Section 16 and remove the steering wheel.
3 Remove the cover under the facia to gain access to the steering column and mountings.
4 Undo the screws which retain the two halves of the steering column shroud. Lift off the left-hand or upper shroud then remove the bulbholder (if equipped) in the right-hand or lower shroud (as applicable). Remove both shrouds from the steering column (photos).
5 On September 1985 models onward, remove the steering column lower panel and louvre along with the demister duct (photo).
6 Disconnect the wiring multi-plugs from the rear of the steering column combination switch. Undo the switch clamp bolt and slide the switch off the column.
7 Disconnect the ignition switch wiring at the harness connector.
8 Undo the pinch-bolt securing the steering column shaft universal joint to the intermediate shaft. Make corresponding marks on both shafts to use as a guide on refitting.

17.4A Steering column shroud retaining screws (arrowed) – pre-September 1985 models

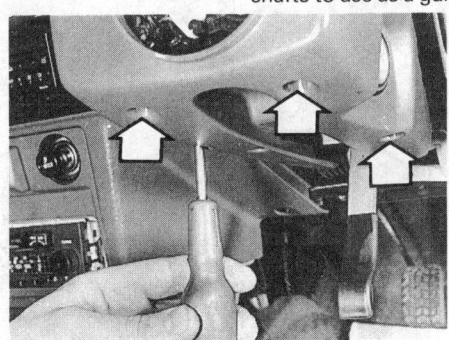
17.4B On September 1985 models onward undo the steering column shroud retaining screws (arrowed)...

17.4C ...then withdraw the shrouds...

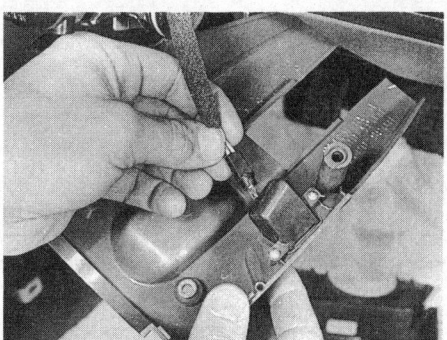
17.4D ...and remove the bulbholder

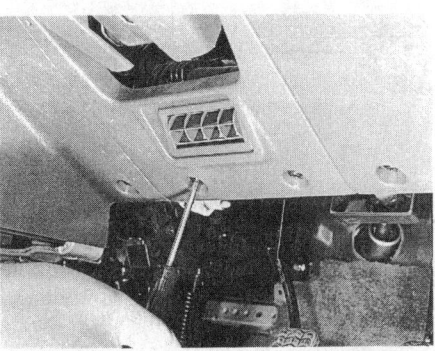
17.5 Removing the steering column lower panel – September 1985 models onward

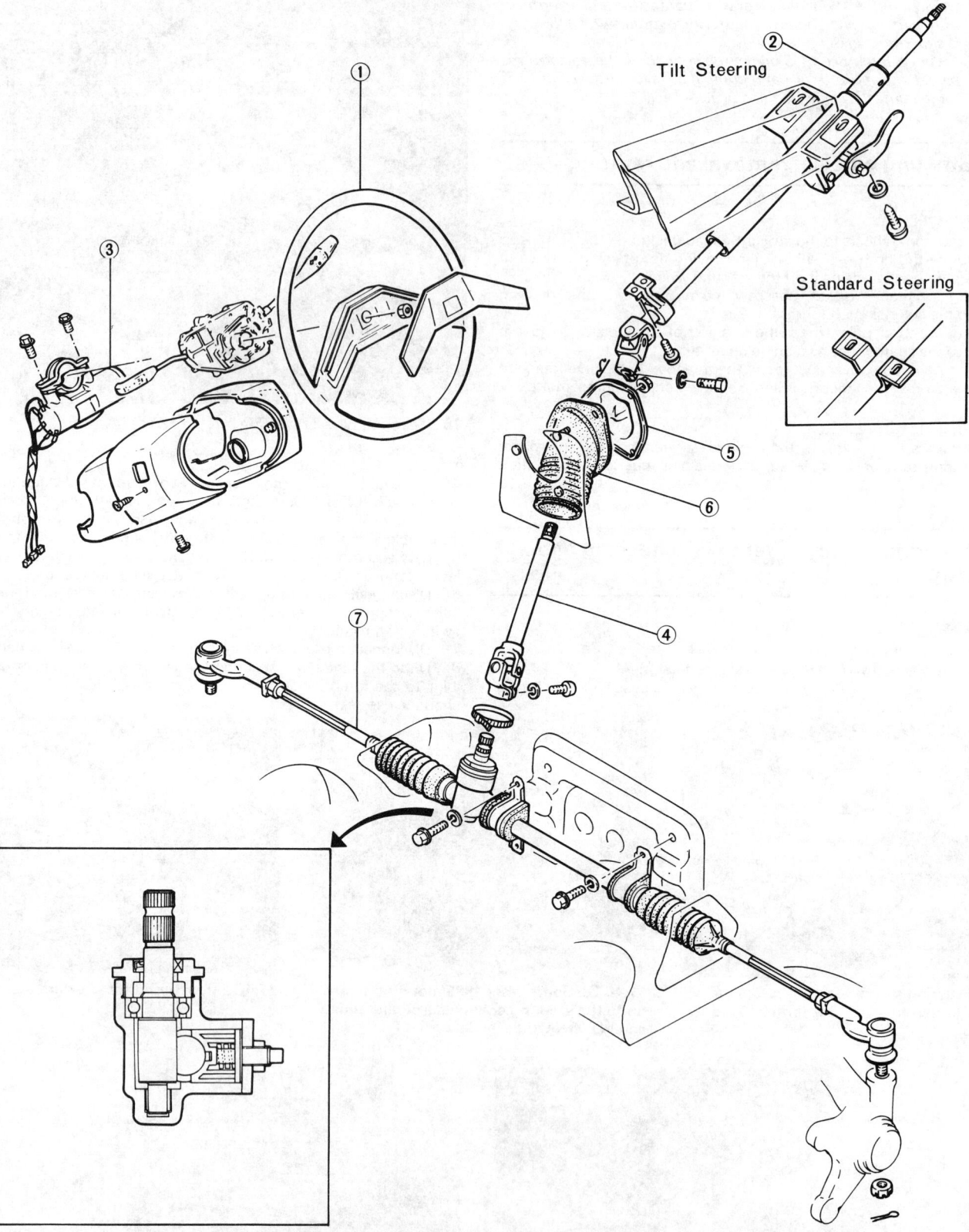

Fig. 10.11 Exploded view of the steering wheel column and assembly – pre-September 1985 models shown (Secs 16 to 21)

1 Steering wheel
2 Steering column and shaft assembly
3 Steering lock/ignition switch
4 Intermediate shaft
5 Gaiter retaining plate
6 Gaiter
7 Steering gear and linkage

Chapter 10 Suspension and steering 237

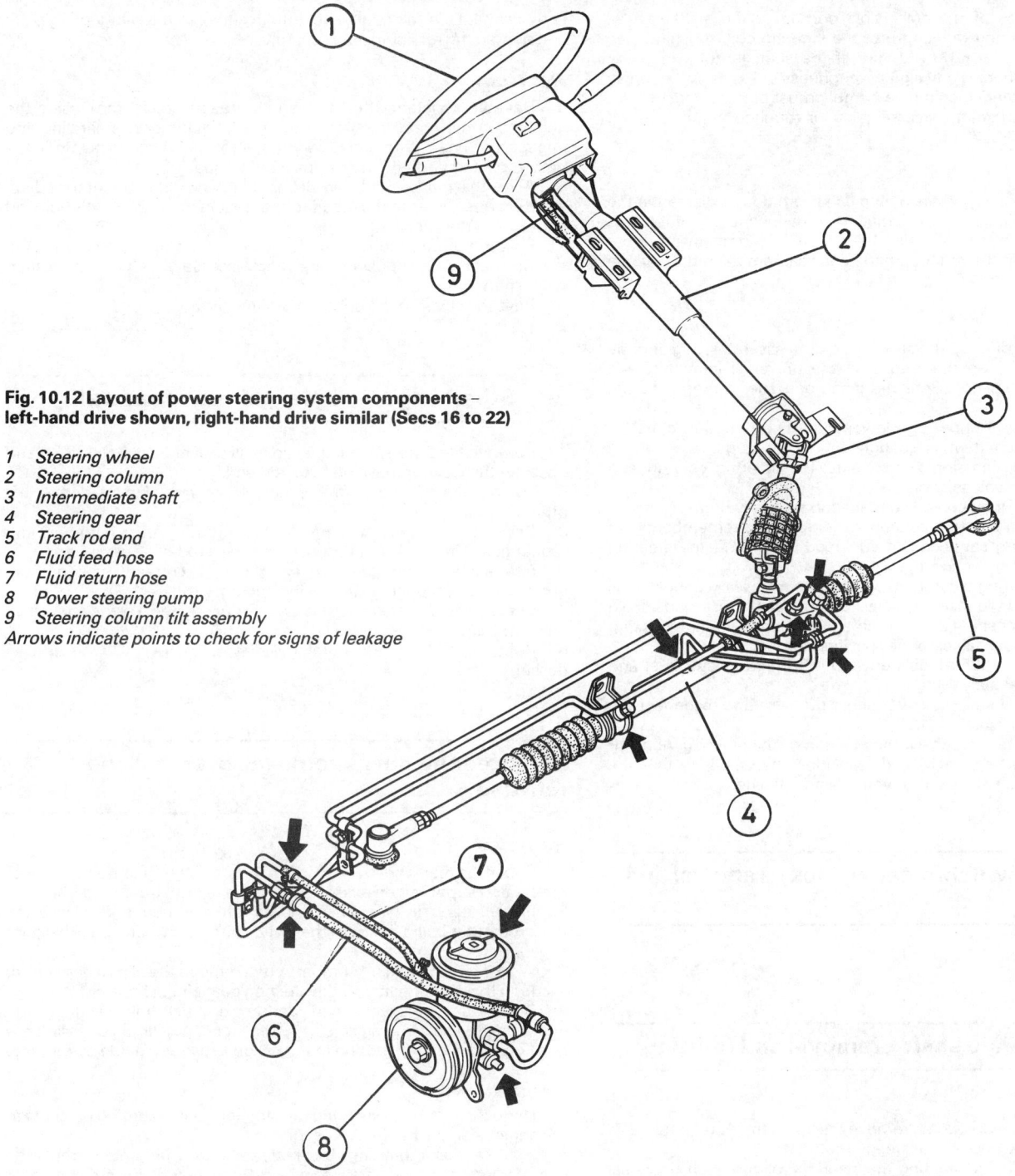

**Fig. 10.12 Layout of power steering system components –
left-hand drive shown, right-hand drive similar (Secs 16 to 22)**

1 Steering wheel
2 Steering column
3 Intermediate shaft
4 Steering gear
5 Track rod end
6 Fluid feed hose
7 Fluid return hose
8 Power steering pump
9 Steering column tilt assembly

Arrows indicate points to check for signs of leakage

9 Undo the upper and lower steering column mounting bracket retaining bolts and nuts (as applicable), withdraw the universal joint from the intermediate shaft and remove the column from the car.

Checking

Pre-September 1985 models

10 Support the steering column in a vice equipped with soft jaws.
11 Insert the ignition key into the switch and turn the switch so that the steering lock is released.
12 Remove the steering column shaft from the base of the column using a twisting motion.
13 Check the steering column shaft for straightness and for signs of impact and damage to the collapsible portion. Similarly check for damage or distortion to the column. Check for wear or roughness in the column bearings and in the intermediate shaft universal joint. If any damage or wear is found, the column and column shaft must be renewed as an assembly. It is possible to obtain the column upper bearing separately, but all other bearings are supplied only as part of the complete column assembly.
14 To renew the column upper bearing prise it out of the column tube using a screwdriver and carefully tap a new bearing into place.
15 To reassemble the steering column, lubricate the column bearings with a lithium based grease then refit the shaft through the base of the column.

September 1985 models onward

16 Undo the pinch-bolt which secures the universal joint to the base of the steering column shaft and pull the joint off the shaft splines. No further dismantling is possible.
17 Check that the steering column shaft turns freely with no traces of roughness. Also check for endfloat and side-play on the steering column

shaft. The condition of the collapsible portion of the shaft can be assessed by measuring the length of the steering column shaft. The shaft length should be 607 ± 1 mm. If the shaft is not within the specified limits or there is any sign of roughness or excessive sideplay and endfloat, the steering column assembly must be renewed. Inspect the steering column shaft universal joint for roughness and renew if necessary.

All models

18 If the column incorporates a tilt mechanism do not dismantle the unit unless it is obviously worn, damaged or malfunctioning. If repair is necessary dismantle the unit using the relevant accompanying figure for reference, obtain the required parts and reassemble in the reverse order of removal.

Refitting

19 On all models offer up the steering column assembly, aligning the marks made on dismantling, and slide the universal joint onto the intermediate shaft splines. Refit the pinch-bolt and tighten it to the specified torque.
20 Refit the column upper and lower mounting nuts and bolts (as applicable) but tighten them finger tight only at this stage.
21 Install the combination switch onto the steering column and tighten its clamping bolt securely.
22 Reconnect the ignition and combination switch wiring.
23 Install the two steering column shroud halves, remembering to reconnect the wiring connector (if equipped), then refit the steering wheel as described in Section 16.
24 Move the steering column up or down as necessary until a small gap exists between the steering wheel and the column shrouds. Hold the column in this position and tighten the upper and lower mounting nuts and bolts (as applicable) to the specified torque.
25 On September 1985 models onward refit the demister duct and lower louvre and panel.
26 On all models install the cover under the facia and reconnect the battery negative terminal.
27 Roadtest the car and check the position of the steering wheel. If necessary, refer to Section 16 and reposition the wheel so that the spokes are level when the car is driven in a straight line.

18 Ignition switch/steering lock – removal and refitting

Refer to Chapter 12, Section 12.

19 Intermediate shaft – removal and refitting

Removal

1 Working from inside the car, remove the cover from under the facia to gain access to the base of the steering column.
2 Undo the pinch-bolt securing the steering column shaft universal joint to the intermediate shaft. Make corresponding marks on both shafts to use as a guide on refitting.
3 Working from within the engine compartment release the clip securing the rubber gaiter to the pinion housing on the steering gear (where fitted). Fold back the gaiter and undo the pinch-bolt securing the intermediate shaft universal joint to the steering gear pinion. Mark the joint in relation to the pinion, then pull the intermediate shaft upwards off the pinion splines and then downwards to release it from the steering column splines. Remove the shaft from the car.
4 Examine the intermediate shaft for signs of wear or damage paying particular attention to the shaft and universal joint splines. Check that the universal joint moves smoothly with no trace of roughness. If the universal joint is worn, or the shaft is damaged, renew the complete shaft assembly. Inspect the intermediate shaft rubber gaiter and renew it if it shows signs of wear, damage or deterioration. To renew the gaiter, undo its three retaining nuts situated inside the car and remove the retaining plate. The gaiter can then be withdrawn from the engine compartment. Fit the new gaiter into position, refit the mounting plate and tighten the retaining nuts securely.

Refitting

5 Have an assistant offer up the intermediate shaft from inside the engine compartment, then align the marks made on dismantling and engage the shaft with the steering column universal joint. Refit the pinch-bolt and tighten it to the specified torque.
6 Align the marks made on dismantling and refit the intermediate shaft universal joint to the steering gear pinion. Refit the pinch-bolt and tighten it to the specified torque.
7 Ensure the intermediate shaft gaiter is correctly located with the steering gear pinion housing and, where necessary, secure it in position with a new clip.
8 Refit the cover to the under side of the facia.

20 Steering gear rubber gaiters – renewal

1 Remove the track rod outer balljoint as described in Section 24, and unscrew the locknut from the track rod end.
2 Using pliers, release the rubber gaiter outer retaining clip and slide it off the track rod end.
3 Remove the inner retaining clip by cutting it, then withdraw the rubber gaiter from the steering gear and track rod.
4 Fit the new rubber gaiter ensuring that it is correctly seated in the grooves in the steering gear housing and track rod.
5 Check that the gaiter is not twisted or dented then secure it in position using new retaining clips.
6 Refit the locknut and outer balljoint onto the track rod end as described in Section 24.

21 Steering gear – removal, overhaul and refitting

Removal

1 Chock the rear wheels, apply the handbrake then jack up the front of the car and support it on axle stands. Remove both front roadwheels.
2 Extract the split pins, unscrew the retaining nuts and release the track rod outer balljoints from the swivel hub using a universal balljoint separator tool (photos).
3 Working in the engine compartment, remove the clip (where fitted) securing the rubber gaiter to the steering gear pinion housing.
4 Disengage the gaiter and undo the pinch-bolt securing the intermediate shaft universal joint to the steering gear pinion. Mark the position of the joint in relation to the pinion to use as a guide on refitting.

Manual steering

5 Undo the four bolts securing the steering gear mounting brackets to the engine compartment bulkhead.
6 Lift off the mounting brackets, release the pinion from the intermediate shaft universal joint and manoeuvre the steering gear sideways and out from under the wheel arch.

Power-assisted steering

7 Undo the fluid feed pipe union bolt, remove the bolt and sealing washers, then slacken the fluid return pipe union nut from the steering gear (photo). Be prepared for fluid spillage and position a suitable container beneath the pipes whilst unscrewing the union nuts. This fluid must be disposed of and new fluid of the type specified in Chapter 1 used when refilling. Plug the pipe ends and steering gear orifices to prevent excessive fluid leakage and the entry of dirt into the hydraulic system.
8 Remove the steering gear unit as described in paragraphs 4 to 6.

Overhaul

9 Complete overhaul of the steering gear is considered to be beyond the scope of the average home mechanic owing to the complexity of the

Chapter 10 Suspension and steering

Fig. 10.13 Power steering gear assembly – left-hand drive shown, right-hand drive similar (Sec 21)

1. Intermediate shaft pinch-bolt
2. Track rod end retaining nut and split pin
3. Track rod
4. Fluid return pipe
5. Fluid feed pipe
6. Mounting bolt
7. Steering gear assembly

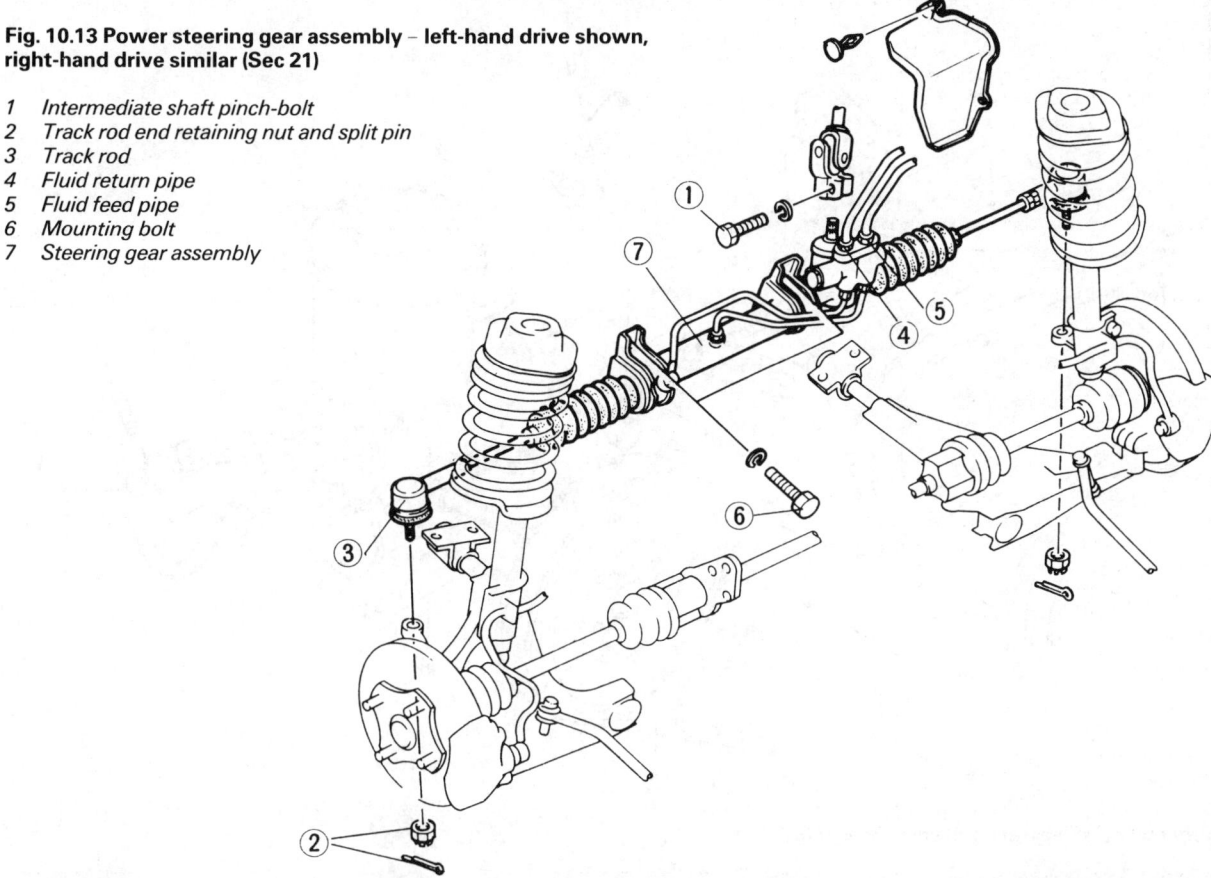

21.2A Track rod outer balljoint retaining nut and split pin

21.2B Using a balljoint separator tool to release the balljoint shank from the steering arm

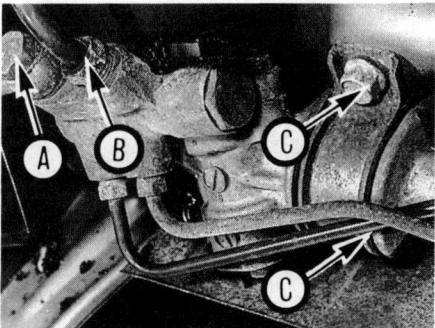

21.7 Fluid feed (A) and return (B) pipes and steering gear mounting bolts (C) on models equipped with power-assisted steering

unit and the need for special tools and expertise to complete the task satisfactorily. If the steering gear is worn to such an extent that overhaul is being considered, it is usually preferable to obtain an exchange reconditioned unit.

10 However, renewal of the steering gear rubber gaiters, track rod outer balljoints and track rods can be carried out reasonably easily and these operations are described in Sections 20, 24 and 25 respectively.

11 Examine the steering gear mounting rubbers for signs of damage or deterioration and renew if necessary.

Refitting

12 Refitting is the reverse of the removal sequence noting the following points.

(a) Tighten all nuts and bolts to the specified torque.
(b) Use new split pins to secure the track rod outer balljoint retaining nuts in position.
(c) After fitting the steering gear, centralize the steering wheel spokes so that they are level with the roadwheels in the straight-ahead setting.
(d) Check the front wheel toe setting as described in Section 26.
(e) On power-assisted steering units top up the level in the pump reservoir using the type of fluid specified in Chapter 1, and bleed the hydraulic system as described in Section 23.

22 Power steering pump – removal and refitting

Removal

1 Slacken the power steering pump mounting bolt and adjuster locknut, then rotate the adjuster bolt until the drivebelt tension is fully slackened (photos).

Chapter 10 Suspension and steering

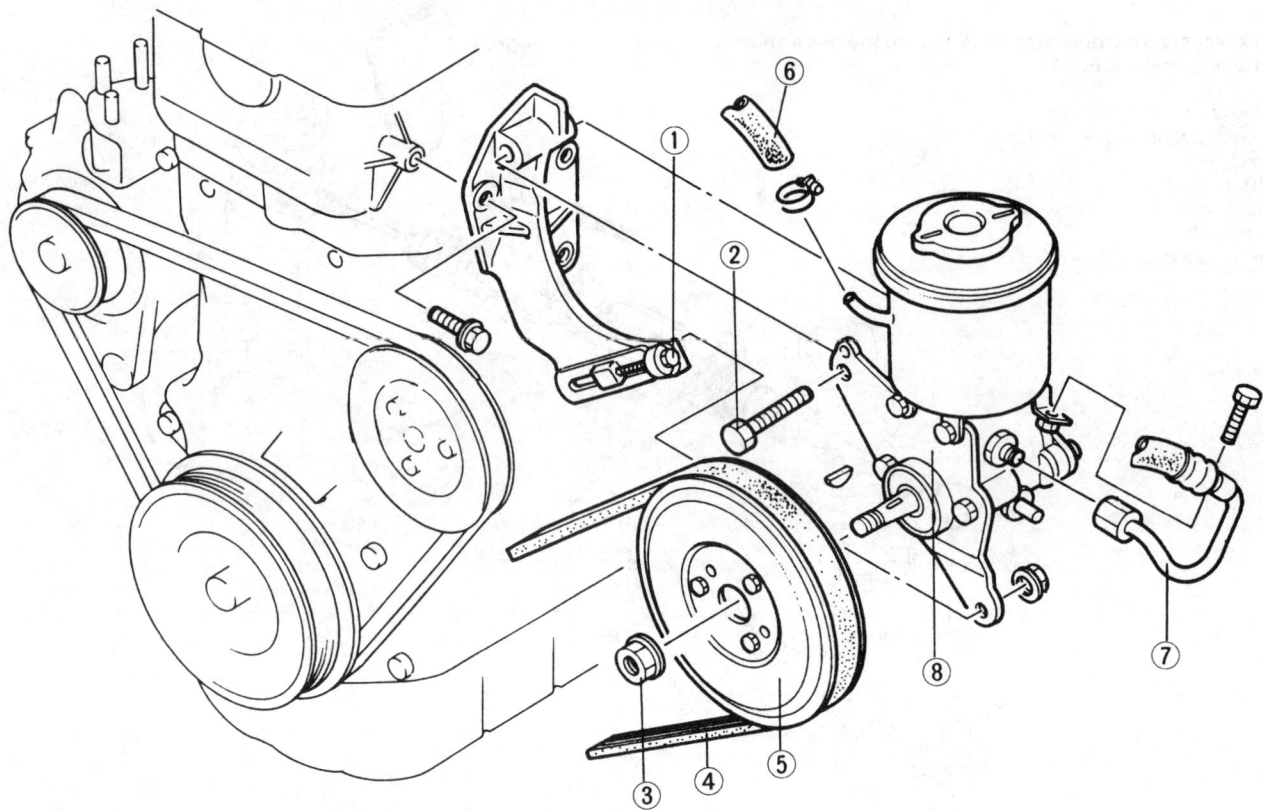

Fig. 10.14 Power steering pump and associated components (Sec 22)

1. Adjuster bolt
2. Mounting bolt
3. Pulley retaining nut
4. Drivebelt
5. Pulley
6. Fluid return hose
7. Fluid feed hose
8. Pump and reservoir assembly

22.1A Slacken the power steering mounting bolt...

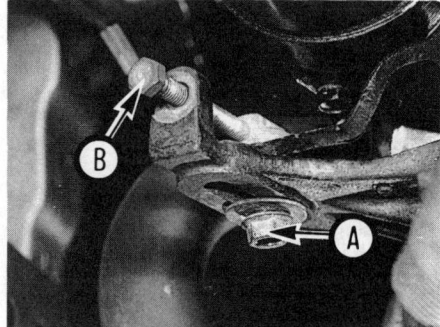

22.1B ...and adjuster locknut (A) then slacken drivebelt tension using the adjuster bolt (B)...

22.2 ...and remove the drivebelt

2 Disengage the drivebelt from the pump and crankshaft pulleys and remove it from the car (photo).

3 Loosen the clip which retains the hydraulic return pipe and disconnect the pipe from the side of the pump reservoir. Undo the feed pipe union nut and disconnect the pipe from the pump. Be prepared for fluid spillage and position a suitable container beneath the pipes as they are removed. Plug the connections to prevent excessive fluid loss and the possible ingress of dirt. Wipe any spilt fluid from the surrounding components.

4 Slacken and remove the pump mounting bolt and lower mounting nut and lift the pump clear (photo).

5 If necessary, undo the pump support bracket retaining bolts and remove the bracket from the engine.

6 Overhaul of the power steering pump is not a task for the average home mechanic and any repairs should therefore be entrusted to a Mazda dealer.

Refitting

7 Install the pump and refit its upper mounting bolt and lower mounting nut. Tighten both the nut and bolt finger tight only at this stage.

Chapter 10 Suspension and steering

22.4 Withdraw the pump mounting bolt and remove pump from the engine

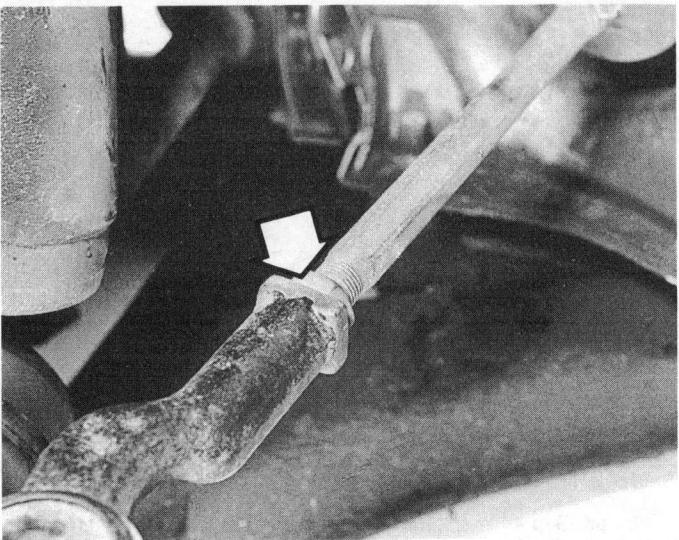

24.2 Track rod outer balljoint locknut (arrowed)

8 Ensure that the hydraulic connections are clean and reconnect the feed and return pipes to the pump assembly. Tighten the feed pipe union nut securely, and secure the return pipe in position with the retaining clip.
9 Refit the drivebelt to the crankshaft and pump pulleys and adjust the drivebelt tension as described in Chapter 1.
10 Top up the level in the pump reservoir with the type of fluid specified in Chapter 1, and bleed the hydraulic system as described in the following Section.

23 Power steering system – bleeding

1 Chock the rear wheels, apply the handbrake then jack up the front of the vehicle and support it on axle stands.
2 Check the fluid level in the power steering reservoir whilst turning the steering from lock to lock. Add more of the fluid specified in Chapter 1 if necessary to maintain the level whilst continuing to turn the steering. When the fluid level stabilizes, lower the vehicle to the ground.
3 Start and run the engine at idle speed, then turn the steering from lock to lock a few times. Check that the fluid level does not drop below the 'L' line on the level dipstick.
4 If when turning the steering an abnormal noise is heard from the fluid lines, it indicates that there is still air in the system. Check this by turning the wheels to the straight-ahead position and switching off the engine. If the fluid level in the reservoir rises, then air is present in the system and further bleeding is necessary.

24 Track rod outer balljoint – removal and refitting

Removal

1 Chock the rear wheels, firmly apply the handbrake then jack up the front of the car and support it on axle stands. Remove the appropriate front roadwheel.
2 Slacken the balljoint locknut by a quarter of a turn (photo).
3 Extract the split pin, then undo the nut securing the balljoint to the swivel hub.
4 Release the tapered shank of the balljoint from the swivel hub using a universal balljoint separator.
5 Using an open ended spanner to hold the track rod, unscrew the balljoint from the track rod end whilst counting the number of turns necessary to remove it.

6 If the locknut is to be removed, note the number of exposed threads behind the nut before unscrewing it from the track rod end.

Refitting

7 If removed, screw the locknut onto the track rod until the correct number of threads noted prior to removal are visible behind it.
8 Screw the balljoint onto the track rod the number of turns noted during removal until the locknut just contacts the balljoint.
9 Refit the balljoint to the swivel hub and refit the retaining nut. Tighten the retaining nut to the specified torque and secure it in position with a new split pin.
10 Tighten the track rod balljoint locknut against the balljoint.
11 Refit the roadwheel and lower the car to the ground.
12 Check the front wheel toe setting as described in Section 26.

25 Track rod – removal and refitting

Removal

1 Remove the steering gear from the car as described in Section 21.
2 Remove the track rod balljoint as described in Section 24.
3 Using pliers, release the rubber gaiter outer retaining clip and then cut off the inner clip. Slide the rubber gaiter off the steering gear and track rod.
4 Turn the steering gear pinion so that the rack is protruding as far as possible on the side on which the track rod is to be removed.
5 Clamp the exposed steering rack in a vice equipped with soft jaws.

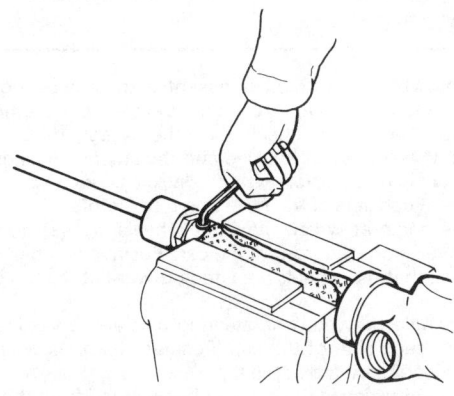

Fig. 10.15 Removing the track rod inner balljoint lock bolt – pre-September 1985 models (Sec 25)

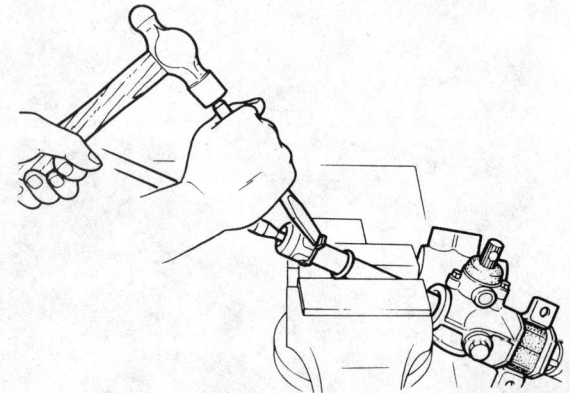

Fig. 10.16 Releasing the track rod balljoint lock washer tabs – September 1985 models onward (Sec 25)

6 On pre-September 1985 models, using a suitable Allen key, unscrew the lock bolt securing the track rod inner balljoint to the rack.
7 On September 1985 models onward, which are fitted with either power steering or variable gear type manual steering, bend the tabs of the lockwasher away from the track rod balljoint using a suitable flat bladed screwdriver and, if necessary, a hammer.
8 On all models the balljoint can then be unscrewed from the rack, using a suitable spanner, and the track rod removed.
9 If necessary, remove the lockwasher from the end of the steering rack.

Refitting

10 Before installing the new track rod assembly liberally lubricate the balljoint with lithium based grease, working it well into the balljoint seat.
11 On pre-September 1985 models, screw the balljoint into the rack and tighten it to the specified torque. Check that the lock bolt holes in the rack and pinion are aligned. Apply a thread locking compound to the lock bolt threads, fit the lock bolt and tighten it securely.
12 On September 1985 models onward fitted with variable gear ratio steering, fit a new lockwasher onto the balljoint, and models equipped with power-assisted steering fit a new lockwasher and damper ring onto the balljoint threads. On all models screw the balljoint into the steering rack and tighten the balljoint to the specified torque. On variable gear and power steering models then stake the balljoint lockwasher into the steering rack groove using a hammer and suitable punch.
13 Refit the rubber gaiter, ensuring that it is correctly located in the grooves on the steering gear housing and track rod. Secure the gaiter in position with new retaining clips.
14 Refit the outer balljoint as described in Section 24.
15 Refit the steering gear to the car as described in Section 21.

26 Wheel alignment and steering angles – general information

1 Accurate front wheel alignment is essential to provide positive steering and prevent excessive tyre wear. Before considering the steering/suspension geometry, check that the tyres are correctly inflated, the front wheels are not buckled and the steering linkage and suspension joints are in good order, without slackness or wear.
2 Wheel alignment consists of the following four factors.
Camber is the angle at which the front wheels are set from the vertical when viewed from the front of the car. 'Positive camber' is the amount (in degrees) that the wheels are tilted outward at the top of the vertical.
Castor is the angle between the steering axis and a vertical line when viewed from each side of the car. 'Positive castor' is when the steering axis is inclined rearward at the top.
Steering axis inclination is the angle (when viewed from the front of the car) between the vertical and an imaginary line drawn through the suspension strut upper mounting and the lower suspension arm balljoint.

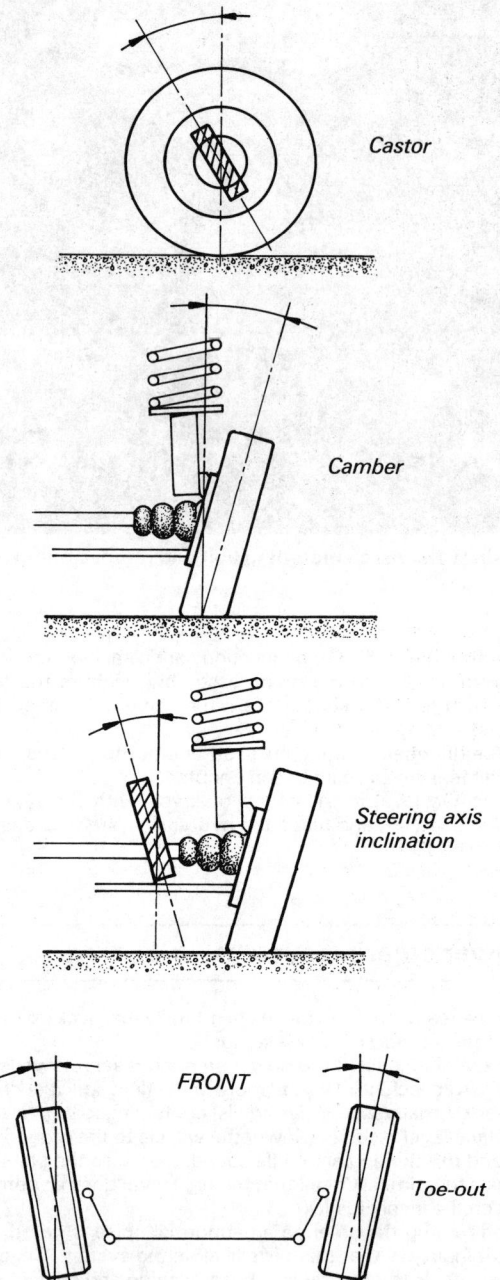

Fig. 10.17 Wheel alignment and steering angles (Sec 26)

Toe setting is the amount by which the distance between the front inside edges of the roadwheels (measured at hub height) differs from the diametrically opposite distance measured between the rear inside edges of the front roadwheels.
3 With the exception of the front and rear toe setting, and front camber angle all other steering angles are set during manufacture and no adjustment is possible. It can be assumed, therefore, that unless the car has suffered accident damage all the preset steering angles will be correct. Should there be some doubt about their accuracy it will be necessary to seek the help of a Mazda dealer, as special gauges are needed to check the steering angles. Where adjustment is necessary proceed as follows.

Front toe setting

4 Two methods are available to the home mechanic for checking the toe setting. One method is to use a gauge to measure the distance between the front and rear inside edges of the roadwheels. The other

Chapter 10 Suspension and steering

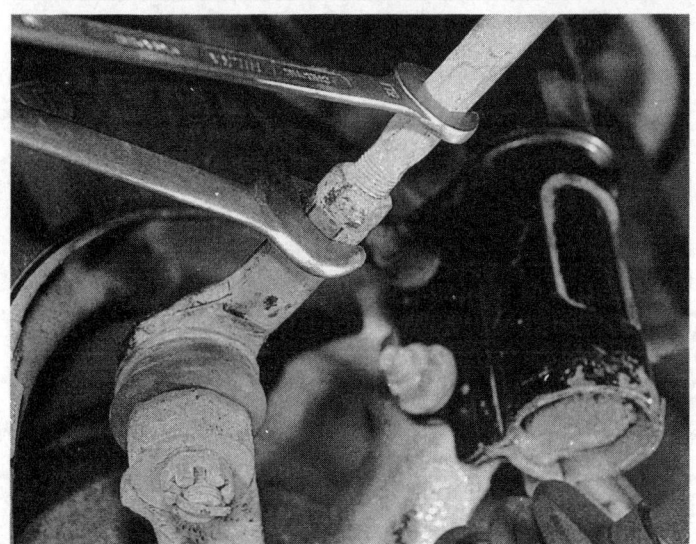

26.7 Adjusting the front wheel toe setting

26.10 Rear lateral link adjuster (A) and locknuts (B) on late Estate models

method is to use a scuff plate in which each front wheel is rolled across a movable plate which records any deviation, or scuff, of the tyre from the straight-ahead position as it moves across the plate. Relatively inexpensive equipment of both types is available from accessory outlets to enable these checks, and subsequent adjustments to be carried out at home.

5 If, after checking the toe setting using whichever method is preferable, it is found that adjustment is necessary, proceed as follows.
6 Turn the steering wheel onto full left lock and record the number of exposed threads on the right-hand track rod end. Now turn the steering onto full right lock and record the number of threads on the left-hand side. If there are the same number of threads visible on both sides then subsequent adjustment can be made equally on both sides. If there are more threads visible on one side than the other it will be necessary to compensate for this during adjustment. *After adjustment there must be the same number of threads visible on each track rod end. This is most important.*
7 To alter the toe setting, slacken the locknut on the track rod end and turn the track rod using a self-grip wrench to achieve the desired setting (photo). When viewed from the side of the car, turning the rod clockwise will increase the toe-in, turning it anti-clockwise will increase the toe-out. Only turn the track rods by a quarter of a turn each time and then recheck the setting using the gauges, or scuff plate.
8 After adjustment tighten the locknuts and reposition the steering gear rubber gaiter to remove any twist caused by turning the track rods.

Rear toe setting

9 The procedure for checking the rear toe setting is the same as described in paragraph 4 except that the equipment is applied to the rear wheels.
10 On later Estate models, to adjust the rear toe setting, loosen off the lateral arm adjuster locknut(s) and then turn the adjuster(s) in the required direction until the required setting is obtained (photo). This is given in the Specifications at the start of this Chapter. The rear lateral link arm lengths must be kept the same on either side. To increase the toe setting, turn the right-hand link adjuster clockwise and the left-hand link adjuster anti-clockwise, and vice versa to reduce the setting. Once the rear toe setting is correct tighten the adjuster locknuts to the specified torque setting.
11 On all other models adjustment of the toe setting of each rear wheel is achieved using an eccentric spacer located behind the rear lateral link to crossmember retaining nut. The washer has a number of notches around its periphery which engage with a lug on the lateral link.
12 If adjustment is necessary, slacken the retaining nut and turn the washer one notch at a time in whichever direction is necessary to obtain the correct setting. Note that one notch of the washer is equal to 2.5 mm of toe setting adjustment. After repositioning the washer, tighten the retaining nut and recheck the setting. Repeat the procedure until the setting is as specified. Tighten the lateral link retaining nut to the specified torque after the final check.

Front camber angle

13 The front camber angle can be set to one of two positions by turning the front suspension strut upper mounting through 180°.
14 If, after checking the front wheel camber using suitable gauges, it is found to be incorrect, adjustment may be carried out as follows.
15 Chock the rear wheels, apply the handbrake then jack up the front of the car and support it on axle stands.
16 From within the engine compartment undo the two strut upper mounting retaining nuts. Lower the strut slightly, turn the mounting 180°, push the strut back up, and refit the two nuts. Tighten the nuts to the specified torque. If the mounting is turned so that the triangular arrow reference mark on the upper face of the mounting rubber is turned from facing out to facing in, the camber alters in the positive direction. If the mark is turned from facing in to facing out, the change is to the negative.
17 Lower the car to the ground, bounce the suspension to settle components in position and recheck the settings.

Chapter 11 Bodywork and fittings

Contents

Body exterior trim strips – general information	26
Bonnet – removal, refitting and adjustment	9
Bonnet lock – removal and refitting	11
Bonnet release cable – removal and refitting	10
Boot lid – removal, refitting and adjustment	18
Boot lid lock – removal and refitting	19
Boot lid lock cylinder – removal and refitting	20
Boot lid/tailgate and fuel filler internal release cables – removal and refitting	21
Centre console – removal and refitting	31
Door – removal, refitting and adjustment	16
Door inner trim panel – removal and refitting	12
Door lock, lock cylinder and handles – removal and refitting	15
Door window glass and regulator – removal and refitting	13
Electric sunroof – general information	30
Exterior mirror and glass – removal and refitting	17
Facia – removal and refitting	32
Front bumper – removal and refitting	6
General information	1
Interior trim panels – general information	28
Maintenance – bodywork and underframe	2
Maintenance – upholstery and carpets	3
Major body damage – repair	5
Minor body damage – repair	4
Quarter window glass (Hatchback models) – removal and refitting	14
Radiator grille – removal and refitting	8
Rear bumper – removal and refitting	7
Seats – removal and refitting	27
Seat belts – removal and refitting	29
Tailgate – removal, refitting and adjustment	22
Tailgate lock and lock cylinder – removal and refitting	24
Tailgate support strut – removal and refitting	23
Underbody and general body check	See Chapter 1
Windscreen and rear window/tailgate glass – general information	25

1 General information

The bodyshell and underframe is of all-steel welded construction, incorporating progressive crumple zones at the front and rear and a rigid centre safety cell. The assembly and welding of the main body unit is completed by computer-controlled robots, and is checked for dimensional accuracy using computer and laser technology.

To facilitate accident damage repair many of the main body panels are supplied as part panel replacements and are bolted rather than welded in place, particularly at the front of the vehicle.

2 Maintenance – bodywork and underframe

The general condition of a vehicle's bodywork is the one thing that significantly affects its value. Maintenance is easy but needs to be regular. Neglect, particularly after minor damage, can lead quickly to further deterioration and costly repair bills. It is important also to keep watch on those parts of the vehicle not immediately visible, for instance the underside, inside all the wheel arches and the lower part of the engine compartment.

The basic maintenance routine for the bodywork is washing – preferably with a lot of water, from a hose. This will remove all the loose solids which may have stuck to the vehicle. It is important to flush these off in such a way as to prevent grit from scratching the finish. The wheel arches and underframe need washing in the same way to remove any accumulated mud which will retain moisture and tend to encourage rust. Paradoxically enough, the best time to clean the underframe and wheel arches is in wet weather when the mud is thoroughly wet and soft. In very wet weather the underframe is usually cleaned of large accumulations automatically and this is a good time for inspection.

Periodically, except on vehicles with a wax-based underbody protective coating, it is a good idea to have the whole of the underframe of the vehicle steam cleaned, engine compartment included, so that a thorough inspection can be carried out to see what minor repairs and renovations are necessary. Steam cleaning is available at many garages and is necessary for the removal of the accumulation of oily grime which sometimes is allowed to become thick in certain areas. If steam cleaning facilities are not available, there are one or two excellent grease solvents available such as Holts Engine Cleaner or Holts Foambrite, which can be brush applied. The dirt can then be simply hosed off. Note that these methods should not be used on vehicles with wax-based underbody protective coating or the coating will be removed. Such vehicles should be inspected annually, preferably just prior to winter, when the underbody should be washed down and any damage to the wax coating repaired using Holts Undershield. Ideally, a completely fresh coat should be applied. It would also be worth considering the use of such wax-based protection for injection into door panels, sills, box sections, etc, as an additional safeguard against rust damage where such protection is not provided by the vehicle manufacturer.

After washing paintwork, wipe off with a chamois leather to give an unspotted clear finish. A coat of clear protective wax polish like the many excellent Turtle Wax polishes, will give added protection against chemical pollutants in the air. If the paintwork sheen has dulled or oxidised, use a cleaner/polisher combination such as Turtle Extra to restore the brilliance of the shine. This requires a little effort, but such dulling is usually caused because regular washing has been neglected. Care needs to be taken with metallic paintwork, as special non-abrasive cleaner/polisher is required to avoid damage to the finish. Always check that the door and ventilator opening drain holes and pipes are completely clear so that water can be drained out. Brightwork should be treated in the same way as paintwork. Windscreens and windows can be kept clear of the smeary film which often appears by the use of proprietary glass cleaner like Holts Mixra. Never use any form of wax or other body or chromium polish on glass.

Chapter 11 Bodywork and fittings

3 Maintenance – upholstery and carpets

Mats and carpets should be brushed or vacuum cleaned regularly to keep them free of grit. If they are badly stained remove them from the vehicle for scrubbing or sponging and make quite sure they are dry before refitting. Seats and interior trim panels can be kept clean by wiping with a damp cloth and Turtle Wax Carisma. If they do become stained (which can be more apparent on light coloured upholstery) use a little liquid detergent and a soft nail brush to scour the grime out of the grain of the material. Do not forget to keep the headlining clean in the same way as the upholstery. When using liquid cleaners inside the vehicle do not over-wet the surfaces being cleaned. Excessive damp could get into the seams and padded interior causing stains, offensive odours or even rot. If the inside of the vehicle gets wet accidentally it is worthwhile taking some trouble to dry it out properly, particularly where carpets are involved. *Do not leave oil or electric heaters inside the vehicle for this purpose.*

4 Minor body damage – repair

Note: *For more detailed information about bodywork repair, the Haynes Publishing Group publish a book by Lindsay Porter called The Car Bodywork Repair Manual. This incorporates information on such aspects as rust treatment, painting and glass-fibre repairs, as well as details on more ambitious repairs involving welding and panel beating.*

The colour bodywork repair photographic sequences between pages 32 and 33 illustrate the operations detailed in the following sub-sections.

Repairs of minor scratches in bodywork

If the scratch is very superficial, and does not penetrate to the metal of the bodywork, repair is very simple. Lightly rub the area of the scratch with a paintwork renovator like Turtle Wax New Color Back, or a very fine cutting paste like Holts Body + Plus Rubbing Compound, to remove loose paint from the scratch and to clear the surrounding bodywork of wax polish. Rinse the area with clean water.

Apply touch-up paint such as Holts Dupli-Color Touch or a paint film like Holts Autofilm, to the scratch using a fine paint brush; continue to apply fine layers of paint until the surface of the paint in the scratch is level with the surrounding paintwork. Allow the new paint at least two weeks to harden, then blend it into the surrounding paintwork by rubbing the scratch area with a paintwork renovator or a very fine cutting paste such as Holts Body + Plus Rubbing Compound or Turtle Wax New Color Back. Finally apply wax polish from one of the Turtle wax range of wax polishes.

Where the scratch has penetrated right through to the metal of the bodywork, causing the metal to rust, a different repair technique is required. Remove any loose rust from the bottom of the scratch with a penknife, then apply rust inhibiting paint such as Turtle Wax Rust Master, to prevent the formation of rust in the future. Using a rubber or nylon applicator fill the scratch with bodystopper paste like Holts Body + Plus Knifing Putty. If required, this paste can be mixed with cellulose thinners such as Holts Body + Plus Cellulose Thinners, to provide a very thin paste which is ideal for filling narrow scratches. Before the stopper-paste in the scratch hardens, wrap a piece of smooth cotton rag around the top of a finger. Dip the finger in cellulose thinners and quickly sweep it across the surface of the stopper-paste in the scratch; this will ensure that the surface of the stopper-paste is slightly hollowed. The scratch can now be painted over as described earlier in this Section.

Repairs of dents in bodywork

When deep denting of the vehicle's bodywork has taken place, the first task is to pull the dent out, until the affected bodywork almost attains its original shape. There is little point in trying to restore the original shape completely, as the metal in the damaged area will have stretched on impact and cannot be reshaped fully to its original contour. It is better to bring the level of the dent up to a point which is about 3 mm below the level of the surrounding bodywork. In cases where the dent is very shallow anyway, it is not worth trying to pull it out at all. If the underside of the dent is accessible, it can be hammered out gently from behind, using a mallet with a wooden or plastic head. Whilst doing this, hold a suitable block of wood firmly against the outside of the panel to absorb the impact from the hammer blows and thus prevent a large area of the bodywork from being 'belled-out'.

Should the dent be in a section of the bodywork which has a double skin or some other factor making it inaccessible from behind, a different technique is called for. Drill several small holes through the metal inside the area – particularly in the deeper section. Then screw long self-tapping screws into the holes just sufficiently for them to gain a good purchase in the metal. Now the dent can be pulled out by pulling on the protruding heads of the screws with a pair of pliers.

The next stage of the repair is the removal of the paint from the damaged area, and from an inch or so of the surrounding 'sound' bodywork. This is accomplished most easily by using a wire brush or abrasive pad on a power drill, although it can be done just as effectively by hand using sheets of abrasive paper. To complete the preparation for filling, score the surface of the bare metal with a screwdriver or the tang of a file, or alternatively, drill small holes in the affected area. This will provide a really good 'key' for the filler paste.

To complete the repair see the Section on filling and respraying.

Repairs of rust holes or gashes in bodywork

Remove all paint from the affected area and from an inch or so of the surrounding 'sound' bodywork, using an abrasive pad or a wire brush on a power drill. If these are not available a few sheets of abrasive paper will do the job most effectively. With the paint removed you will be able to judge the severity of the corrosion and therefore decide whether to renew the whole panel (if this is possible) or to repair the affected area. New body panels are not as expensive as most people think and it is often quicker and more satisfactory to fit a new panel than to attempt to repair large areas of corrosion.

Remove all fittings from the affected area except those which will act as a guide to the original shape of the damaged bodywork (eg headlamp shells etc). Then, using tin snips or a hacksaw blade, remove all loose metal and any other metal badly affected by corrosion. Hammer the edges of the hole inwards in order to create a slight depression for the filler paste.

Wire brush the affected area to remove the powdery rust from the surface of the remaining metal. Paint the affected area with rust inhibiting paint such as Turtle Wax Rust Master; if the back of the rusted area is accessible treat this also.

Before filling can take place it will be necessary to block the hole in some way. This can be achieved by the use of aluminium or plastic mesh, or aluminium tape.

Aluminium or plastic mesh or glass fibre matting, such as the Holts Body + Plus Glass Fibre Matting, is probably the best material to use for a large hole. Cut a piece to the approximate size and shape of the hole to be filled, then position it in the hole so that its edges are below the level of the surrounding bodywork. It can be retained in position by several blobs of filler paste around its periphery.

Aluminium tape should be used for small or very narrow holes. Pull a piece off the roll and trim it to the approximate size and shape required, then pull off the backing paper (if used) and stick the tape over the hole; it can be overlapped if the thickness of one piece is insufficient. Burnish down the edges of the tape with the handle of a screwdriver or similar, to ensure that the tape is securely attached to the metal underneath.

Bodywork repairs – filling and respraying

Before using this Section, see the Sections on dent, deep scratch, rust holes and gash repairs.

Many types of bodyfiller are available, but generally speaking those proprietary kits which contain a tin of filler paste and a tube of resin hardener are best for this type of repair like Holts Body + Plus or Holts No Mix which can be used directly from the tube. A wide, flexible plastic or nylon applicator will be found invaluable for imparting a smooth and well contoured finish to the surface of the filler.

Mix up a little filler on a clean piece of card or board – measure the hardener carefully (follow the maker's instructions on the pack) otherwise the filler will set too rapidly or too slowly. Alternatively, Holts No Mix can be used straight from the tube without mixing, but daylight is required to cure it. Using the applicator apply the filler paste to the prepared area; draw the applicator across the surface of the filler to achieve the correct contour and to level the surface. As soon as a contour that approximates to the correct one is achieved, stop working the paste – if you carry on too long the paste will become sticky and begin to 'pick-up' on the applicator. Continue to add thin layers of filler paste at twenty minute intervals until the level of the filler is just proud of the surrounding bodywork.

Once the filler has hardened, excess can be removed using a metal plane or file. From then on, progressively finer grades of abrasive paper should be used, starting with a 40 grade production paper and finishing with a 400 grade wet-and-dry paper. Always wrap the abrasive paper around a flat rubber, cork, or wooden block – otherwise the surface of the filler will not be completely flat. During the smoothing of the filler surface the wet-and-dry paper should be periodically rinsed in water. This will ensure that a very smooth finish is imparted to the filler at the final stage.

At this stage the 'dent' should be surrounded by a ring of bare metal, which in turn should be encircled by the finely 'feathered' edge of the good paintwork. Rinse the repair area with clean water, until all of the dust produced by the rubbing-down operation has gone.

Spray the whole area with a light coat of primer, either Holts Body+Plus Grey or Red Oxide Primer – this will show up any imperfections in the surface of the filler. Repair these imperfections with fresh filler paste or bodystopper, and once more smooth the surface with abrasive paper. If bodystopper is used, it can be mixed with cellulose thinners to form a really thin paste which is ideal for filling small holes. Repeat this spray and repair procedure until you are satisfied that the surface of the filler, and the feathered edge of the paintwork are perfect. Clean the repair area with clean water and allow to dry fully.

The repair area is now ready for final spraying. Paint spraying must be carried out in a warm, dry, windless and dust free atmosphere. This condition can be created artificially if you have access to a large indoor working area, but if you are forced to work in the open, you will have to pick your day very carefully. If you are working indoors, dousing the floor in the work area with water will help to settle the dust which would otherwise be in the atmosphere. If the repair area is confined to one body panel, mask off the surrounding panels; this will help to minimise the effects of a slight mis-match in paint colours. Bodywork fittings (eg chrome strips, door handles etc) will also need to be masked off. Use genuine masking tape and several thicknesses of newspaper for the masking operations.

Before commencing to spray, agitate the aerosol can thoroughly, then spray a test area (an old tin, or similar) until the technique is mastered. Cover the repair area with a thick coat of primer; the thickness should be built up using several thin layers of paint rather than one thick one. Using 400 grade wet-and-dry paper, rub down the surface of the primer until it is really smooth. While doing this, the work area should be thoroughly doused with water, and the wet-and-dry paper periodically rinsed in water. Allow to dry before spraying on more paint.

Spray on the top coat using Holts Dupli-color Autospray, again building up the thickness by using several thin layers of paint. Start spraying in the centre of the repair area and then, using a circular motion, work outwards until the whole repair area and about 2 inches of the surrounding original paintwork is covered. Remove all masking material 10 to 15 minutes after spraying on the final coat of paint.

Allow the new paint at least two weeks to harden, then, using a paintwork renovator or a very fine cutting paste such as Turtle Wax New Color Back or Holts Body+Plus Rubbing Compound, blend the edges of the paint into the existing paintwork. Finally, apply wax polish.

Plastic components

With the use of more and more plastic body components by the vehicle manufacturers (eg bumpers, spoilers, and in some cases major body panels), rectification of more serious damage to such items has become a matter of either entrusting repair work to a specialist in this field, or renewing complete components. Repair of such damage by the DIY owner is not really feasible owing to the cost of the equipment and materials required for effecting such repairs. The basic technique involves making a groove along the line of the crack in the plastic using a rotary burr in a power drill. The damaged part is then welded back together by using a hot air gun to heat up and fuse a plastic filler rod into the groove. Any excess plastic is then removed and the area rubbed down to a smooth finish. It is important that a filler rod of the correct plastic is used, as body components can be made of a variety of different types (eg polycarbonate, ABS, polypropylene).

Damage of a less serious nature (abrasions, minor cracks etc) can be repaired by the DIY owner using a two-part epoxy filler repair material such as Holts Body+Plus or Holts No Mix which can be used directly from the tube. Once mixed in equal proportions (or applied directly from the tube in the case of Holts No Mix), this is used in similar fashion to the bodywork filler used on metal panels. The filler is usually cured in twenty to thirty minutes, ready for sanding and painting.

If the owner is renewing a complete component himself, or if he has repaired it with epoxy filler, he will be left with the problem of finding a suitable paint for finishing which is compatible with the type of plastic used. At one time the use of a universal paint was not possible owing to the complex range of plastics encountered in body component applications. Standard paints, generally speaking, will not bond to plastic or rubber satisfactorily, but Holts Professional Spraymatch paints to match any plastic or rubber finish can be obtained from dealers. However, it is now possible to obtain a plastic body parts finishing kit which consists of a pre-primer treatment, a primer and coloured top coat. Full instructions are normally supplied with a kit, but basically the method of use is to first apply the pre-primer to the component concerned and allow it to dry for up to 30 minutes. Then the primer is applied and left to dry for about an hour before finally applying the special coloured top coat. The result is a correctly coloured component where the paint will flex with the plastic or rubber, a property that standard paint does not normally posses.

5 Major body damage – repair

Where serious damage has occurred, or large areas need renewal due to neglect, it means that complete new panels will need welding in, and this is best left to professionals. If the damage is due to impact, it will also be necessary to check completely the alignment of the bodyshell, and this can only be carried out accurately by a Mazda dealer using special jigs. If the body is left misaligned, it is primarily dangerous as the car will not handle properly, and secondly, uneven stresses will be imposed on the steering, suspension and possibly transmission, causing abnormal wear, or complete failure, particularly to such items as the tyres.

6 Front bumper – removal and refitting

Removal

Pre-September 1985 models

1 Undo the nuts accessed from the rear of the bumper, and remove the four upper and four lower bolts which secure the bumper to its mounting brackets.

2 Release the bumper wrap-around sections from the side mountings and lift the bumper away from the car.

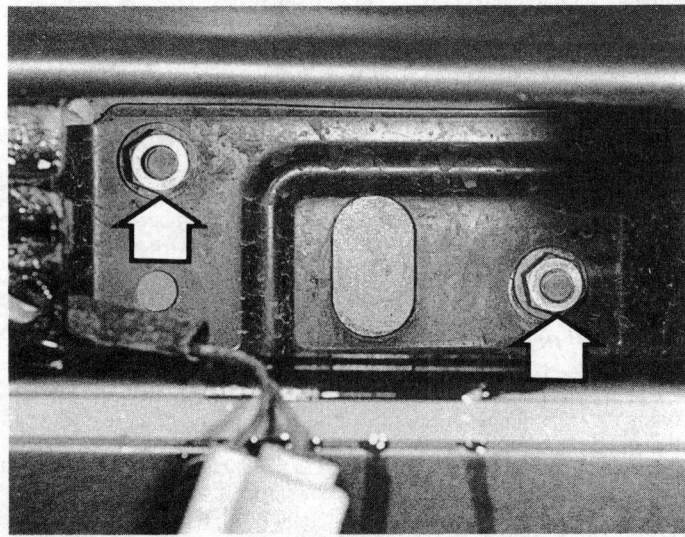

6.6 Front bumper mounting nuts (arrowed) – September 1985 models onward

Chapter 11 Bodywork and fittings

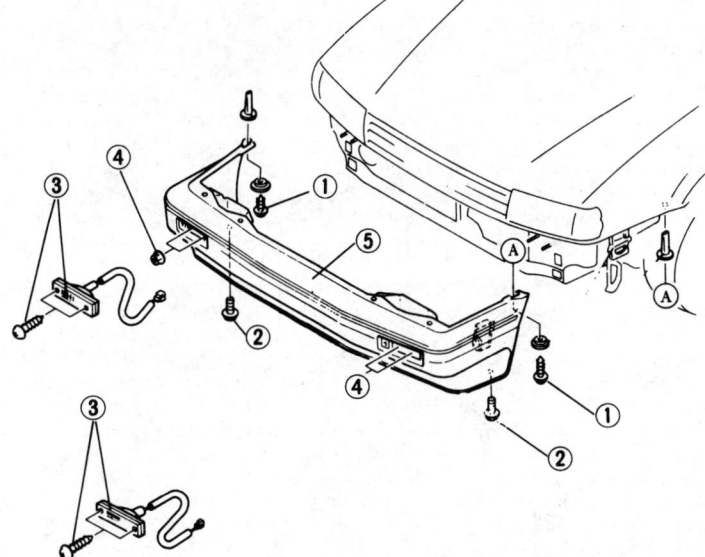

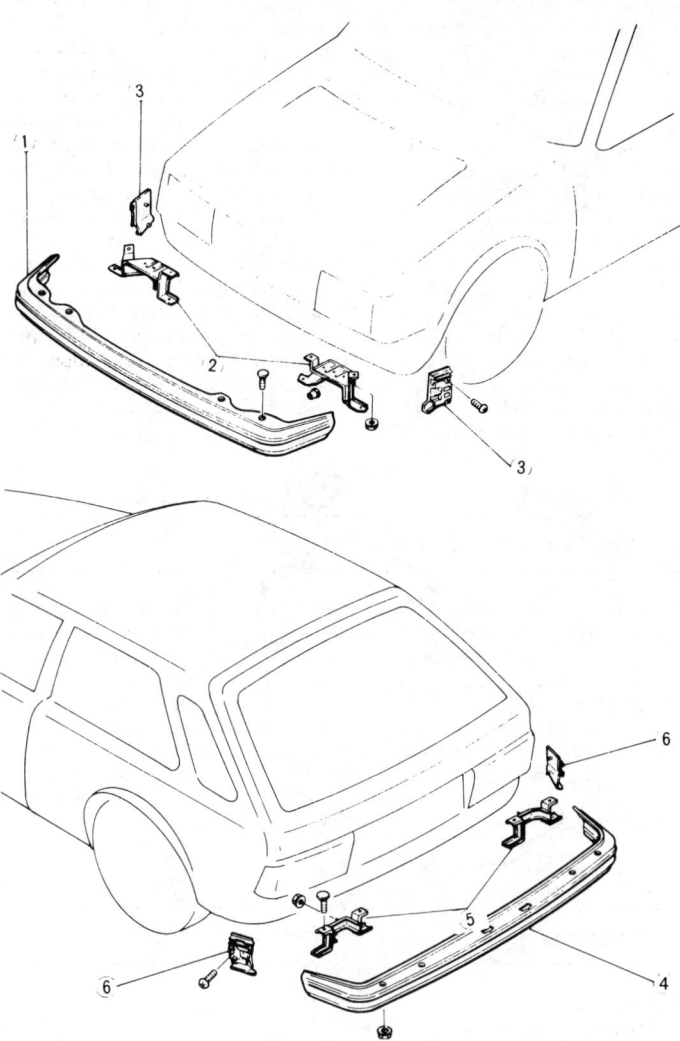

Fig. 11.1 Front and rear bumper fittings – pre-September 1985 models (Secs 6 and 7)

1. Front bumper
2. Mounting brackets
3. Side mountings
4. Rear bumper
5. Mounting brackets
6. Side mountings

3 If necessary, undo the nuts and bolts and remove the bumper mounting brackets from the vehicle.

September 1985 models onward

4 Undo the screws securing the left and right-hand side wrap-around sections of the bumper to the wings. Screws are accessed from the underside of the bumper.
5 Undo the screws securing the front turn signal lamps to the bumper. Partially withdraw the turn signal lamps then disconnect the wiring connectors and remove both lamps.
6 Undo the four nuts securing the bumper mounting brackets to the car, these can be accessed through the turn signal lamp cutouts (photo).
7 Release the bumper wrap-around sections from the side mountings and remove the bumper from vehicle.
8 If necessary, undo the nuts and remove the mounting brackets from the bumper.

Refitting

9 Refitting is a reversal of the removal procedure.

Fig. 11.2 Front bumper fittings – September 1985 models onward (Sec 6)

1. Screw
2. Screw
3. Turn signal lamps and retaining screws
4. Bumper retaining nuts
5. Front bumper

7 Rear bumper – removal and refitting

Removal

Pre-September 1985 models

1 Prise the number plate lamp out of the rear bumper, then disconnect the wiring connector and remove the lamp.
2 Remove the rear bumper using the information given in Section 6, paragraphs 1 to 3.

September 1985 onward Saloon and Hatchback models

3 Undo the screws securing the left and right-hand side wrap-around sections of the bumper to the wings. Screws are accessed from the underside of the bumper.
4 Open up the boot/tailgate, and from inside the car undo the four nuts securing the bumper mountings to the vehicle body.
5 Release the bumper wrap-around sections from the side mountings and remove the assembly from the vehicle, remembering to disconnect the number plate lamp wiring connector.

May 1986 to October 1989 Estate models

6 Chock the front wheels, jack up the rear of the car and support it on axle stands.
7 Undo the screws securing the left and right-hand side wrap-around sections of the bumper to the body. Screws are accessed from the underside of the bumper.
8 From underneath the car undo the four retaining bolts securing the bumper mountings to the underside of the vehicle.
9 Release the bumper wrap-around sections from the side mountings and remove the assembly from the vehicle. If necessary dismantle the bumper assembly as follows.
10 Undo the four nuts securing each mounting to the bumper and remove both mountings. Carefully prise out the bumper trim, then undo the four nuts securing each wrap-around side section to the bumper centre section. Separate the bumper sections noting the correct positions of the mounting plates which retain the side sections. Reassemble the bumper by reversing the dismantling sequence.

248 Chapter 11 Bodywork and fittings

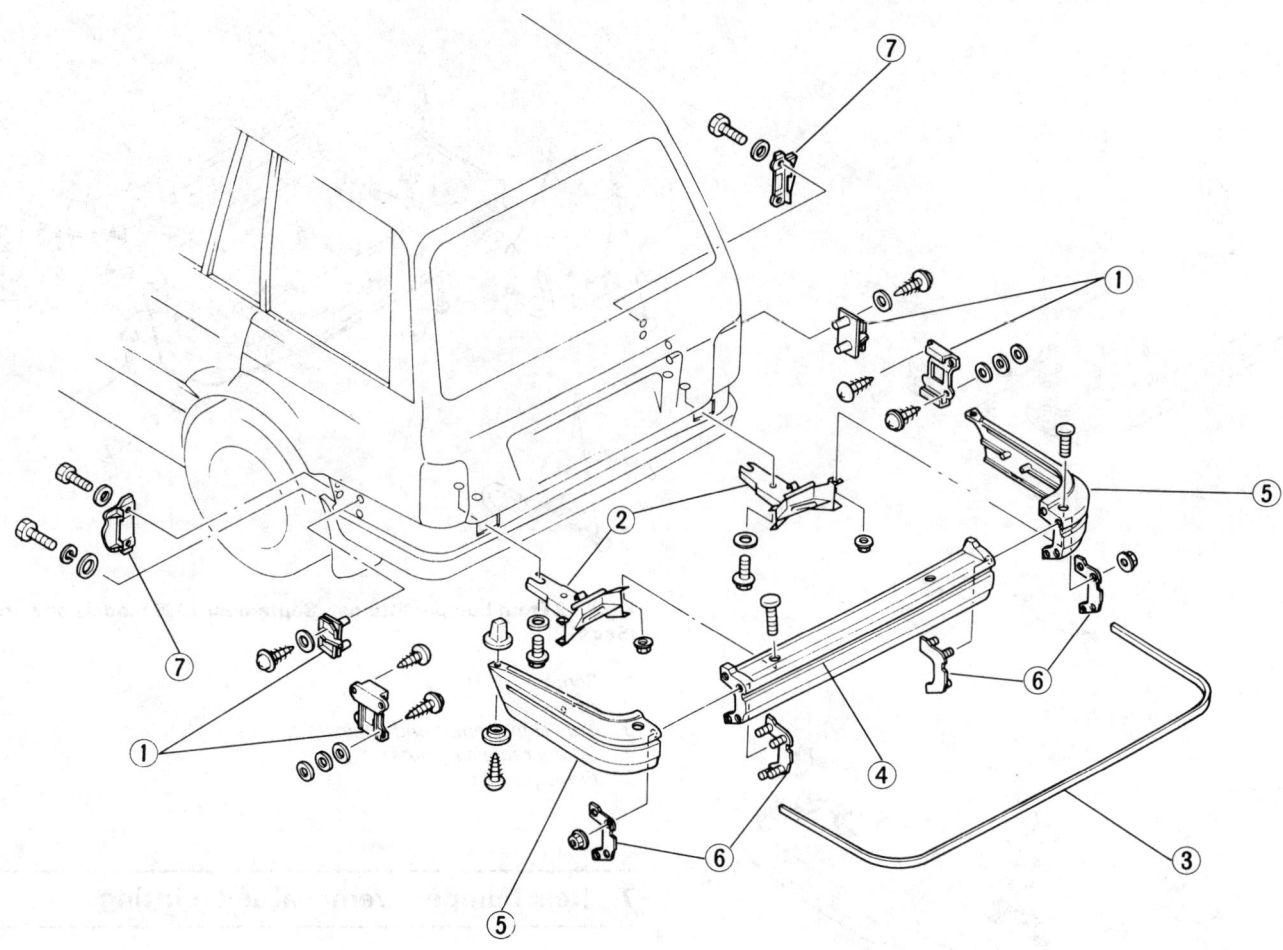

Fig. 11.3 Exploded view of the rear bumper and fittings – pre-October 1989 Estate models (Sec 7)

| 1 Side mountings | 3 Bumper trim | 5 Side sections | 7 Trim panel bracket |
| 2 Bumper mounting brackets | 4 Centre section | 6 Mounting plates and nuts | |

October 1989 onward Estate models

11 Chock the front wheels, jack up the rear of the car and support it on axle stands.
12 Undo the four screws securing the bumper wrap-around sections to the body. These screws are accessed from the underside of the bumper.
13 From underneath the car undo the four retaining bolts securing the bumper mountings to the underside of the vehicle.
14 Remove the bumper assembly from the vehicle and, if necessary, dismantle as follows.
15 Undo the two bolts securing the support bar to the left and right-hand mounting brackets. Undo the centre mounting bracket nut and remove the support bar. The mounting brackets, and lower skirt can then be removed from the bumper once their retaining nuts or bolts (as applicable) have been undone. Reassemble the bumper by reversing the dismantling sequence.

Refitting

16 Refitting is a reversal of the removal procedure.

8 Radiator grille – removal and refitting

Removal

1 Where fitted, remove the single screw securing the centre of the grille to the body.
2 Insert a screwdriver into the centre of one of the grille retaining clips, and carefully spread the clip ears as you pull outward on the grille. Repeat this procedure on all the remaining clips until the grille comes free and can be lifted away.

Refitting

3 Gently push the grille, with all its retaining clips, inwards until all the clips are engaged and the grille has snapped into place.
4 Refit the grille retaining screw (where fitted) and tighten it securely.

9 Bonnet – removal, refitting and adjustment

Removal

Pre-September 1985 models

1 With the bonnet open, undo the two bolts securing the stay to the bonnet inner panel (photo).
2 With an assistant supporting the bonnet, extract the two hinge pin retaining clips and remove the washers (photo).
3 Withdraw the two hinge pins and remove the bonnet from the car. Store the bonnet out of the way in a safe place (photo).

September 1985 models onward

4 Raise the bonnet and get an assistant to support it, then using a If necessary, loosen of the hinge bolts and realign the bonnet to suit,

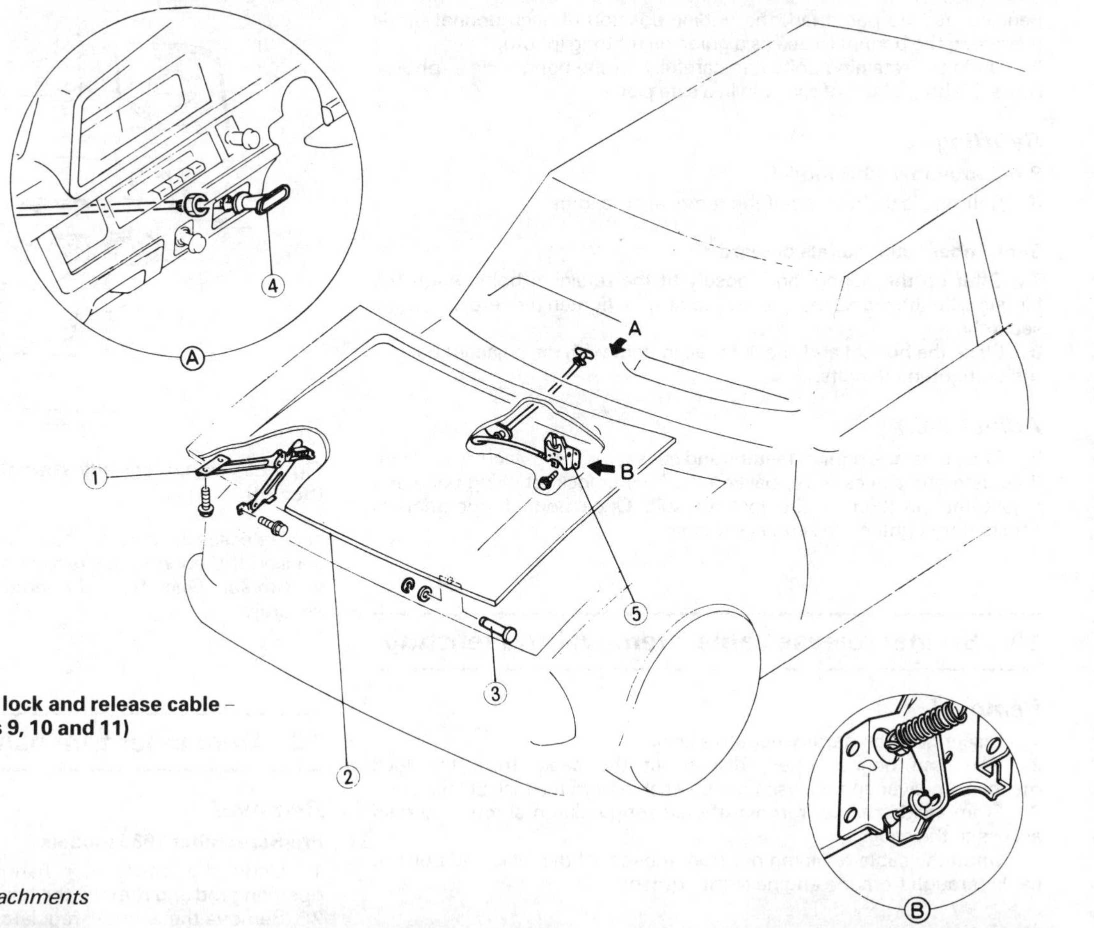

Chapter 11 Bodywork and fittings

Fig. 11.4 Layout of bonnet, bonnet lock and release cable – pre-September 1985 models (Secs 9, 10 and 11)

1 Bonnet stay
2 Bonnet
3 Hinge pin
4 Release cable knob
5 Bonnet lock

Inset A and B show release cable attachments

9.1 Remove the bonnet stay retaining bolts

9.2 Extract the hinge pin retaining clip (arrowed)...

9.3 ...then withdraw the hinge pin and remove the bonnet

9.4 Support the bonnet with the help of an assistant...

9.5 ...then undo the bonnet retaining bolts

250 Chapter 11 Bodywork and fittings

pencil or felt tip pen, mark the outline position of each bonnet hinge relative to the bonnet to use as a guide on refitting (photo).
5 Undo the retaining bolts and carefully lift the bonnet clear (photo). Store the bonnet out of the way in a safe place.

Refitting

Pre-September 1985 models

6 Refitting is the reverse of the removal procedure.

September 1985 models onward

7 Offer up the bonnet and loosely fit the retaining bolts. Align the hinges with the marks made on removal then tighten the retaining bolts securely.
8 Close the bonnet and check for alignment with the adjacent panels, then retighten the bolts.

Adjustment

9 Check that the bonnet fastens and releases in a satisfactory manner. If adjustment is necessary, slacken the bonnet lock retaining bolts and adjust the position of the lock to suit. Once bonnet operation is satisfactory tighten the screws securely.

10 Bonnet release cable – removal and refitting

Removal

1 Disconnect the battery negative lead.
2 With the bonnet open, disconnect the cable from the lock mechanism lever and release the outer cable from the lock bracket.
3 From inside the car remove the steering column shrouds to gain access to the cable.
4 Undo the cable retaining nut from the rear of the facia and pull the cable through from the engine compartment.

Refitting

5 Refitting is the reverse of the removal procedure.

11 Bonnet lock – removal and refitting

Removal

1 With the bonnet open, disconnect the lock mechanism lever and free the outer cable from the lock bracket.
2 Undo the bolts securing the bonnet lock to the body and remove the lock from the engine compartment.

Refitting

3 Refit the lock and tighten its retaining bolts securely.
4 Reconnect the cable to the lock and check that the bonnet fastens

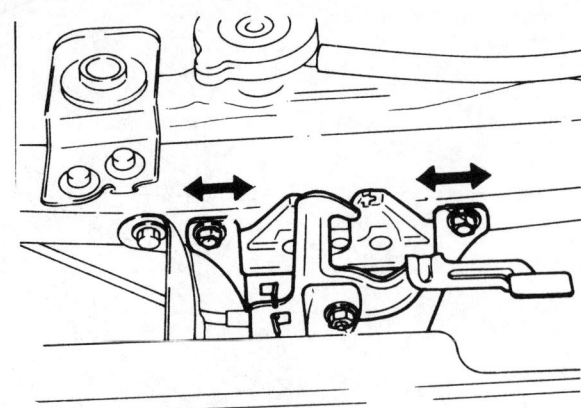

Fig. 11.5 Bonnet lock adjustment – September 1985 models onward (Sec 11)

and releases in a satisfactory manner. If adjustment is necessary, slacken the bonnet lock retaining bolts and adjust the position of the lock to suit. Once bonnet operation is satisfactory tighten the screws securely.

12 Door inner trim panel – removal and refitting

Removal

Pre-September 1985 models

1 Undo the door inner handle retaining screw, disengage the operating rod and remove the handle from the door (photo).
2 Remove the window regulator handle horseshoe clip by hooking it out with a screwdriver or a bent piece of wire, then pull the handle off the spindle.
3 Extract the trim caps (where applicable) over the armrest retaining screws, then undo the screws and remove the armrest (photos).
4 Release the door trim panel studs by carefully levering between the panel and door with a suitable flat bladed screwdriver (photo). When all the studs are released lift the panel upwards and away from the door.
5 If necessary, carefully peel back the polythene watershield to gain access to the window regulator and door lock components (photo).

September 1985 models onward

6 Undo the door inner handle retaining screw, disengage the operating rod and remove the handle from the door.
7 On models with manual windows, remove the window regulator handle horseshoe clip by hooking it out with a screwdriver or bent piece of wire, then pull the handle off the spindle.
8 Where an armrest is fitted, extract the trim caps (where applicable) over the armrest retaining screws then undo the screws and remove the

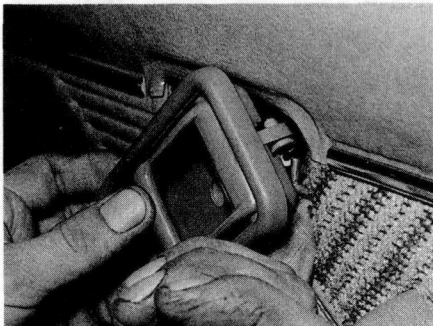

12.1 Disengage the operating rod from the door inner handle

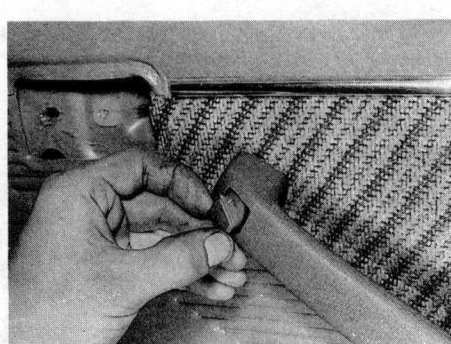

12.3A Extract the trim caps...

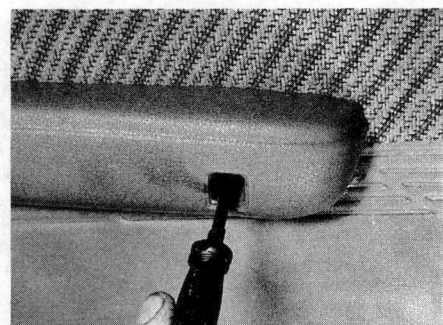

12.3B ...then undo the armrest retaining screws

Chapter 11 Bodywork and fittings

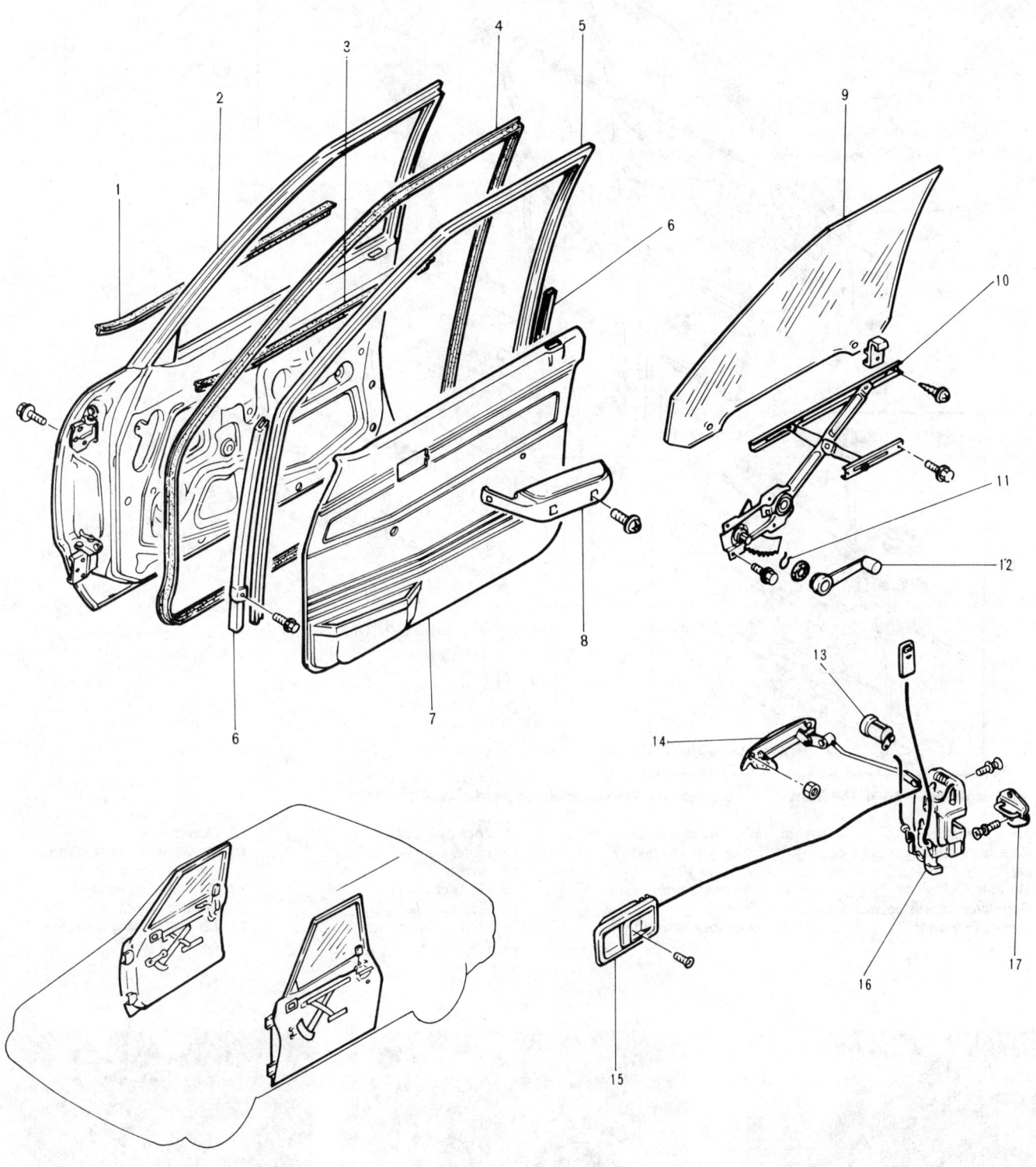

Fig. 11.6 Exploded view of the front door – pre-September 1985 models (Secs 12 to 16)

1 Outer sealing strip
2 Door frame
3 Inner sealing strip
4 Door sealing strip
5 Glass channel
6 Glass guide
7 Inner trim panel
8 Armrest
9 Window glass
10 Window regulator
11 Regulator handle horseshoe clip
12 Regulator handle
13 Lock cylinder
14 Outer handle
15 Inner handle
16 Door lock
17 Striker

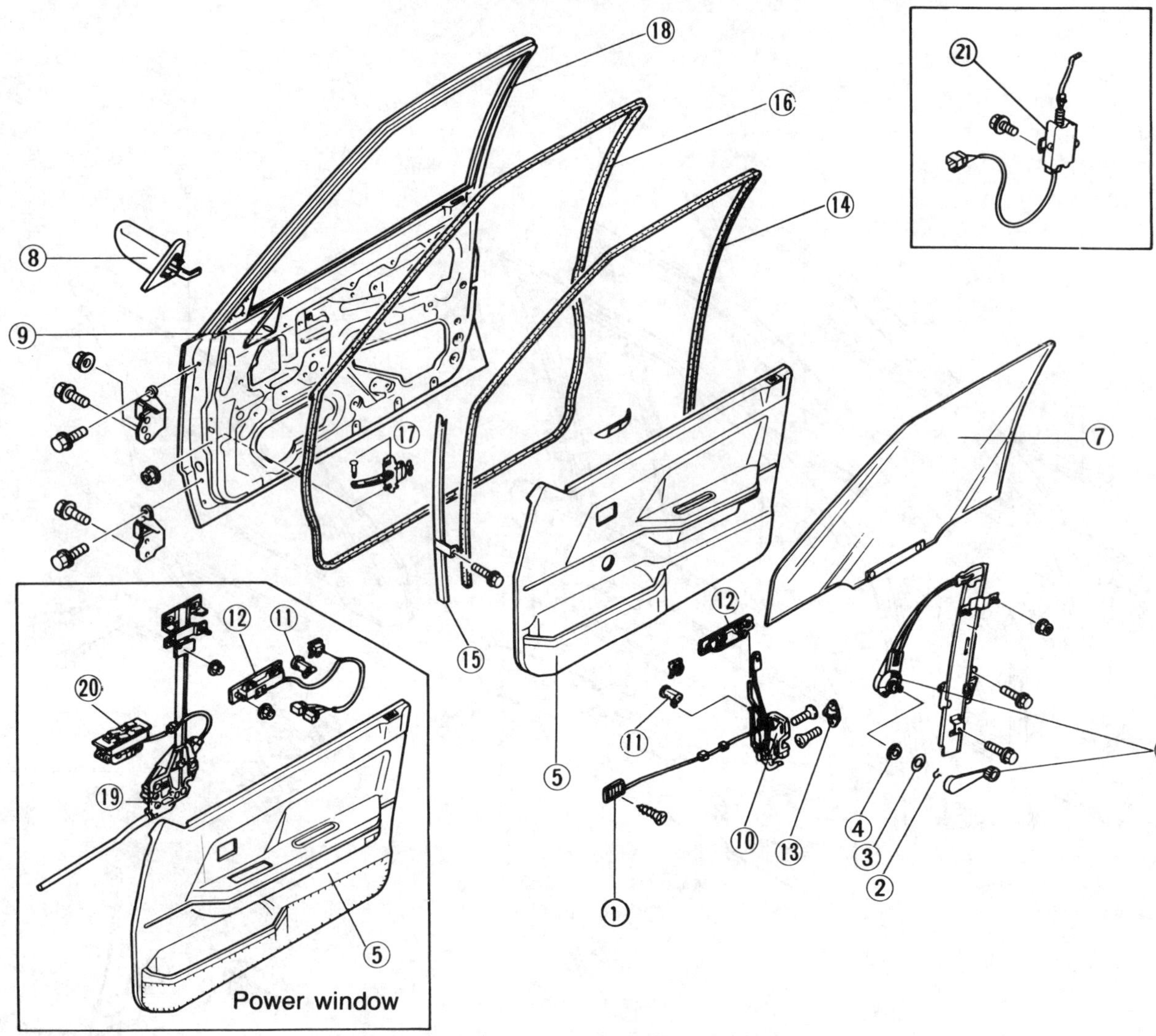

Chapter 11 Bodywork and fittings

Fig. 11.7 Exploded view of the front door – September 1985 models onward (Sec 12 to 16)

1. Inner handle cover
2. Regulator handle horseshoe clip
3. Washer
4. Regulator handle bezel
5. Inner trim panel
6. Window regulator assembly and handle
7. Window glass
8. Mirror assembly
9. Trim
10. Door lock assembly
11. Lock cylinder
12. Outer handle
13. Striker
14. Glass channel
15. Glass guide
16. Door sealing strip
17. Check link
18. Door frame
19. Power window regulator (where fitted)
20. Power window switch (where fitted)
21. Central locking actuator (where fitted)

12.4 Release all the trim panel retaining studs

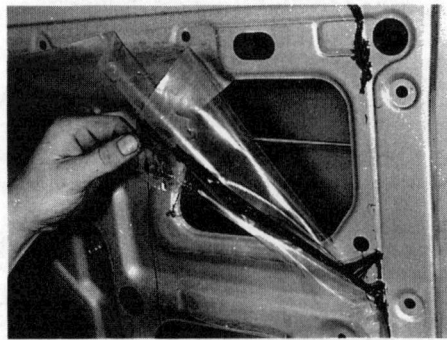

12.5 Peel back the watershield to gain access to the door components

12.8 Remove the screw from the bottom of the armrest pocket

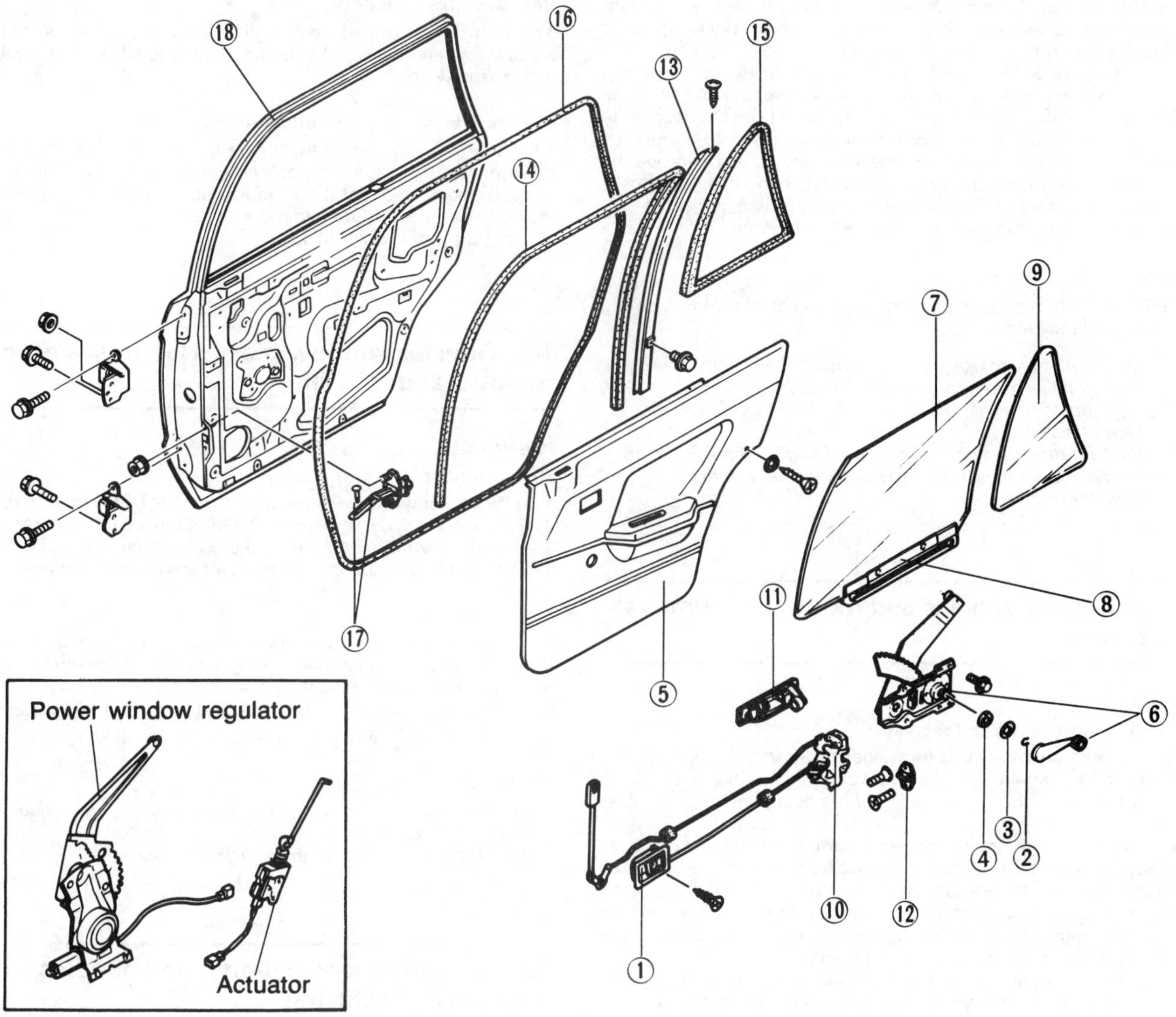

Chapter 11 Bodywork and fittings

Fig. 11.8 Exploded view of the rear door – September 1985 models onward (Secs 12 to 16)

1. Inner handle cover
2. Regulator handle horseshoe clip
3. Washer
4. Regulator handle bezel
5. Inner trim panel
6. Window regulator assembly and handle
7. Window glass
8. Window glass holder
9. Fixed window glass
10. Door lock assembly
11. Outer handle
12. Striker
13. Guide channel support
14. Glass channel
15. Fixed window sealing strip
16. Door sealing strip
17. Check link
18. Door frame
19. Power window regulator (where fitted)
20. Central locking actuator (where fitted)

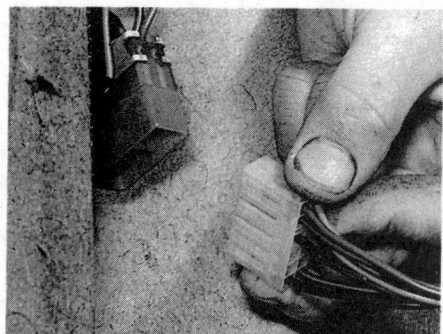

12.10A If necessary, disconnect the electric window switch wiring connector...

12.10B ...and remove the door panel warning lamp bulbholder

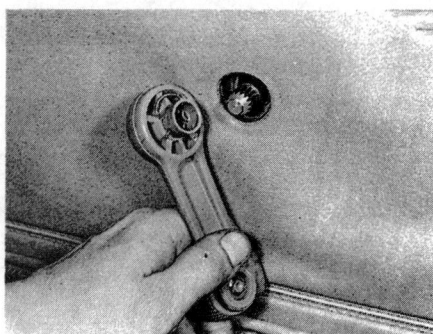

12.12 Refit the clip to the regulator handle before fitting the handle to the spindle

254 Chapter 11 Bodywork and fittings

armrest. On models where the armrest is an integral part of the door trim, remove the screw from the bottom of the armrest pocket (photo).
9 Undo the door trim panel retaining screws.
10 Release the door trim panel studs by carefully levering between the panel and door with a suitable flat bladed screwdriver. When all the studs are released lift the panel upwards and away from the door. Note that on models with electric windows it will be necessary to disconnect the switch wiring connector as the panel is removed. Also remove the bulbholder from the door panel warning lamps (where fitted) (photos).
11 If necessary, carefully peel back the polythene watershield to gain access to the window regulator and door lock components.

Refitting
12 Refitting the trim panel is the reverse sequence of removal, noting the following points (photo).

 (a) Apply a suitable mastic between the polythene watershield and the door panel to ensure a watertight seal.
 (b) Check the trim panel retaining clips for breakage and renew them as necessary.
 (c) When refitting the window regulator handle (where fitted), fit the clip to the handle first then push the handle onto the regulator spindle.

13 Door window glass and regulator – removal and refitting

Removal
Front door window glass and regulator
1 Remove the front door inner trim panel as described in Section 12.
2 Release all the retaining clips, and carefully prise the two sealing strips out of position from the top edge of the door panel and the glass channel.
3 Temporarily refit the regulator handle or reconnect the switch (as applicable) and position the window glass so its retaining screws or nuts can be accessed through the cutaway in the door.
4 With an assistant supporting the glass, undo the screws or nuts securing the window glass to the regulator mechanism, then lift the glass upwards and manoeuvre it out of the door.
5 Undo the bolts securing the regulator mechanism to the door panel and manoeuvre the assembly out through the door panel cutaway.

Rear door fixed window
6 Lower the door window glass to its lowest position.
7 Remove the rear door inner trim panel as described in Section 12.
8 Carefully prise out the door sealing strip and glass channel from the top of the door frame, to gain access to the centre guide channel support upper retaining screw. Undo this screw and the lower retaining bolt and withdraw the support from the door.
9 Ease the fixed window out of position and remove it from the door.

Rear door window glass and regulator
10 Remove the fixed window as described in paragraphs 6 to 9.
11 Carefully release all the retaining clips, and prise out the two sealing strips from the top edge of the door panel and the glass channel.
12 Lower the rear portion of the glass and disengage the regulator arm roller from the glass holder.
13 Lift the glass upwards and remove it from the door.
14 Undo the bolts securing the regulator to the door panel and manoeuvre the regulator out through the cutaway in the door.

Refitting
Front door window glass and regulator
15 Refitting is the reverse of the removal procedure, noting that on pre-September 1985 models, the regulator must be adjusted as follows before installing the trim panel.
16 Raise the glass to its highest position using the regulator. Adjust the position of the regulator guide so that the glass closes fully and moves up and down smoothly. Once the window operation is satisfactory tighten the regulator retaining bolts securely.

Rear door fixed window
17 Refitting is the reverse of the removal procedure, noting that installation can be eased by lubricating the fixed window sealing strip with soapy water.

Rear door window glass and regulator
18 Refitting is the reverse of the removal procedure. Prior to refitting the trim, slacken the lower bolt and adjust the centre guide channel support position so that the window glass is well supported but slides freely. Once window operation is satisfactory tighten the centre guide channel support lower bolt securely.

14 Quarter window glass (Hatchback models) – removal and refitting

Removal
3-door models
1 Using a screwdriver carefully prise off the front hinge covers.
2 Slacken the screws securing the front hinges to the body.
3 Open the window and undo the screw securing the rear retaining catch to the body and remove the quarter window from the car.

5-door models
4 From inside the car, free the rear quarter window inner trim panel from the door and window sealing strips, then carefully release the retaining clips and remove the panel.
5 Undo the three retaining nuts and remove the quarter window assembly from the car.

Refitting
6 Before refitting, check that the window seal is in a good condition, renewing it if necessary.
7 Refitting is a reverse of the removal procedure.

15 Door lock, lock cylinder and handles – removal and refitting

Removal
Door lock
1 Remove the door inner trim panel as described in Section 12.
2 Working through the cutaway in the door panel, disconnect the lock cylinder (where applicable), inner handle and inner lock button operating rods by releasing their plastic clips (photo). On models with central locking also disconnect the solenoid operating rod.
3 Undo the three screws securing the lock assembly to the door and remove the lock through the door cutaway (photo).

Lock cylinder
4 Remove the door trim inner panel as described in Section 12.
5 Working through the cutaway in the door panel, disconnect the door lock operating rod from the lock by releasing the plastic retaining clip (photo).
6 Where applicable, disconnect the wiring connector from the door lock central locking switch.
7 Extract the wire lock retaining clip and withdraw the lock from the door.

Interior handle
8 Undo the inner handle retaining screw then disengage the lock from the operating rod and remove the handle.

Exterior handle
9 Remove the door inner trim panel as described in Section 12.
10 Working through the cutaway in the door panel, disconnect the

Chapter 11 Bodywork and fittings

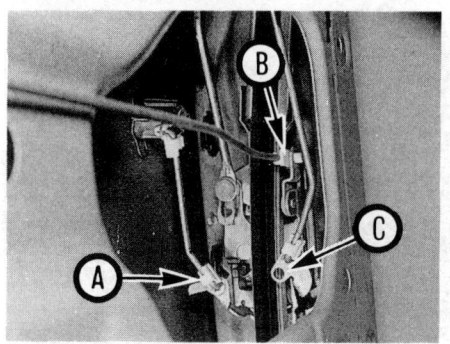

15.2 Lock cylinder (A), inner handle (B) and inner lock button (C) control rod retaining clips

15.3 Door lock retaining screws (arrowed)

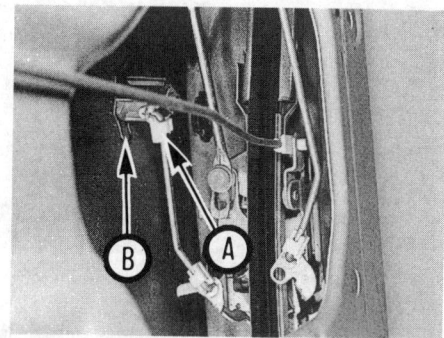

15.5 Lock cylinder control rod clip (A) and lock cylinder retaining clip (B)

operating rod from the handle and remove the retaining nuts securing the handle to the door panel (photo).
11 Remove the handle from door.

Refitting

12 Refitting is the reverse of the removal sequence ensuring that all operating rods are securely held in position by their retaining clips. Apply grease to all lock and operating rod pivot points.

16 Door – removal, refitting and adjustment

Removal

1 On models equipped with electric windows and/or central locking, trace the wiring back from the front edge of the door, and disconnect it from the main harness at the wiring connector(s). If required, remove any necessary trim panels to gain access to the wiring connector.
2 Using a pencil or felt tip pen, mark the outline position of each door hinge relative to the door to use as a guide on refitting.
3 Remove the retaining clip and extract the pin securing the door check link to the door pillar (photo).
4 Have an assistant support the door, and undo the nuts and/or bolts which secure the upper and lower hinges to the door pillar (photo).
5 Remove the door from the car noting any shims which may be fitted between the hinges and door pillar.

Refitting and adjustment

6 The door is refitted by reversing the removal sequence aligning the hinges with the marks made on removal.
7 On completion check that the door is correctly aligned with all surrounding bodywork with an equal clearance all around. If necessary, adjustment can be made by slackening the hinge bolts and moving the door. Once the door is positioned correctly tighten the hinge bolts securely.
8 Once the door is correctly aligned, check that the door closes easily

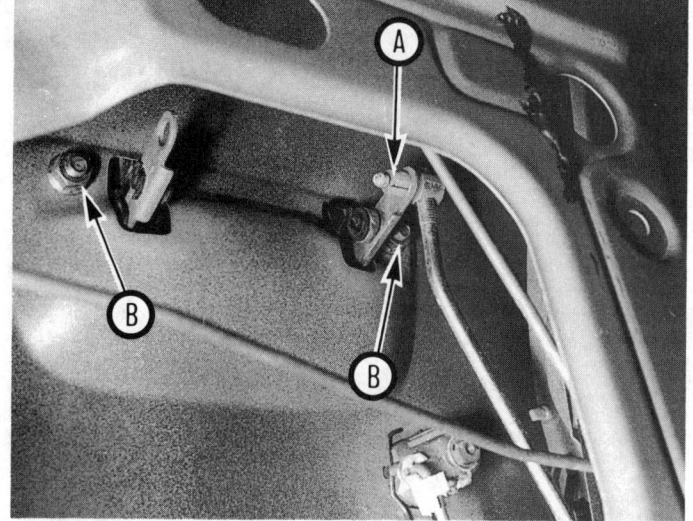

15.10 Exterior handle operating rod clip (A) and retaining nuts (B)

and does not rattle when closed. If not, slacken the door striker retaining screws and reposition the striker (photo). Once the door operation is satisfactory tighten the striker retaining screws securely.

17 Exterior mirror and glass – removal and refitting

Removal

Mirror glass

1 Using a wooden wedge inserted between the glass and mirror frame, carefully lever the glass out of position.

16.3 Door check link retaining pin and clip (arrowed)

16.4 Rear door upper hinge retaining nut and bolts

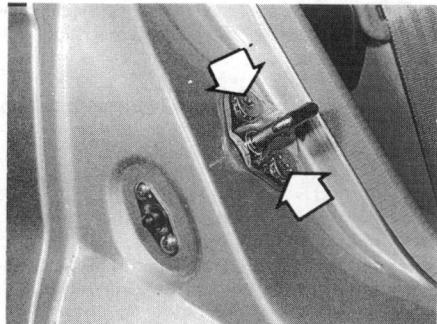

16.8 Front door striker retaining screws

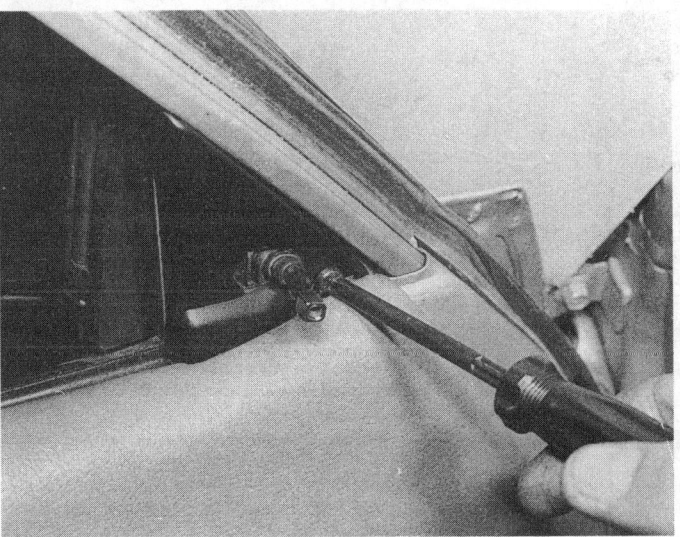

17.4 Remove the trim panel retaining screw

17.5 Undo the mirror retaining screws and remove the mirror from the door

Mirror assembly

2 On models with electrically operated exterior mirrors, remove the door inner trim panel as described in Section 12 and disconnect the mirror wiring connector.
3 On models with manually operated exterior mirrors, undo the adjusting lever retaining screw (where fitted) and remove the lever from the mirror.
4 Undo the retaining screw (where fitted) and remove the inner trim panel (photo).
5 Undo the mirror retaining screws and lift the mirror assembly away from the door (photo).

Refitting

Mirror glass

6 Align the mirror glass with its retaining clips and push the glass in until the clips snap into place.

Mirror assembly

7 Refitting is the reverse of the removal sequence.

18 Boot lid – removal, refitting and adjustment

Removal

1 Remove the boot lid lock as described in Section 19, then detach the lock release cable from its retaining clips on the boot lid.
2 Using a pencil or felt tip pen, mark the relative position of the hinges on the boot lid.
3 Have an assistant support the boot lid, then undo the hinge retaining bolts and lift the boot lid away from the car.

Refitting

4 Refitting is the reverse of the removal sequence aligning the hinges with the marks made during removal.
5 Once fitted check that the boot lid is correctly aligned with all surrounding bodywork with an equal clearance all around. If necessary, adjustment can be made by slackening the hinge nuts and moving the boot lid.
6 Once the boot lid is correctly aligned ensure that it closes without slamming and is securely retained. If not slacken the boot lid striker retaining screws and reposition the striker. Once the boot lid operation is satisfactory tighten the striker retaining screws securely.

19 Boot lid lock – removal and refitting

Removal

1 Disconnect the lock cylinder operating rod from the lid lock.
2 Disconnect the lid lock release cable from the lock lever and free the cable.
3 Undo the lock retaining bolts and remove the lock from the boot lid.

Refitting

4 Refitting is the reverse of the removal sequence.

20 Boot lid lock cylinder – removal and refitting

Removal

1 Disconnect the operating rod from the lock cylinder.
2 Extract the large horseshoe shaped clip securing the lock cylinder to the boot lid.
3 Withdraw the lock cylinder and sealing washer from the boot lid.

Refitting

4 Refitting is the reverse of the removal sequence.

21 Boot lid/tailgate and fuel filler internal release cables – removal and refitting

Removal

Boot lid/tailgate release cable

1 Undo the two retaining screws and remove the cover from the operating lever (photo).
2 Undo the retaining bolt and detach the boot lid release cable from the operating lever inside the car. Tie a long piece of string around the end of the cable.
3 Disconnect the release cable from the boot lid lock assembly and withdraw the cable from the boot. Untie the cable from the string and leave the string in position in the vehicle.

Fuel filler release cable

4 Undo the two retaining screws and remove the cover from the operating lever.
5 Undo the retaining bolt and detach the fuel filler release cable from the operating lever inside the car. Tie a long piece of string around the end of the cable.

Chapter 11 Bodywork and fittings

21.1 Undo the retaining screws and remove the cover from the operating lever

21.6 Fuel filler lid release mechanism retaining nut and screws

6 Open the fuel filler lid and undo the fuel filler lid release mechanism retaining nut and screws (photo).
7 Disconnect the cable from the release mechanism and withdraw the cable from the car. Untie the cable from the string and leave the string in position in the vehicle.

Refitting

8 Refitting is the reverse of the removal procedure. Tie the string onto the cable and, working from inside the car, use the string to pull the cable back through until it appears from the operating lever aperture. On completion ensure that the release mechanism operates satisfactorily.

22 Tailgate – removal, refitting and adjustment

Removal

1 Disconnect the battery negative lead.

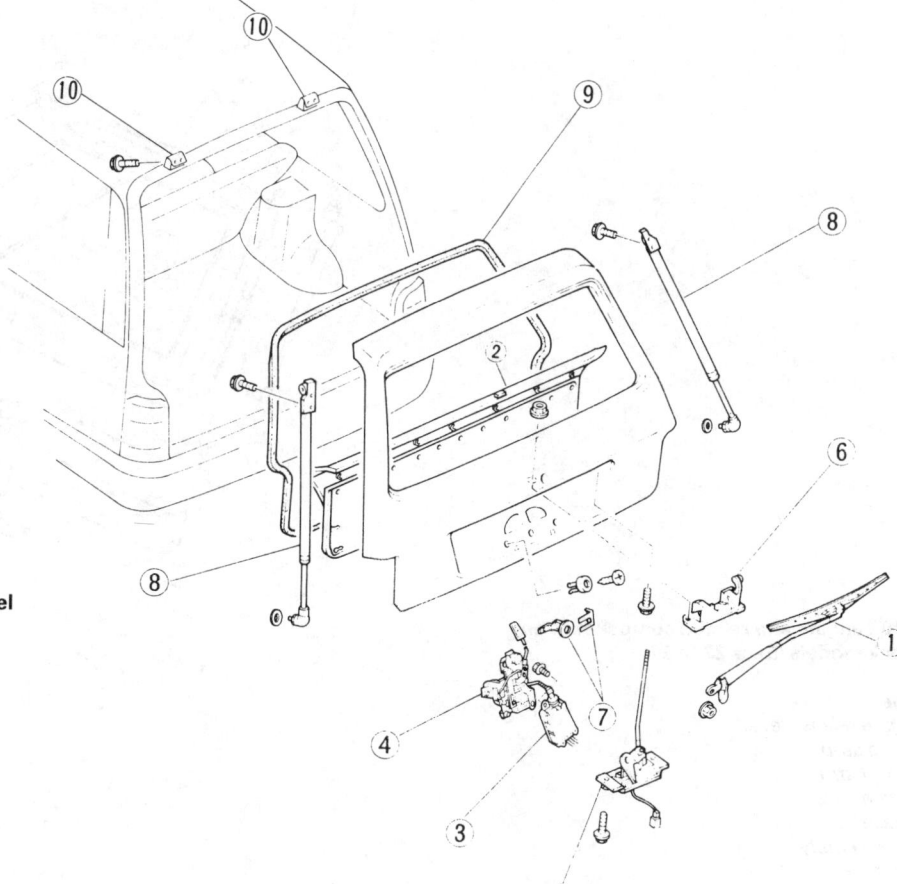

Fig. 11.9 Exploded view of the Estate model tailgate (Secs 22 to 24)

1 Rear wiper arm and blade
2 Tailgate trim
3 Lock controller
4 Lock unit
5 Tailgate lock assembly
6 Outer handle
7 Lock cylinder and retainer
8 Support struts
9 Sealing strip
10 Hinges

Chapter 11 Bodywork and fittings

22.6 Tailgate hinge retaining bolts (Estate model shown)

22.8 Tailgate striker retaining bolts (early Hatchback model shown)

2 Remove the tailgate interior trim retaining clips and remove the trim panel.
3 Disconnect the hose from the tailgate washer and withdraw the hose from the tailgate.
4 Disconnect the wiring from the tailgate wiper motor, rear window demister and central locking unit (as applicable) and withdraw the wiring loom from the tailgate.
5 Have an assistant support the tailgate then undo the two bolts securing the upper support strut brackets to the body.
6 Undo the hinge retaining nuts or bolts and carefully remove the tailgate from the car (photo).

Refitting and adjustment

7 Refitting is a reversal of the removal procedure. Prior to refitting the tailgate interior trim, shut the tailgate and check that it is correctly aligned with all surrounding bodywork with an equal clearance all

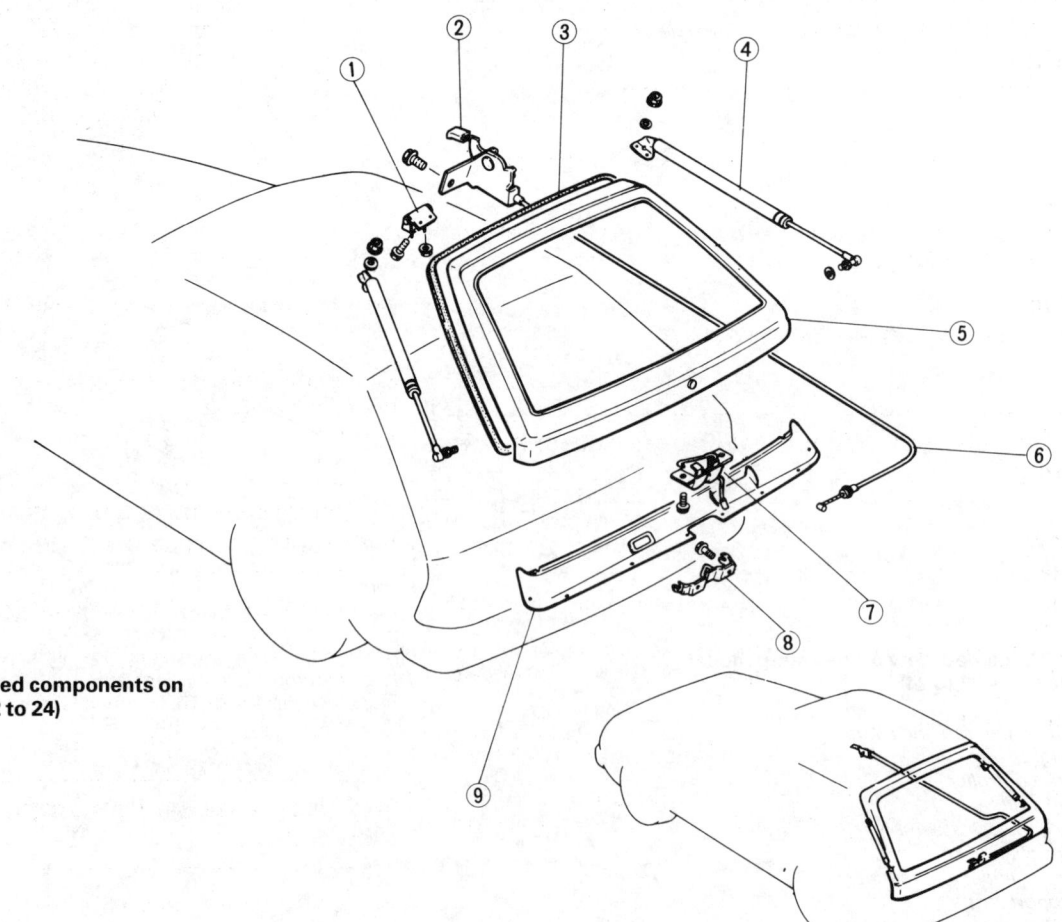

Fig. 11.10 Tailgate and related components on Hatchback models (Secs 22 to 24)

1 Hinge
2 Tailgate release lever
3 Sealing strip
4 Support strut
5 Tailgate
6 Release cable
7 Lock assembly
8 Striker
9 Trim panel

Chapter 11 Bodywork and fittings

23.2A Undo the tailgate strut retaining bolts...

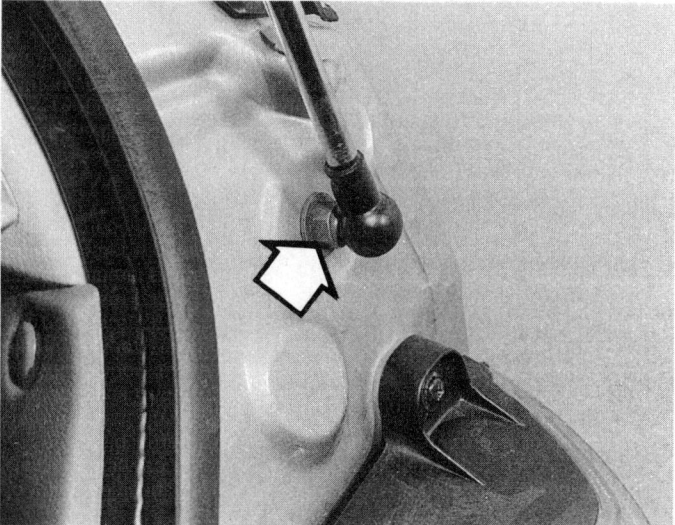

23.2B ...and unscrew the lower end fitting (arrowed)

around. If necessary, adjustment can be made by slackening the hinge bolts and moving the tailgate. Once the tailgate is positioned correctly tighten the hinge bolts securely.

8   On completion, check that the tailgate closes easily and does not rattle when closed. If not slacken the tailgate striker retaining bolts and reposition the striker (photo). Once the tailgate operation is satisfactory tighten the striker retaining bolts securely.

23 Tailgate support strut – removal and refitting

Removal

1   Support the tailgate in the open position using a stout piece of wood, or with the help of an assistant.
2   Undo the two bolts securing the support strut bracket to the body, then unscrew the lower support strut end fitting from the tailgate and remove the strut (photos).

Refitting

3   Refitting is the reverse sequence of removal.

24 Tailgate lock and lock cylinder – removal and refitting

Removal

Tailgate lock

1   Remove the tailgate interior trim panel retaining clips and remove the panel.
2   Disconnect the operating rod(s) from the tailgate lock lever (photo).
3   Undo the two retaining bolts and remove the lock from the tailgate.

Tailgate lock cylinder

4   Remove the tailgate interior trim panel retaining clips and remove the panel.
5   Disconnect the operating rod from the lock cylinder.
6   Extract the horseshoe shaped retaining clip and withdraw the lock cylinder from the tailgate.

Refitting

7   Refitting is the reverse of the removal procedure. When refitting the lock, check that the striker enters the lock centrally when the tailgate is

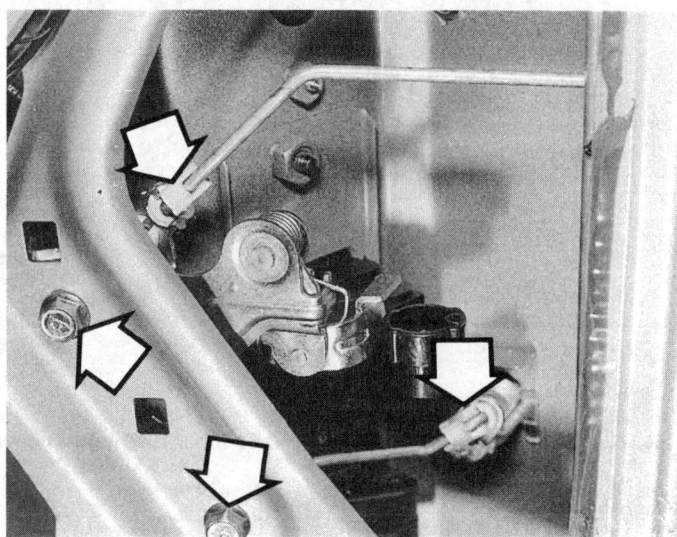

24.2 Tailgate lock operating rods and retaining bolts (arrowed) on Estate model

closed, and if necessary re-position the striker by loosening the mounting screws.

25 Windscreen and rear window/tailgate glass – general information

The windscreen and rear window/tailgate glass are bonded in place with special mastic, and special tools are required to cut free the old units and fit the new units together with cleaning solutions and primers. It is therefore recommended that this work is entrusted to a Mazda dealer or windscreen replacement specialist.

26 Body exterior trim strips – general information

The exterior door trim strips are held in position with a special adhesive tape. Removal requires the trim to be cut away from the door surface. Due to the high risk of damage to the vehicles paintwork during

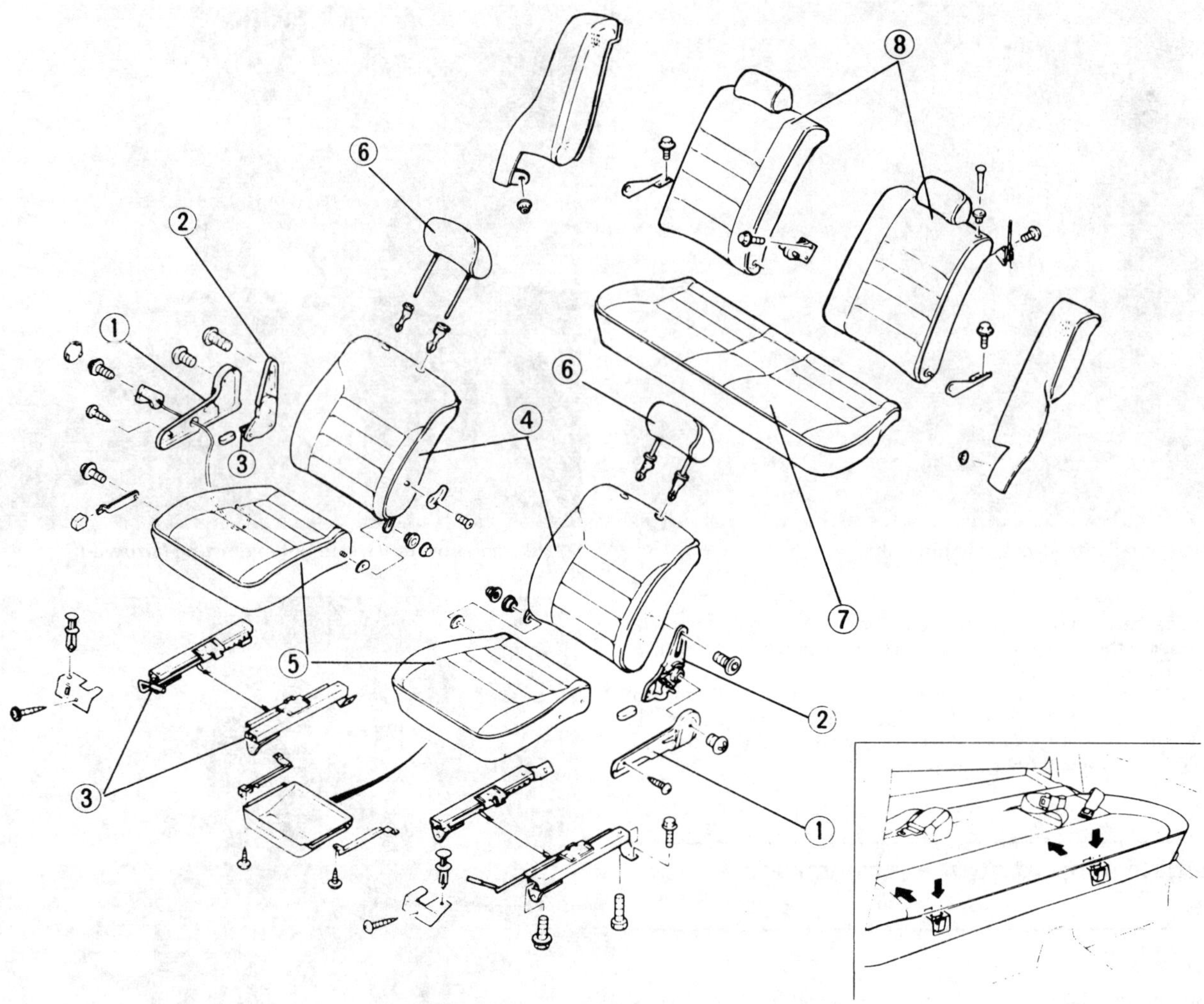

Fig. 11.11 Exploded view of the seats – September 1985 models onward shown (Sec 27)

1 Covers
2 Reclining joints
3 Seat adjuster rails
4 Front seat back
5 Front seat cushion
6 Headrest
7 Rear seat cushion
8 Rear seat back

this operation it is recommended that the work should be entrusted to a Mazda dealer. The front wing and rear panel exterior trim strips are secured to the wheel arches by rivets in addition to the adhesive tape.

27 Seats – removal and refitting

Removal

Front seats

1 Move the seat forwards to gain access to the rear seat rail retaining bolts. Undo the two bolts and move the seat backwards.
2 Undo the two front seat rail retaining bolts and remove the seat from the car.

Rear seats

3 Detach the rear seat back from the striker catch and push the seat rearward.
4 Remove the rear seat back fasteners and, if necessary, remove the mat.
5 Undo the seat back retaining bolts and remove the seat back from the car (photo).
6 Where fitted, undo the bolts securing the rear of the seat cushion to the floor.

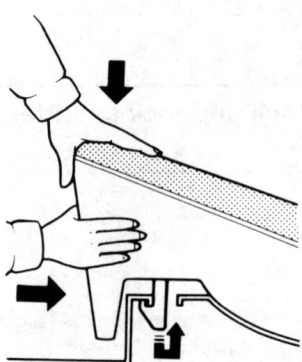

Fig. 11.12 Removing the rear seat cushion (Sec 27)

7 Push the front of the seat cushion downwards and rearwards to release the cushion front catches, then lift up the front of the cushion and remove it from the car.

Refitting

8 Refitting is the reverse of the removal sequence.

Chapter 11 Bodywork and fittings

27.5 Seat back retaining bolts (arrowed)

3 To remove the front pillar trim carefully prise the trim out from behind the windscreen and door sealing strips using a suitable screwdriver. The front lower trim panel is removed in similar manner once its retaining screw has been removed.
4 To remove the rear door pillar trim on Saloon and Hatchback models, prise the trim out from behind the rear window/tailgate and rear door/quarter window sealing strips (as applicable), free the trim fastener clips and remove the panel. On Estate models undo the seat belt from the rear pillar and remove the panel retaining screws. Prise the trim out from behind the rear door and tailgate sealing strips, extract the panel fasteners and remove the panel.
5 To remove the roof lining trim sections first remove the front and rear pillar trim panels. Undo the trim retaining screws then carefully prise the roof lining trim out of position.

28 Interior trim panels – general information

1 The interior trim panels are held in position with a mixture of screws and plastic retaining clips. The plastic clips break easily and it is recommended that they are renewed on refitting to ensure secure fitting.
2 To remove the centre pillar trim, carefully peel back the door sealing strips and prise the lower panel out of position. Remove the seat belt mounting from the centre pillar, then free the upper trim panel and remove it from the car.

29 Seat belts – removal and refitting

Removal
Front seat belts
1 To remove the seat belt stalks on pre-September 1985 undo the bolts securing the stalk assembly to the floor. On models since September 1985 the stalks can be removed individually once their retaining bolts have been removed.
2 To remove the inertia reel and belt, prise the cap off the belt upper mounting. Undo the mounting bolt and recover the washers. Remove the centre pillar lower trim panel, and undo the inertia reel retaining bolts. Where necessary remove the bolt securing the lower end of the belt to the floor. Remove the belt and inertia reel assembly from the car.

Rear seat belts
3 Remove the rear seat as described in Section 27.

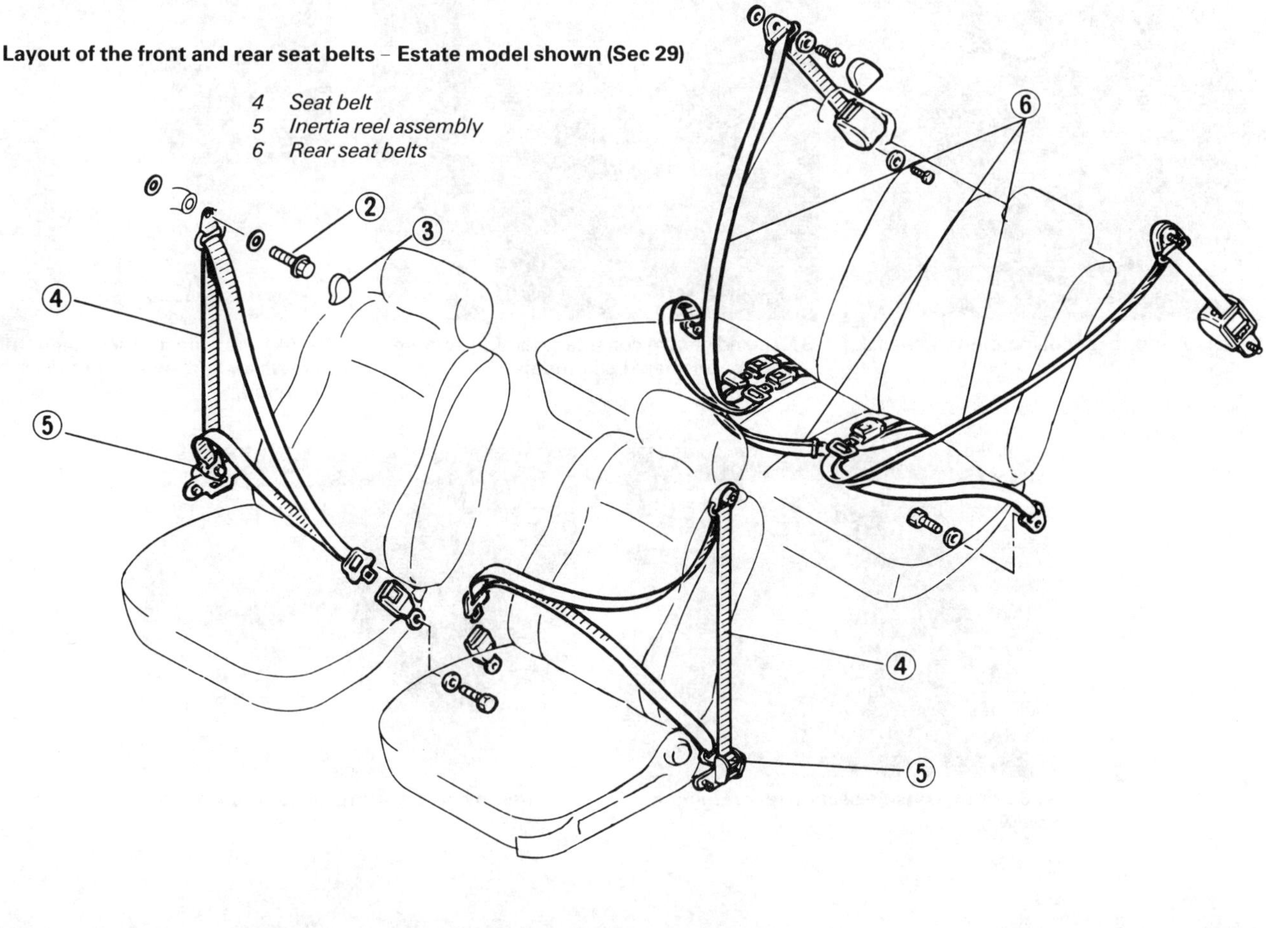

Fig. 11.13 Layout of the front and rear seat belts – Estate model shown (Sec 29)

1 Bolt
2 Bolt
3 Cover
4 Seat belt
5 Inertia reel assembly
6 Rear seat belts

4 On Saloon and Hatchback models undo the mounting bolt from each end of the belt and remove the belts from the car.
5 On Estate models, prise the cap off the belt upper mounting. Undo the mounting bolt and recover the washers. Undo the inertia reel retaining bolts and the bolt securing the lower end of the belt to the floor. Then remove the belt and inertia reel assembly from the car.

Refitting

6 Refitting is a reverse of the removal procedure.

30 Electric sunroof – general information

An electrically operated sunroof is fitted on some models. The sunroof being operated by two switches in the overhead console. Considerable expertise is needed to repair or replace sunroof components successfully. Any problems should therefore be referred to a Mazda dealer.

If the sunroof motor fails to operate, first check the appropriate fuse. If the fault can not be traced and rectified, the sunroof can be opened and closed manually using a suitable Allen key inserted into the bolt behind the overhead console cover (photo). A suitable key was supplied with the vehicle and should be found in the glove compartment.

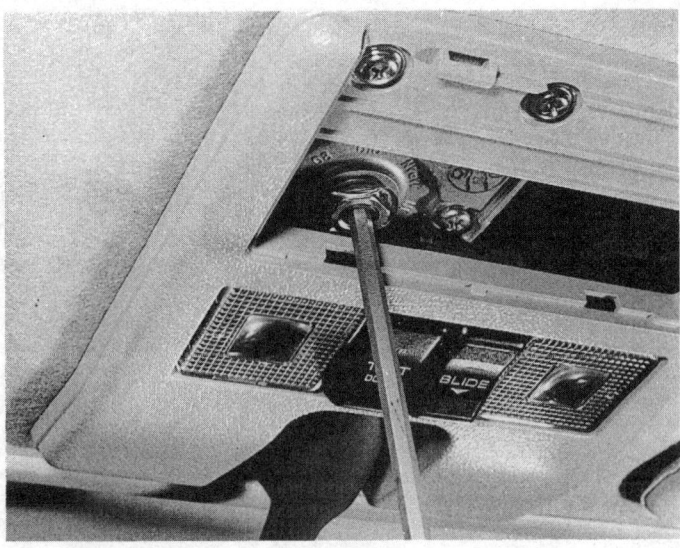

30.1 Using an Allen key to operate the electric sunroof

31 Centre console – removal and refitting

Removal

Pre-September 1985 models

1 Carefully prise the upper panel out of its locating catches, lift it up over the gear lever (or selector lever) and remove it (photo).
2 Undo the two screws in the centre, and the two screws at the rear sides securing the console in position (photo). Lift the console up and remove it from the car.

September 1985 models onward

3 Remove the four screws securing the front console section in position, and remove the single retaining screw from both the left and right-hand front console sidewall sections (photos).
4 Lift up the rear of the front console section and remove the front console and sidewalls as an assembly (photo).
5 Undo the two screws from the rear of the rear console section and remove the rear section from the car, if necessary.

Refitting

6 Refitting is a reversal of the removal procedure.

31.1 Console upper panel locating catches...

31.2 ...and centre console retaining screws – pre-September 1985 models

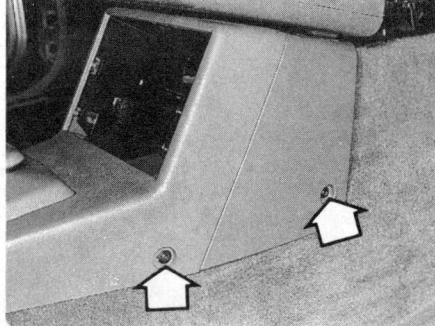

31.3A Front console section front retaining screws and sidewall retaining screws (arrowed)

31.3B Front console section rear retaining screw

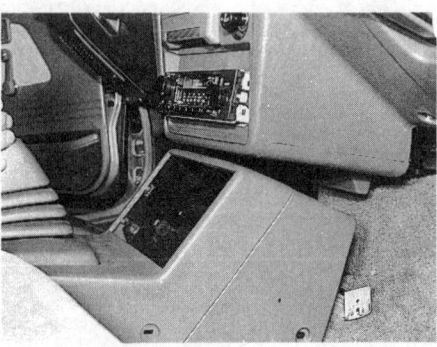

31.4 Removing the front console assembly

Chapter 11 Bodywork and fittings

32 Facia – removal and refitting

Removal

Pre-September 1985 models

1 Refer to Chapter 12 and remove the instrument panel.
2 Undo the retaining screws and remove the oddments tray from under the centre of the facia (photo).
3 Remove the radio control knobs and retaining nuts.
4 Undo the choke knob retaining grub screw and pull off the knob.
5 Undo the knurled retaining nut securing the choke cable to the switch panel (photo).
6 Undo the screws along the upper edge securing the left and right-hand halves of the switch panel (photos).
7 Pull the upper part of the switch panel outwards then disengage the lower catches. Remove the left-hand panel, disconnect the switch wiring and remove the right-hand panel.
8 Undo the two screws securing the radio to the facia. Withdraw the radio, disconnect the wiring and aerial connections and remove the radio (photo).
9 Undo the two clock retaining screws. Disconnect the wiring and remove the clock.
10 Undo the screws securing the glovebox lid striker and remove the striker.
11 Undo the screws securing the glovebox lid and glovebox to the facia. Lift off the lid and remove the glovebox (photos).
12 Undo the nuts securing the heater controls to the facia frame and separate the controls and frame (photo).
13 Undo the fusebox retaining screws and separate the fusebox from the facia.
14 Undo the steering column mounting bolts and lower the column clear of the facia.
15 Refer to Fig. 11.15 and undo the bolts securing the facia and frame

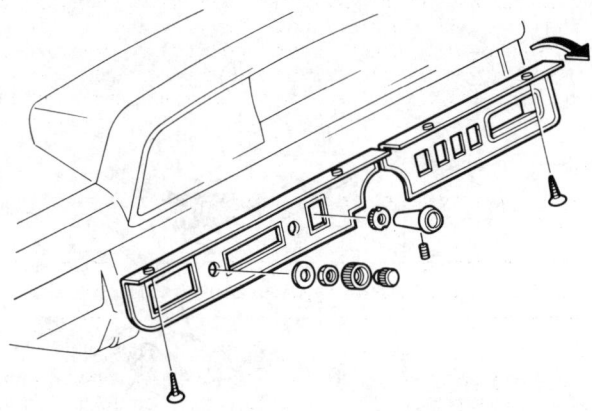

Fig. 11.14 Facia switch panel removal – pre-September 1985 models (Sec 32)

to the bulkhead, noting that some bolts may be concealed behind trim caps (photos). Ease the facia away from its location and, as access improves, disconnect the heater ducts and any remaining wiring. Remove the facia from the car.

September 1985 models onward

16 Remove the instrument panel as described in Chapter 12.
17 Remove the centre console as described in Section 31 of this Chapter.

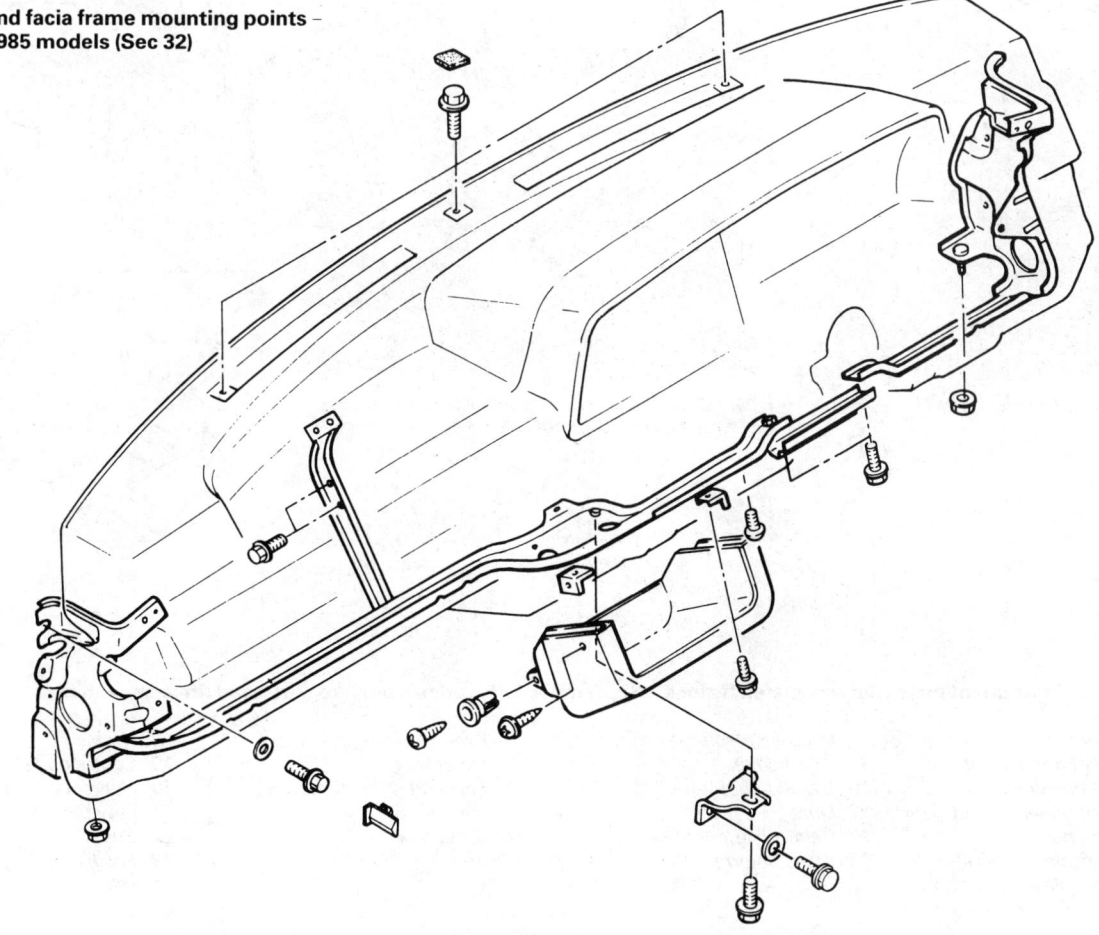

Fig. 11.15 Facia and facia frame mounting points – pre September 1985 models (Sec 32)

264 Chapter 11 Bodywork and fittings

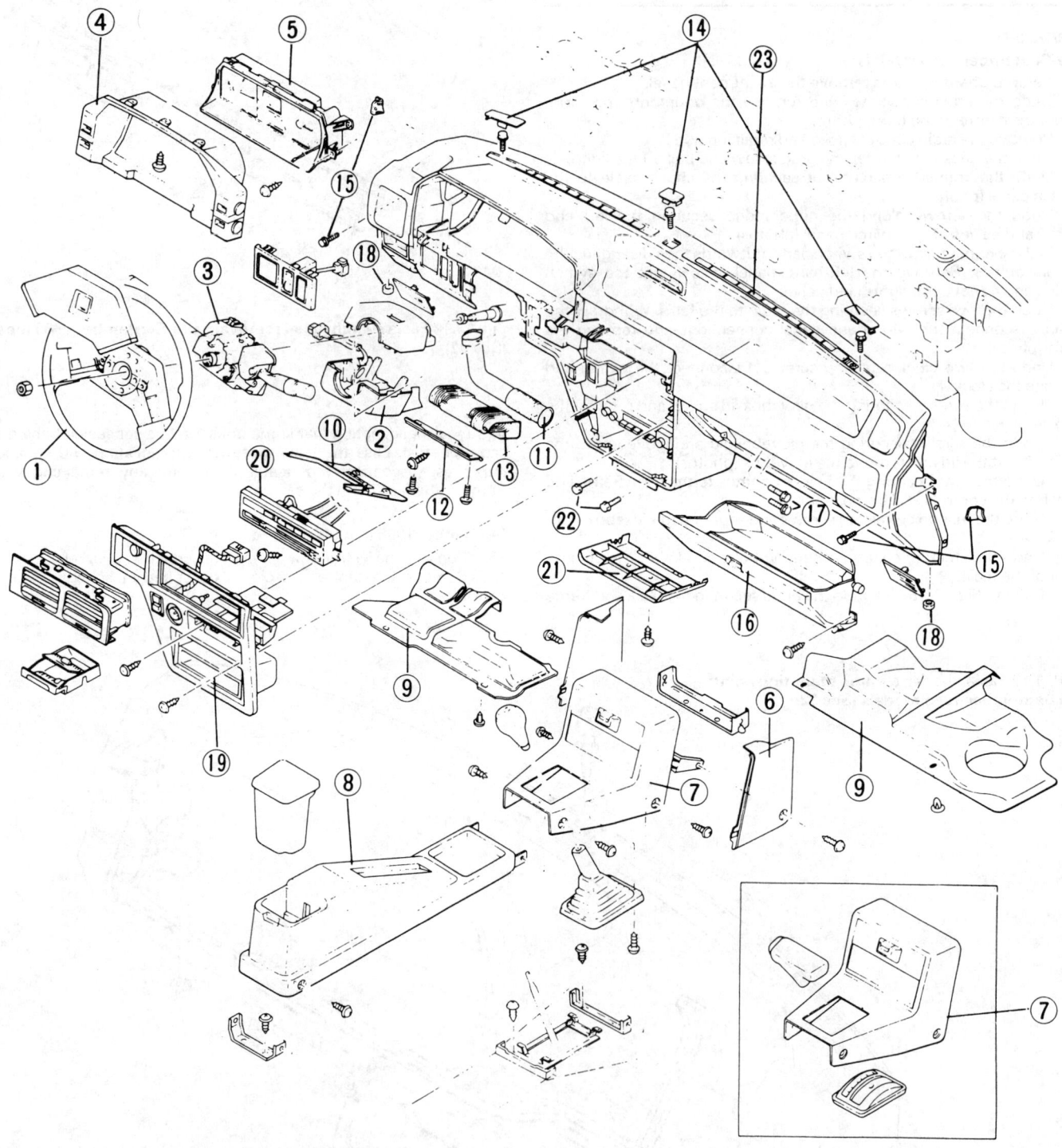

Fig. 11.16 Facia, instrument panel and associated fittings – September 1985 models onward (left-hand drive shown) (Sec 32)

1. Steering wheel
2. Steering column shroud
3. Combination switch
4. Instrument panel shroud
5. Instrument panel
6. Centre console side wall
7. Centre console front section
8. Centre console rear section
9. Undercover
10. Lower panel
11. Duct
12. Reinforcement plate
13. Lower louvre
14. Facia retaining bolts and covers
15. Facia retaining bolts and covers
16. Glovebox
17. Bolts
18. Nuts
19. Centre panel
20. Heater control panel
21. Lower cover
22. Bolts
23. Facia panel

Chapter 11 Bodywork and fittings 265

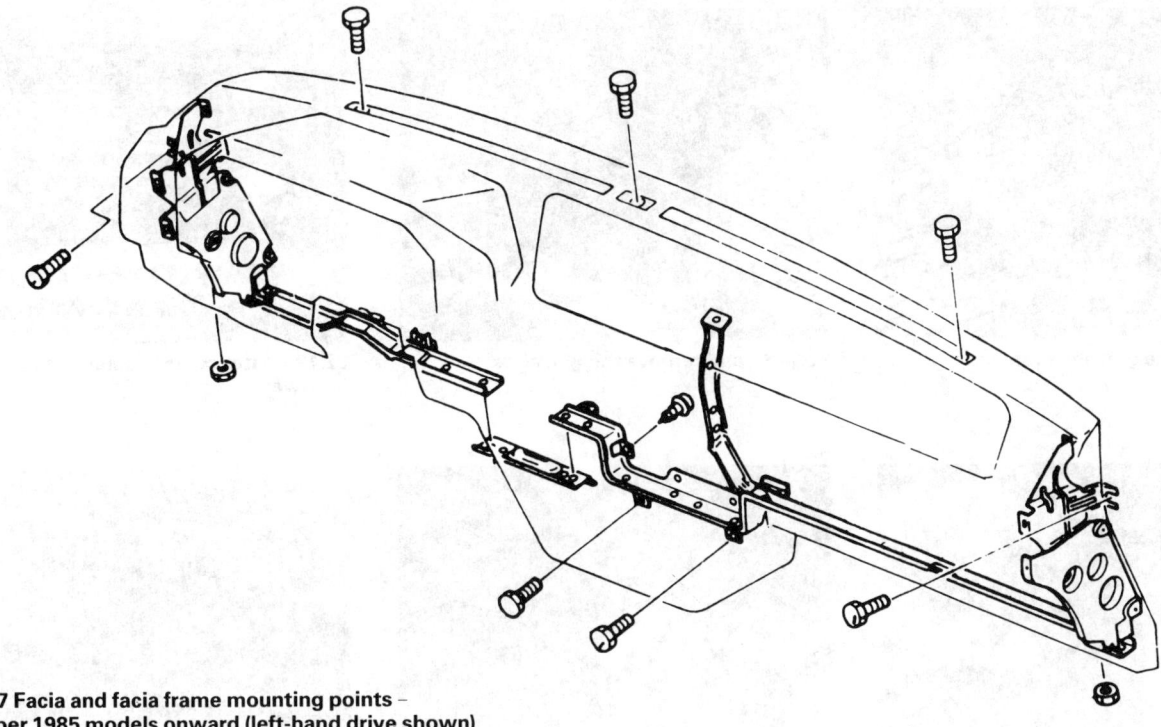

Fig. 11.17 Facia and facia frame mounting points –
September 1985 models onward (left-hand drive shown)
(Sec 32)

32.2 Oddments tray retaining screws

32.5 Choke cable knurled retaining ring

32.6A Undo the screws securing the left-hand...

32.6B ...and right-hand halves of the switch panel

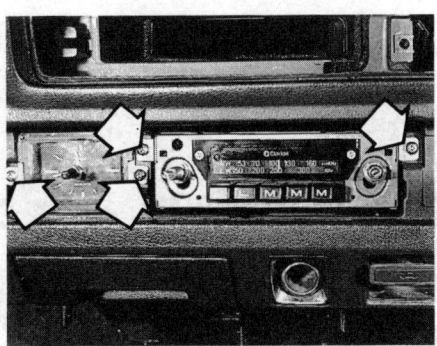

32.8 Clock and radio retaining screws (arrowed)

32.11A Undo the glovebox retaining screws...

32.11B ...and remove the glovebox

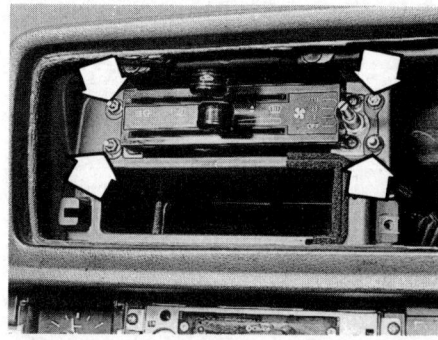

32.12 Heater control panel retaining nuts (arrowed)

32.15A Remove the trim caps to gain access to the facia upper...

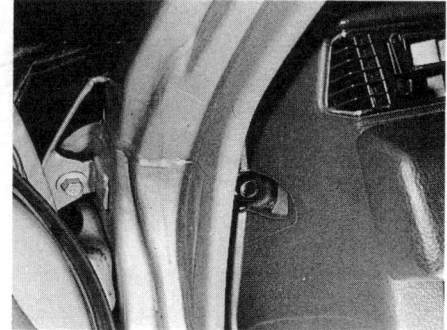

32.15B ...and side retaining bolts

32.18A Remove both the left-...

32.18B ...and right-hand facia undercovers by releasing their retaining clips (arrowed)

32.20 Remove the coin holder box...

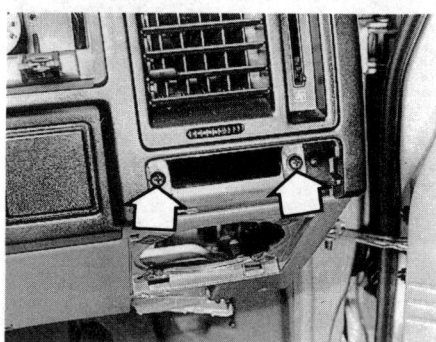

32.21 ...to gain access to two right-hand vent/switch panel retaining screws (arrowed)

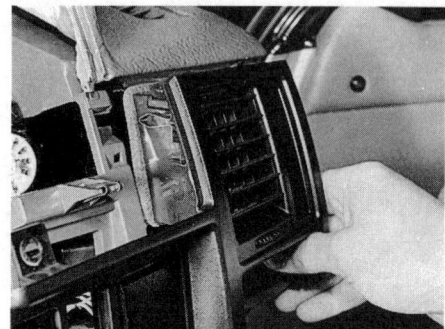

32.22 Withdrawing the right hand vent/switch panel

32.24A Undo the panel retaining screws...

32.24B ...and prise out the side retaining tabs (arrowed) to remove lower central panel (radio removed)

Chapter 11 Bodywork and fittings

32.25A Glovebox retaining screw

32.25B Removing the black metal upper panel from the glovebox recess

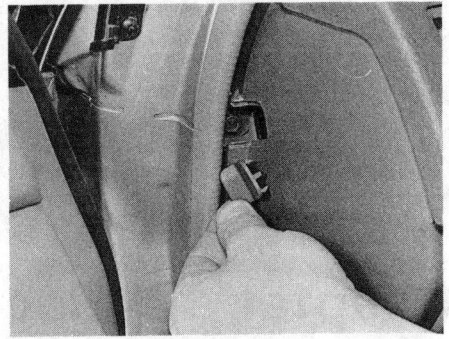

32.26 Where necessary, remove the trim caps to gain access to the facia mounting bolts

18 Remove both the left and right-hand facia undercovers (photos).
19 Undo the screws securing the reinforcement plate, situated directly under the steering column, to the facia and disconnect the heater duct and lower louvre.
20 Remove the coin holder box by prising it free at its outside edge and pull the knob off the right-hand vent control lever (photo).
21 Undo the two retaining screws accessed through the coin box recess and the single screw on the left side (photo).
22 Withdraw the right-hand vent/switch panel, disconnect the switch wiring and remove the panel from the car (photo).
23 Remove the heater control panel as described in Chapter 3.
24 Remove the ashtray to gain access to the two lower central panel retaining screws. Undo the screws and carefully prise the panel retaining tabs out using a flat bladed screwdriver to free them (photos). Withdraw the panel and disconnect the wiring and aerial connections from the radio/cassette player and cigar lighter.
25 Open the glovebox then unscrew the both the left and right-hand hinge screws and remove the glovebox. Undo the retaining screws and remove the black metal upper panel from inside the glovebox recess (photos).
26 Refer to Fig. 11.17 and undo the nuts and bolts securing the facia and frame to the bulkhead, noting that some bolts will be hidden behind trim caps (photo). These can be prised out of position using a screwdriver. Ease the facia away from its location and, as access improves, disconnect the heater ducts and any remaining wiring. Remove the facia from the car.

Refitting

27 Refitting is a reverse of the removal sequence ensuring that all wiring is correctly connected. On completion reconnect the battery, switch on the ignition and check that all electrical components and switches function correctly.

Chapter 12 Electrical system

Contents

Alternator – overhaul	7
Alternator – removal and refitting	6
Alternator drivebelt check, adjustment and renewal	See Chapter 1
Battery – removal and refitting	4
Battery – testing and charging	3
Battery check and maintenance	See Chapter 1
Bulbs (exterior lamps) – renewal	13
Bulbs (interior lamps) – renewal	14
Charging system – testing	5
Cigar lighter – removal and refitting	18
Clock – removal and refitting	19
Electrical fault finding – general information	2
Exterior lamp units – removal and refitting	15
Fuses and relays – general information	11
General information and precautions	1
Headlamp beam alignment check	See Chapter 1
Horn – removal and refitting	20
Instrument panel – removal and refitting	16
Instrument panel components – removal and refitting	17
Radio aerial – removal and refitting	28
Radio/cassette player – removal and refitting	26
Speakers – removal and refitting	27
Speedometer drive cable – removal and refitting	21
Starter motor – removal and refitting	9
Starter motor brushes – checking and renewal	10
Starting system – testing	8
Switches – removal and refitting	12
Tailgate/rear window washer system components – removal and refitting	25
Tailgate/rear window wiper motor and linkage – removal and refitting	23
Windscreen/headlamp washer system check and adjustment	See Chapter 1
Windscreen/headlamp washer system components – removal and refitting	24
Windscreen/tailgate wiper blades and arms check and renewal	See Chapter 1
Windscreen wiper motor and linkage – removal and refitting	22
Wiring diagrams – explanatory notes	29

Specifications

System type
12 volt, negative earth

Battery
Type: Low maintenance
Capacity: 33, 50 or 60 Ah

Alternator
Type: Mitsubishi
Output: 50, 55 or 60 amps
Regulated voltage: 14 to 15 volts

Starter motor
Type: Mitsubishi pre-engaged
Brush length:
  New: 17.0 mm
  Minimum: 11.5 mm

Bulbs
	Wattage
Headlamp	60/55
Front sidelamp	5
Front direction indicator	21
Front repeater	5
Rear sidelamp	5
Rear direction indicator	21
Reversing lamp	21
Stop lamp	21
Rear foglamp	21
Number plate lamp	5

Bulbs (continued)
Interior lamp
Map reading lamps
Instrument panel illumination

Torque wrench settings
Alternator pivot mounting bolt
Alternator adjusting arm bolt
Starter motor mounting bolts

Wattage
10
6
1.4 or 3.4

Nm	lbf ft
19 to 30	14 to 22
43 to 61	32 to 45
31 to 46	23 to 34

1 General information and precautions

Warning: *Before carrying out any work on the electrical system, read through the precautions given in Safety First! at the beginning of this manual*

The electrical system is of 12 volt negative earth type, and consists of a battery, alternator, starter motor and related electrical accessories, components and wiring.

The battery, charged by the alternator which is belt-driven from the crankshaft pulley, provides a steady amount of current for the ignition, starting, lighting and other electrical circuits.

The starter motor is of the pre-engaged type incorporating an integral solenoid. On starting, the solenoid moves the drive pinion into engagement with the flywheel ring gear before the starter motor is energised. Once the engine has started, a one-way clutch prevents the motor armature being driven by the engine until the pinion disengages from the flywheel.

It is necessary to take extra care when working on the electrical system to avoid damage to semi-conductor devices (diodes and transistors), and to avoid the risk of personal injury. In addition to the precautions given in *Safety first!* at the beginning of this manual, observe the following when working on the system.

Always remove rings, watches, etc before working on the electrical system. Even with the battery disconnected, capacitive discharge could occur if a component live terminal is earthed through a metal object. This could cause a shock or nasty burn.

Do not reverse the battery connections. Components such as the alternator, fuel and ignition control units, or any other having semi-conductor circuitry could be irreparably damaged.

If the engine is being started using jump leads and a slave battery, connect the batteries positive-to-positive and negative-to-negative. This also applies when connecting a battery charger.

Never disconnect the battery terminals, any electrical wiring or any test instruments, when the engine is running.

Never use an ohmmeter of the type incorporating a hand-cranked generator for circuit or continuity testing.

Always ensure that the battery negative lead is disconnected when working on the electrical system.

2 Electrical fault finding – general information

1 A typical electrical circuit consists of an electrical component, any switches, relays, motors, fuses, fusible links or circuit breakers related to that component and the wiring and connectors that link the component to both the battery and the chassis. To help you pinpoint an electrical circuit problem, wiring diagrams are included at the end of this manual.
2 Before tackling any troublesome electrical circuit, first study the appropriate wiring diagrams to get a complete understanding of what components are included in that individual circuit. Trouble spots, for instance, can be narrowed down by noting if other components related to the circuit are operating properly. If several components or circuits fail at one time, the problem is probably in a shared fuse or earth connection, because several circuits are often routed through the same connections.
3 Electrical problems usually stem from simple causes, such as loose or corroded connections, a blown fuse, a melted fusible link or a faulty relay. Visually inspect the condition of all fuses, wires and connections in a problem circuit before testing the components. Use the diagrams to note which terminal connections will need to be checked in order to pinpoint the trouble spot.
4 The basic tools needed for electrical fault finding include a circuit tester or voltmeter (a 12-volt bulb with a set of test leads can also be used), a continuity tester, a battery and set of test leads, and a jumper wire, preferably with a circuit breaker incorporated, which can be used to bypass electrical components. Before attempting to locate a problem with test instruments, use the wiring diagram to decide where to make the connections.

Voltage checks
5 Voltage checks should be performed if a circuit is not functioning properly. Connect one lead of a circuit tester to either the negative battery terminal or a known good earth. Connect the other lead to a connector in the circuit being tested, preferably nearest to the battery or fuse. If the bulb of the tester lights, voltage is present, which means that the part of the circuit between the connector and the battery is problem free. Continue checking the rest of the circuit in the same fashion. When you reach a point at which no voltage is present, the problem lies between that point and the last test point with voltage. Most problems can be traced to a loose connection. **Note:** *Bear in mind that some circuits are only live when the ignition switch is switched to a particular position.*

Finding a short circuit
6 One method of finding a short circuit is to remove the fuse and connect a test light or voltmeter to the fuse terminals with all the relevant electrical components switched off. There should be no voltage present in the circuit. Move the wiring from side to side while watching the test light. If the bulb lights up, there is a short to earth somewhere in that area, probably where the insulation has rubbed through. The same test can be performed on each component in the circuit, even a switch.

Earth check
7 Perform an earth test to check whether a component is properly earthed. Disconnect the battery and connect one lead of a self-powered test light, known as a continuity tester, to a known good earth point. Connect the other lead to the wire or earth connection being tested. If the bulb lights up, the earth is good. If the bulb does not light up, the earth is not good.

Continuity check
8 A continuity check is necessary to determine if there are any breaks in a circuit. With the circuit off (ie no power in the circuit), a self-powered continuity tester can be used to check the circuit. Connect the test leads to both ends of the circuit (or to the positive end and a good earth), and if the test light comes on, the circuit is passing current properly. If the light does not come on, there is a break somewhere in the circuit. The same procedure can be used to test a switch, by connecting the continuity tester to the switch terminals. With the switch turned on, the test light should come on.

Finding an open circuit
9 When checking for possible open circuits, it is often difficult to locate them by sight because oxidation or terminal misalignment are hidden by the connectors. Merely moving a connector on a sensor or in the wiring harness may correct the open circuit condition. Remember this when an open circuit is indicated when fault finding in a circuit. Intermittent problems may also be caused by oxidized or loose connections.

General
10 Electrical fault finding is simple if you keep in mind that all electrical circuits are basically electricity flowing from the battery, through the wires, switches, relays, fuses and fusible links to each electrical

component (light bulb, motor, etc.) and to earth, from which it is passed back to the battery. Any electrical problem is an interruption in the flow of electricity from the battery.

3 Battery – testing and charging

1 Where a conventional battery is fitted, the electrolyte level of each cell should be checked and if necessary topped up with distilled or de-ionized water at the intervals given in Chapter 1. On some batteries the case is translucent, and incorporates minimum and maximum level marks. The check should be made more often if the car is operated in high ambient temperature conditions.
2 Where a low-maintenance battery is fitted, it is not usually possible to check the electrolyte level.
3 Periodically disconnect and clean the battery terminals and leads. After refitting them, smear the exposed metal with petroleum jelly.
4 When the battery is removed for whatever reason, it is worthwhile checking it for cracks and leakage.
5 If frequent topping-up is required, and the battery case is not fractured, the battery is being over-charged, and the voltage regulator will have to be checked.
6 If the car covers a very small annual mileage, it is worthwhile checking the specific gravity of the electrolyte every three months to determine the state of charge of the battery. Use a hydrometer to make the check, and compare the results with the following table.

	Normal climates	Tropics
Discharged	1.120	1.080
Half charged	1.200	1.160
Fully charged	1.280	1.230

7 If the battery condition is suspect, first check the specific gravity of electrolyte in each cell. A variation of 0.040 or more between any cells indicates loss of electrolyte or deterioration of the internal plates.
8 A further test can be made using a battery heavy discharge meter. The battery should be discharged for a maximum of 15 seconds at a load of three times the ampere-hour capacity (at the 20 hour discharge rate). Alternatively, connect a voltmeter across the battery terminals and operate the starter motor with the HT king lead from the ignition coil earthed with a suitable wire, and the headlamps, heated rear window and heater blower switched on. If the voltmeter reading remains above 9.6 volts, the battery condition is satisfactory. If the voltmeter reading drops below 9.6 volts, and the battery has already been charged, it is faulty.
9 In winter when heavy demand is placed on the battery (starting from cold and using more electrical equipment), it is a good idea to have the battery fully charged from an external source occasionally at a rate of 10% of the battery capacity (ie 3.3 amps for a 33 Ah battery).
10 Both battery terminal leads must be disconnected before connecting the charger leads (disconnect the negative lead first). Continue to charge the battery until no further rise in specific gravity is noted over a four-hour period.
11 Alternatively, a trickle charger, charging at a rate of 1.5 amps can safely be used overnight.

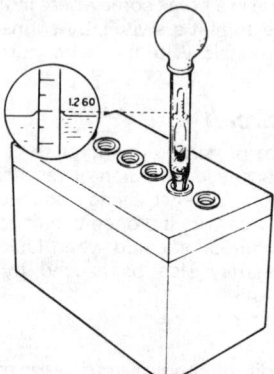

Fig. 12.1 Checking battery condition with a hydrometer (Sec 3)

4 Battery – removal and refitting

Removal

1 The battery is located on the left-hand side of the engine compartment. First check that all electrical components are switched off in order to avoid a spark occurring as the negative lead is disconnected. Note also that if the radio has a security coding, it will be necessary to insert this code when the battery is re-connected.
2 Slacken the negative (–) lead clamp bolt and lift the clamp and lead from the terminal and place it to one side. This is the lead to disconnect before working on any electrical component on the car. If the clamp is tight, carefully ease it off by moving it from side to side.
3 Lift the plastic cover off the positive (+) lead, slacken the clamp bolt and lift the clamp and lead off the battery terminal.
4 Undo the retaining nuts and lift off the battery clamp bolt.
5 Lift the battery from the tray keeping it upright and taking care not to touch clothing.
6 If necessary, undo the retaining bolts and remove the battery tray from the car.
7 Clean the battery terminal posts, clamps, tray and battery casing. If the engine compartment around the battery area is rusted as a result of battery acid spilling onto it, clean it thoroughly and re-paint with reference to Chapter 11.

Refitting

8 Refitting is a reversal of removal, but always connect the positive (+) lead first and the negative (–) lead last.

5 Charging system – testing

1 If the alternator warning lamp fails to illuminate when the ignition is switched on, first check the wiring connections at the rear of the alternator for security. If satisfactory, check that the warning lamp bulb has not blown and is secure in its holder. If the lamp still fails to illuminate check the continuity of the warning lamp feed wire from the alternator to the bulb holder. If all is satisfactory, the alternator is at fault and should be renewed or taken to an automobile electrician for testing and repair.
2 If the alternator warning lamp illuminates when the engine is running, ensure that the drivebelt is correctly tensioned (see Chapter 1), and that the connections on the rear of the alternator are secure. If the fault still persists, the alternator should be taken to an automobile electrician for testing and repair, or renewed.
3 If the alternator output is suspect even though the warning lamp functions correctly, the regulated voltage may be checked as follows.
4 Connect a voltmeter across the battery terminals and then start the engine.
5 Increase the engine speed until the reading on the voltmeter remains steady. This should be between 13.5 and 14.8 volts.
6 Switch on as many electrical accessories as possible and check that the alternator maintains the regulated voltage at between 13.5 and 14.8 volts.
7 If the regulated voltage is not as stated, the fault may be due to a faulty regulator, a faulty diode, a severed phase winding or worn brushes, springs or commutator. The alternator should be taken to an automobile electrician for testing and repair, or renewed.

6 Alternator – removal and refitting

Removal

1 Disconnect the battery negative lead.
2 For greater access, apply the handbrake, chock the rear wheels and jack up the front of the car. Support the car on axle stands and remove the vehicle undertrays as necessary.
3 Undo the nut and disconnect the battery lead from the alternator. Unplug the wiring connector(s) from the rear of the alternator (photo).
4 Slacken the alternator pivot mounting bolt and the adjusting arm bolt, then move the alternator towards the engine. Slip the drivebelt off the alternator pulley (photo).

Chapter 12 Electrical system

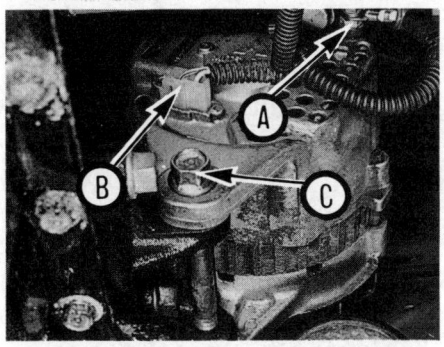

6.3 Undo the battery lead retaining nut (A) and disconnect the wiring connector (B), then slacken the pivot mounting bolt (C)...

6.4 ...and the adjusting arm bolt

6.5 Removing the alternator from the engine compartment

5 Remove the adjusting arm bolt and pivot mounting bolt, then carefully manoeuvre the alternator out of the engine compartment (photo).

Refitting

6 Refitting is the reverse sequence of removal. Adjust the drivebelt tension as described in Section 11 of Chapter 1 before tightening the mounting and adjusting arm bolts to the specified torque.

7 Alternator – overhaul

For reference purposes an exploded view of the alternator is shown in either Fig. 12.2 or 12.3 (as applicable). However, owing to the specialist knowledge and equipment which is required to dismantle and repair the alternator, it is recommended that if performance is suspect, the alternator should be taken to an automobile electrician who will have the facilities and experience to carry out such work. A practical alternative may be to replace the alternator with an exchange reconditioned unit.

8 Starting system – testing

1 If the starter motor fails to operate, first check the condition of the battery by switching on the headlamps. If they glow brightly then gradually dim after a few seconds, the battery is in a discharged condition.
2 If the battery is satisfactory, check the starter motor main terminal

Fig. 12.2 Exploded view of the alternator – pre-September 1985 models (Sec 7)

1 Through-bolt
2 Front bracket and rotor
3 Pulley retaining nut
4 Rear bracket and stator
5 Pulley and fan
6 Brush holder assembly
7 Rectifier

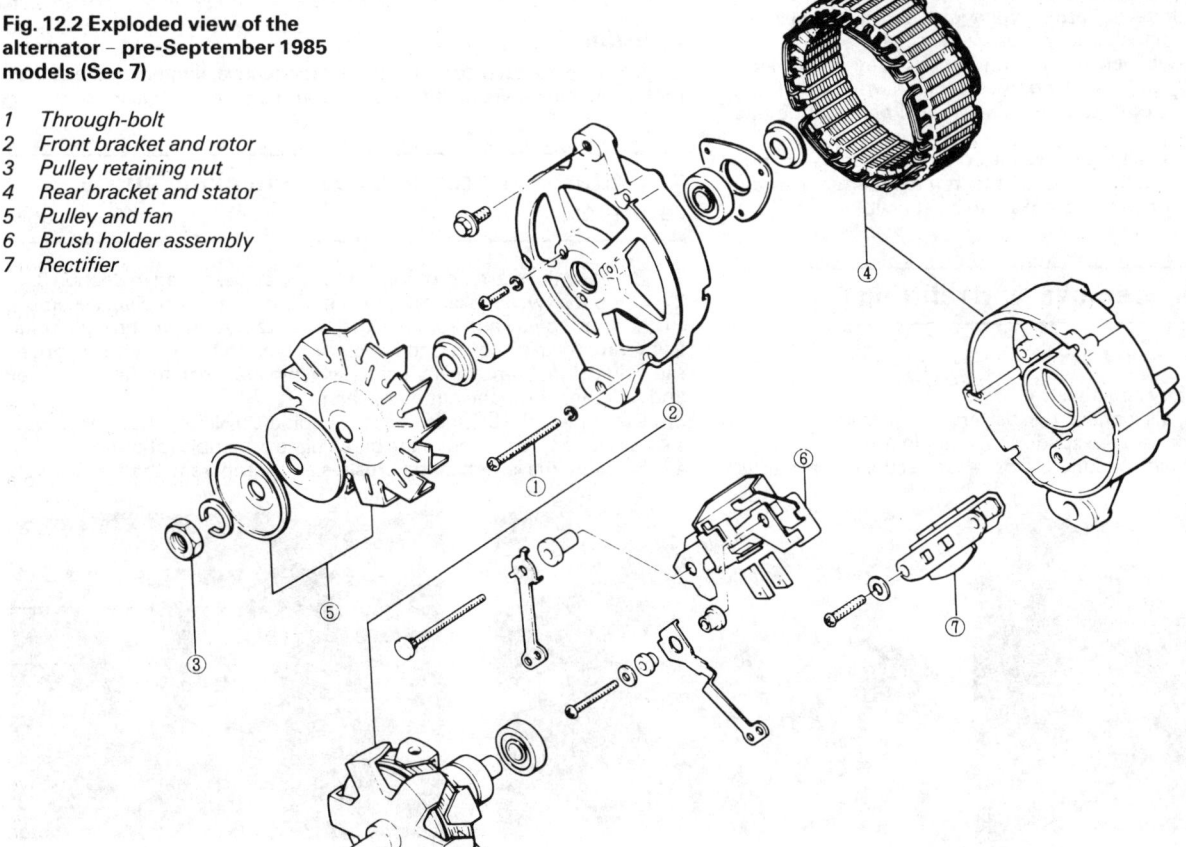

Chapter 12 Electrical system

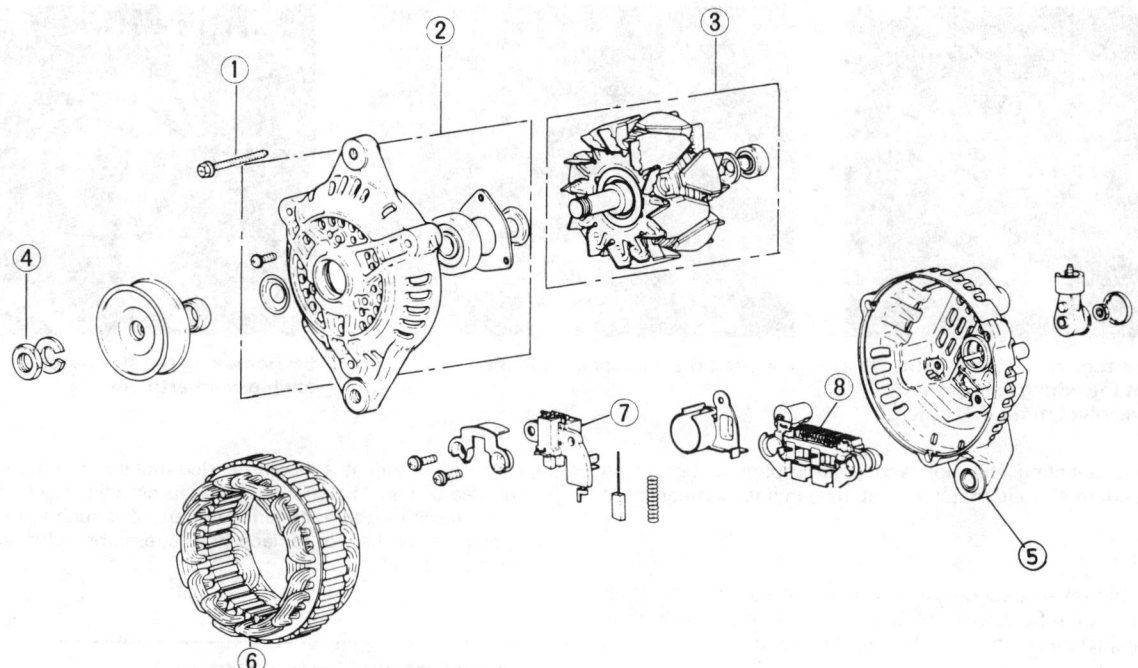

Fig. 12.3 Exploded view of the alternator – September 1985 models onward (Sec 7)

1 Through-bolt
2 Front bracket and bearing assembly
3 Rotor and fan assembly
4 Pulley retaining nut
5 Rear bracket
6 Stator
7 Brush holder assembly
8 Rectifier

and the engine earth cable for security. Check the terminal connections on the solenoid, located on the starter motor.
3 If the starter still fails to turn, use a voltmeter, or 12 volt test light and leads, to ensure that there is battery voltage at the solenoid main terminal (ie the cable from the battery positive terminal).
4 With the ignition switched on and the ignition key in the start position check that voltage is reaching the solenoid terminal with the spade connector, and also the starter main terminal beneath the end cover.
5 If there is no voltage reaching the spade connector there is a wiring or ignition switch fault. If voltage is available, but the starter does not operate, then the starter or solenoid is likely to be at fault.

9 Starter motor – removal and refitting

Removal

1 Disconnect the battery negative lead.
2 Undo the nut and disconnect the battery cable from the main solenoid terminal. Disconnect the spade connector (photo).
3 Undo the three retaining nuts and/or bolts securing the starter motor to the gearbox housing and, where necessary, the single bolt securing the rear of the starter motor to the block (photo).
4 Remove the starter motor from the engine (photo).

Refitting

5 Refitting is a reverse of the removal sequence, tightening the starter motor retaining bolts to the specified torque.

10 Starter motor brushes – checking and renewal

1 Remove the starter motor from the car as described in Section 9.
2 Undo the two small screws which secure the brush plate assembly to the rear cover, and remove the two starter motor through-bolts. Withdraw the rear cover and release it from the rubber grommet over the field cable. Remove any shims which have stuck to the rear cover and refit them onto the armature shaft.
3 Disconnect the brushes from their holders by lifting the springs with a screwdriver, then remove the brush plate assembly (photo).
4 Measure the length of the brushes and compare with the minimum

9.2 Starter solenoid main terminal nut and spade connector (arrowed)

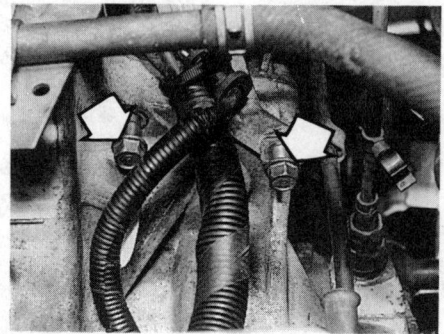

9.3 Undo the starter motor retaining bolts (upper two arrowed)...

9.4 ...and remove the starter motor from the engine compartment

Chapter 12 Electrical system

Fig. 12.4 Exploded view of the starter motor – pre-September 1985 model shown (Sec 10)

1 Solenoid and spacers
2 Through-bolt
3 Rear cover
4 Brush holder assembly
5 Yoke
6 Drive-end housing
7 Drive pinion and thrust collar
8 Armature

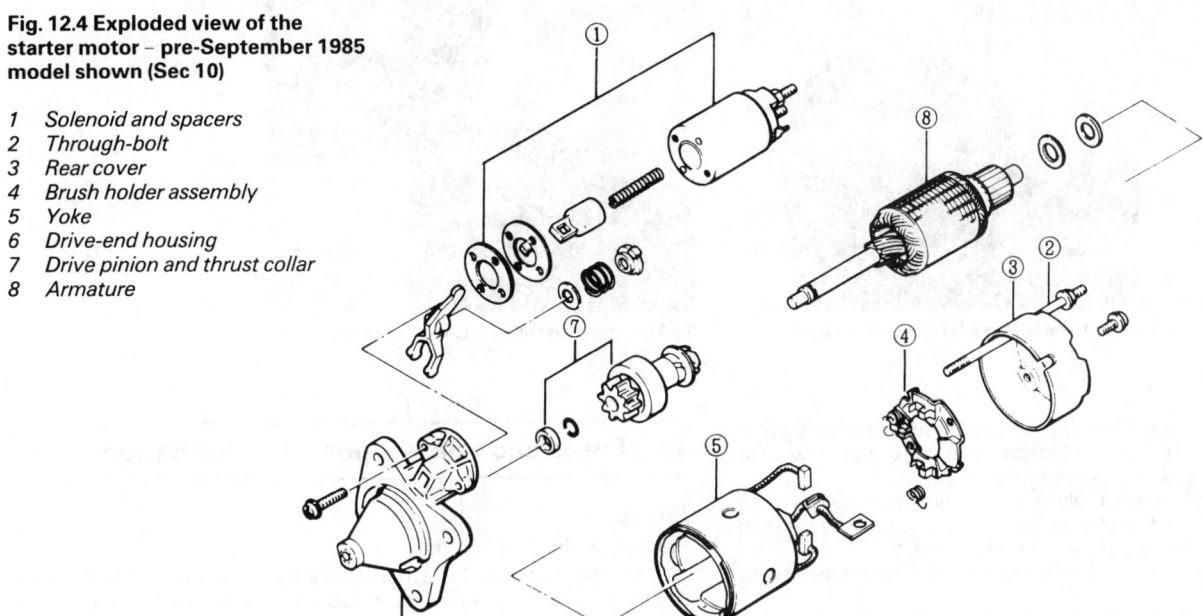

10.3 Lift the springs with a screwdriver and remove the brushes from the holders

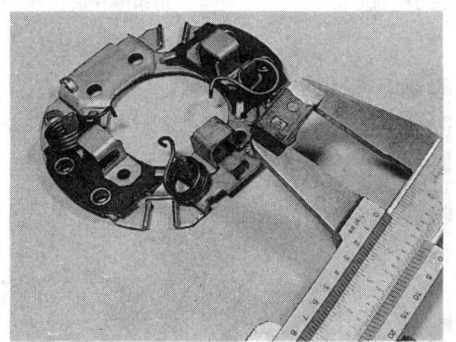

10.4 Using vernier calipers to measure the brush length

10.5 Using fine glass paper to clean the commutator

10.7A Fit the brush plate assembly over the commutator...

10.7B ...and check that the springs (arrowed) are seated correctly in the brush grooves

10.8 Locate the grommet (arrowed) in the slot and refit the rear cover

Chapter 12 Electrical system

10.9 Refit the brush plate retaining screws...

10.10 ...then install the through-bolts

dimension given in the Specifications (photo). If either brush is worn below this amount, renew all of the brushes. Note that new field brushes must be soldered to the existing leads.

5 Clean the brush holder assembly and wipe the commutator with a petrol-moistened cloth. If the commutator is dirty, it may be cleaned with fine glass paper, then wiped with the cloth (photo).
6 Fit the new brushes into their holders and check that they are able to slide freely.
7 Fit the brush plate assembly over the commutator and check that the springs are correctly seated on the brushes (photos).
8 Ensure that any necessary shims are in position on the armature shaft, then locate the field coil wire grommet in the rear cover and refit the cover (photo).
9 Align the cover holes with those in the brush holder then refit the small brush plate retaining screws and tighten them securely (photo).
10 Refit the starter motor through-bolts and tighten them securely (photo).
11 Refit the starter motor to the car as described in Section 9.

11 Fuses and relays – general information

Fuses

Pre-September 1985 models

1 On pre-September 1985 models the fusebox is situated on the front of the facia just to the right of the steering column. Lift the lid of the fusebox to gain access to the fuses.
2 To remove a fuse from its location, fit the tweezers supplied to the fuse then pull it directly out of the holder (photo). Slide the fuse sideways to remove it from the tweezers. The wire within the fuse is clearly visible and it will be broken if the fuse is blown.
3 Always renew a fuse with one of an identical rating. Never renew a fuse more than once without tracing the source of the trouble. The fuse rating is stamped on top of the fuse. The circuits protected by each fuse are shown on the fusebox.
4 In addition to the fuses, the headlamps are protected by two fusible

11.2 Using the tweezers provided to remove a fuse on pre-September 1985 models

11.4 Fusible links are located in the engine compartment

11.6 Circuit fuse box location on September 1985 models onward

11.7A Main fuses are located in the engine compartment

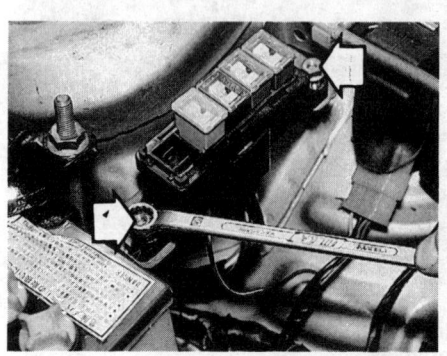

11.7B Remove the fuse box retaining nuts (arrowed)...

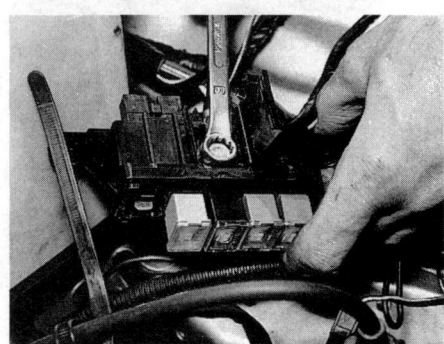

11.7C ...then undo the fuse terminal bolts...

Chapter 12 Electrical system

11.7D ...and remove the main 80 amp fuse

11.7E Smaller 30 and 40 amp fuses can simply be pulled out of position

11.10 Blower motor circuit breaker is located directly above the circuit fuse box

links which are located in the engine compartment in between the ignition coil and battery (photo).

5 If the headlamps fail to operate and the relevant fuses are not blown, check the fusible links. If either of the links has melted it must be replaced. Note however that if this happens there may be a fault in the circuit and thorough check should be carried out before fitting the new fusible link.

September 1985 models onward

6 On September 1985 models onward the fusebox, which contains the smaller circuit fuses, is located beneath the facia just to the right of the accelerator pedal. To remove a fuse from its location remove the fuse box cover, then fit the tweezers supplied to the fuse and pull it directly out of the holder (photo). Slide the fuse sideways to remove it from the tweezers.

7 A further fusebox which contains the main fuses is situated next to the battery in the engine compartment. To remove the main (80 amp) fuse, undo the two fusebox retaining nuts and lift up the fuse box. Open the covers on the side of the fusebox to gain access to the main fuse terminals. Undo the bolts, then disconnect the terminals and remove the main fuse. On refitting tighten the terminal screws and fusebox retaining nuts securely. The smaller 30 and 40 amp fuses can simply be pulled out of position (photos).

8 The wire within the fuse is clearly visible and it will be broken if the fuse is blown. Always renew a fuse with one of an identical rating. Never renew a fuse more than once without tracing the source of the trouble. The fuse rating is stamped on top of the fuse. The circuits protected by each fuse are shown on the fusebox.

9 In addition to the fuses, the heater blower motor is protected by a circuit breaker.

10 If the heater blower motor fails to operate and the relevant fuses are not blown, check the circuit breaker which is located in the junction box directly above the circuit fusebox (photo). If the circuit breaker has been tripped, the red button will have popped out. If this is the case, it is likely that there is a fault in the blower motor circuit. Thoroughly check the blower motor circuit before resetting the circuit breaker by pushing the button in.

Relays

11 On pre-September 1985 models the cooling fan and horn relay are situated in the left-hand corner of the engine compartment. All other relays are located behind the instrument panel which must first be removed to gain access to them.

12 On September 1985 models onward all the relays, with the exception of those which operate the electric sunroof (where fitted), are either located in the left-hand corner of the engine compartment or in the relay junction box which is situated behind the right-hand end of the facia (photo). The electric sunroof relays are situated behind the roof trim panel where they are just in front of the sunroof itself. To gain access to the relay junction box remove the facia as described in Chapter 11.

13 If a system controlled by a relay becomes inoperative and the relay is suspect, operate the system and if the relay is functioning it should be possible to hear it click as it is energized. If this is the case the fault lies with the components or wiring of the system. If the relay is not being energized then the relay is not receiving a main supply voltage or a switching voltage, or the relay is faulty.

12 Switches – removal and refitting

Ignition switch/steering lock

1 Remove the steering column as described in Chapter 10.
2 With the column assembly on the bench, use a hammer and chisel to cut a slot in the round heads of the clamp plate retaining shear bolts (Fig. 12.5).
3 Unscrew the shear bolts using a flat bladed screwdriver applied to the chisel slot, then lift off the clamp plate and remove the ignition switch.
4 To refit the lock, place the lock in position on the column then fit the clamp plate. Screw in two new shear bolts tightening them finger tight only at this stage.
5 Using the ignition key check the operation of the lock mechanism. If the lock operation is satisfactory tighten the shear bolts until their heads break off.
6 Refit the steering column as described in Chapter 10.

Steering column combination switch

7 Disconnect the battery negative lead.
8 Remove the steering wheel as described in Chapter 10.
9 Undo the screws which retain the two halves of the steering column shroud. Lift off the left-hand or upper shroud and remove the bulbholder (if equipped) in the right-hand or lower shroud (as applicable). Remove both shrouds from the steering column.
10 Disconnect the wiring connectors from the rear of the steering column combination switch. Undo the switch clamp bolt and slide the switch off the column (photo). If necessary the switches can be removed from the main body after undoing the retaining screws at the rear.

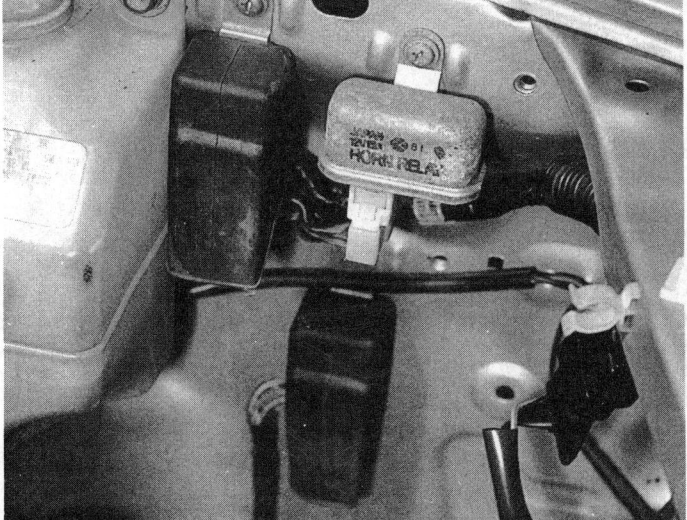

11.12 Engine compartment relays on September 1985 models onward

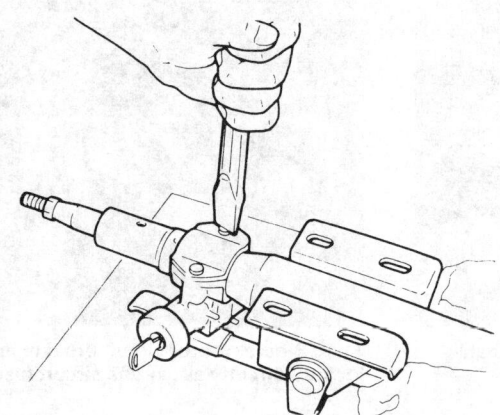

Fig. 12.5 Use a chisel to cut slots in the ignition switch shear bolts (Sec 12)

11 Refitting is a reverse of the removal procedure.

Facia switches

12 Disconnect the battery negative lead.

Pre-September 1985 models

13 Undo the screws along the upper edge securing the right-hand half of the switch panel to the facia.
14 Pull the upper part of the switch panel outwards then disengage the lower catches. Disconnect the switch wiring and remove the panel (photo).
15 Depress the switch retaining tangs and withdraw the switches from the panel.
16 To remove the switches on the auxiliary panel adjacent to the fusebox, first remove the facia undercover. Disconnect the wiring and release the relevant switch by reaching up behind the facia and releasing its retaining tangs.
17 Refitting is the reverse sequence of removal.

September 1985 models onward

18 Undo the three upper screws, and the two lower screws securing the instrument panel shroud to the facia and withdraw the shroud (photos).
19 Disconnect the wiring from the instrument panel shroud switches and remove the panel (photo).
20 Undo the cluster switch retaining screws and remove them from the shroud (photo). If necessary, each individual switch unit can be removed from the rear of the assembly by carefully prising off the knob (where applicable) and freeing the switch retaining tangs at the rear of the switch as shown in Fig. 12.6.
21 To remove the switches from the right-hand vent panel, referring to Chapter 11 if necessary, remove the coin box holder and pull the knob off the right-hand vent control lever. Remove the three vent/switch panel retaining screws, disconnect the switch wiring and remove the panel. Compress the retaining tangs and remove the switches from the panel.
22 Refitting is a reverse of the removal procedure.

Door switches

23 Remove the door inner trim panel as described in Chapter 11.
24 Compress the switch retaining tangs and remove the switch from the panel.
25 Refitting is a reverse of removal.

Courtesy lamp switches

26 With the door open, undo the two screws securing the switch to the body. Pull out the switch and tie a piece of string to the wiring to prevent it from dropping into the body.
27 Disconnect the switch and remove it from the vehicle.
28 Refitting is a reverse of removal.

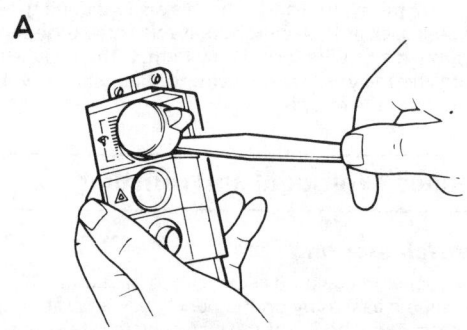
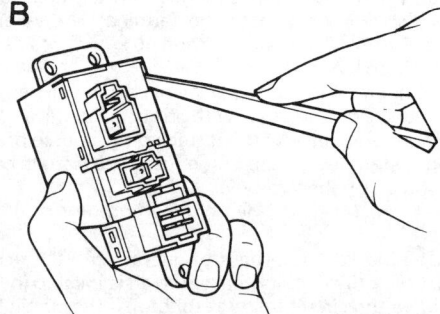

Fig. 12.6 Instrument panel shroud switch disassembly – September 1985 models onward (Sec 12)

A Prise free the knob (where fitted)
B Carefully prise switch free from the rear face of the panel

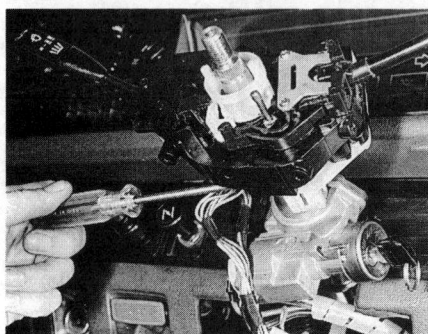

12.10 Undo the combination switch clamp and slide the assembly off the steering column

12.14 Disconnect the switch wiring and remove the switch panel – pre-September 1985 models

12.18A On September 1985 models onward undo the instrument panel shroud upper...

Chapter 12 Electrical system

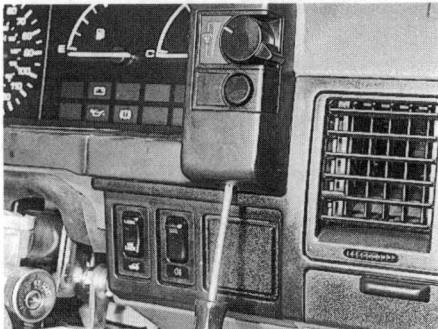

12.18B ...and lower retaining screws...

12.18C ...then withdraw the shroud...

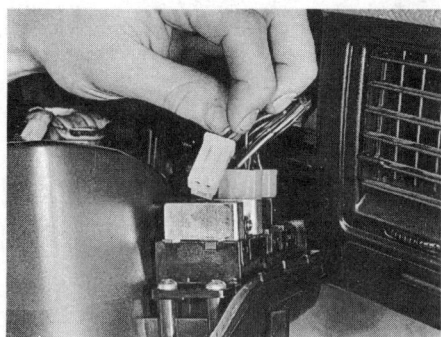

12.19 ...and disconnect the switch wiring connectors

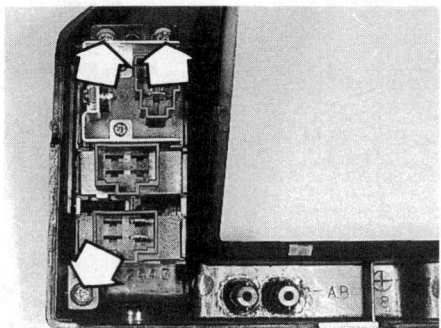

12.20 Cluster switch retaining screws (arrowed)

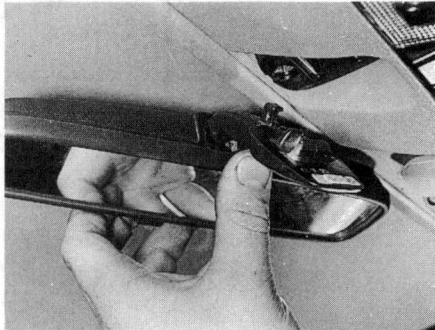

12.33 Unclip the rear view mirror...

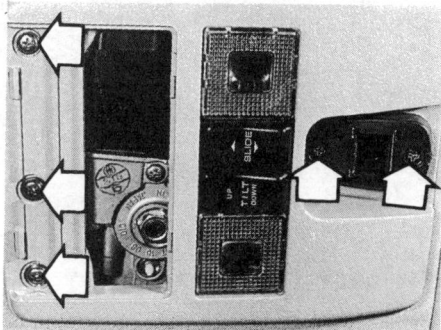

12.35 ...and remove the overhead console retaining screws (arrowed)

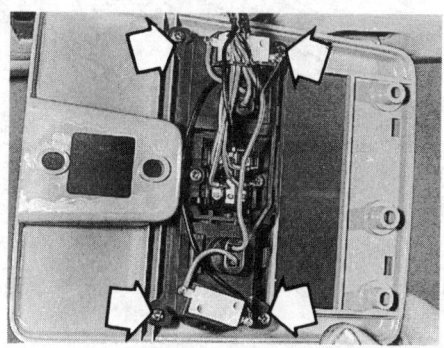

12.36 Switch assembly retaining screws (arrowed)

Handbrake warning lamp switch

29 Remove the centre console as described in Chapter 11.
30 Disconnect the wire from the handbrake switch, then undo the retaining bolt and remove the switch from the handbrake lever.
31 On refitting, reconnect the wire and tighten the switch retaining bolt finger tight only. Switch on the ignition switch and check that the warning lamp on the instrument panel illuminates when the handbrake lever is applied by one click of the handbrake ratchet mechanism. If not, reposition the switch until the operation is correct then tighten the retaining bolt securely.
32 Refit the centre console as described in Chapter 11.

Overhead console switches

33 Unclip the rear view mirror by pushing it towards the front of the car (photo).
34 On models fitted with a manual sunroof carefully prise off the sunroof handle. On models fitted with an electric sunroof remove the sliding panel from the rear of the console.

35 Undo all the console retaining screws and free the console from the roof (photo). Disconnect the switch wiring and remove the console.
36 Undo the switch assembly retaining screws and separate the switch assembly and console (photo).
37 Refitting is the reverse of the removal procedure.

13 Bulbs (exterior lamps) – renewal

Headlamp

1 From within the engine compartment pull off the wiring connector from the rear of the headlamp bulb (photo).
2 Slip off the rubber cover then either, remove the bulb retaining ring, or release the bulb retaining spring clip and pivot the clip clear (as applicable). Withdraw the bulb. Take care not to touch the bulb glass with your fingers – if touched, clean the bulb with methylated spirits (photos).

Chapter 12 Electrical system

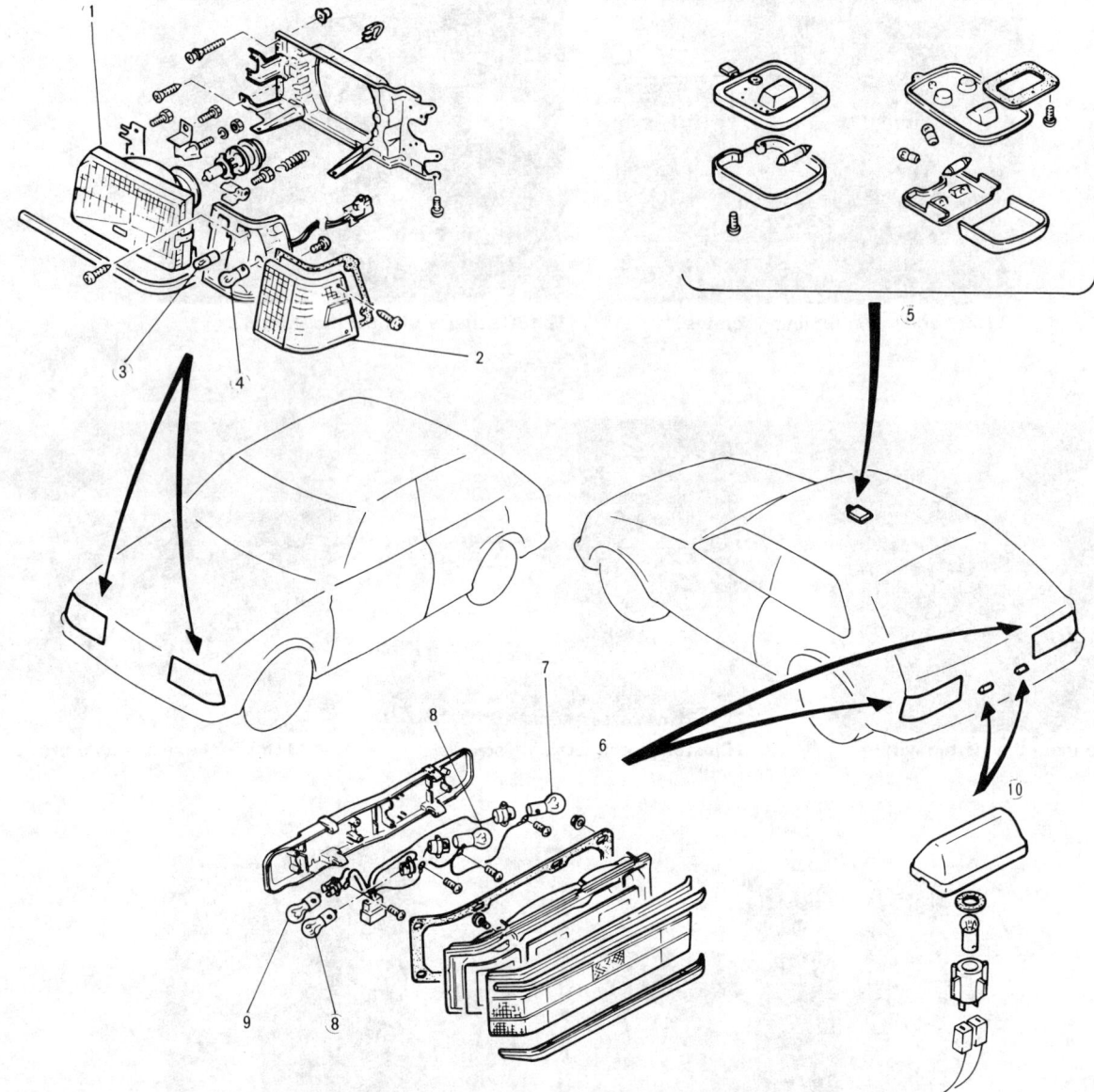

Fig. 12.7 Exploded view of the vehicle lighting assemblies – pre-September 1985 models (Secs 13 to 15)

1 Headlamp lens unit	3 Sidelamp bulb	6 Rear lamp clusters	9 Direction indicator bulb
2 Direction indicator/sidelamp lens	4 Direction indicator bulb	7 Reversing lamp bulb	10 Number plate lamp
	5 Interior courtesy lamp	8 Stop/tail lamp bulb	

3 Fit the new bulb using a reversal of the removal procedure, ensuring that the tabs on the bulb support are correctly located in the lens assembly cutouts.

Front sidelamp

4 Undo the screws and remove the sidelamp lens.
5 Remove the bulb from the holder by turning it anti-clockwise.
6 Refit the bulb by reversing the removal sequence. Do not overtighten the lens retaining screws as the lens is easily cracked.

Front direction indicator

7 Undo the screws and remove the direction indicator lens, which is integral with the sidelamp lens on early models, or located in the front bumper on later models (photo).
8 Remove the bulb from the holder by turning it anti-clockwise (photo).
9 Refit the bulb by reversing the removal sequence. Do not overtighten the lens retaining screws as the lens is easily cracked.

Front direction indicator repeater

10 Undo the two screws and remove the lens from the side of the wing (photo).
11 Withdraw the push fit bulb from the holder (photo).
12 Refitting is the reverse of removal.

Rear lamp cluster

Hatchback and Saloon models

13 Working in the luggage compartment, where necessary release the retaining clip and open up the rear lamp trim panel cover.
14 Depress the catches and withdraw the bulb panel from the lens unit (photo).
15 The relevant bulb can then be removed from the panel by twisting it anti-clockwise.
16 Refitting is the reverse of removal.

Estate models

17 Open the tailgate, then undo the two cluster unit retaining screws (photo).

Chapter 12 Electrical system

13.1 Pull the wiring connector off the headlamp bulb...

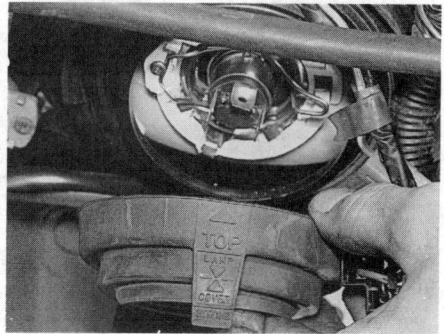

13.2A ...and remove the rubber cover

13.2B Release the bulb retaining clip...

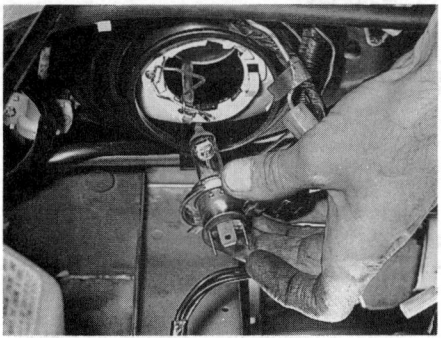

13.2C ...and withdraw the bulb

13.7 On later models undo the direction indicator lens retaining screws...

13.8 ...and remove the lens followed by the bulb

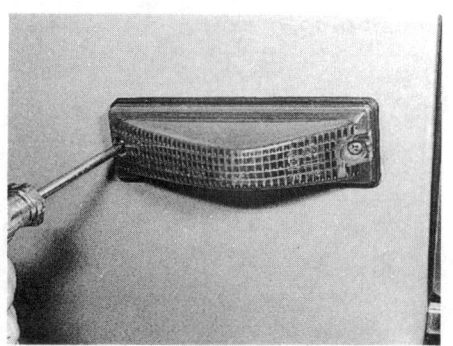

13.10 Undo the side repeater lens retaining screws then remove the lens...

13.11 ...and withdraw the push fit bulb

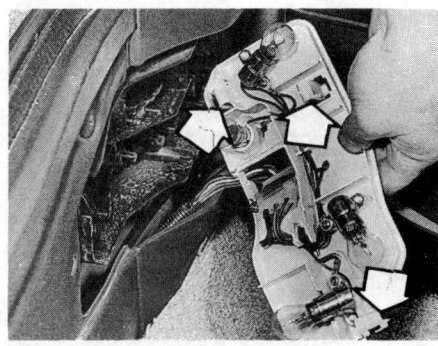

13.14 Release the retaining catches (arrowed) and remove the rear lamp cluster bulb panel from the lens unit (late Hatchback model shown)

13.17 On Estate models undo the rear lamp cluster retaining screws...

13.18 ...then withdraw the assembly and release the bulbholders by twisting them in an anti-clockwise direction

13.24A On Saloon and Hatchback models remove the number plate lamp from the bumper...

13.24B ...and release the bulbholder from the rear of the unit

13.25 Twist the bulb anti-clockwise and remove it from the holder

18 Withdraw the cluster unit, then twist the bulbholder(s) in an anti-clockwise direction to free them from the rear of the assembly (photo).
19 Remove the bulb from the holder by twisting it in an anti-clockwise direction.
20 Refitting is the reverse of removal.

Rear foglamp (Estate models)
21 Undo the two lens retaining screws and remove the lens.
22 Remove the bulb by twisting it in an anti-clockwise direction.
23 Refitting is the reverse of removal.

Number plate bulbs
Hatchback and Saloon models
24 Carefully prise the light unit out of the bumper then turn the bulbholder anti-clockwise to remove it from the unit (photos).
25 Remove the bulb from its holder by turning it anti-clockwise (photo).

26 Refitting is the reverse of removal.

Estate models
27 Raise the tailgate, undo the two retaining screws and remove the lens.
28 Twist the bulb anti-clockwise and remove it from its holder.
29 Refitting is the reverse of removal.

14 Bulbs (interior lamps) – renewal

Courtesy lamps
1 Carefully prise the lens off the light unit, then remove the festoon bulb from its edge contacts.
2 Fit the new bulb using a reversal of the removal procedure, but check the tension of the spring contacts and if necessary bend them so that they firmly grip the bulb end caps (photo).

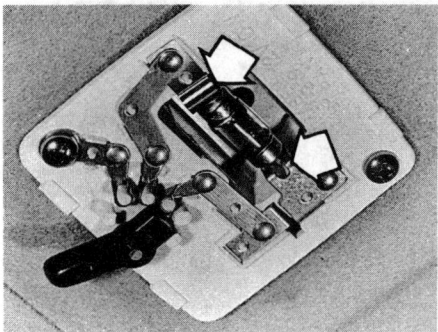

14.2 Ensure spring contacts (arrowed) grip the bulb firmly

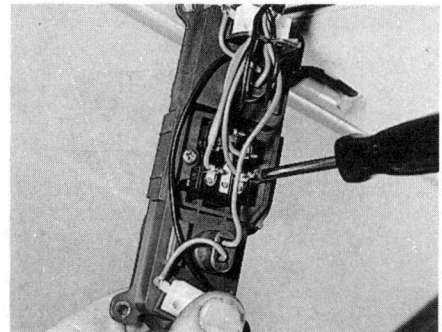

14.4A Loosen the switch retaining screws...

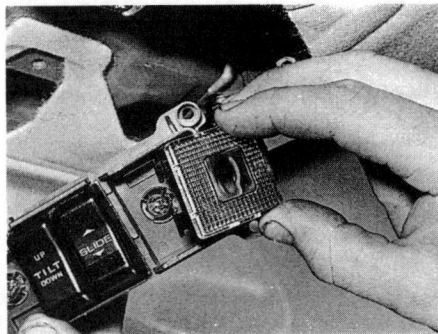

14.4B ...and remove the map reading lamp lenses

14.4C Bulbs are a push fit in their holders

14.11 Removing an instrument panel bulb

Chapter 12 Electrical system

Fig. 12.8 Exploded view of headlamp, sidelamp and front direction indicator lamps – September 1985 models onward (Sec 15)

1 Sidelamp assembly
2 Front direction indicator lamp assembly
3 Headlamp assembly

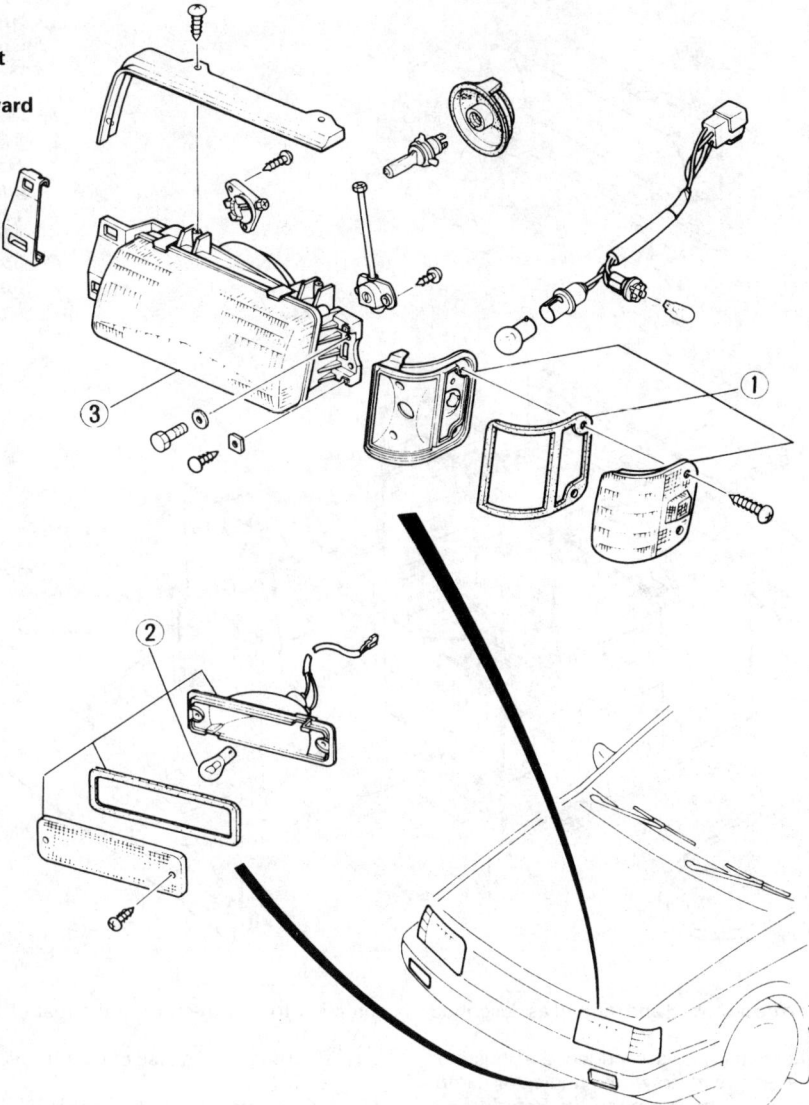

Map reading lamps

3 Remove the overhead console switch assembly as described in Section 12.
4 Loosen the two screws on the rear of the assembly and remove the map reading lamp lenses. The bulbs can then be pulled out of position (photos).
5 Refitting is a reversal of the removal procedure.

Luggage compartment lamp

6 Prise out the lamp using a small screwdriver.
7 Release the festoon type bulb from the spring contacts or pull the bulb out of its holder (as applicable).
8 Fit the new bulb using a reversal of the removal procedure. If a festoon bulb is fitted check the tension of the spring contacts and if necessary bend them so that they firmly grip the bulb end caps.

Instrument panel

9 Remove the instrument panel as described in Section 16.
10 The bulbholders are secured to the rear of the instrument panel by a bayonet fitting and can be removed by twisting them anti-clockwise.
11 Most bulbs can be removed from their holders by twisting them in an anti-clockwise direction (photo). **Note:** *Some bulbs are an integral part of the bulbholder and cannot be removed. These must be renewed as a unit.*
12 Refitting is a reversal of the removal procedure.

15 Exterior lamp units – removal and refitting

Headlamp

1 Remove the radiator grille as described in Chapter 11.
2 From within the engine compartment disconnect the wiring from the rear of the headlamp bulb.
3 Remove the front sidelamp as described in paragraphs 7 to 9 inclusive.
4 Undo the two inner headlamp retaining screws and remove the headlamp unit.
5 Refitting is a reversal of the removal procedure. On completion adjust the headlamp aim as described in Chapter 1, Section 11.

Front sidelamp

6 Working from inside the engine compartment disconnect the sidelamp or sidelamp/indicator assembly wiring at the wiring connector.
7 Undo the screws and remove the sidelamp assembly lens.
8 Undo the two retaining screws and remove the assembly from the car.
9 Refitting is a reverse of the removal procedure.

Front direction indicator

10 On early models the front direction indicator is an integral part of

Chapter 12 Electrical system

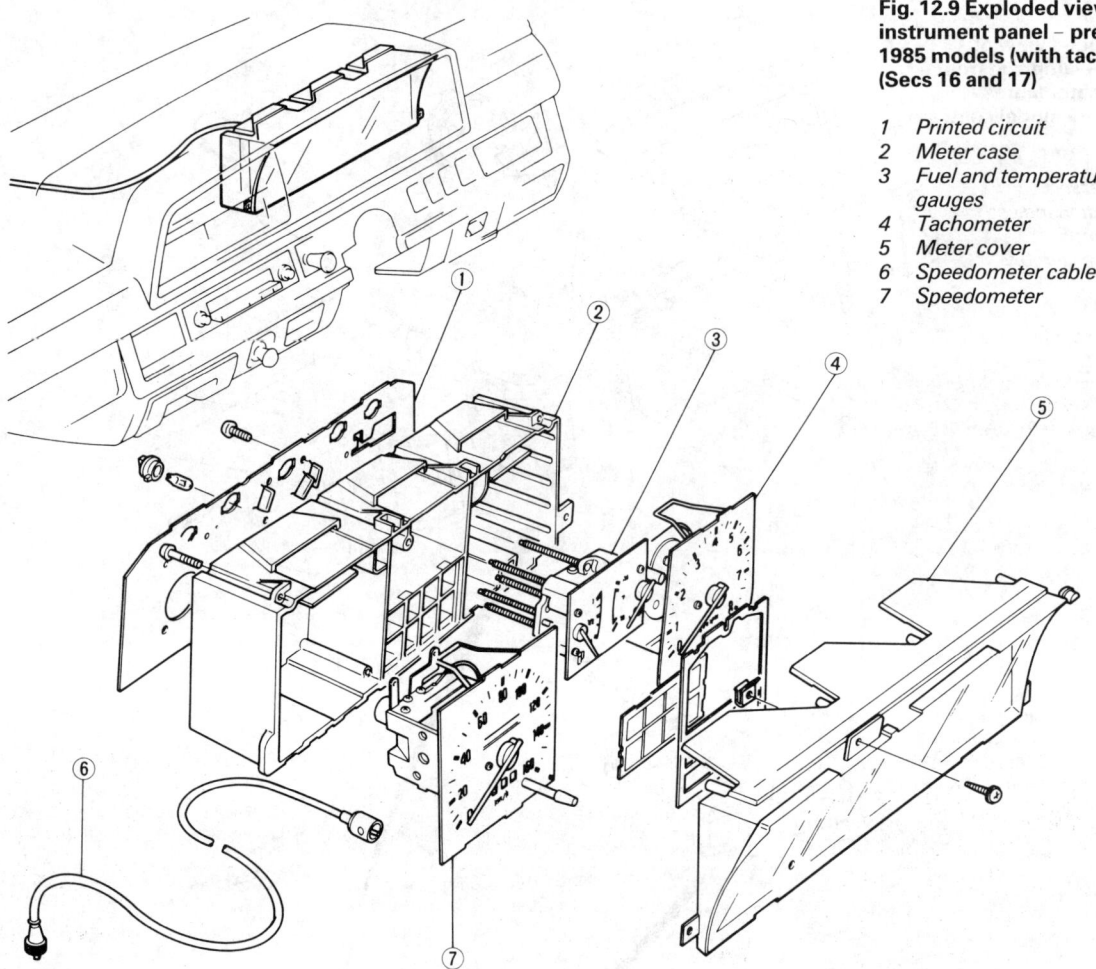

Fig. 12.9 Exploded view of the instrument panel – pre-September 1985 models (with tachometer) (Secs 16 and 17)

1. Printed circuit
2. Meter case
3. Fuel and temperature gauges
4. Tachometer
5. Meter cover
6. Speedometer cable
7. Speedometer

the sidelamp assembly which can be removed and refitted as described in paragraphs 7 to 10 inclusive.
11 On later models, undo the two screws securing the indicator assembly to the front bumper and withdraw the indicator lamp. Disconnect the wiring and remove the lamp from the car. Refitting is a reversal of removal.

Front direction indicator repeater

12 Undo the two screws securing the direction indicator lamp to the wing. Withdraw the lamp assembly, disconnect the wiring and remove the lamp from the car along with the gasket.
13 Refitting is a reverse of the removal procedure, noting that the gasket should be renewed if damaged.

Rear lamp cluster

Hatchback and Saloon models

14 Working in the luggage compartment, where necessary release the retaining clip and open up the rear lamp trim panel cover.
15 Depress the catches and withdraw the bulb panel from the lens unit.
16 Undo the lens unit retaining nuts or screws (as applicable) and remove it from the car along with its gasket.
17 Examine the gasket for signs of damage or deterioration and renew if necessary. Refit the lamp by reversing the removal procedure.

Estate models

18 Open the tailgate then undo the two cluster unit retaining screws and withdraw the cluster unit.
19 Carefully prise the cluster wiring grommet out of position then withdraw the rear lamp wiring until the wiring connector appears. Disconnect the wiring connector and remove the lamp cluster assembly from the car. Alternatively, twist the bulbholders in an anti-clockwise

direction to free them from the rear of the assembly then remove the lens unit.
20 Refitting is a reverse of the removal procedure.

Rear foglamp (Estate models)

21 Working in the luggage compartment, lift up the carpet to gain access to the foglamp mountings and wiring.
22 Disconnect the wiring at the wiring connector then displace the grommet from the floor panel and feed the wiring out through the floor.
23 Undo the fog lamp retaining bolts and remove them from the car.
24 Refitting is a reverse of the removal procedure.

Number plate lamps

25 On Hatchback and Saloon models, carefully prise the lamp unit out of the bumper then disconnect the wiring and remove it from the car. Refitting is the reverse of removal.
26 On Estate models open up the tailgate then undo the two retaining screws. Disconnect the wiring connector and remove the lamp from the car along with the gasket. Refitting is a reversal of the removal procedure noting that the gasket should be renewed if damaged.

16 Instrument panel – removal and refitting

Removal

1 Remove the steering column combination switch as described in Section 12 of this Chapter.
2 On pre-September 1985 models, pull the knob off the heater blower motor switch then undo the four upper retaining screws securing the instrument panel shroud to the facia. Push down on the flat surface of

Chapter 12 Electrical system

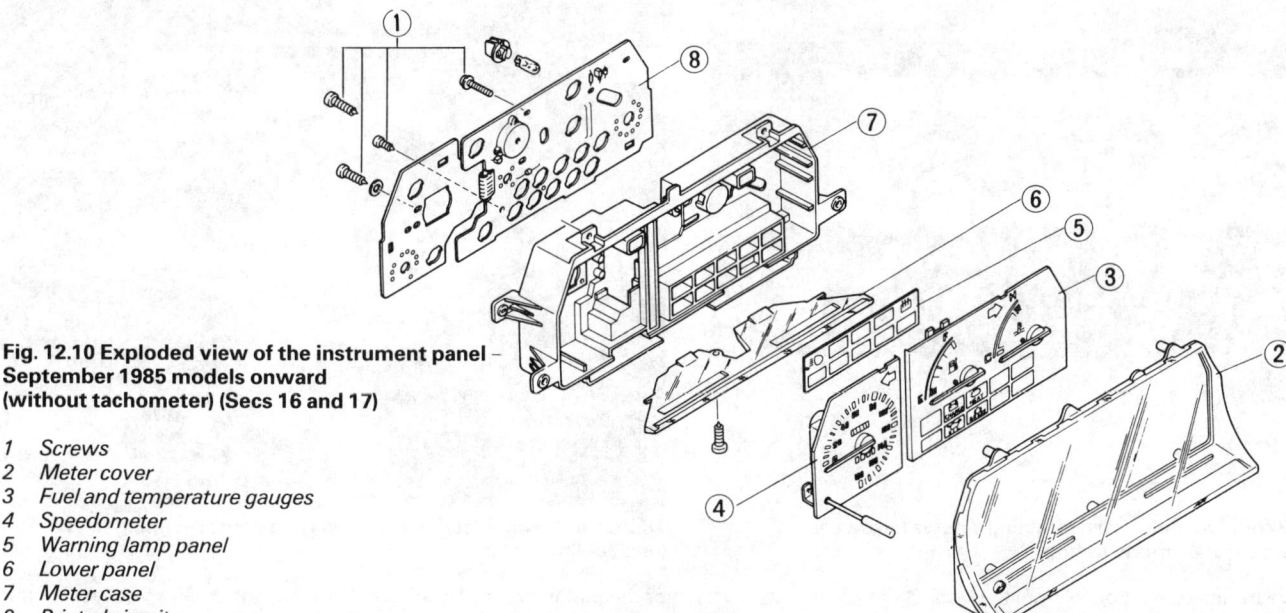

Fig. 12.10 Exploded view of the instrument panel – September 1985 models onward (without tachometer) (Secs 16 and 17)

1 Screws
2 Meter cover
3 Fuel and temperature gauges
4 Speedometer
5 Warning lamp panel
6 Lower panel
7 Meter case
8 Printed circuit

16.2A On pre-September 1985 models remove the retaining screws...

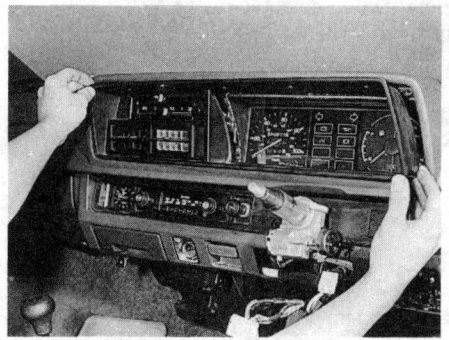

16.2B ...and remove the instrument panel shroud

16.4A Instrument panel retaining screws – pre-September 1985 models

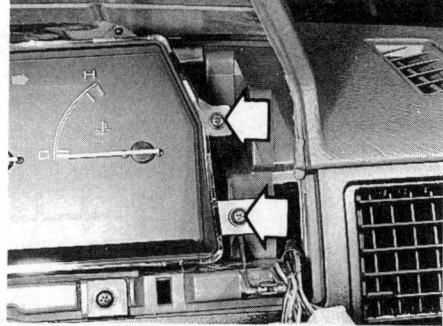

16.4B Instrument panel retaining screws – September 1985 models onward

16.5 Disconnect the wiring from the rear of the panel

the shroud below the instrument panel and pull the left-hand side of the shroud outwards. Withdraw the shroud from the facia (photos).
3   On September 1985 models onward, undo the three upper screws, and the two lower screws securing the instrument panel shroud to the facia and withdraw the shroud. Disconnect the wiring from the instrument panel shroud switches and remove the panel.
4   Undo the four retaining screws securing the instrument panel to the facia frame (photos).

5   Pull the instrument panel out slightly then carefully reach behind the panel and disconnect the speedometer cable by either unscrewing its knurled retaining ring or releasing its retaining clip (as applicable). **Note:** *If necessary, disconnect the speedometer cable at the transmission to allow the instrument panel to be withdrawn sufficiently to gain access to the rear of the panel.* Make a note of the location of the wiring connectors, to use as a guide on refitting, then disconnect them from the rear of the panel (photo).

17.2 Meter cover to meter case retaining screws (arrowed) (pre-September 1985 models shown)

18.5 Cigar lighter location showing facia panel retaining tabs (arrowed)

6 Remove the instrument panel from the facia.

Refitting

7 Refitting is a reverse of the removal procedure. On completion check the operation of all the panel warning lamps and switches (as applicable) to ensure that they are functioning correctly.

17 Instrument panel components – removal and refitting

Removal

1 Remove the instrument panel as described in Section 16 of this Chapter.
2 Undo the retaining screws from the rear of the assembly, then release the lower tangs and separate the meter cover from the meter case (photo).
3 To remove the speedometer, undo the two retaining screws from the rear of the case and withdraw the meter. Where fitted, the tachometer is removed in the same way.
4 To remove the water temperature gauge/fuel gauge assembly, undo the nuts securing the gauge terminals to the printed circuit then remove the gauge assembly from the meter case.
5 To remove the voltage regulator, disconnect the lead at the spade terminal then undo the retaining screw and remove the regulator from the back of the meter case.
6 To remove the printed circuit first remove the tachometer, speedometer, temperature/fuel gauge and regulator (as applicable) as described in paragraphs 3 to 5. Remove all the bulbholders from the rear

of the meter case by twisting them in an anti-clockwise direction, then undo the small retaining screws and remove the printed circuit from the meter case.

Refitting

7 Refitting is a reversal of the removal procedure.

18 Cigar lighter – removal and refitting

Removal

1 Disconnect the battery negative lead.

Pre-September 1985 models

2 Release the retaining clip and withdraw the ashtray from the facia.
3 Reaching in through the ashtray aperture in the facia, release the cigar lighter retaining tangs and push the cigar lighter out from the facia.
Note: *If it is not possible to release the cigar lighter retaining clips through the ashtray aperture it will be necessary to remove the facia panel as described in Chapter 11.*
4 Disconnect the wiring and remove the cigar lighter.

September 1985 models onward

5 Remove the ashtray to gain access to the two lower central facia panel retaining screws. Undo the screws and carefully prise the panel retaining tabs out using a flat bladed screwdriver (photo). Withdraw the panel and disconnect the wiring and aerial connections from the radio/cassette player.
6 Remove the bulbholder from the cigar lighter housing and disconnect the wiring from the cigar lighter (photo).

18.6 Remove the bulbholder and disconnect the wiring connector

18.7 Remove the cigar lighter housing...

18.8 ...and compress the retaining tang (arrowed) to free the cigar lighter from the facia panel

Chapter 12 Electrical system

7 Undo the retaining screw and remove the cigar lighter housing from the rear of the panel (photo).
8 Release the cigar lighter retaining tangs and withdraw the lighter from the facia panel (photo).

Refitting
9 Refitting is a reversal of the removal procedure.

19 Clock – removal and refitting

Removal
1 Disconnect the battery negative lead.

Pre-September 1985 models
2 Remove the radio control knobs and retaining nuts.
3 Undo the choke knob retaining grub screw and pull off the knob.
4 Undo the knurled retaining nut securing the choke cable to the switch panel.
5 Undo the screws along the upper edge that secure the left-hand half of the switch panel to the facia.
6 Pull the upper part of the switch panel outwards then disengage the lower catches. Remove the panel.
7 Undo the two clock retaining screws and withdraw it from the facia.
8 Disconnect the wiring and remove the clock.

September 1985 models onward
9 Remove the facia as described in Chapter 11.
10 Release the clock retaining tangs and remove it from the facia.

Refitting
11 Refitting is a reversal of the removal procedure.

20 Horn – removal and refitting

Removal
1 The horns are located on the body front valance behind the front bumper. To remove a horn, first apply the handbrake then jack up the front of the car and support on axle stands.
2 Disconnect the battery negative lead, then reach up and disconnect the horn supply wires.
3 Unscrew the nut securing the horn to the mounting bracket and remove the horn from the car.

Refitting
4 Refitting is a reversal of removal.

21 Speedometer drive cable – removal and refitting

Removal
1 Refer to Section 16 and remove the instrument panel sufficiently to allow the speedometer cable to be disconnected.
2 Release the grommet from the bulkhead and pull the cable through into the engine compartment.
3 Unscrew the knurled retaining ring securing the cable to the speedometer drive on the transmission housing and disconnect the cable at its lower end.
4 Release all the relevant cable clips and ties and remove the cable from the car.

Refitting
5 Refitting is a reverse of the removal procedure ensuring that the cable is correctly routed and retained by all the relevant clips and ties.

22 Windscreen wiper motor and linkage – removal and refitting

Removal
1 Operate the wiper motor then switch it off so that it returns to its rest position.
2 Disconnect the battery negative lead.
3 Lift up the hinged covers and unscrew the wiper arm retaining nuts (photo). Carefully prise the wiper arms off the spindles.

Pre-September 1985 models
4 Undo the screw securing the windscreen washer nozzle to the motor access cover on the engine compartment bulkhead (photo).
5 Undo the access cover retaining screws and withdraw the cover (photo).
6 Lift off the covers and undo the nuts securing the wiper spindles to the scuttle. Lift off the rubber grommets and push the spindles through the scuttle.
7 Disconnect the wiper motor at the wiring harness.
8 Undo the screws securing the motor mounting plate to the body and withdraw the motor mounting plate through the opening.
9 Manipulate the assembly out of the opening and remove it from the car.
10 If necessary the motor can be removed from the mounting plate after undoing the retaining screws and removing the primary cranking arm. Before disturbing the crank arm mark its position in relation to the motor body to ensure that it is refitted in its original position on reassembly.

22.4 Washer nozzle retaining screw (arrowed) – pre-September 1985 models

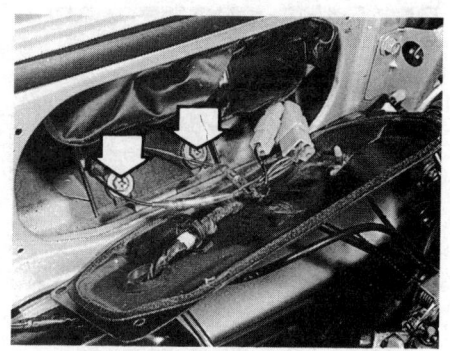

22.5 Withdraw the cover to gain access to the wiper motor retaining bolts (arrowed)

22.3 Lift up the hinged covers to gain access to wiper arm nuts

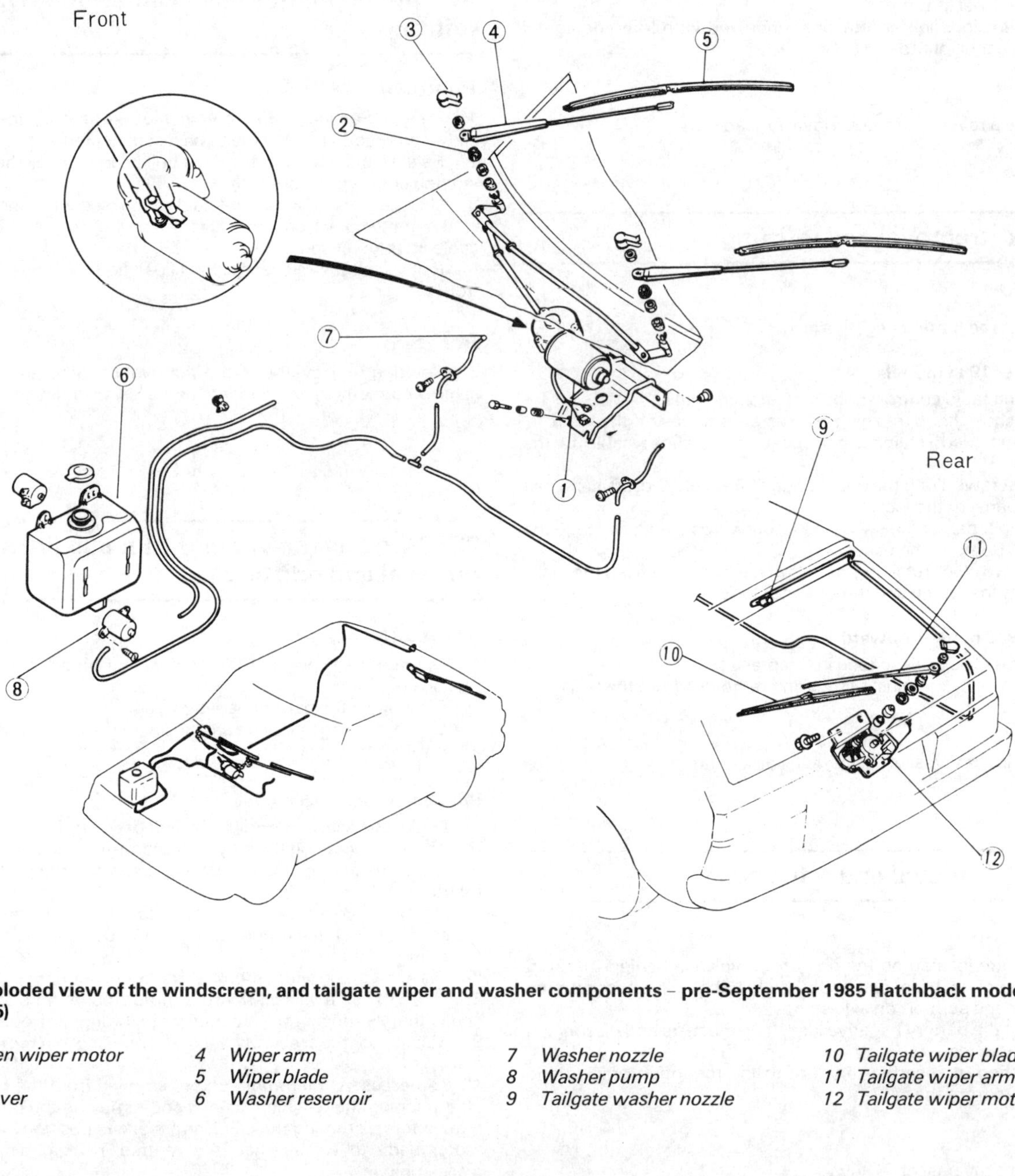

Chapter 12 Electrical system

Fig. 12.11 Exploded view of the windscreen, and tailgate wiper and washer components – pre-September 1985 Hatchback models (Secs 22 to 25)

1 Windscreen wiper motor
2 Seal
3 Hinged cover
4 Wiper arm
5 Wiper blade
6 Washer reservoir
7 Washer nozzle
8 Washer pump
9 Tailgate washer nozzle
10 Tailgate wiper blade
11 Tailgate wiper arm
12 Tailgate wiper motor

22.14 On September 1985 models onward remove the wiper motor cover...

22.15 ...and separate the wiper linkage balljoint

22.16 Wiper motor retaining bolts and wiring connector – September 1985 models onward

Chapter 12 Electrical system 287

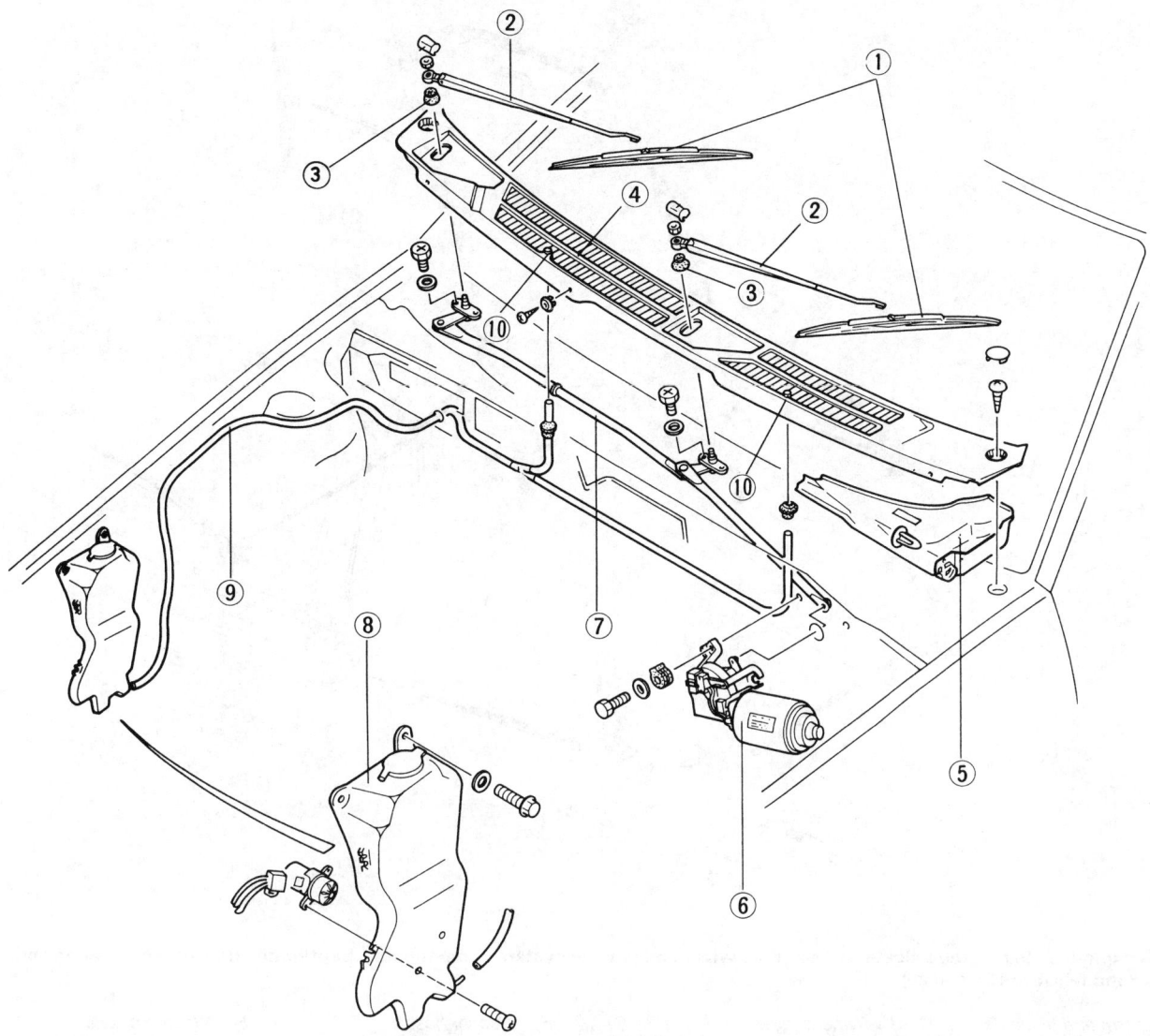

Fig. 12.12 Exploded view of the windscreen wiper and washer components – September 1985 models onward (Secs 22 and 24)

1 Wiper blade
2 Wiper arm
3 Rubber cover
4 Ventilation grille
5 Cover
6 Wiper motor
7 Wiper linkage
8 Washer reservoir assembly
9 Washer hose
10 Washer nozzle

September 1985 models onward
11 Carefully prise the rubber covers off the wiper spindles.
12 Prise out the caps from the left and right-hand ends of the ventilation grille to gain access to the two grille retaining screws and undo the screws. Undo all the retaining screws along the front edge of the grille and lift up the grille.
13 Disconnect the windscreen washer hoses from the underside of the grille and remove the grille from the car.
14 Release the retaining clips and remove the wiper motor cover to gain access to the wiper motor crank arm (photo).
15 Using a large flat bladed screwdriver, carefully lever the wiper linkage arm off the wiper motor crank arm balljoint (photo). **Note:** *Do not separate the motor and crank arm unless absolutely necessary. If removal is necessary, mark the position of the arm in relation to the body to ensure that it is positioned correctly on refitting.*
16 Remove the four wiper motor retaining bolts and washers and remove the wiper motor (photo).
17 Undo the four bolts securing the wiper arm pivots to the car and remove the wiper linkage assembly.

Refitting
18 Refitting is the reverse of the removal procedure.

23 Tailgate/rear window wiper motor and linkage – removal and refitting

Removal
1 Operate the wiper then switch it off so that it returns to its rest position.
2 Disconnect the battery negative lead.
3 Lift up the hinged covers and unscrew the wiper arm retaining nuts. Carefully prise the wiper arms off the spindles.
4 On Hatchback and Estate models remove the retaining clips and remove the tailgate inner trim panel to gain access to the wiper motor.

288 Chapter 12 Electrical system

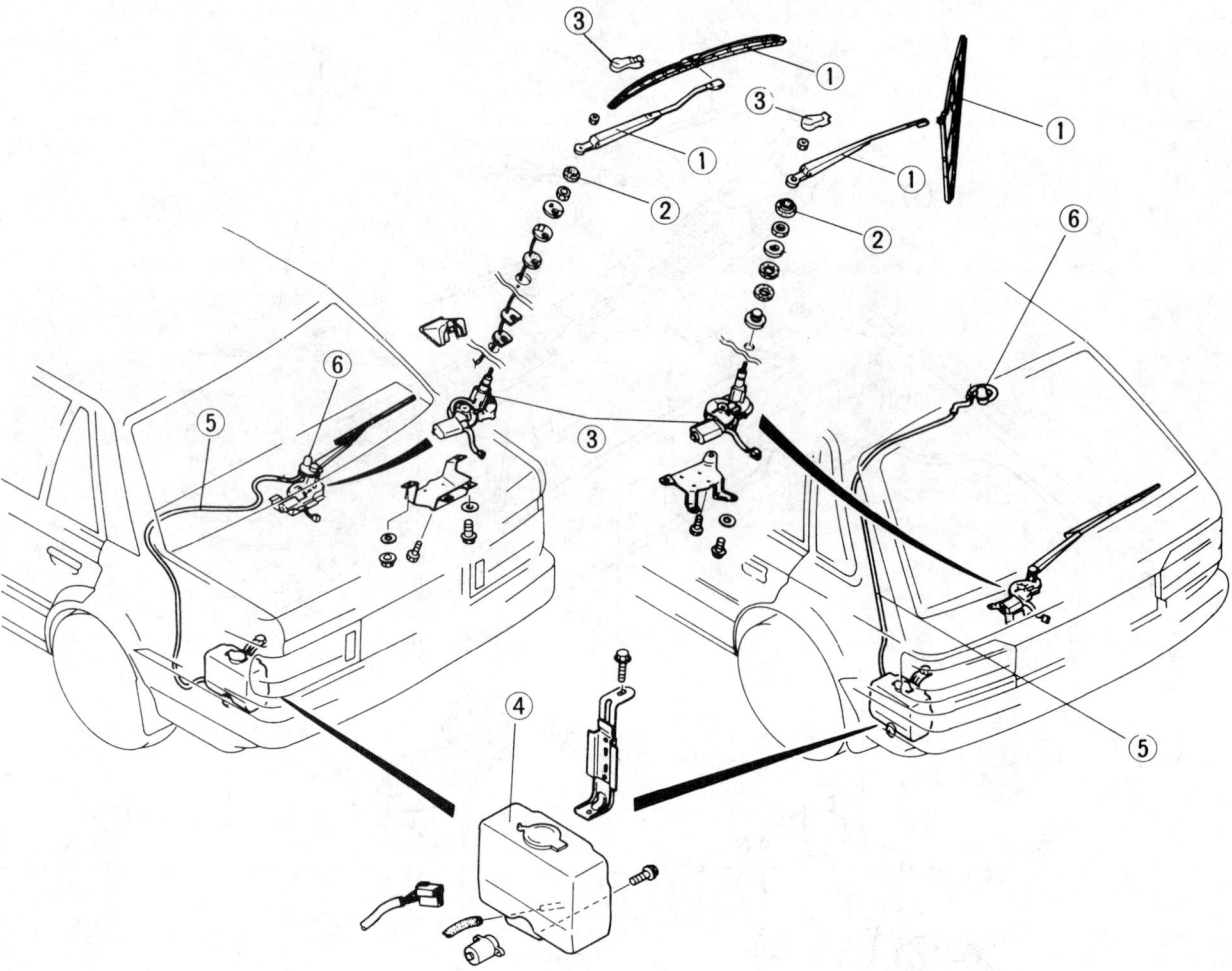

Fig. 12.13 Exploded view of the tailgate/rear window wiper and washer system components – September 1985 onward Saloon and Hatchback models (Secs 23 and 25)

1 Wiper arm and blade
2 Rubber cover
3 Wiper motor
4 Washer reservoir
5 Washer hose
6 Washer nozzle

On Saloon models access to the wiper motor can be gained from inside the boot.
5 Lift off the rubber cover and undo the nut securing the wiper spindle to the tailgate/body.
6 Disconnect the motor wiring at the connector then undo the four retaining bolts.
7 Withdraw the wiper motor from the tailgate/boot.

Refitting
8 Refitting is a reversal of the removal procedure.

24 Windscreen/headlamp washer system components – removal and refitting

Removal
1 To remove the windscreen washer reservoir and pump, unscrew the mounting bolts and lift the reservoir from the front right-hand corner of the engine compartment. On July 1987 models onward the reservoir is also used for the headlamp washer system which is supplied by a second pump fitted on the underside of the reservoir.

2 Disconnect the wiring connector(s) from the pump(s), then disconnect the plastic tubing from the reservoir and remove the assembly from the car.
3 Empty the reservoir of any remaining fluid then undo the retaining screws and separate the pump(s) and reservoir.
4 If necessary the washer nozzles can carefully be prised out of the ventilation grille and disconnected from the tubing.

Refitting
5 Refitting is a reversal of removal.

25 Tailgate/rear window washer system components – removal and refitting

Removal
1 To remove the washer reservoir and pump, unscrew the mounting bolts and lift the reservoir from the right (Estate) or left-hand (Hatchback and Saloon) side of the luggage compartment/boot.
2 Disconnect the wiring connector from the pump, then disconnect

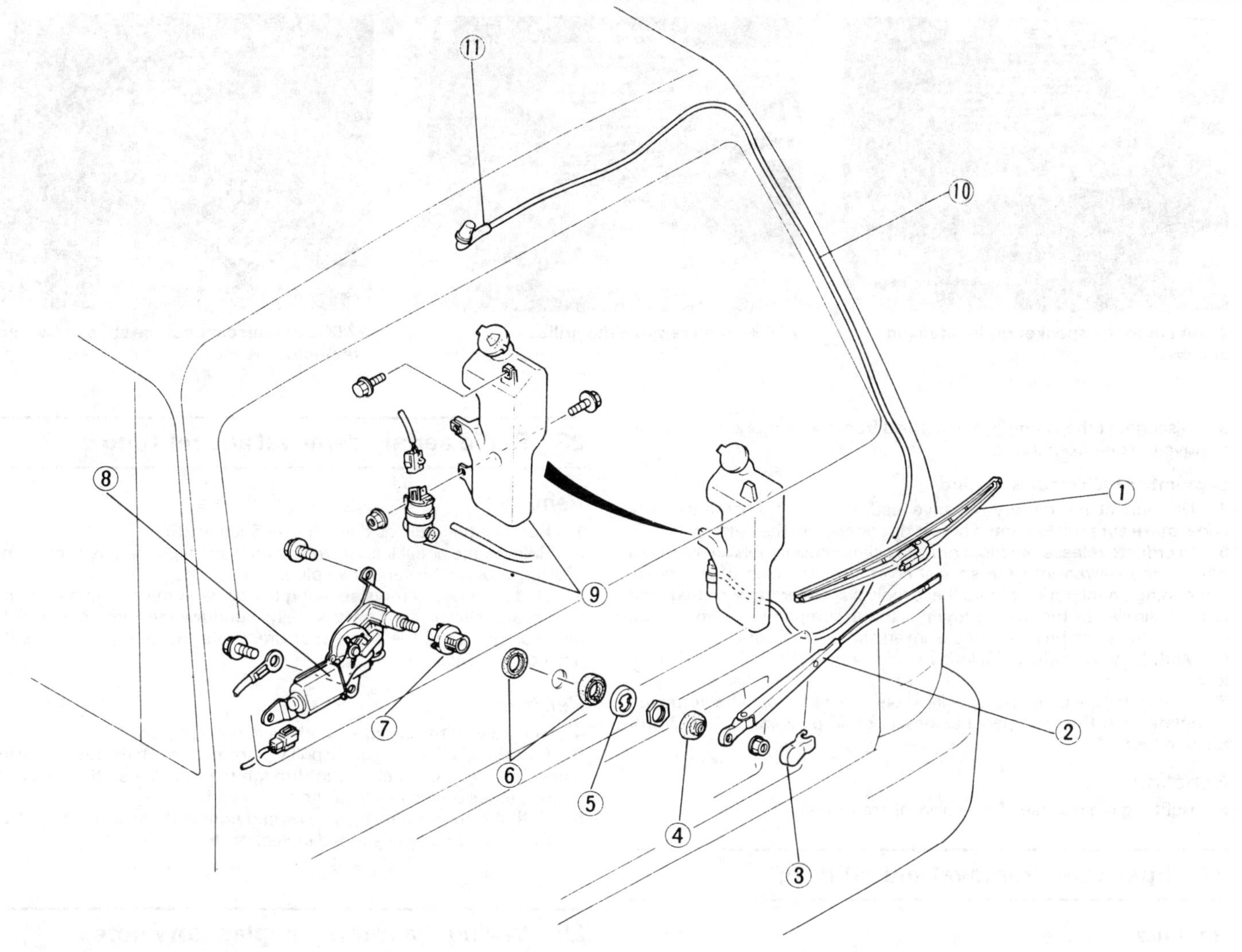

Chapter 12 Electrical system   289

Fig. 12.14 Exploded view of the tailgate wiper and washer system components – Estate models (Secs 23 and 25)

1 Wiper blade
2 Wiper arm
3 Hinged cover
4 Rubber cover
5 Outer bush
6 Seals
7 Inner bush
8 Wiper motor
9 Washer reservoir and pump
10 Washer hose
11 Washer nozzle

the plastic tubing from the reservoir and remove the assembly from the car.
3 Empty the reservoir of any remaining fluid then undo the two retaining screws and separate the pump and reservoir.
4 On Hatchback and Estate models the washer nozzle can be carefully prised out of the tailgate and disconnected from the tubing.
5 On Saloon models it will be necessary to remove the wiper arm as described in Section 23 before the washer nozzle can be removed.

Refitting
6 Refitting is a reverse of the removal procedure.

26 Radio/cassette player – removal and refitting

Removal
Pre-September 1985 models
1 Perform the operations listed in paragraphs 1 to 7 of Section 19.
2 Undo the two screws securing the radio to the facia and withdraw the radio.

26.5 Using special tools to remove the radio cassette player – September 1985 models onward

Chapter 12 Electrical system

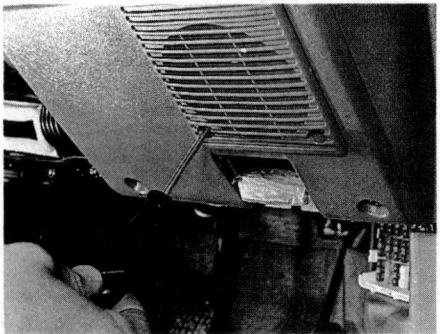
27.6A Undo the speaker grille retaining screws...

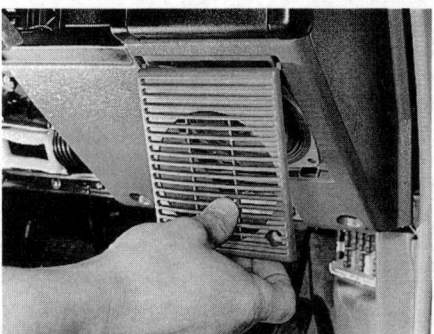
27.6B ...then remove the grille...

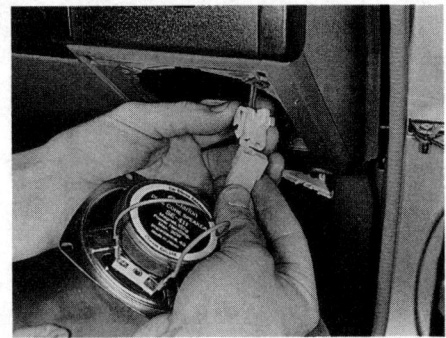
27.6C ...and disconnect the speaker wiring connector

3 Disconnect the wiring and aerial lead from the rear of the radio and remove the unit from the car.

September 1985 models onward

4 Disconnect the battery negative lead. If the radio has a security code, make sure this is known before disconnecting the battery.
5 In order to release the radio retaining clips special tools are required which are inserted into the special holes on each side of the radio. Depending on the type of radio these tools will either be two U-shaped rods or similar to the ones shown in the accompanying photograph (photo). These can be purchased from an audio specialist.
6 Withdraw the radio sufficiently to disconnect the wiring and aerial lead.
7 In the absence of the special tools the radio can be removed complete with the facia panel as described in paragraphs 5 and 6 of Section 18.

Refitting

8 Refitting is a reverse of the removal procedure.

27 Speakers – removal and refitting

Removal

Pre-September 1985 models

1 Depending on the audio system the car is equipped with either one, two or four speakers. On models equipped with a single speaker the speaker is located in the facia panel, behind the glovebox. Models fitted with four speakers have the front speakers fitted in the left and right-hand trim panels, just below the facia panel, and the rear speakers on the right and left-hand side of the parcel shelf. Models equipped with two speakers are just equipped with the two front speakers in the left and right-hand trim panels.
2 To remove the facia mounted speaker, open the glovebox and undo the speaker retaining screws. Disconnect the wiring from the rear of the speaker and remove it from the facia panel.
3 To remove a front speaker which is fitted in the left or right-hand trim panel, undo the four speaker grille retaining screws and remove the grille. Withdraw the speaker then disconnect the wiring at the connector and remove the speaker from the car.
4 Access to the rear speakers is gained from the luggage compartment/boot. From inside the luggage compartment/boot, disconnect the speaker wiring then undo the four retaining bolts and remove the speaker from the car.

September 1985 models onward

5 On later models the front speakers are located on the left and right-hand underside of the facia panel and the rear speakers, where fitted, are on the left and right-hand side of the parcel shelf (Saloon and Hatchback) or rear quarter panel (Estate).
6 To remove a speaker, undo the speaker grille retaining screws and remove the grille. Withdraw the speaker then disconnect the wiring connector and remove the speaker from the car (photos).

Refitting

7 Refitting is the reverse of the removal procedure.

28 Radio aerial – removal and refitting

Removal

1 Remove the radio as described in Section 26.
2 Release the aerial lead from any relevant clips then tie a long piece of string around the aerial lead plug.
3 Undo the two screws securing the aerial to the roof and withdraw the aerial assembly (photo). Carefully withdraw the aerial lead until the plug comes out of the aerial aperture. Untie the string and leave it in position in the car.

Refitting

4 Securely tie the string around the aerial lead plug.
5 From inside the car, gently pull the string through the radio aperture whilst feeding the aerial lead in through the roof. When the aerial lead plug emerges on the inside of the car untie the string.
6 Refit the aerial retaining screws and tighten them securely.
7 Refit the radio as described in Section 26.

29 Wiring diagrams – explanatory notes

The wiring diagrams included at the end of this Chapter are of both conventional type and the current flow type where each wire is shown in the simplest line form without crossing over other wires.

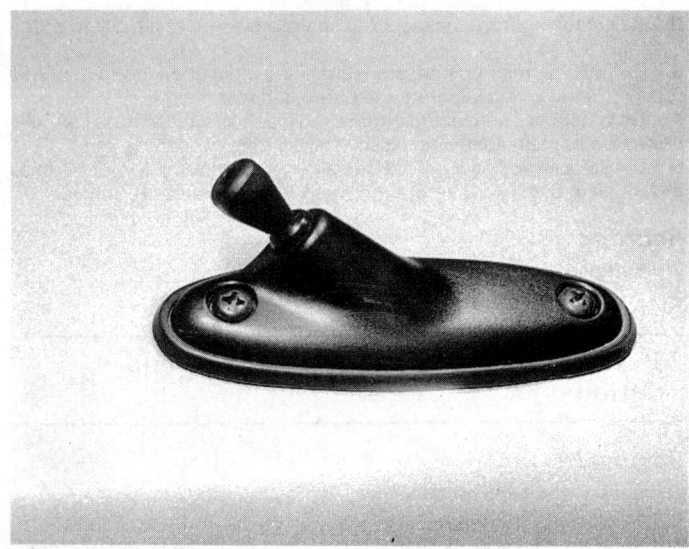
28.3 Aerial retaining screws

Chapter 12 Electrical system

Battery	Ground (Harness, Body)	Fuse (Holder, Box)	Fusible link	Motor
Coil solenoid	Resistance	Variabel resistance	Thermister	Diode
Condenser	Transistor	Pump	Lamp	Horn
Speaker	Cigar lighter	Heater	Illuminated Diode	Zener Diode

12.15 Symbols used in the wiring diagrams

Chapter 12 Electrical system

DESCRIPTION OF HARNESS	SYMBOL	DESCRIPTION OF HARNESS	SYMBOL
Front harness	[F]	No. 1 Door harness	[Dr1]
Engine harness	[E]	No. 2 Door harness	[Dr2]
Instrument panel harness	[I]	No. 3 Door harness	[Dr3]
Rear harness	[R]	No. 4 Door harness	[Dr4]
Interior light harness	[In]	Air-Cond harness	[A]
Floor harness	[Fr]	Others	

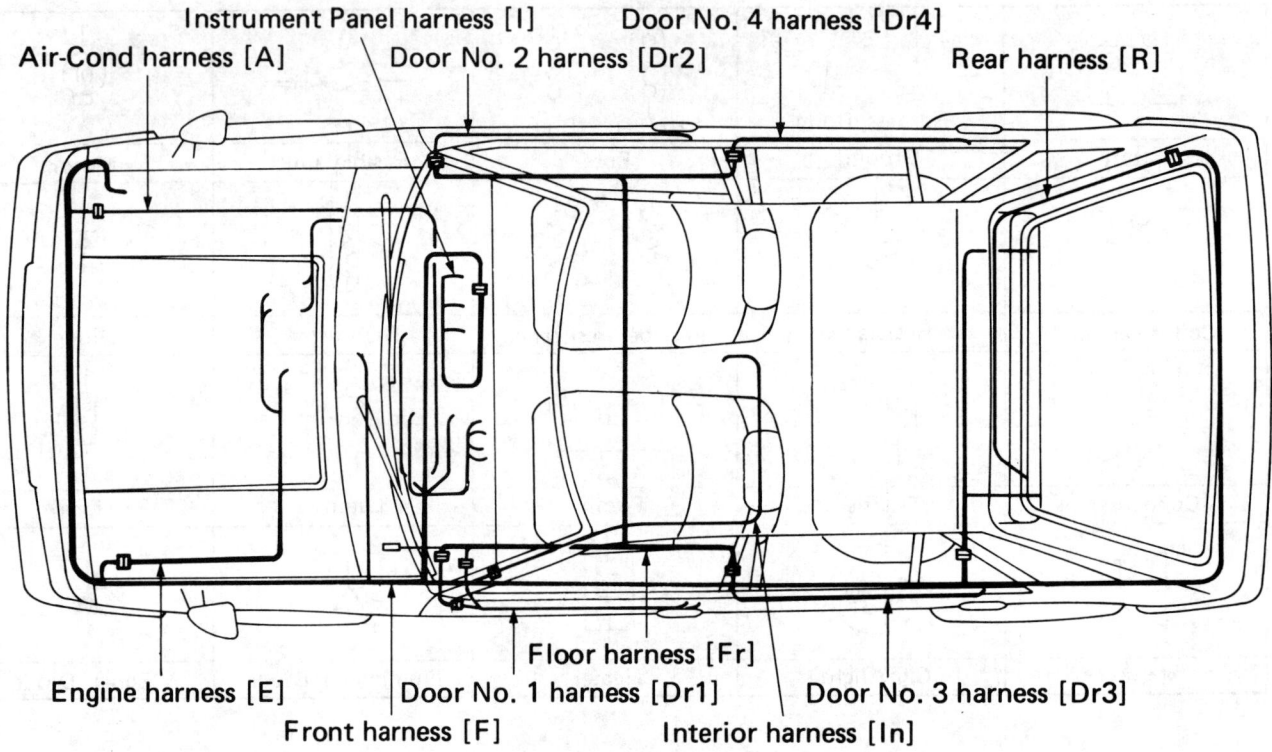

Abbr.	Term	Abbr.	Term
St	Start	A	Ampere
IG	Ignition	W	Watt
ACC	Accessory	R	Resistance
AS	Auto stop	Tr	Transistor
INT	Intermittent	M	Motor
Lo	Low	SW	Switch
Mi	Middle	Sq	Square per millimeter
Hi	High		
R.H.	Right hand	A/T	Automatic transmission
L.H.	Left hand		
F.R	Front righ	M/T	Manual transmission
F.L	Front left		
R.R	Rear right	NO	Normal opened
R.L	Rear left	NC	Normal closed
V	Volt		

CODE	COLOR	CODE	COLOR
B	Black	Lg	Light green
Br	Brown	O	Orange
G	Green	R	Red
Gy	Gray	W	White
L	Blue	Y	Yellow
Lb	Light blue		

12.16 Wiring harness locations, symbols, abbreviations and colour codes

Chapter 12 Electrical system

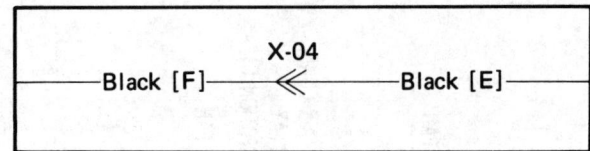

12.17 Wiring diagram explanation

In the illustration shown Black indicates the wire colour (which may appear in the diagrams as a letter indicating the colour or two letters where a tracer colour is used). The letter in brackets indicates the wiring harness where this is shown. In this case (F) is the front harness and (E) is the engine harness. The arrow symbol indicates a harness connector and X-04 is the connector number

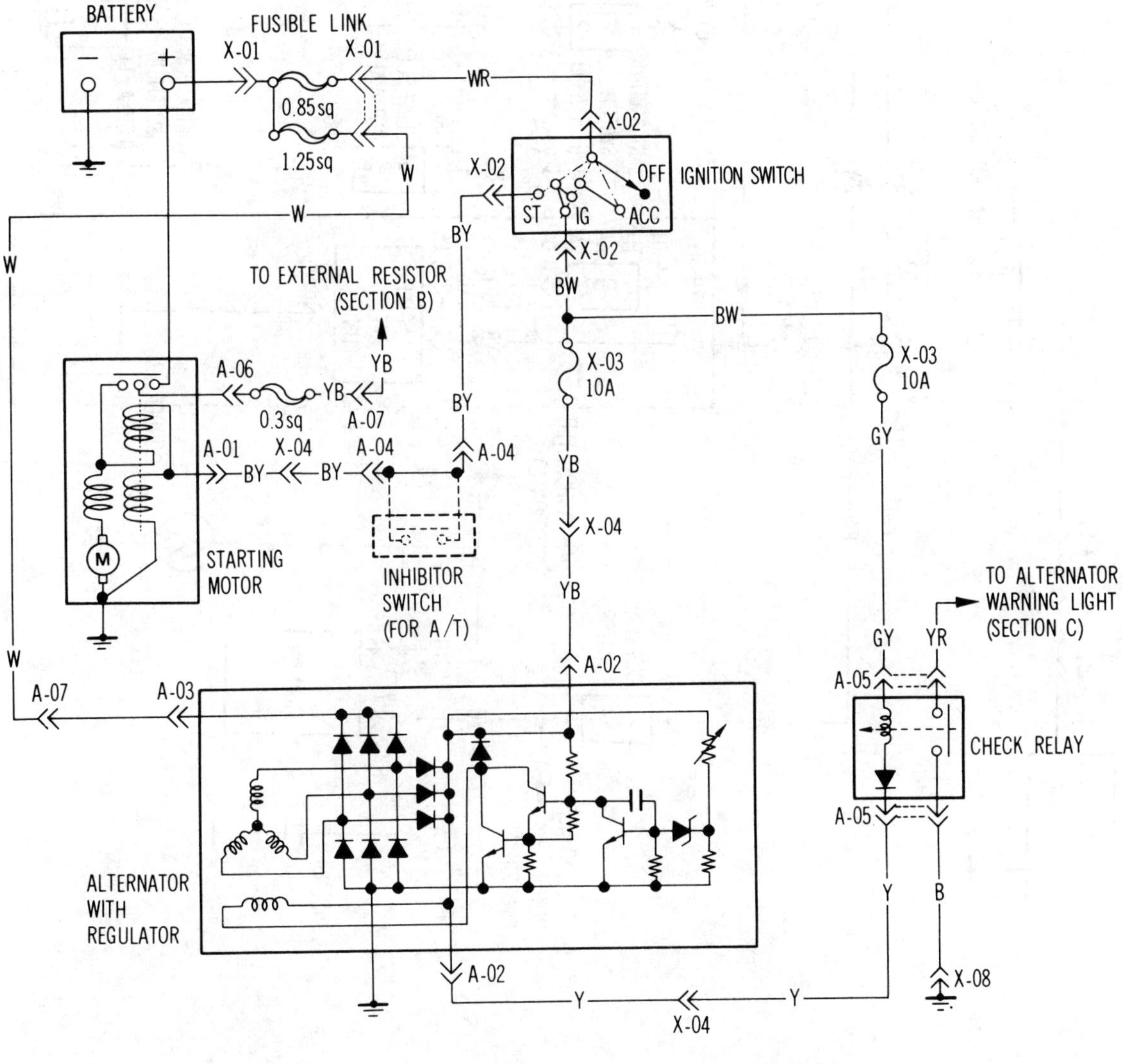

12.18 Charging and starting system wiring diagram (1981 models)

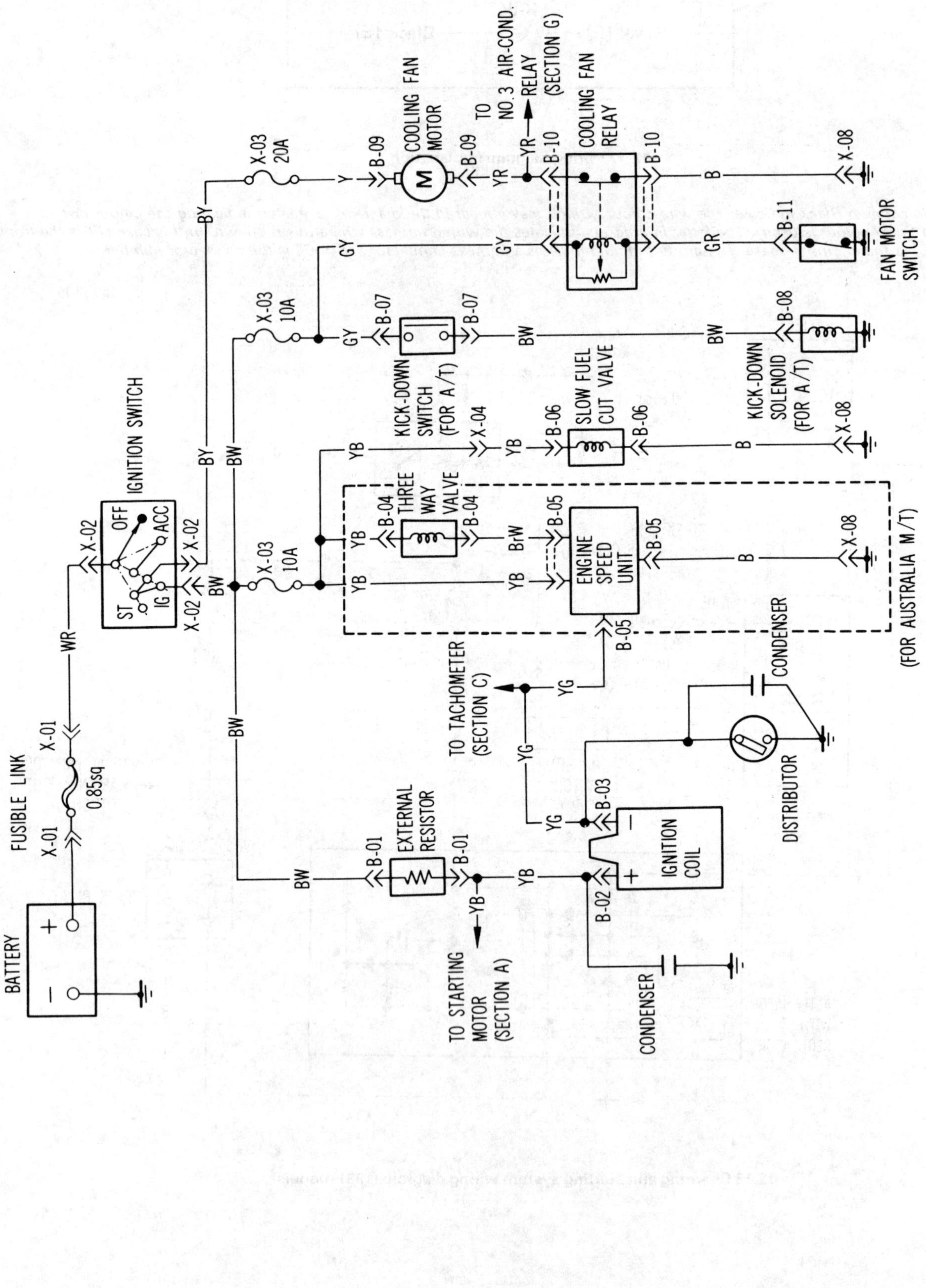

12.19 Fuel, ignition, cooling fan and automatic transmission kickdown system wiring diagram (1981 models)

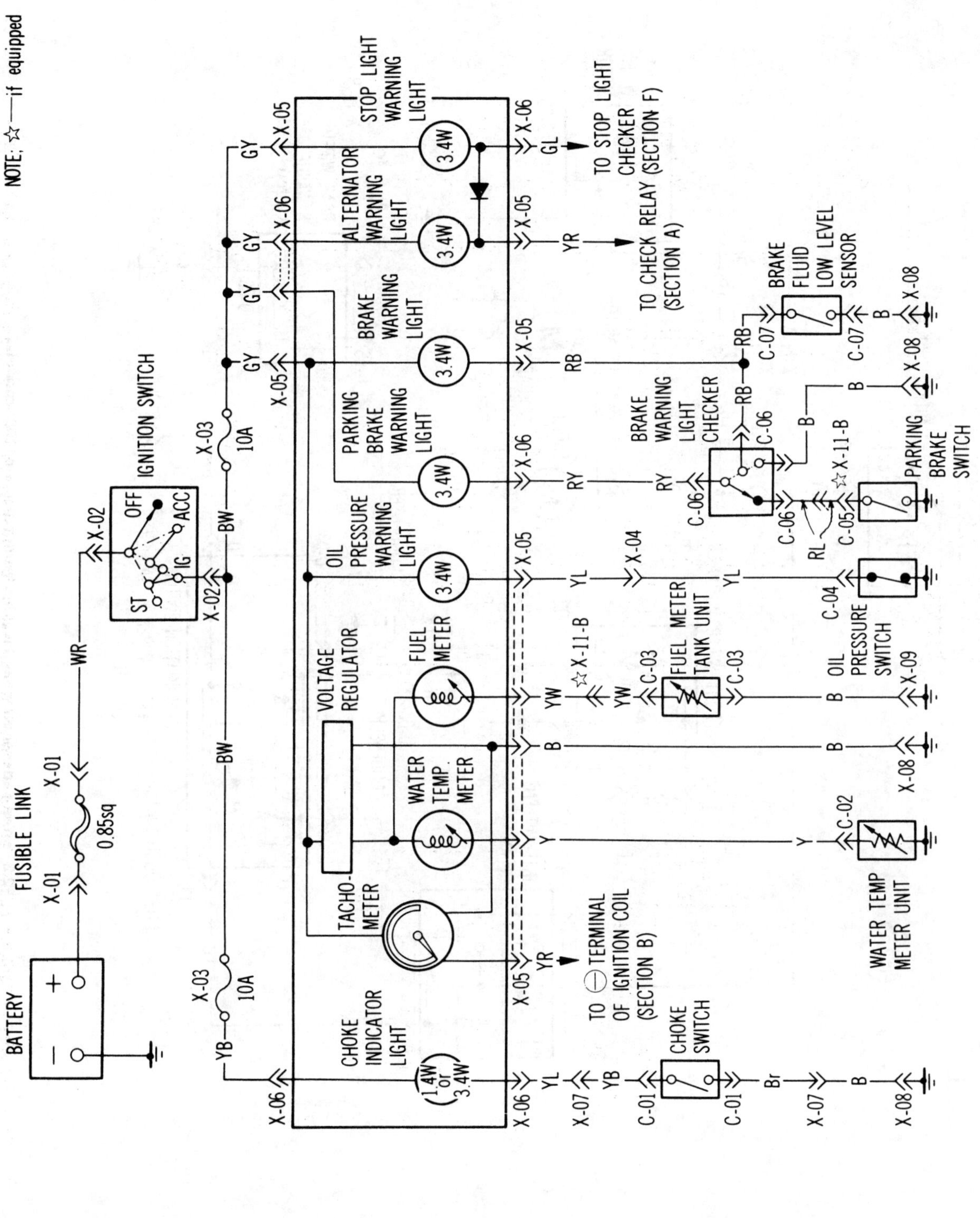

Chapter 12 Electrical system

12.20 Instrument and warning light wiring diagram (1981 models)

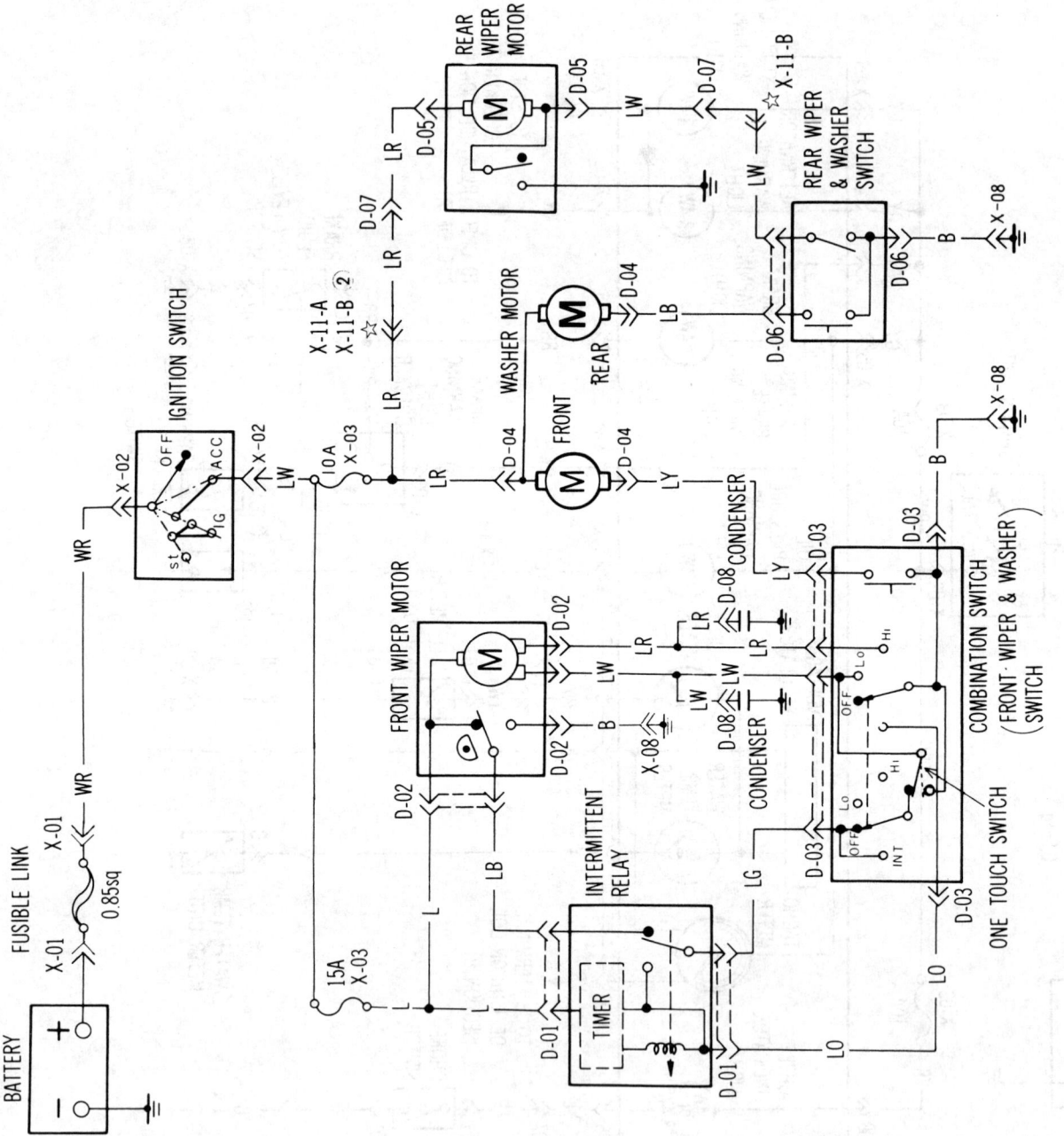

12.21 Front and rear wiper and washer wiring diagram (1981 models)

Chapter 12 Electrical system

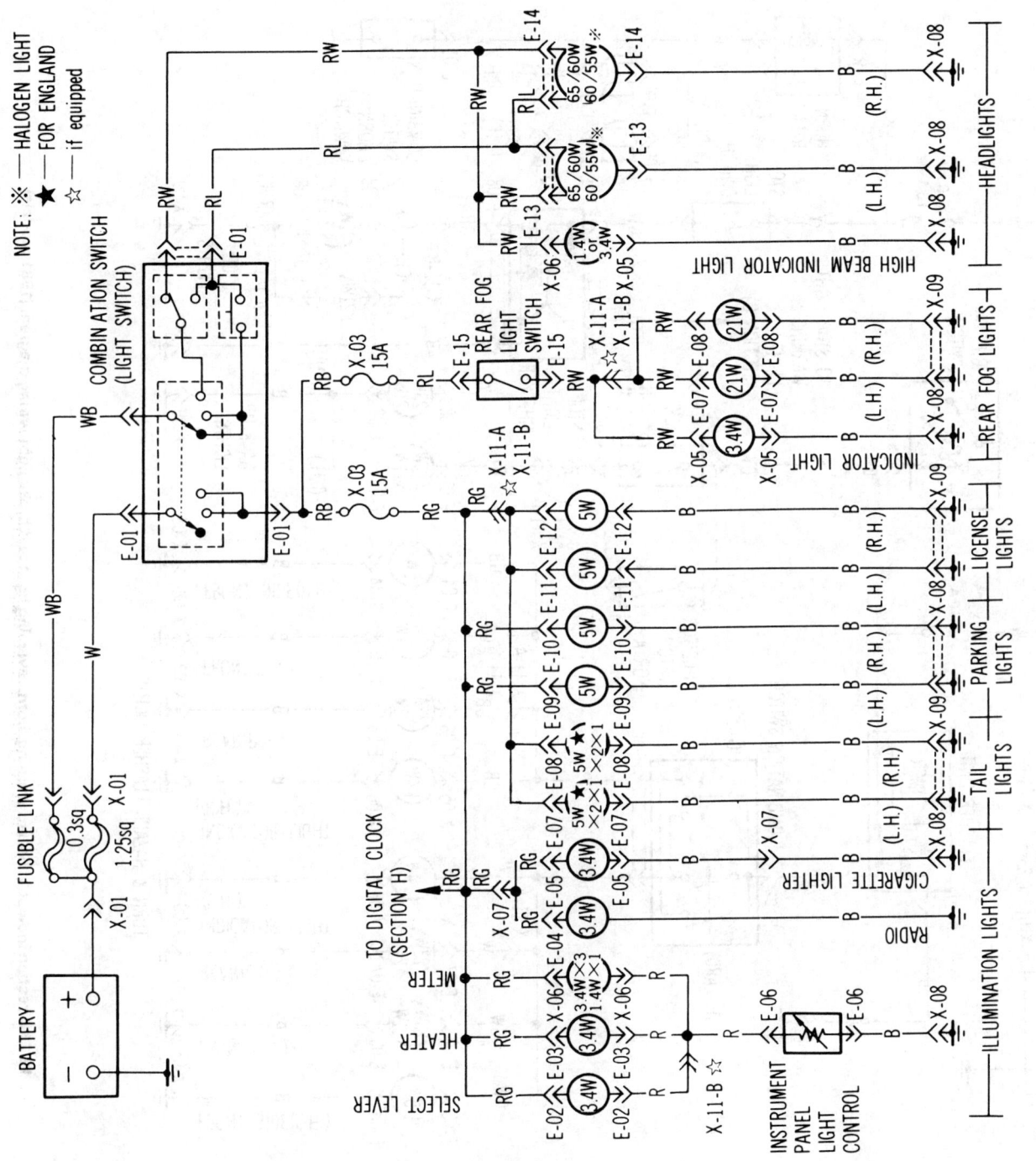

12.22 Headlight, sidelight, stop/tail light, number plate light and interior light wiring diagram (1981 models)

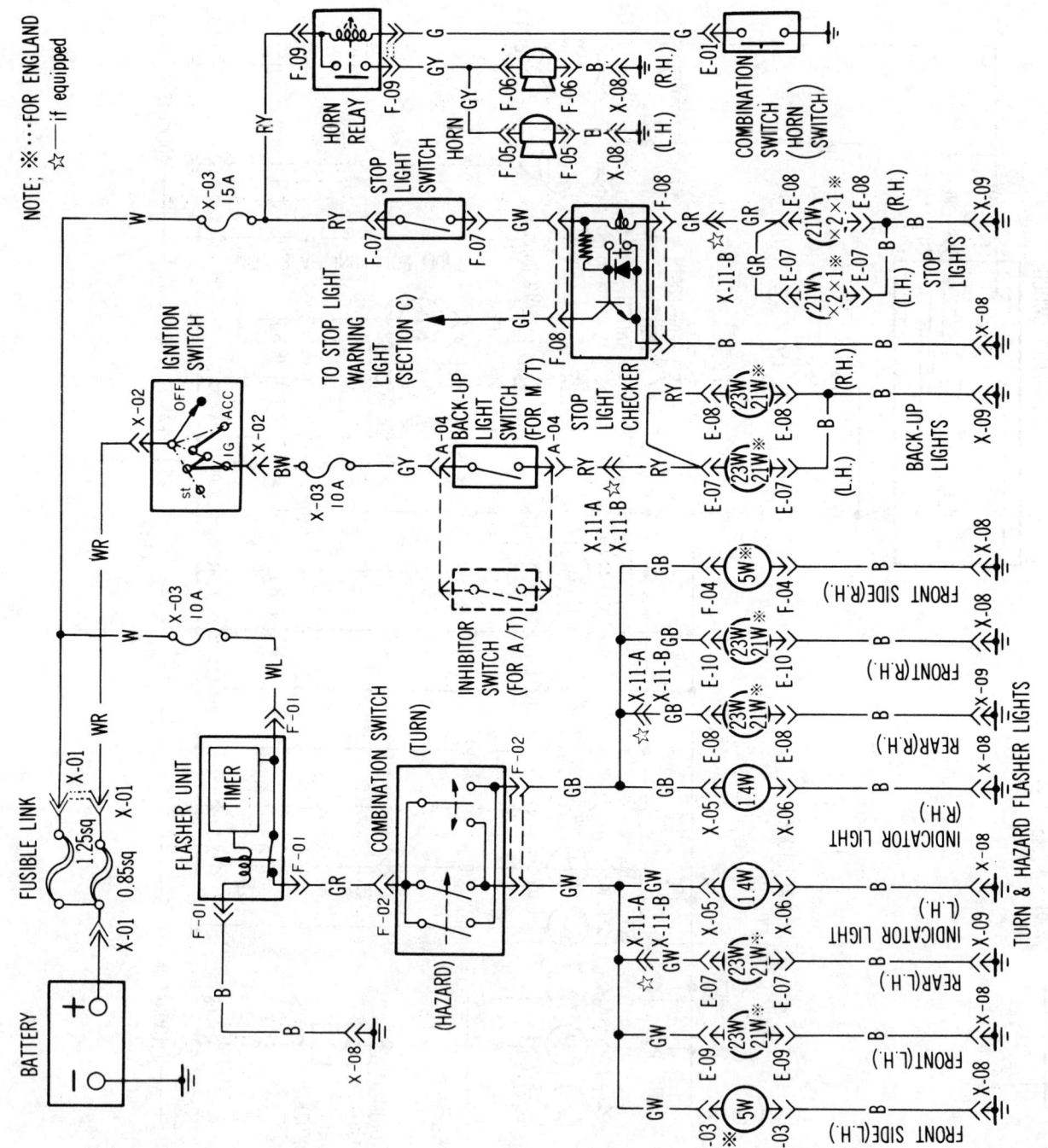

12.23 Direction indicator, hazard warning, horn, reversing light and brake light wiring diagram (1981 models)

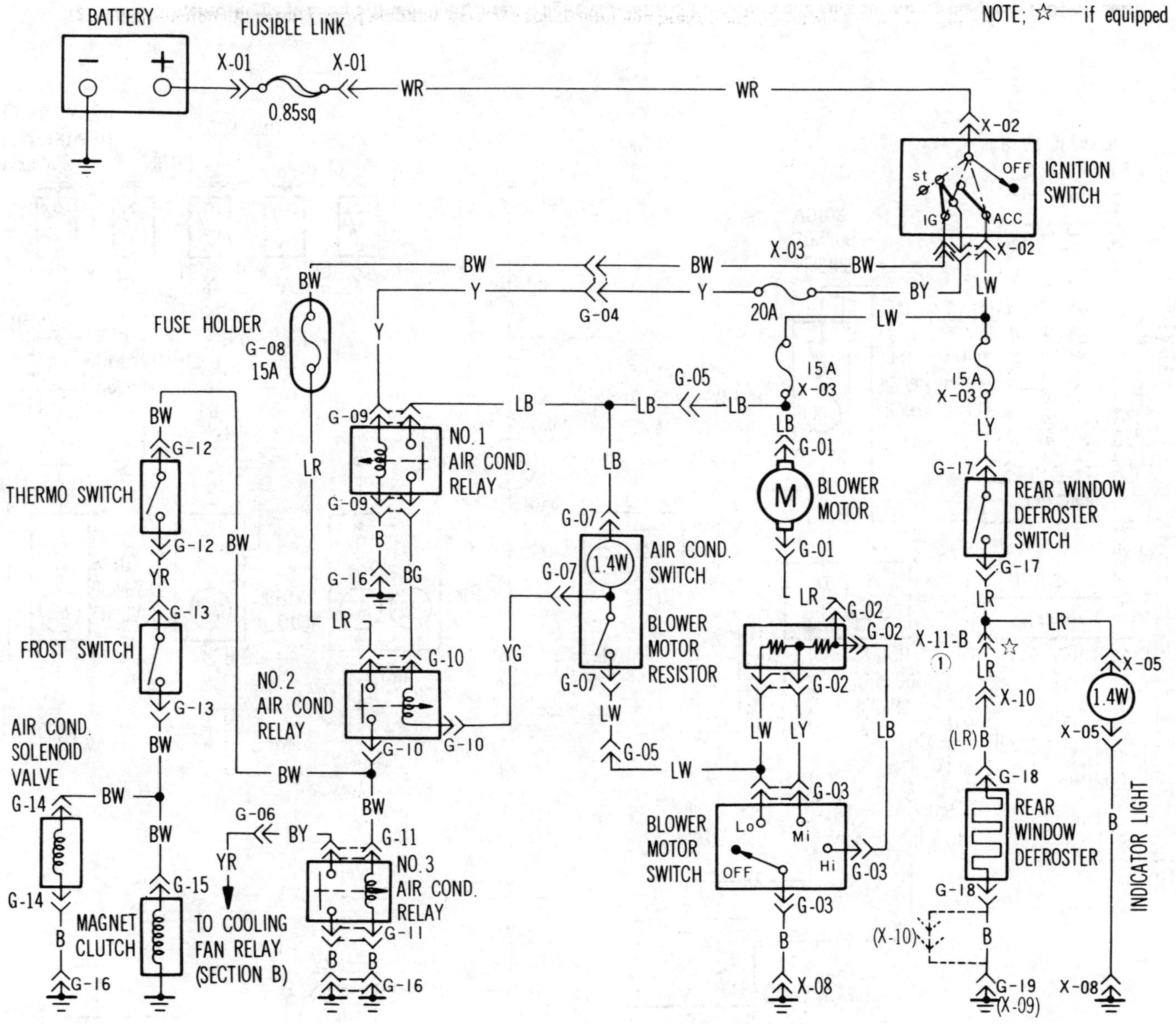

Chapter 12 Electrical system

12.24 Heater, air conditioner and rear screen demister wiring diagram (1981 models)

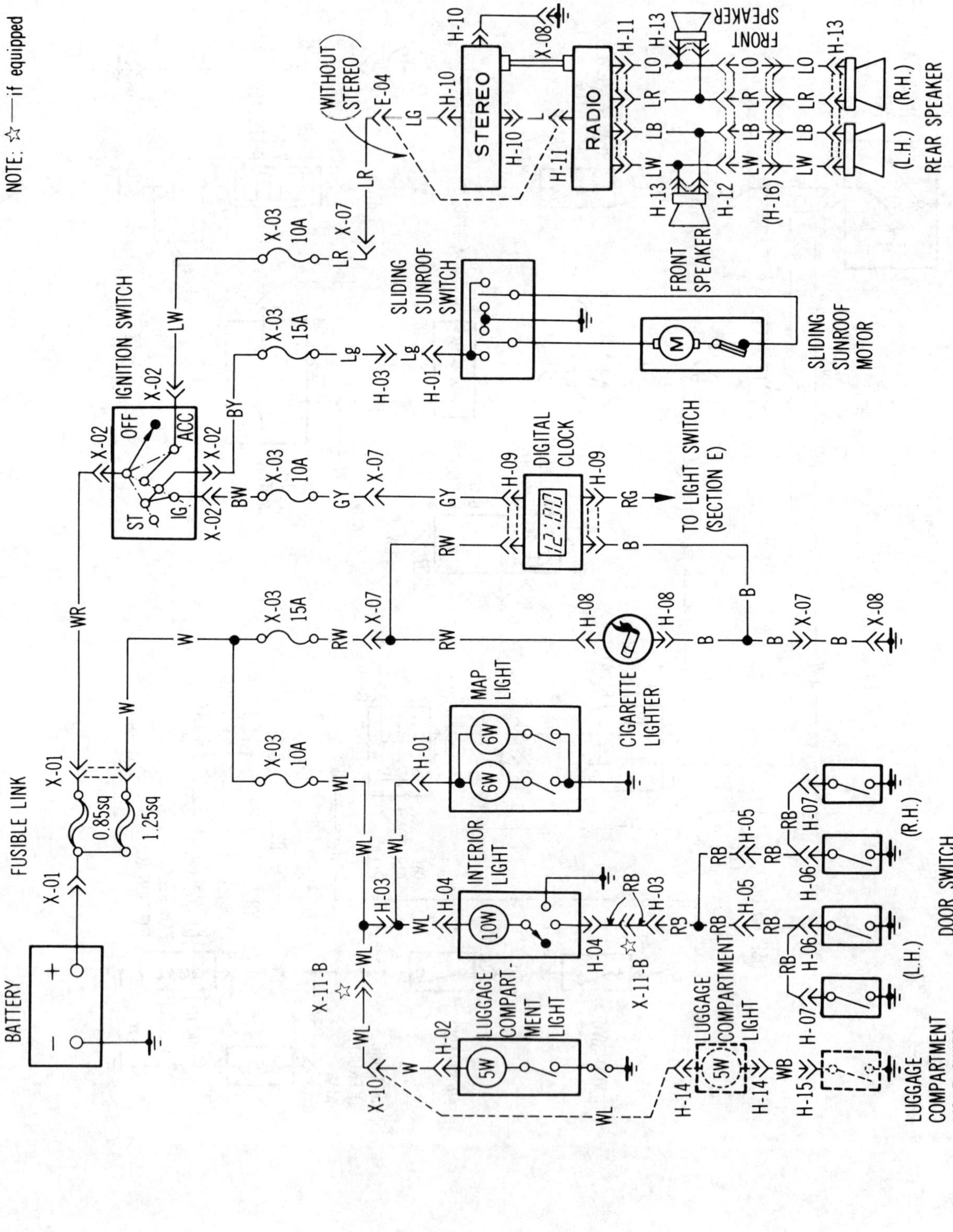

12.25 Radio, stereo, luggage compartment light, interior light, cigarette lighter, clock and sunroof wiring diagram (1981 models)

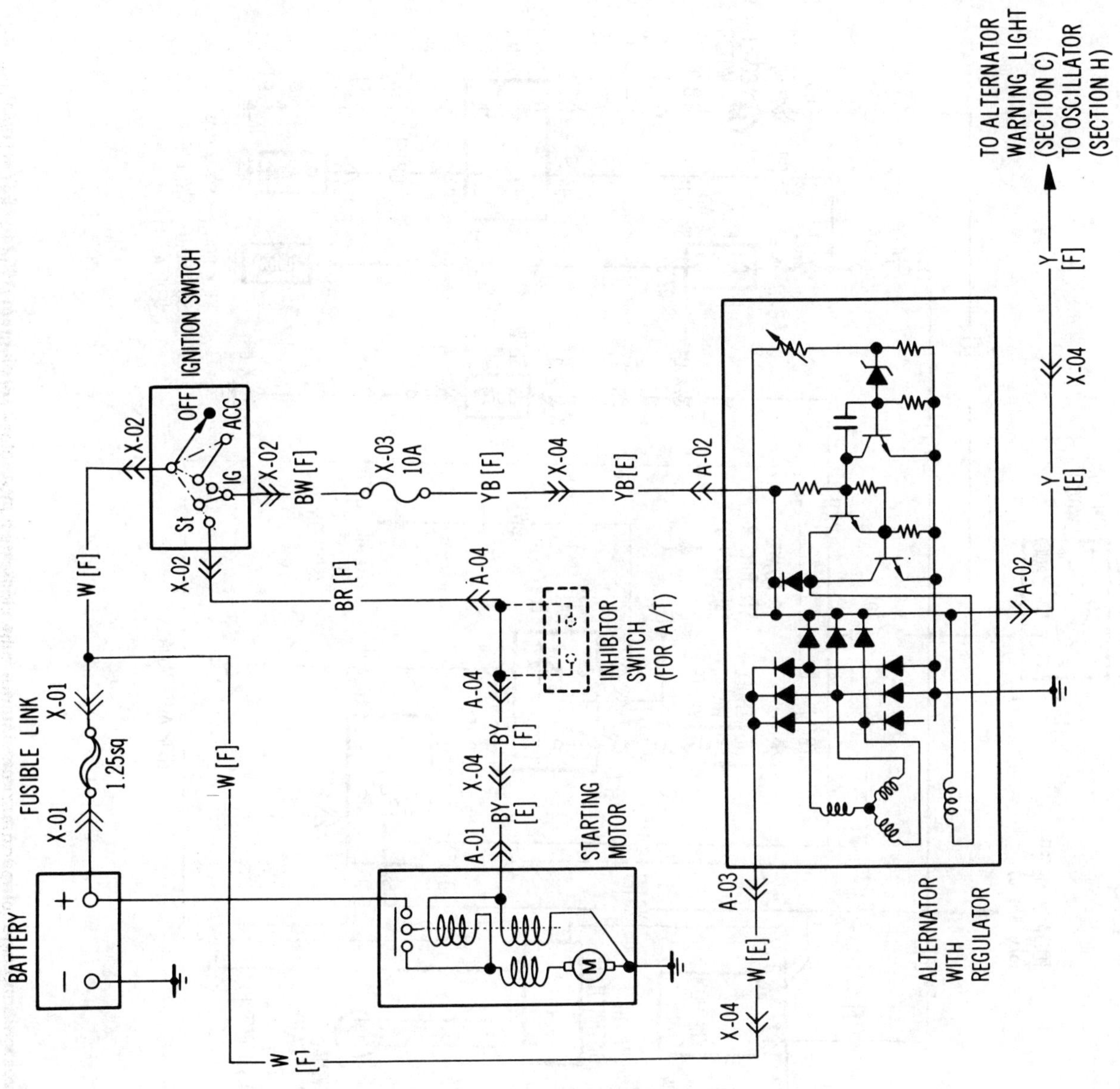

Chapter 12 Electrical system

12.26 Charging and starting system wiring diagram (1982 to 1987 models)

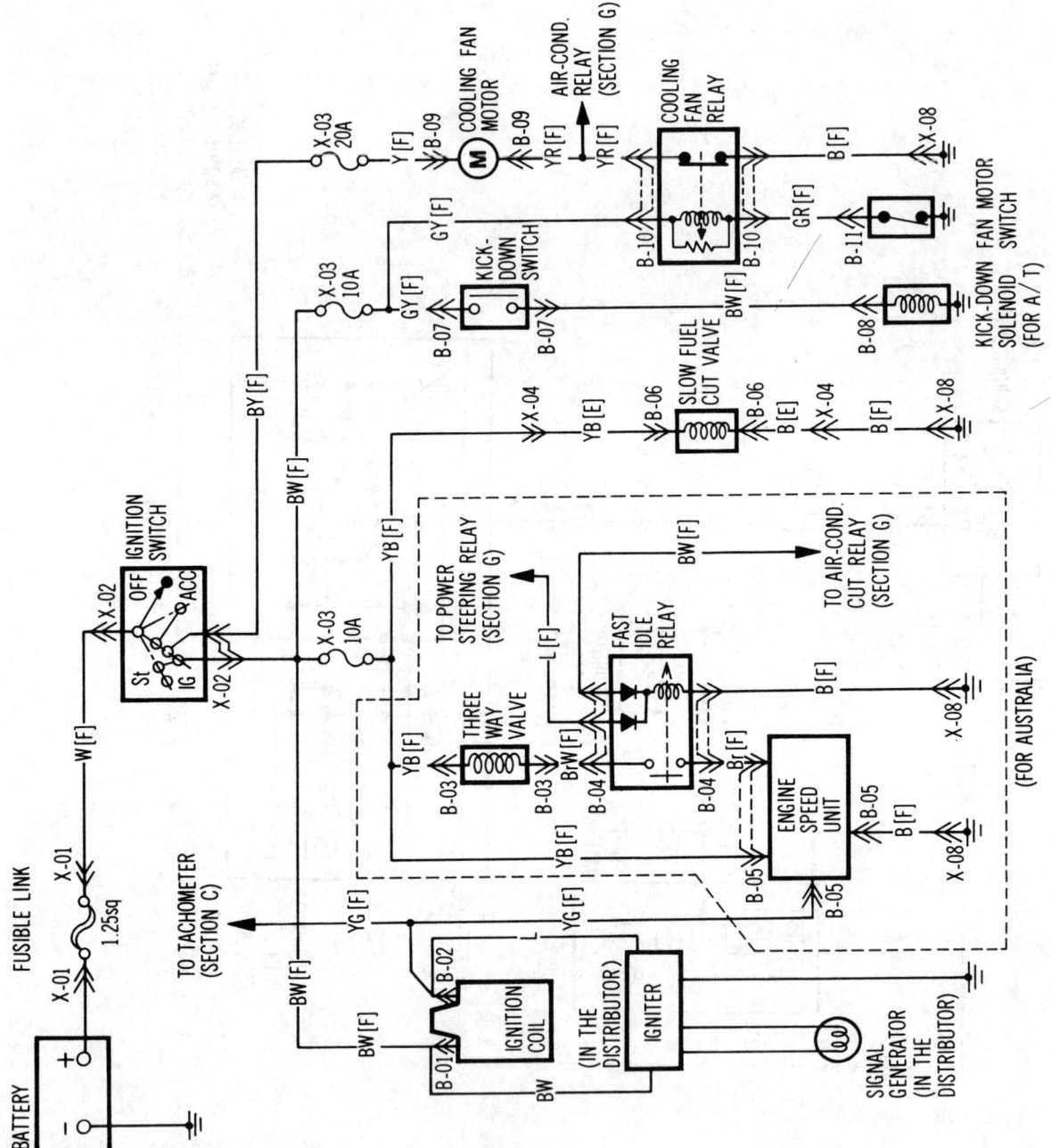

12.27 Fuel, ignition, cooling fan and automatic transmission kickdown system wiring diagram (1982 to 1987 models)

Chapter 12 Electrical system

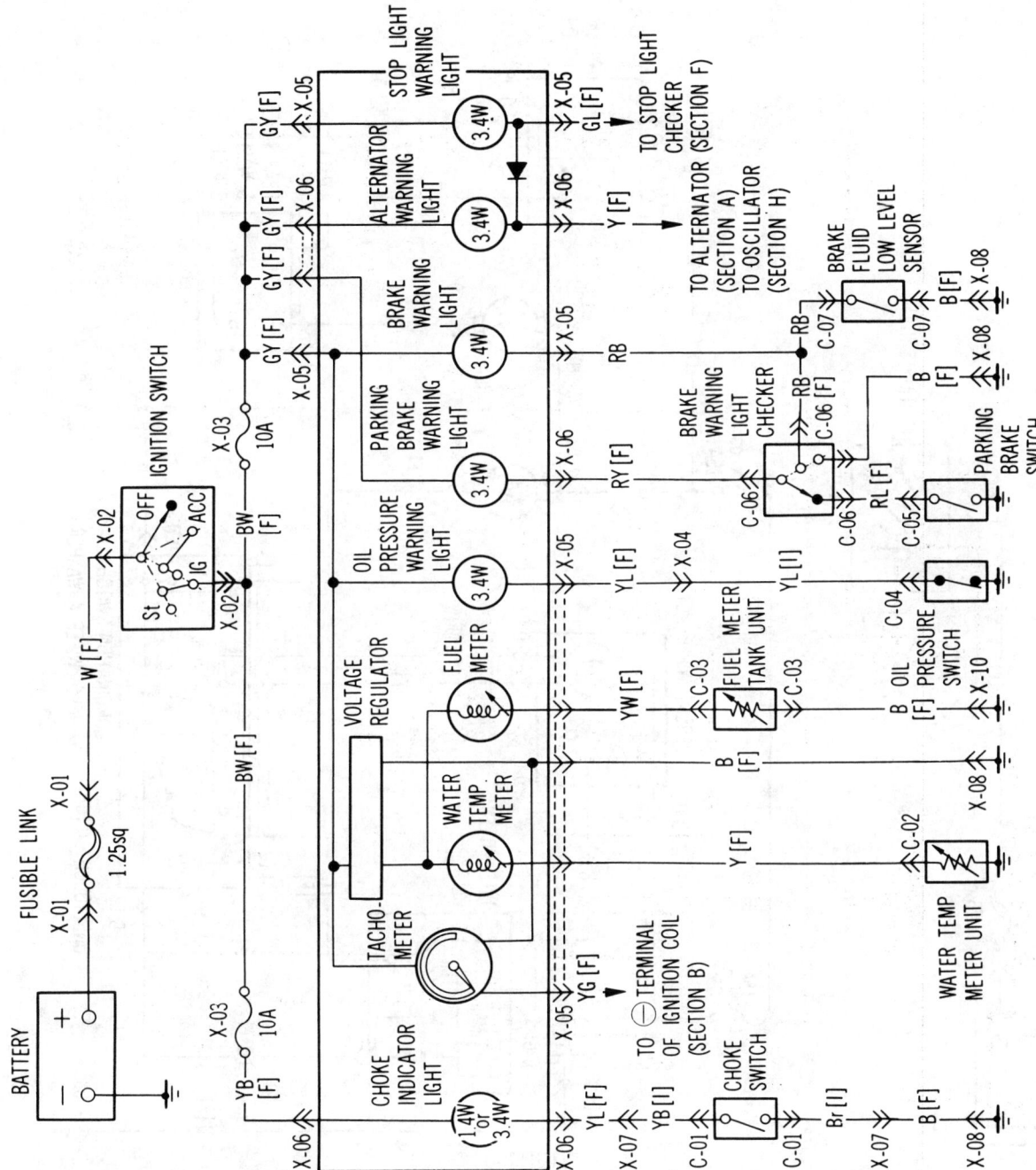

12.28 Instrument and warning light wiring diagram (1982 to 1987 models)

Chapter 12 Electrical system

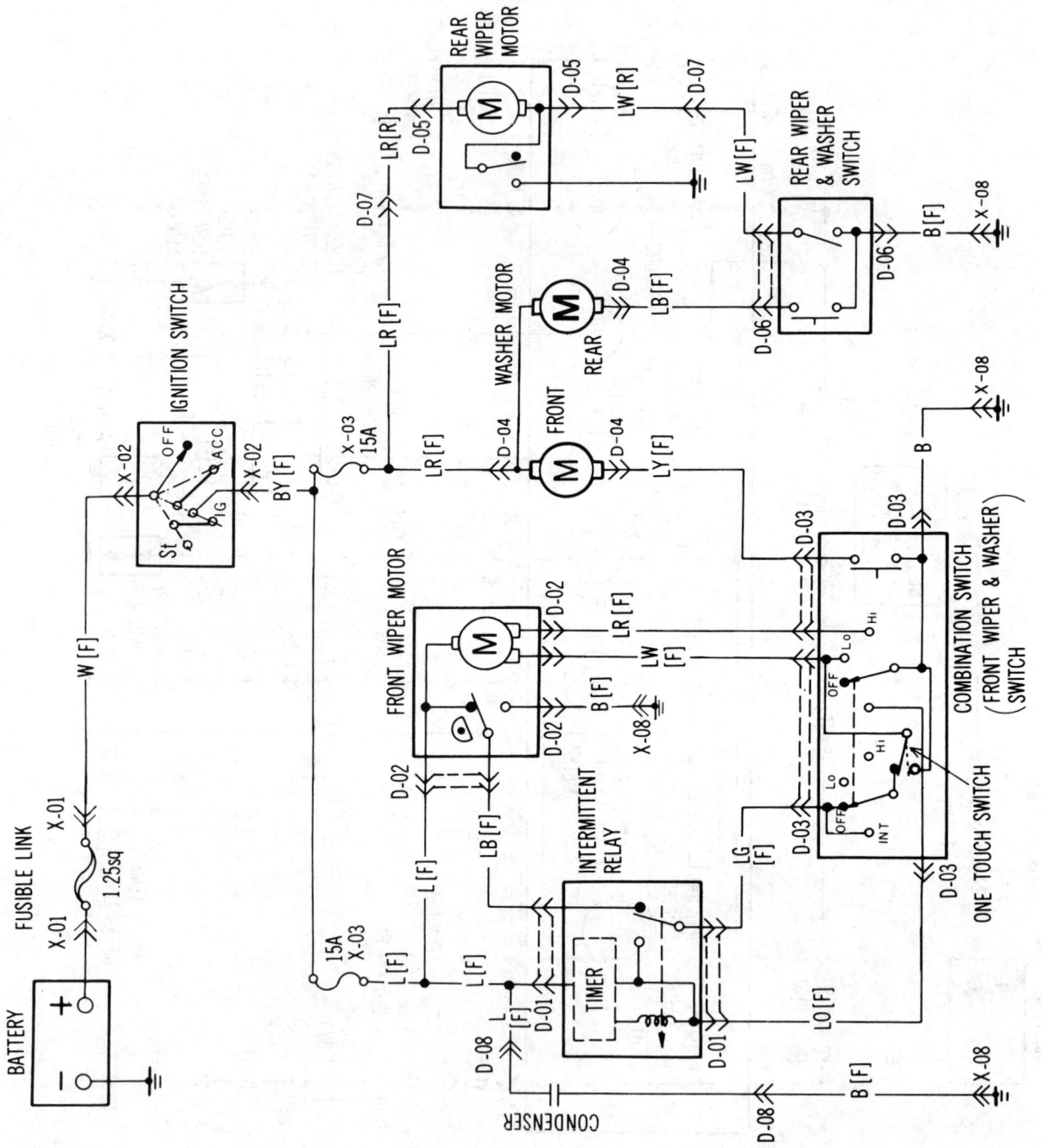

12.29 Front and rear wiper and washer wiring diagram (1982 to 1987 models)

Chapter 12 Electrical system

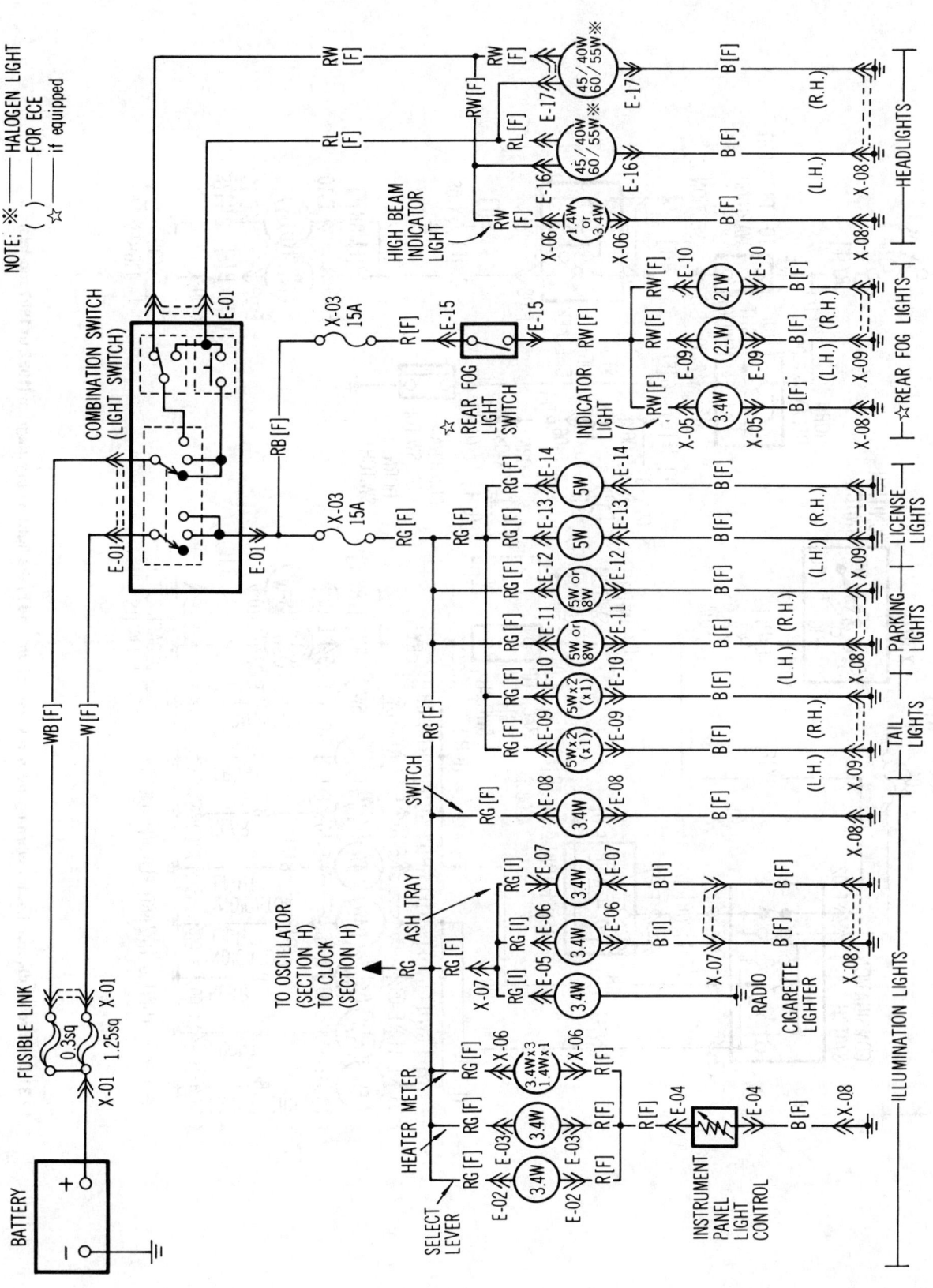

12.30 Headlight, sidelight, tail light, number plate light and interior light wiring diagram (1982 to 1987 models)

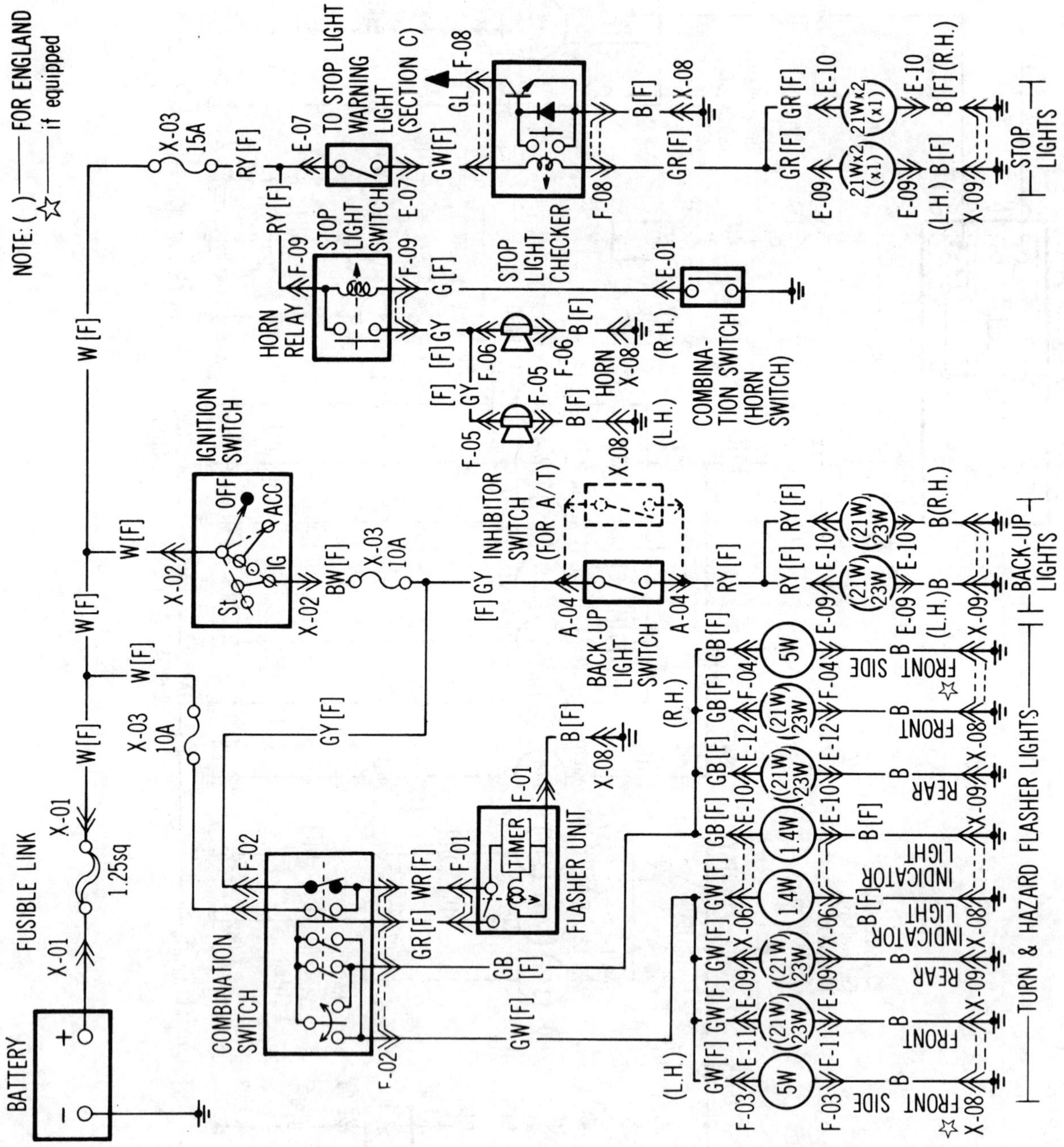

Chapter 12 Electrical system

12.31 Direction indicator, hazard warning, horn, reversing light and brake light wiring diagram (1982 to 1987 models)

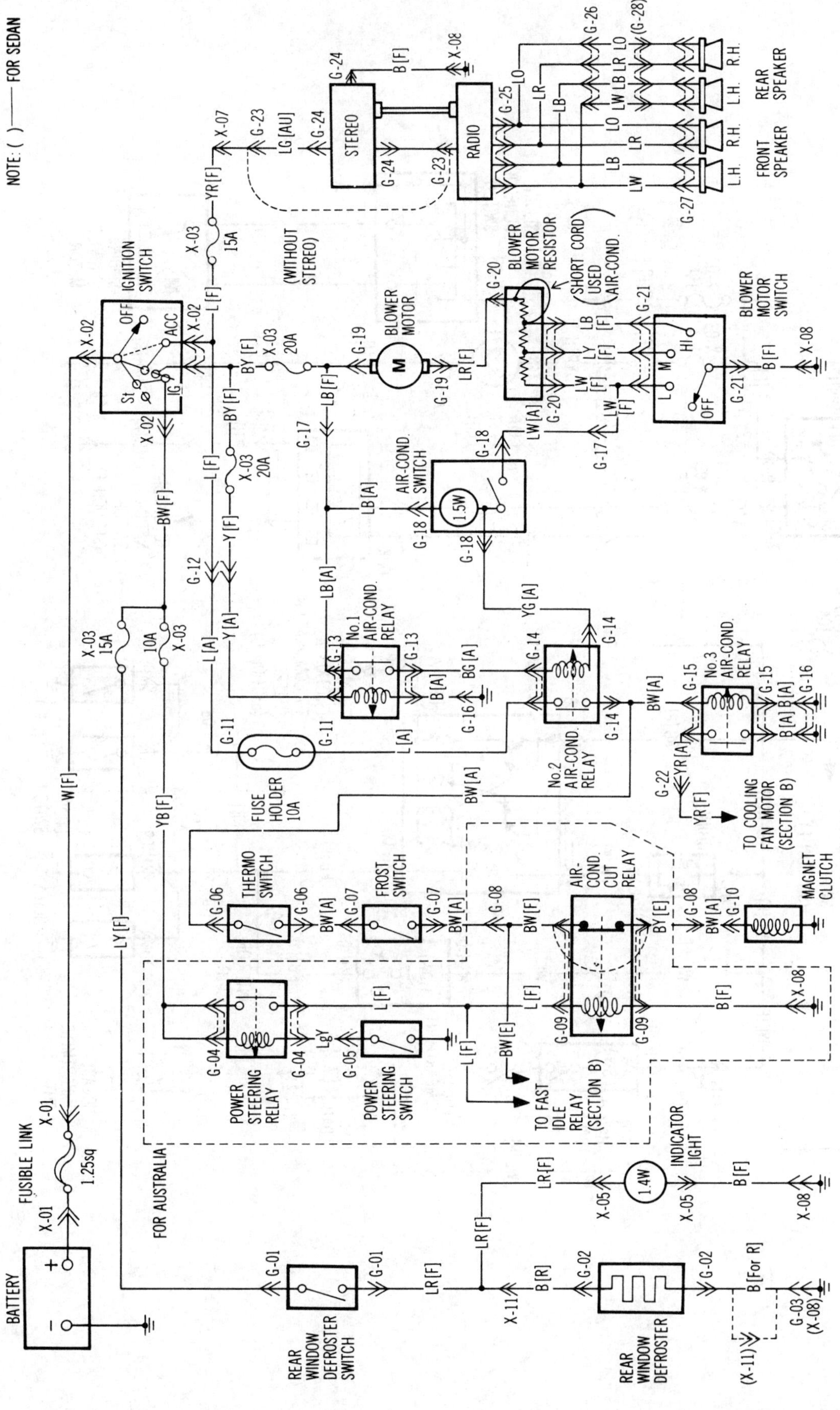

Chapter 12 Electrical system

12.32 Radio, stereo, air conditioner and rear screen demister wiring diagram (1982 to 1987 models)

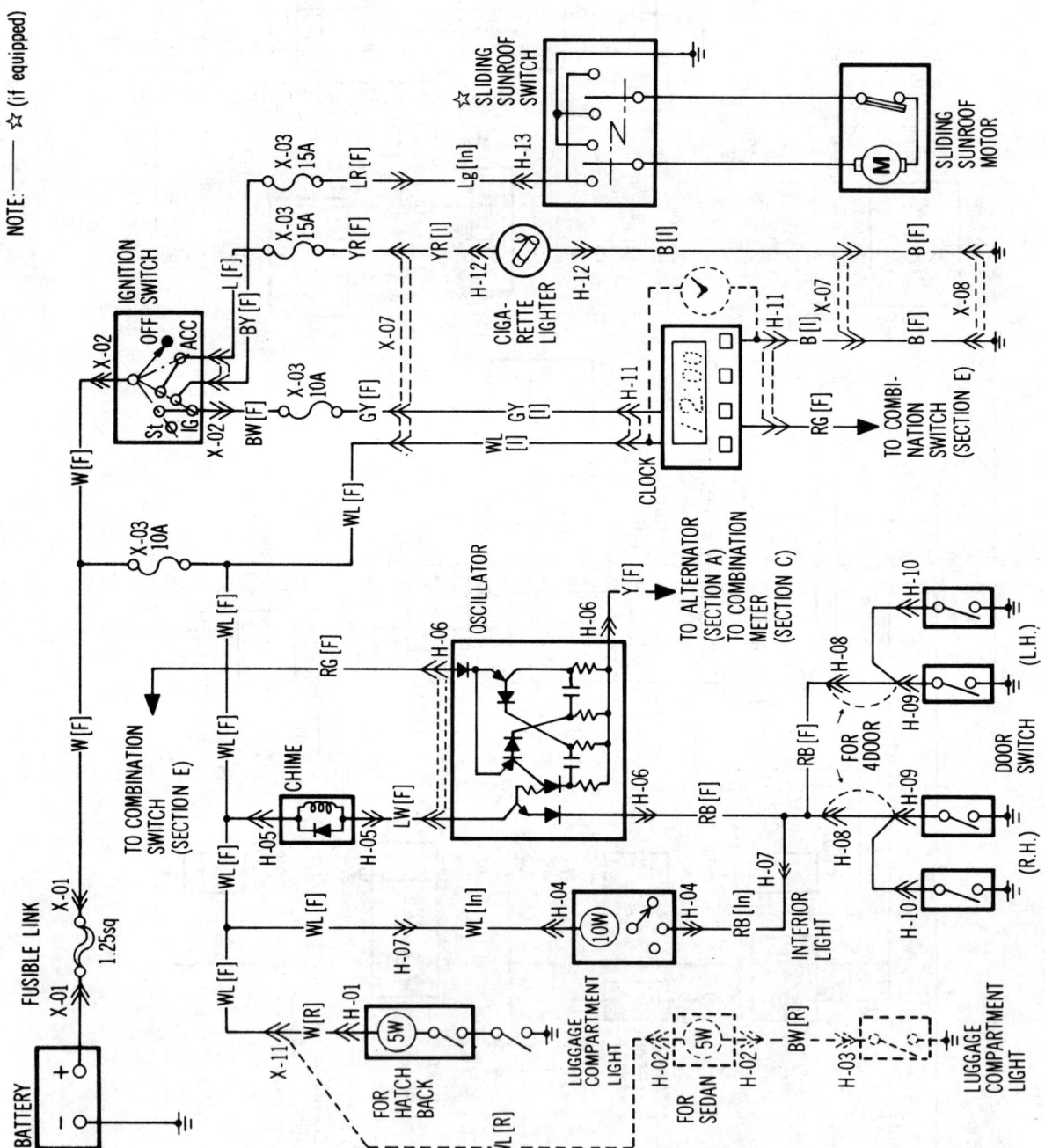

12.33 Luggage compartment light, interior light, cigarette lighter, and sunroof wiring diagram (1982 to 1987 models)

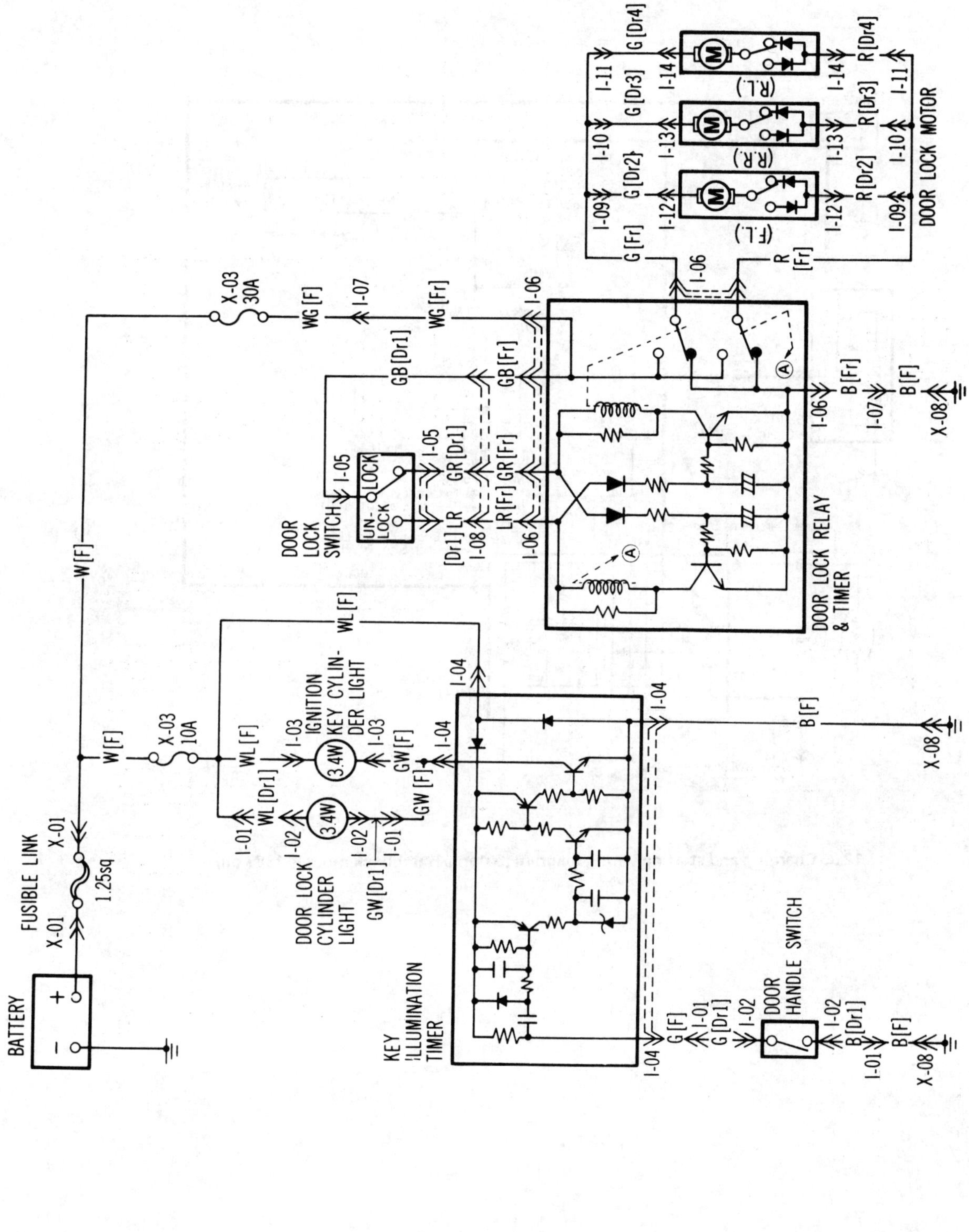

12.34 Ignition key light, door lock light and central locking wiring diagram (1982 to 1987 models)

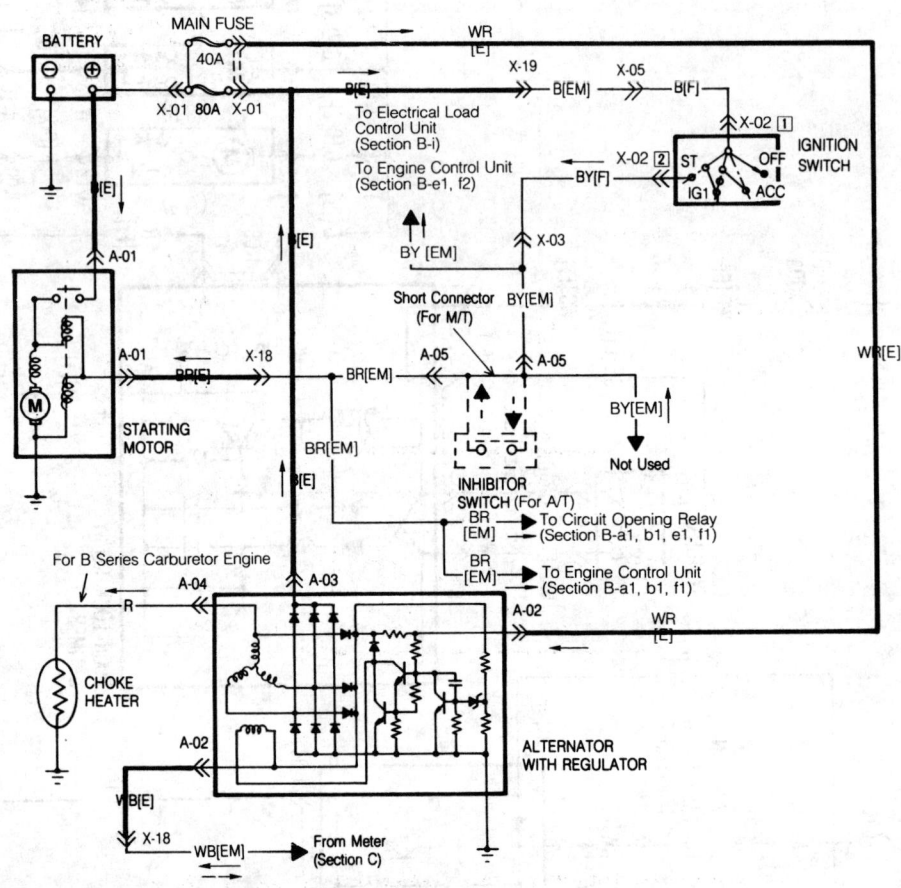

12.35 Charging and starting wiring diagram (Saloon/Hatchback models 1987 on)

Chapter 12 Electrical system

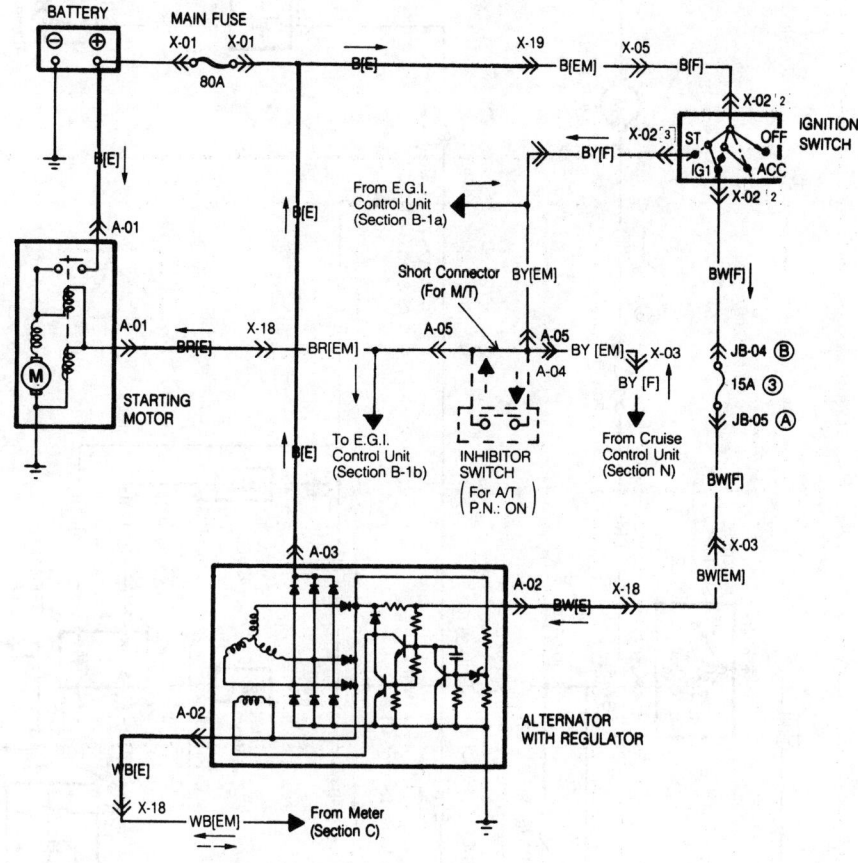

12.36 Charging and starting wiring diagram (Estate models 1987 on)

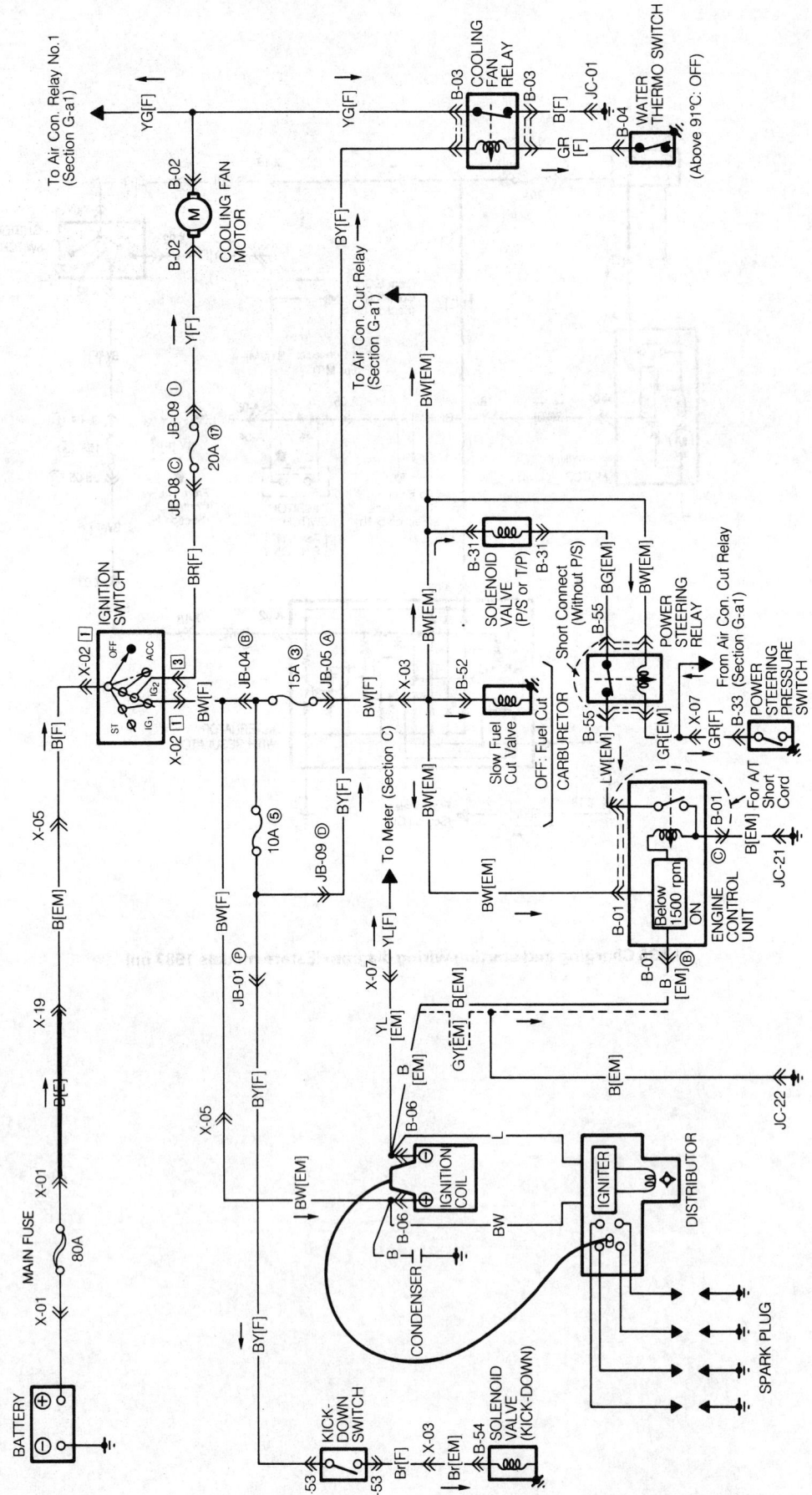

12.37 Cooling fan, ignition system, engine control system and kickdown system (automatic transmission) wiring diagram (E series carburettor engine fitted to Saloon/Hatchback models 1987 on)

Chapter 12 Electrical system

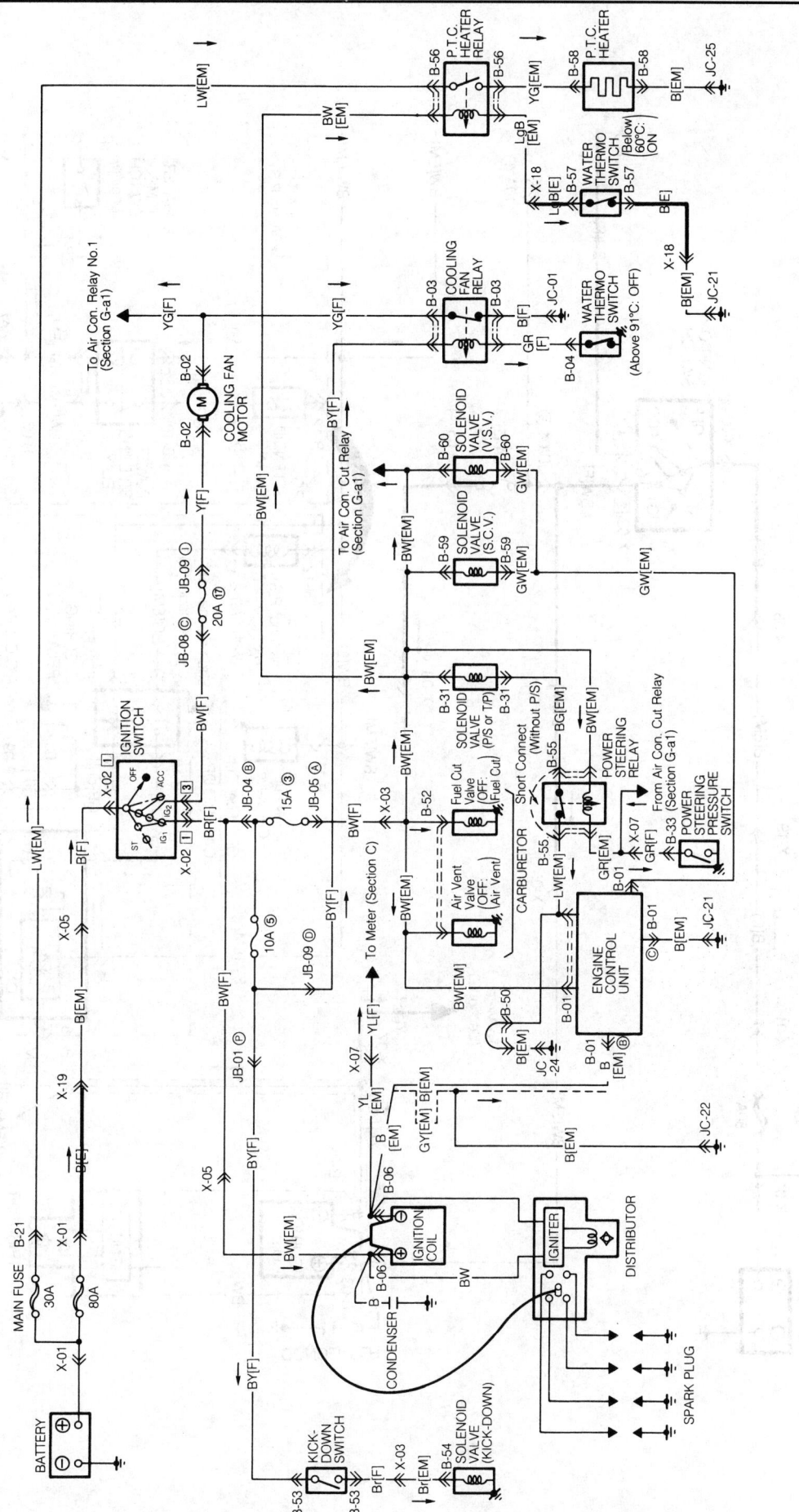

12.38 Cooling fan, ignition system, engine control system, kickdown system (automatic transmission) and PTC heater wiring diagram (B series carburettor engine fitted to Saloon/Hatchback models 1987 on)

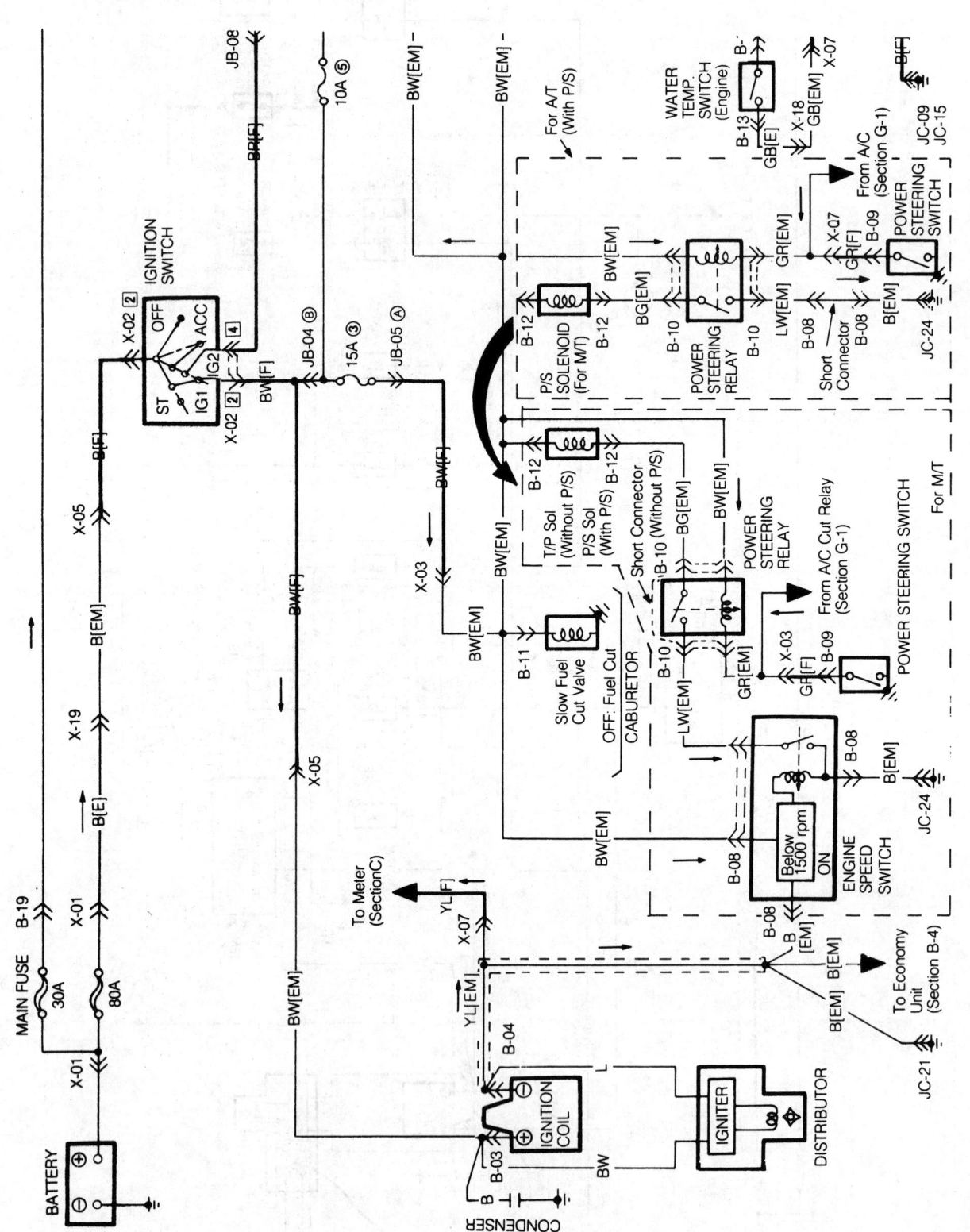

12.39 Cooling fan, ignition system, engine control system, kickdown system (automatic transmission) and PTC heater wiring diagram (Estate models 1987 on)

Chapter 12 Electrical system

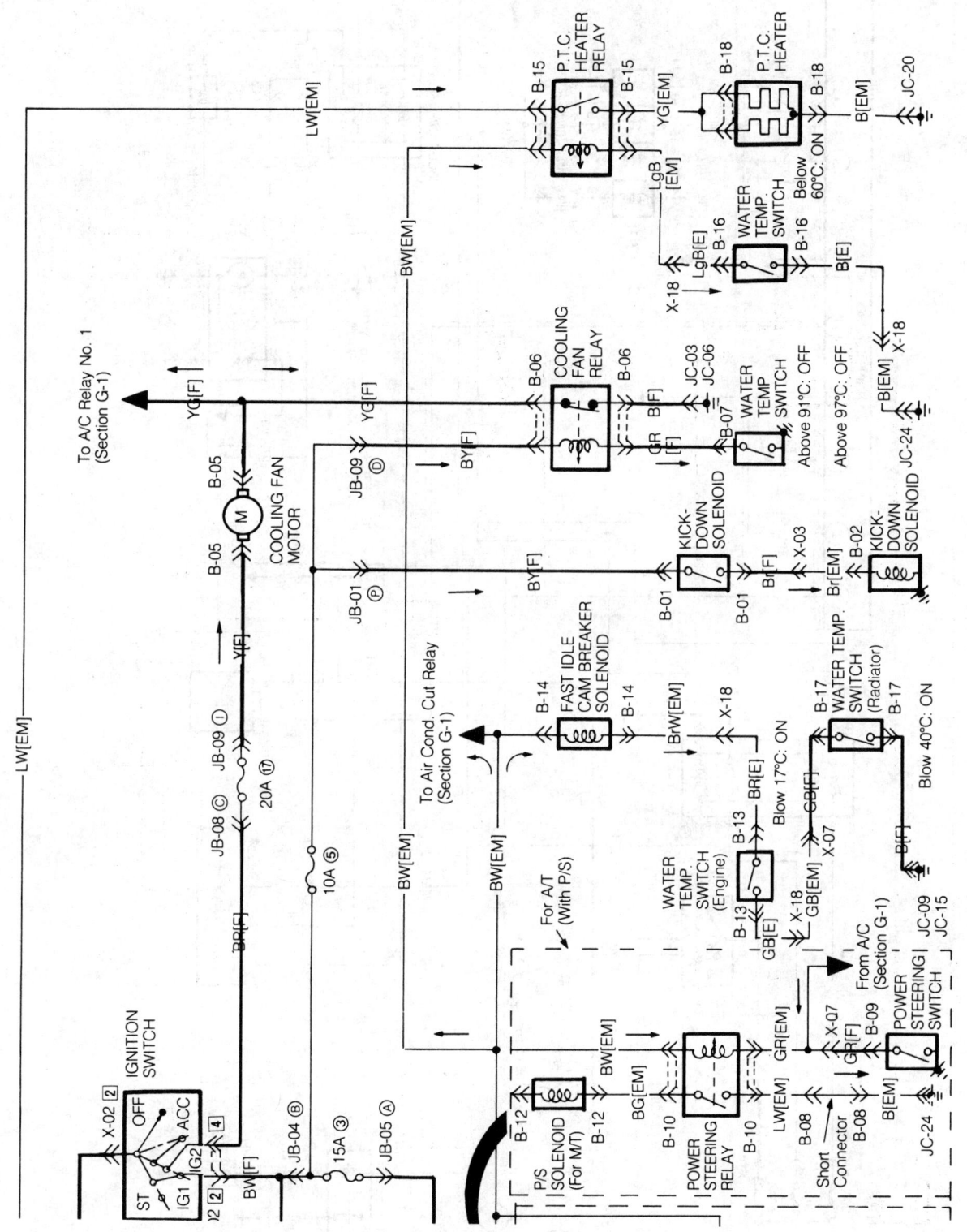

12.39 Cooling fan, ignition system, engine control system, kickdown system (automatic transmission) and PTC heater wiring diagram (Estate models 1987 on) (continued)

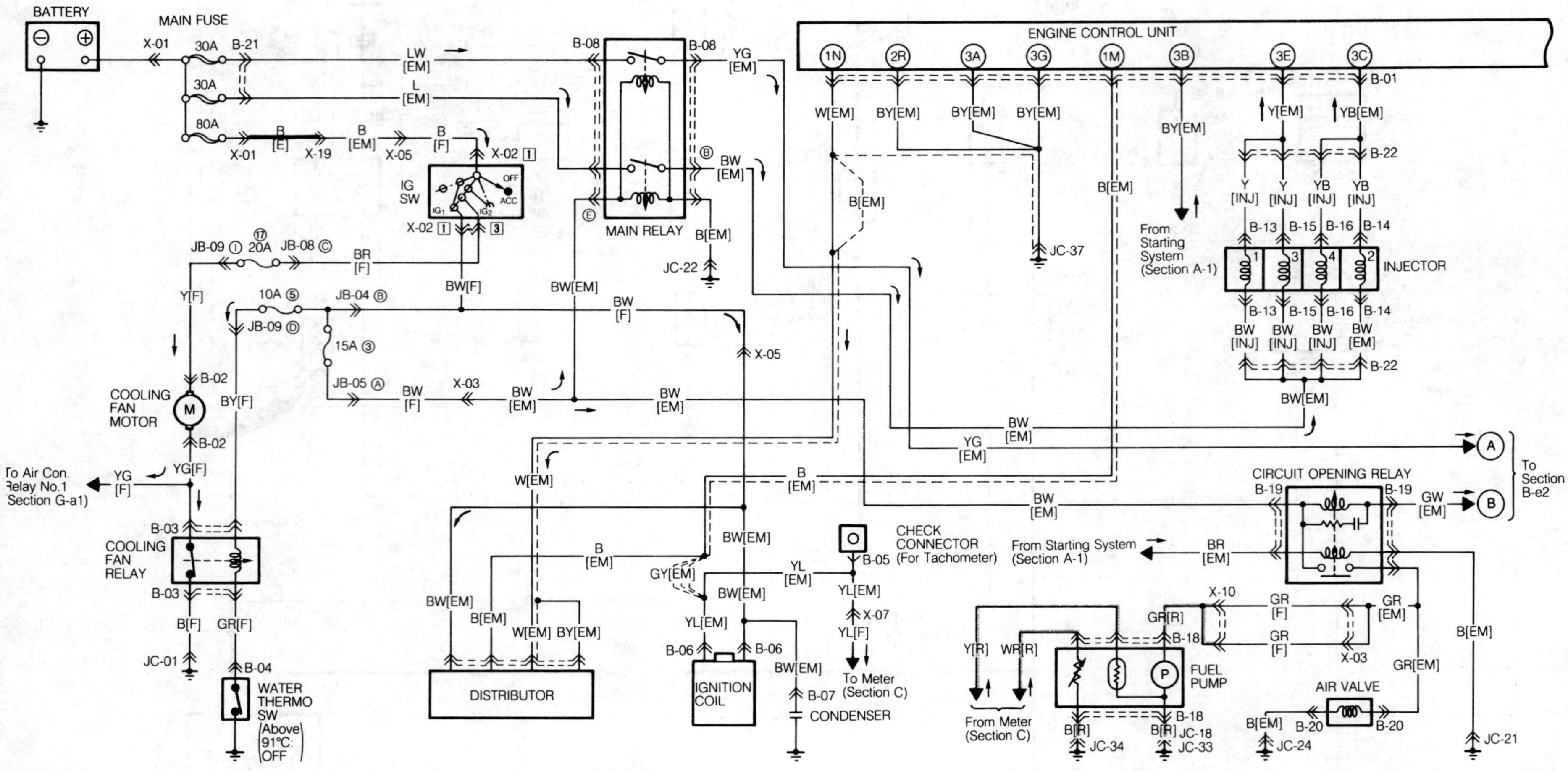

12.40 Cooling fan, ignition system, fuel and engine control system wiring diagram (1600i models)

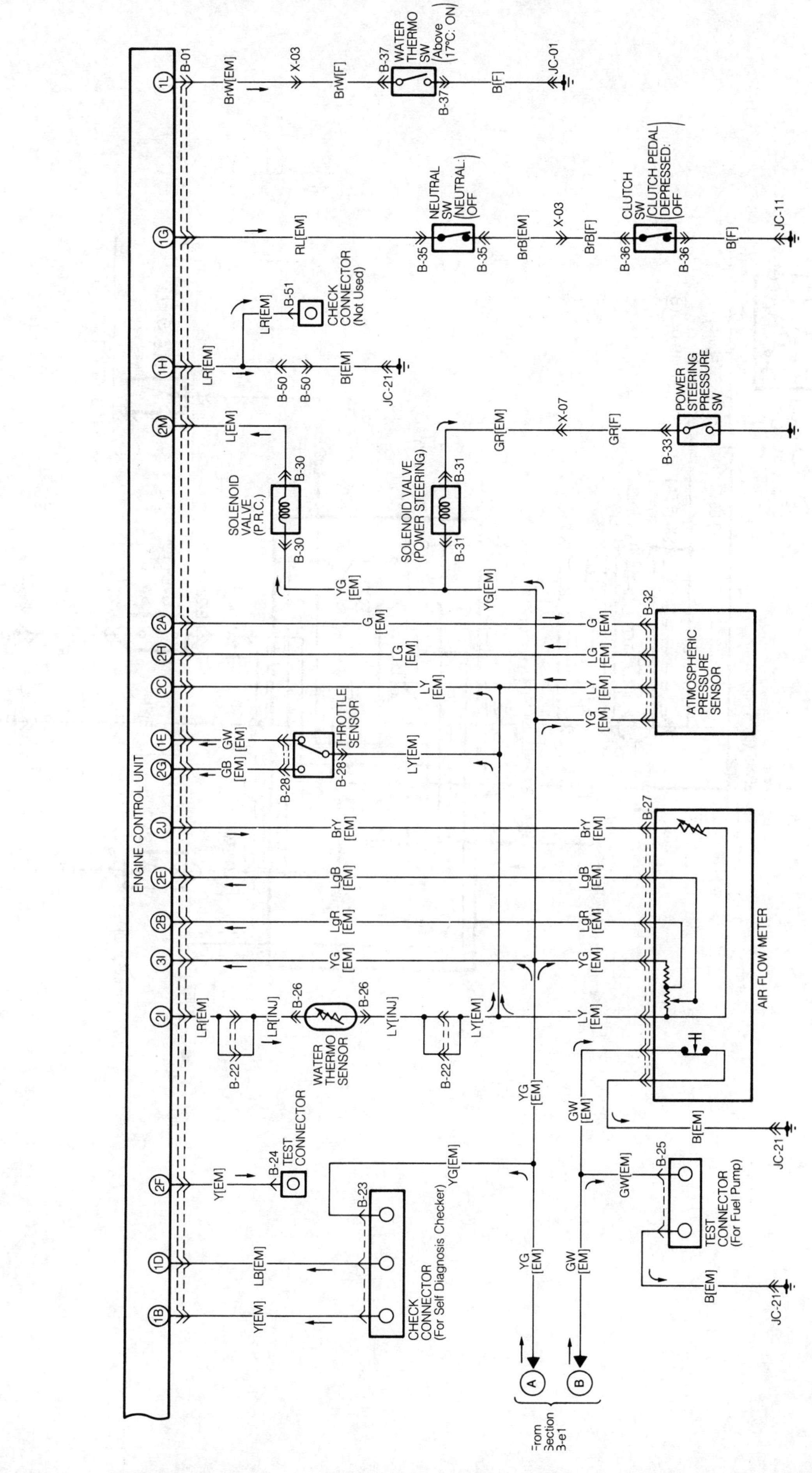

Chapter 12 Electrical system

12.40 Cooling fan, ignition system, fuel and engine control system wiring diagram (1600i models) (continued)

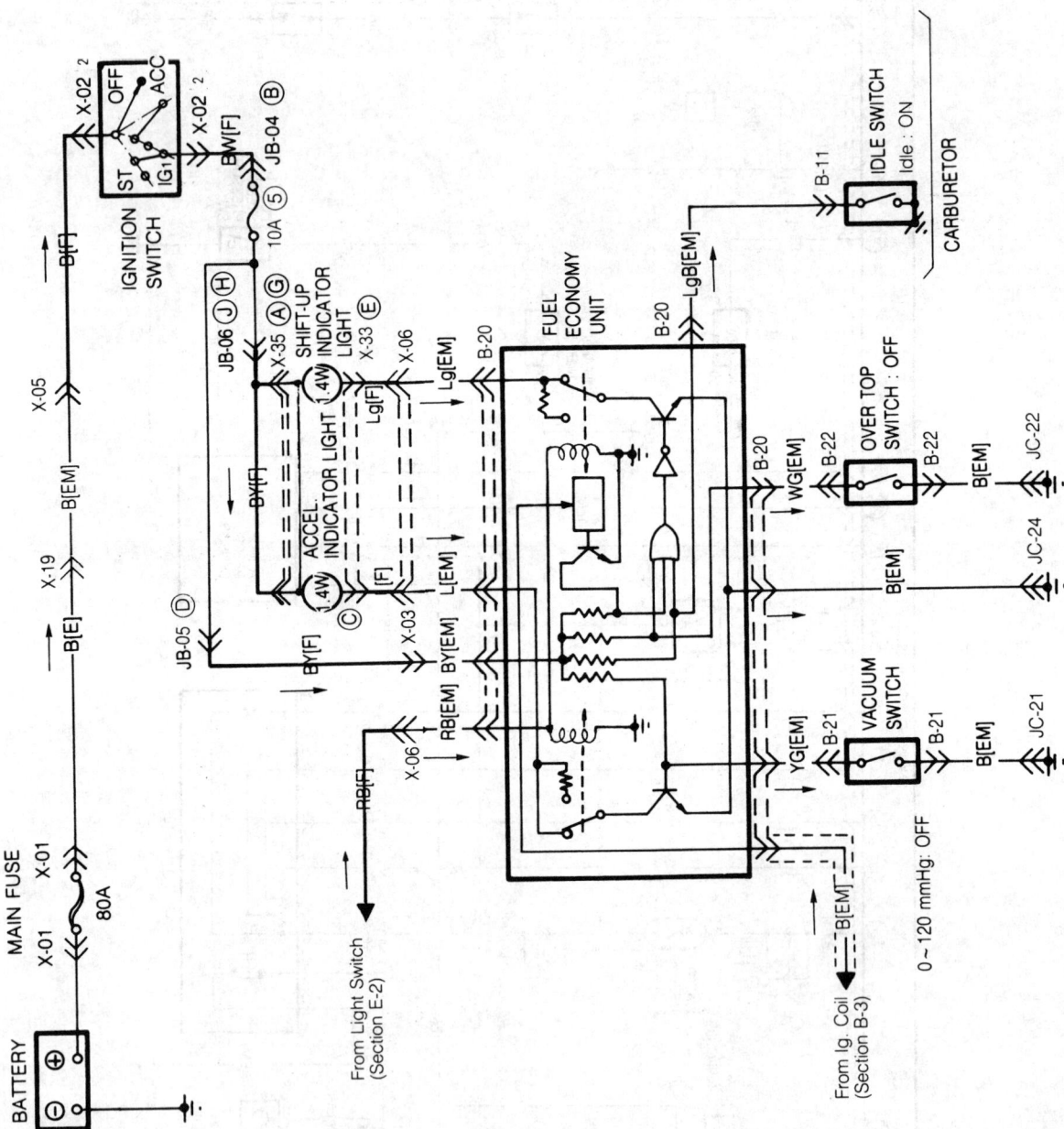

12.41 Fuel economy indicator system wiring diagram (Estate models 1987 on)

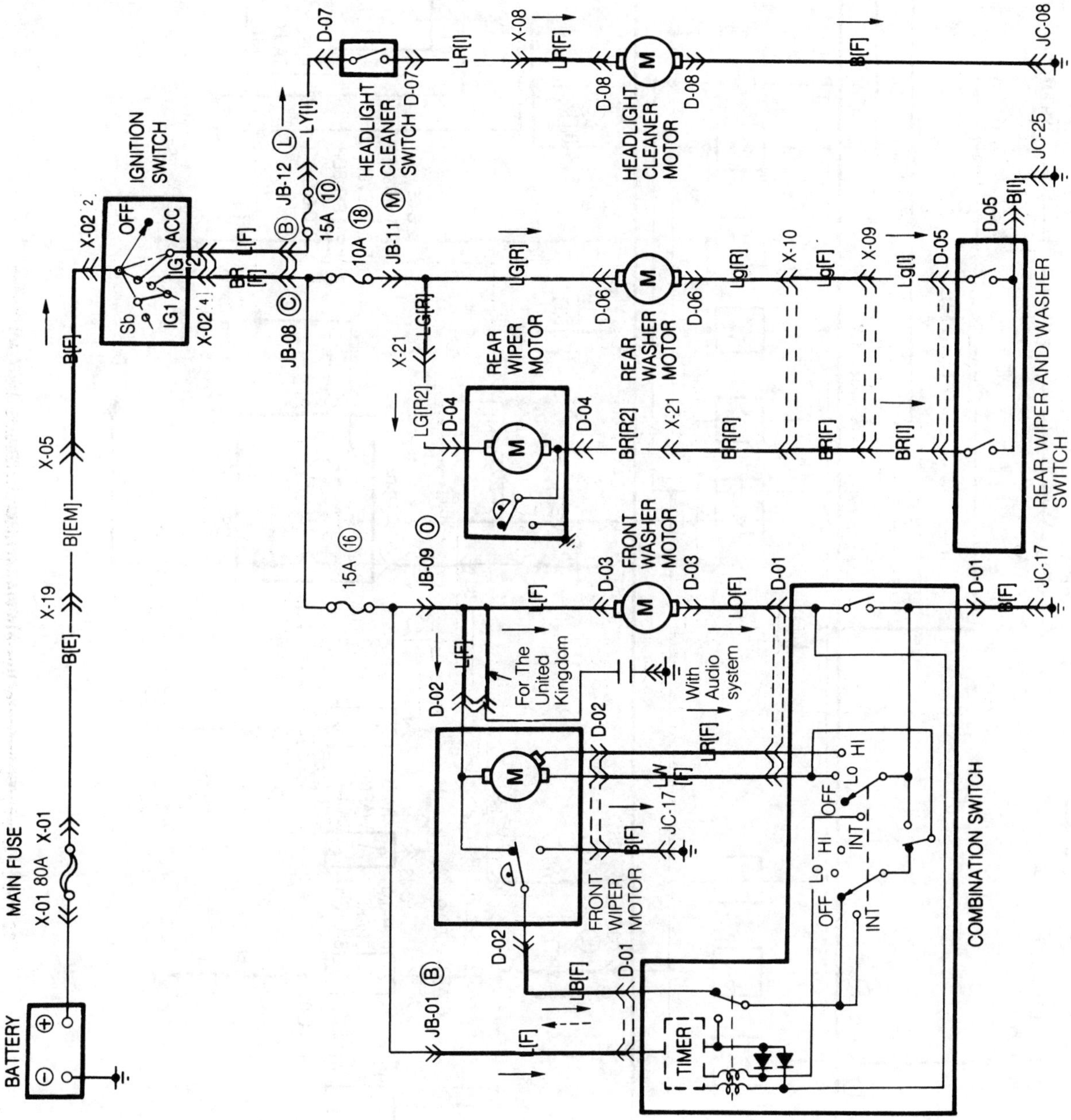

Chapter 12 Electrical system

12.42 Typical wiring diagram – Wipers, washers and headlight cleaners (1987 on)

Chapter 12 Electrical system

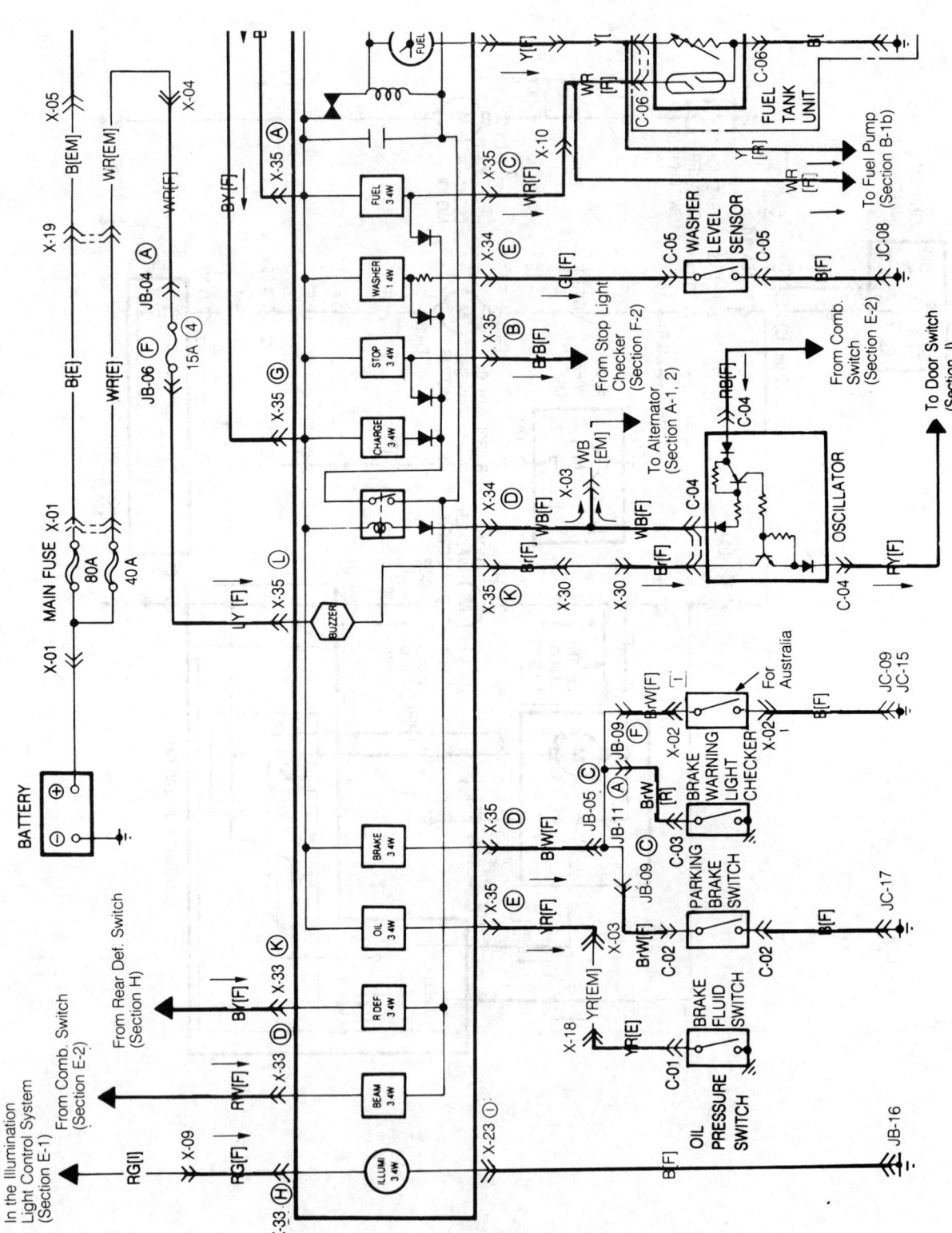

12.43 Typical wiring diagram – Meters and warning lights (1987 on)

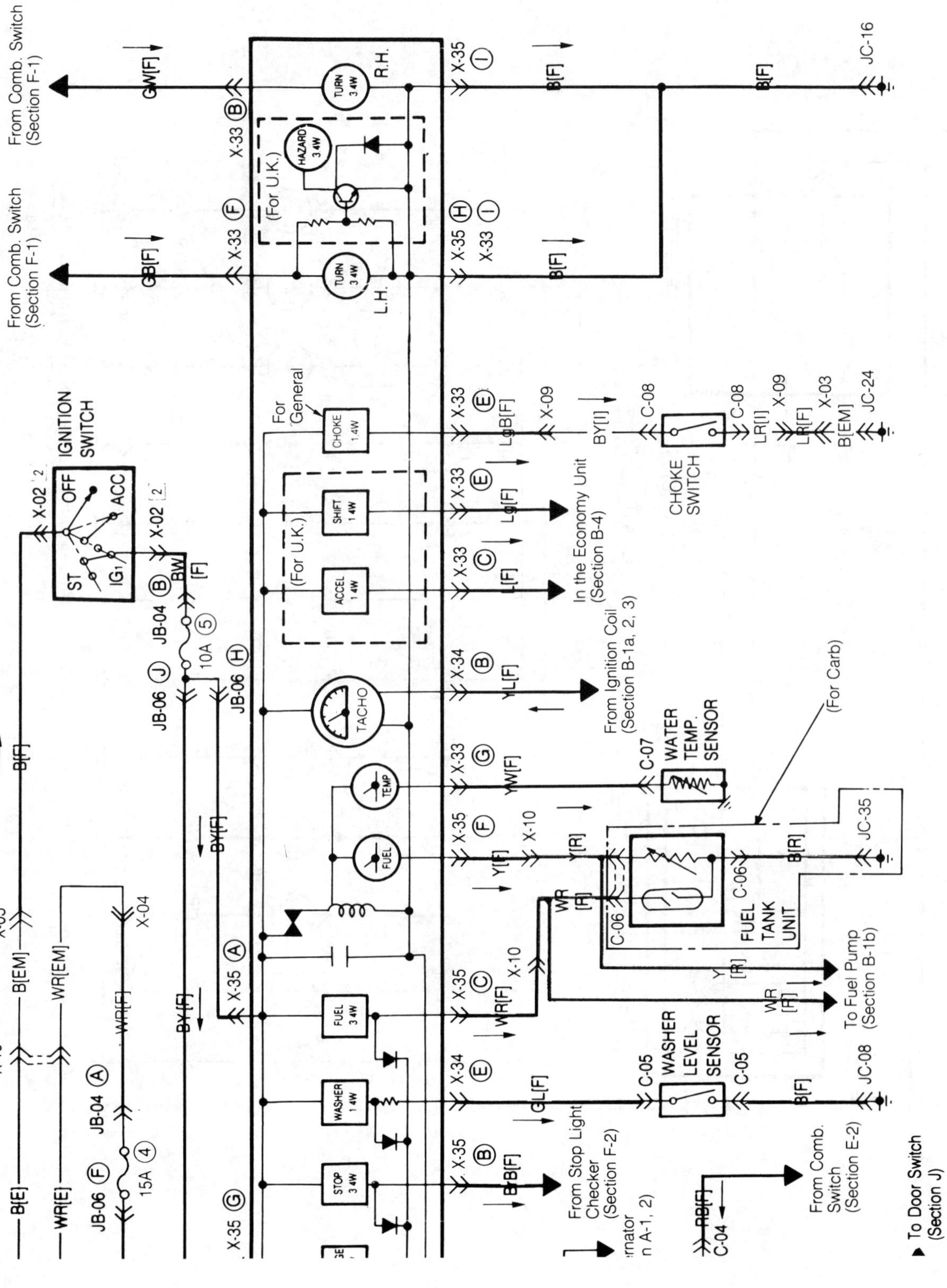

Chapter 12 Electrical system

12.43 Typical wiring diagram – Meters and warning lights (1987 on) (continued)

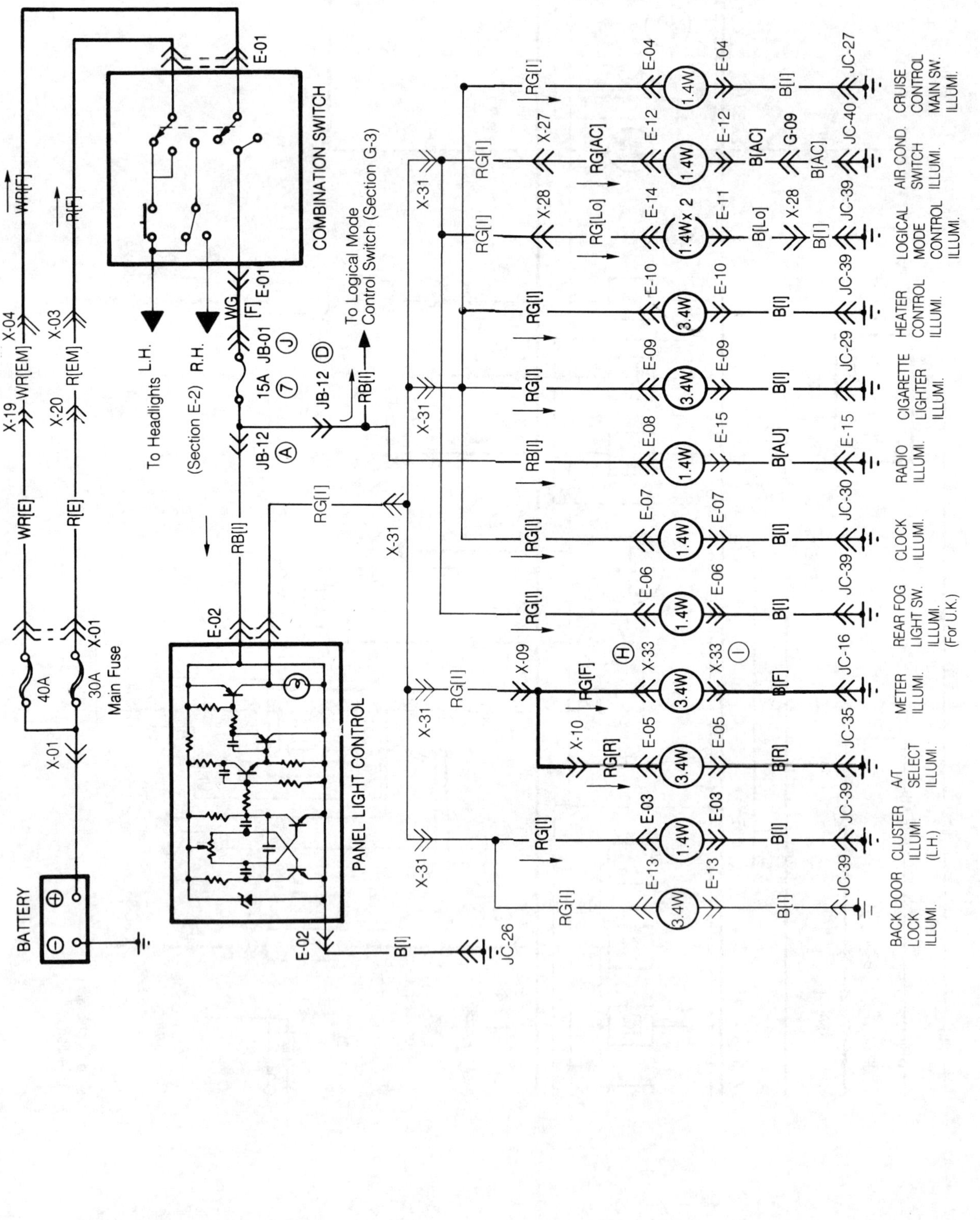

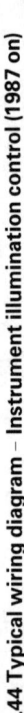

12.44 Typical wiring diagram – Instrument illumination control (1987 on)

Chapter 12 Electrical system

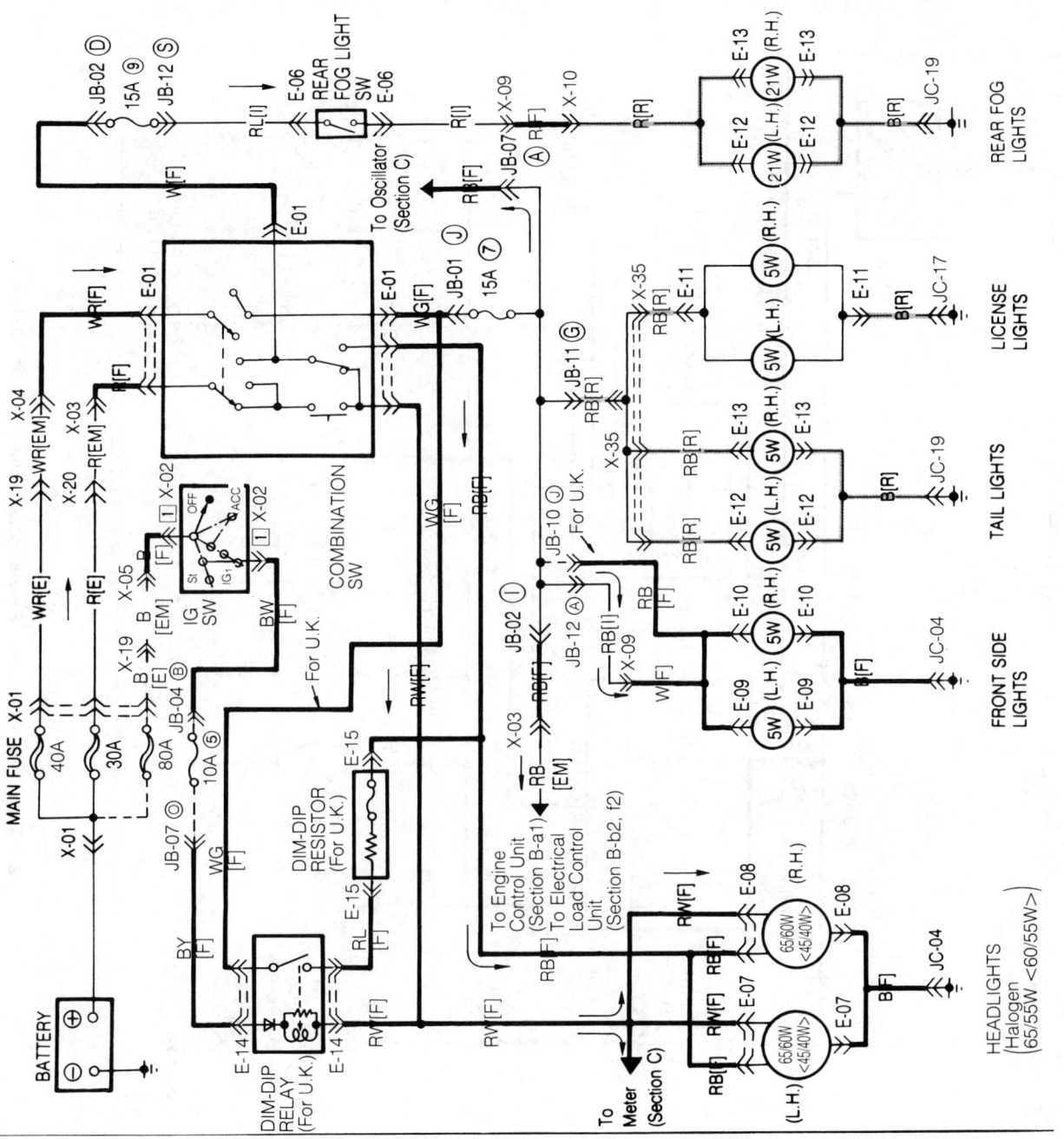

12.45 Typical wiring diagram – Sidelights, headlights, tail lights number plate and rear fog lights (1987 on)

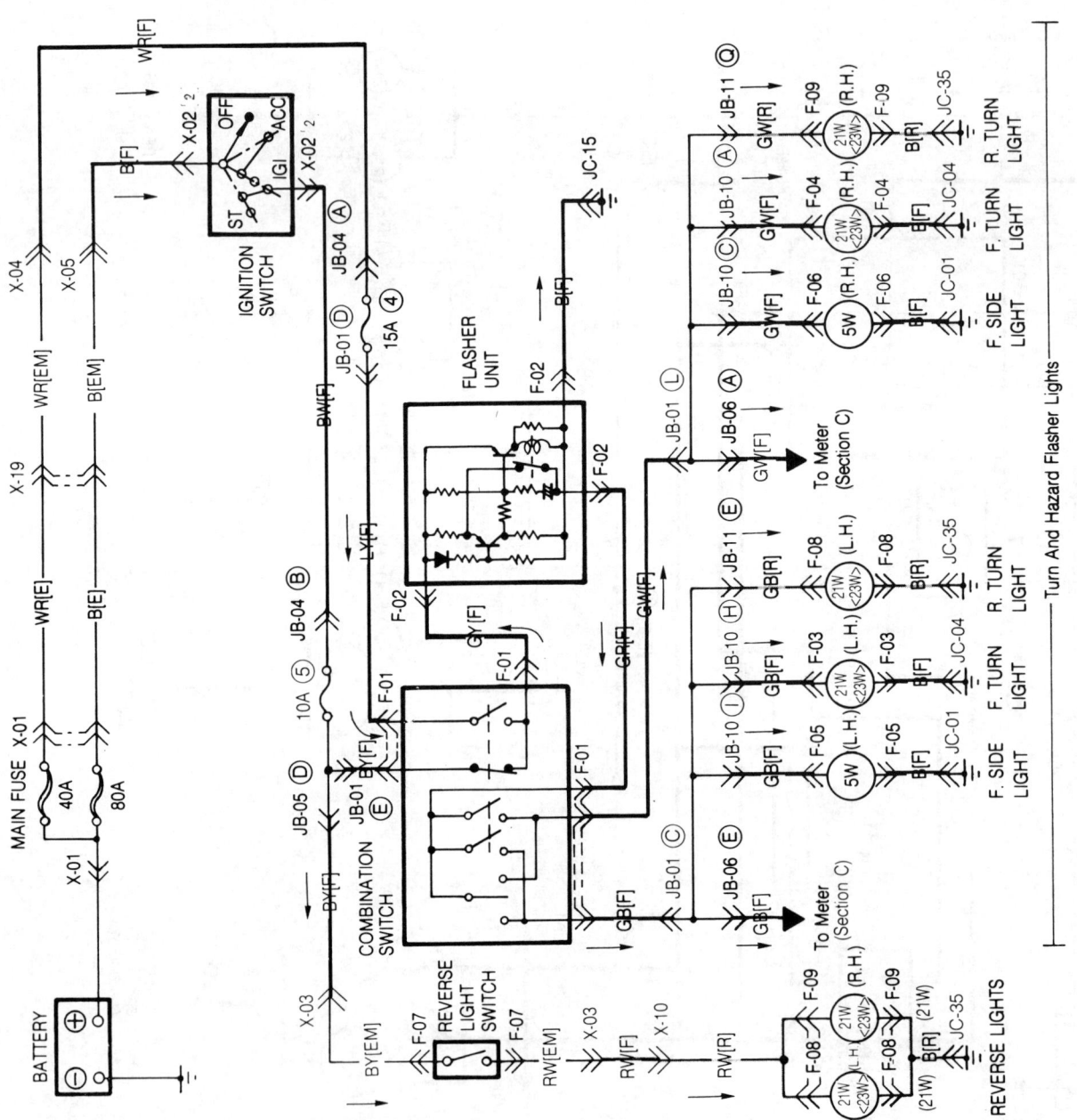

12.46 Typical wiring diagram – Reversing lights and indicators (1987 on)

Chapter 12 Electrical system 325

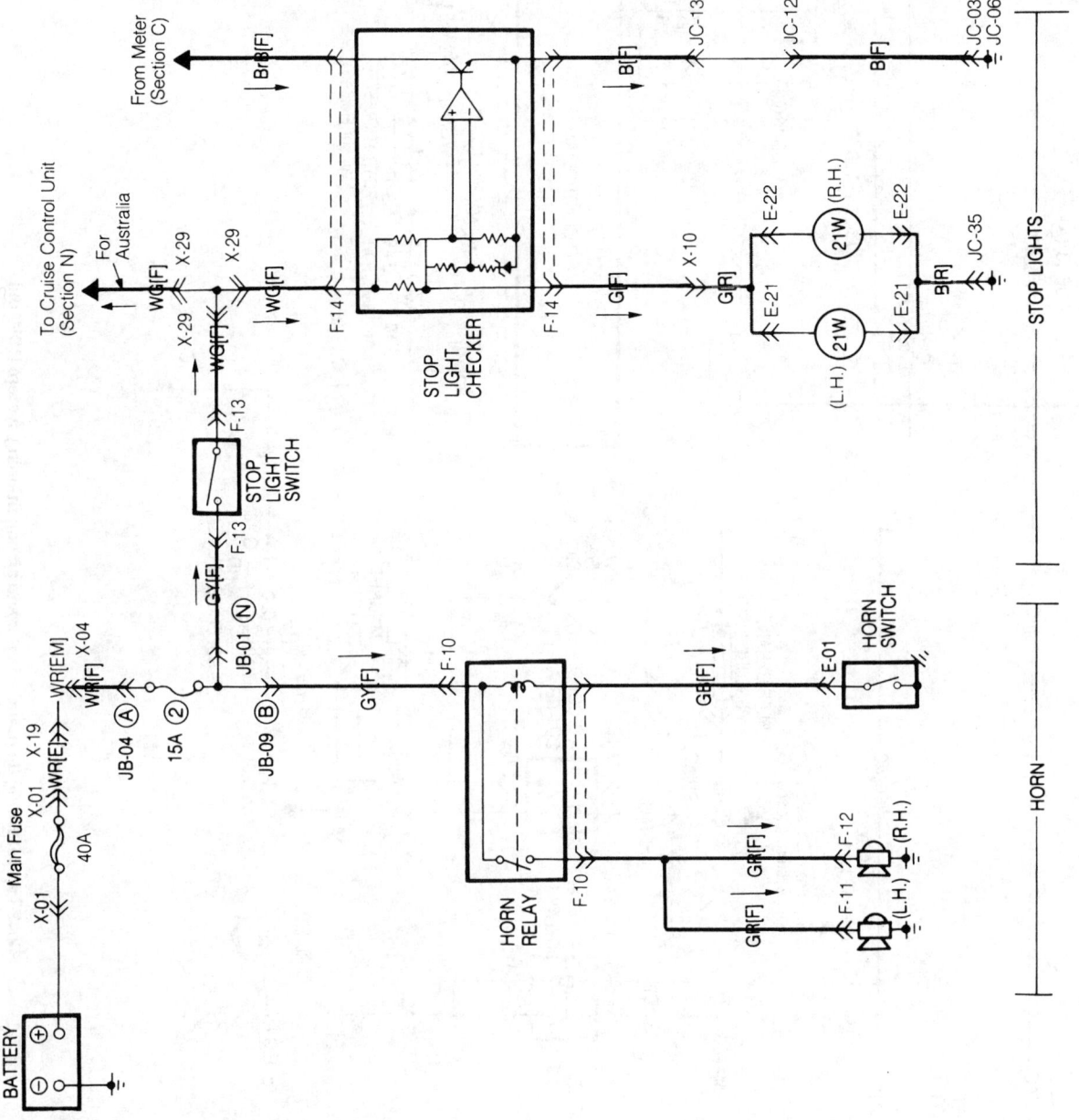

12.47 Typical wiring diagram – Horn and brake lights (1987 on)

326 Chapter 12 Electrical system

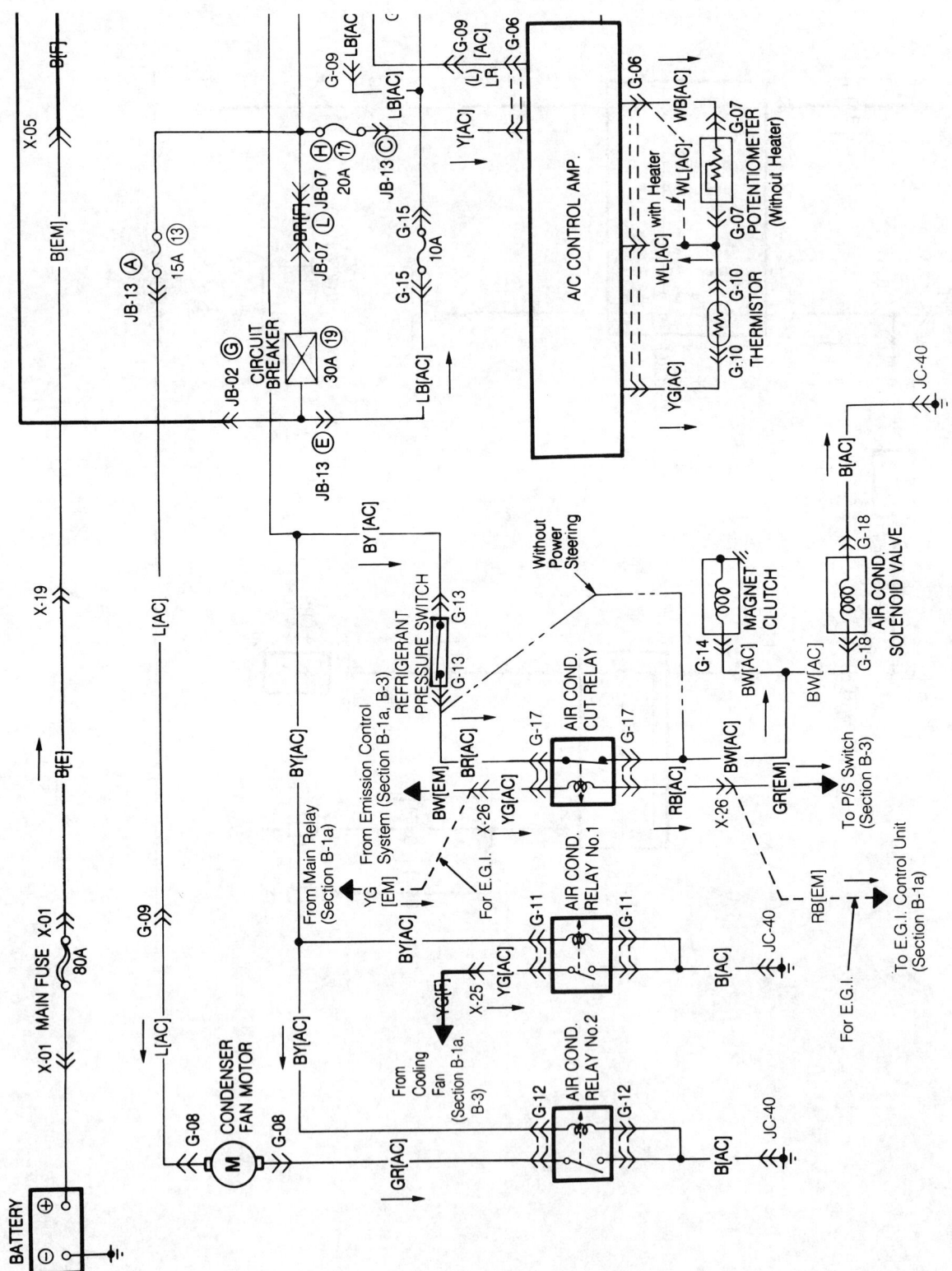

12.48 Typical wiring diagram – Heater and air conditioning system (1987 on)

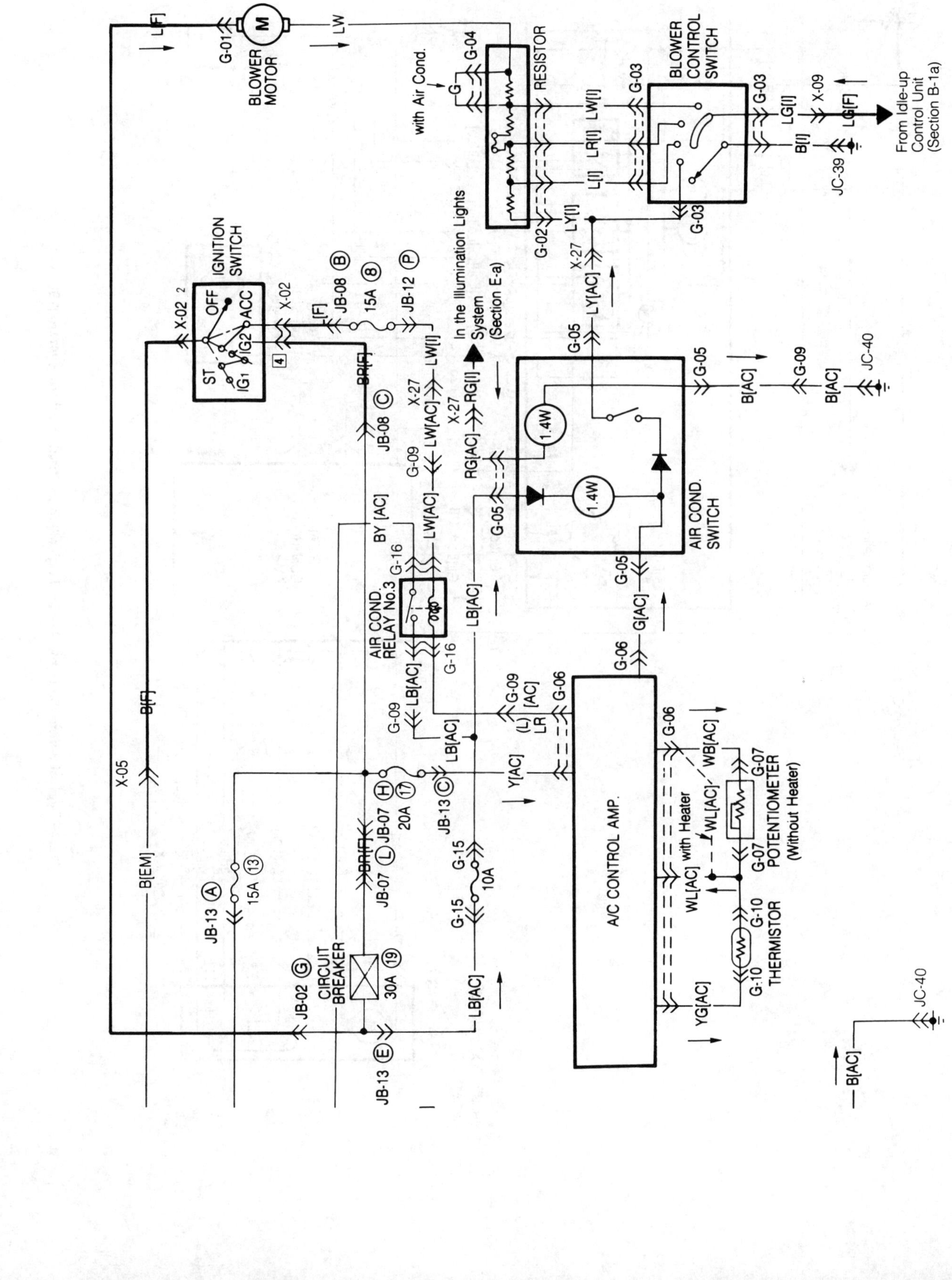

Chapter 12 Electrical system

12.48 Typical wiring diagram – Heater and air conditioning system (1987 on) (continued)

Chapter 12 Electrical system

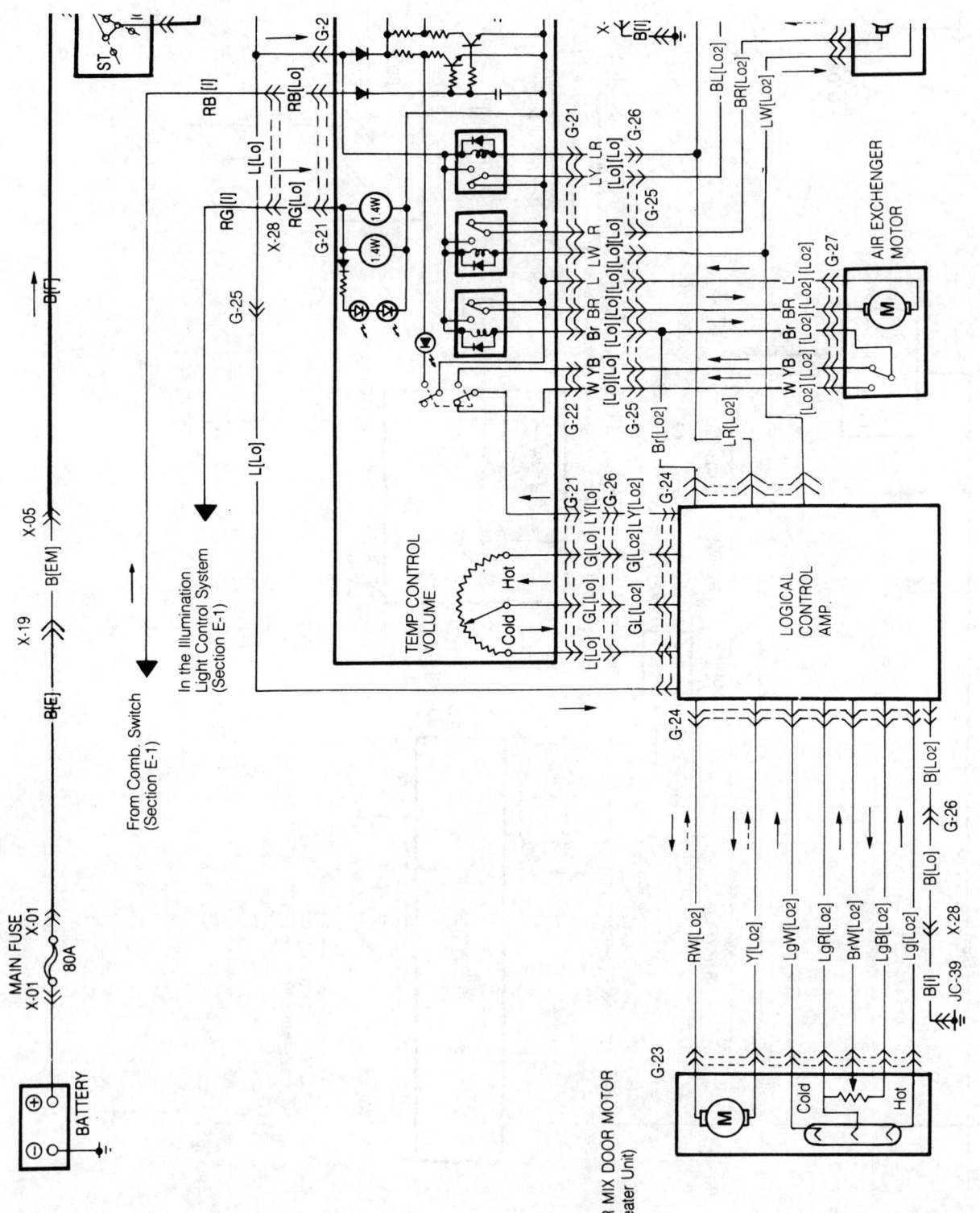

12.49 Typical wiring diagram – Logic ('Logical') mode heater control system (1987 on)

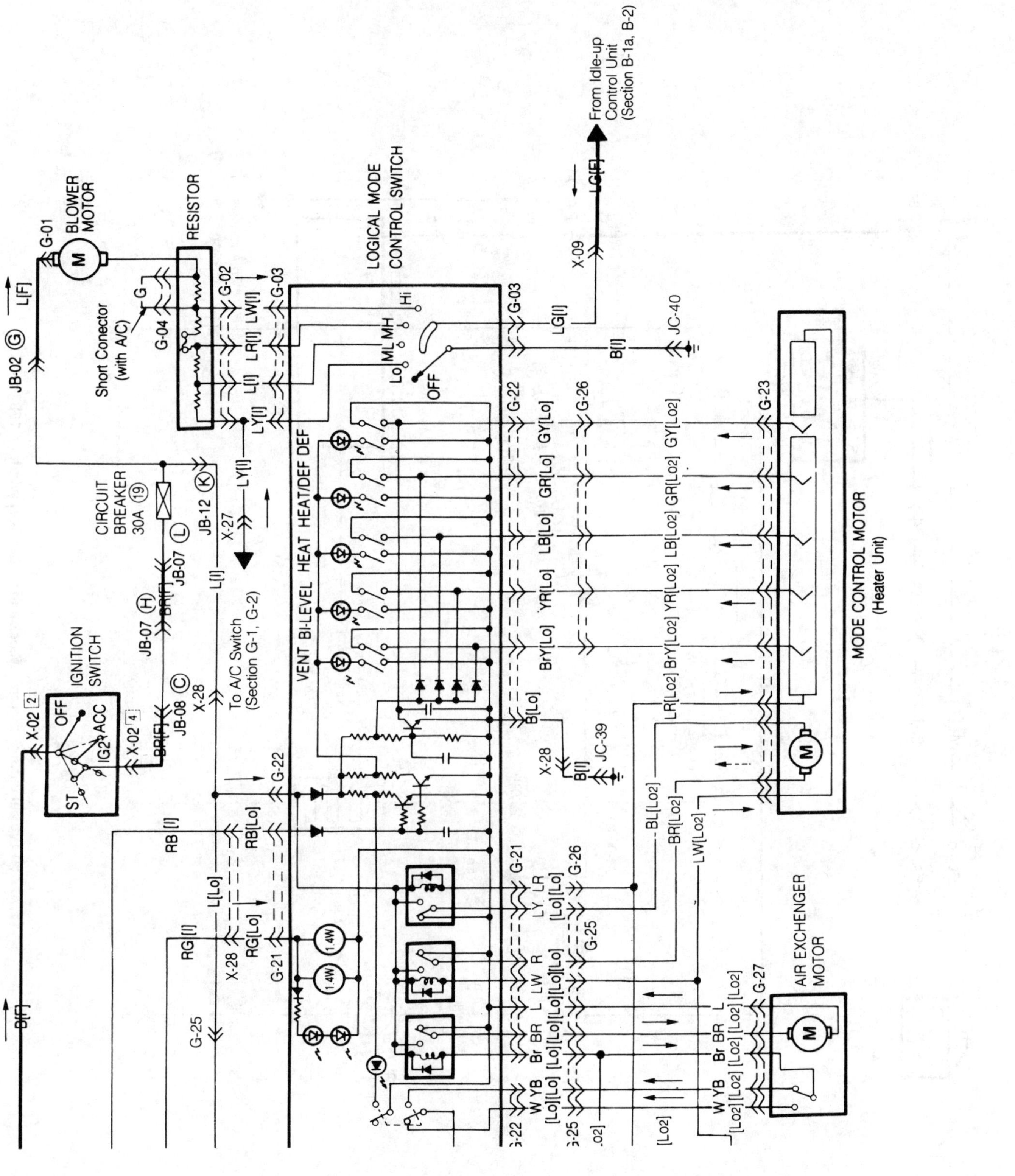

12.49 Typical wiring diagram – Logic ("Logical") mode heater control system (1987 on) (continued)

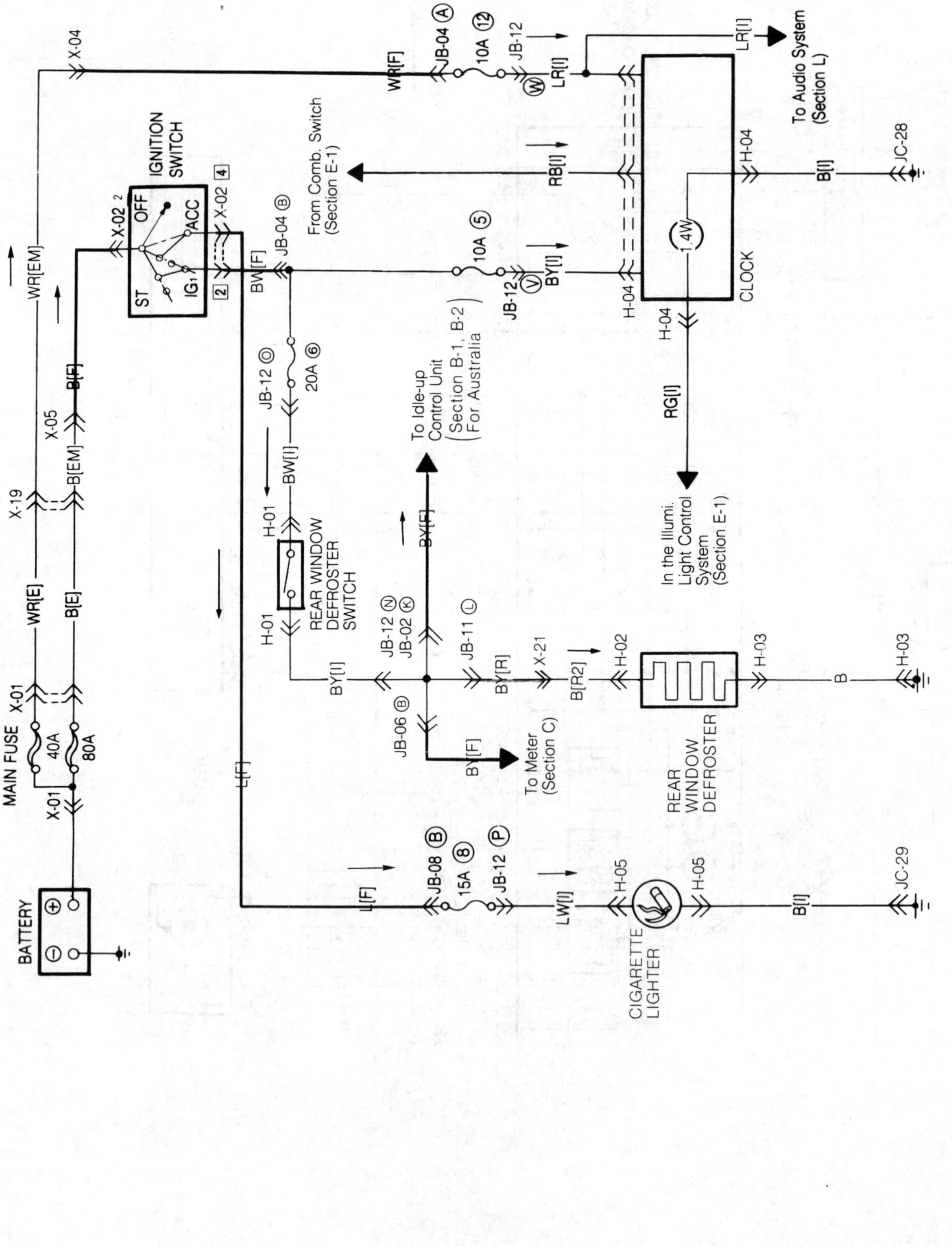

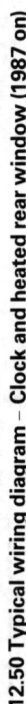

12.50 Typical wiring diagram – Clock and heated rear window (1987 on)

Chapter 12 Electrical system

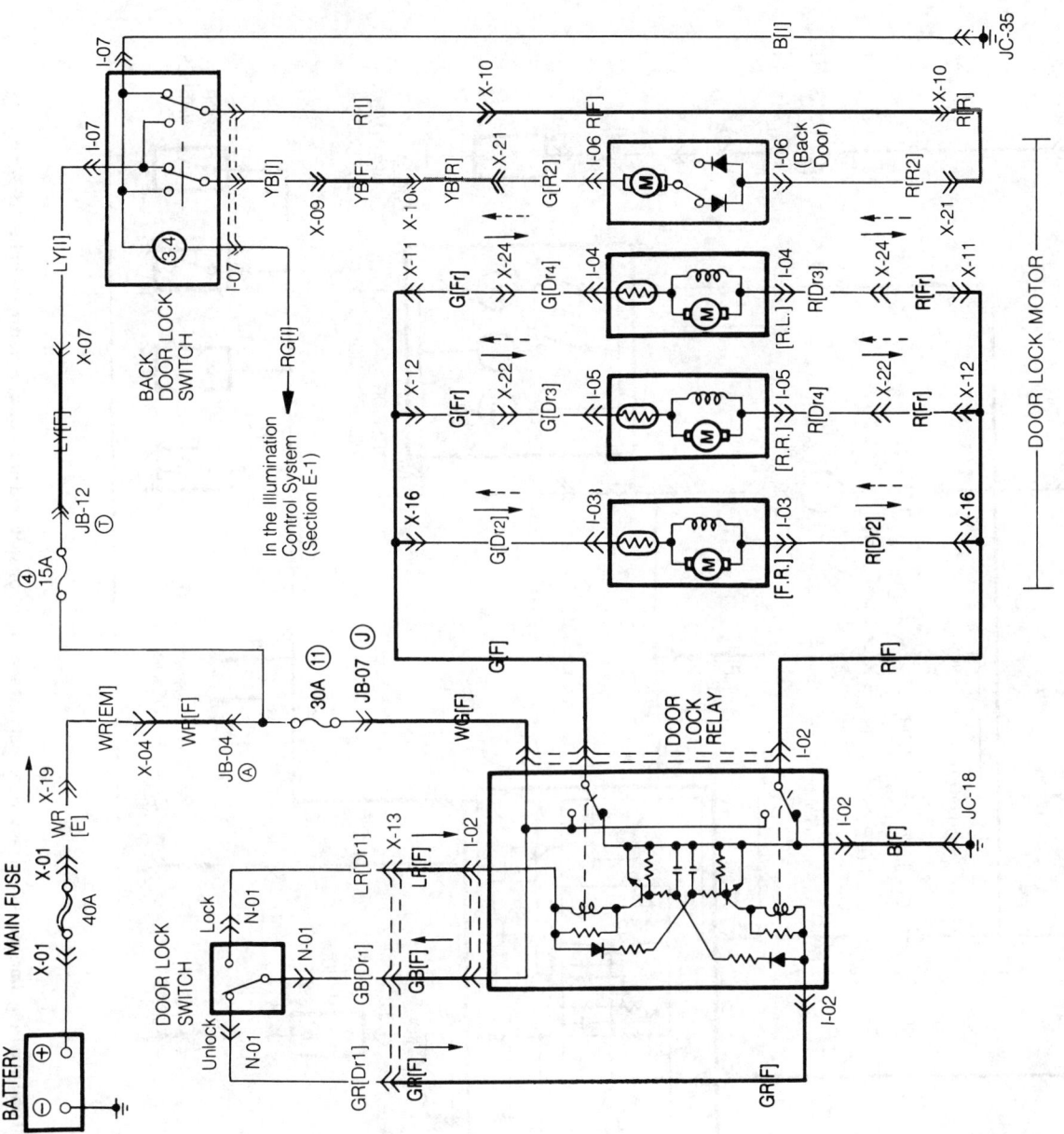

12.51 Typical wiring diagram – Central locking system (1987 on)

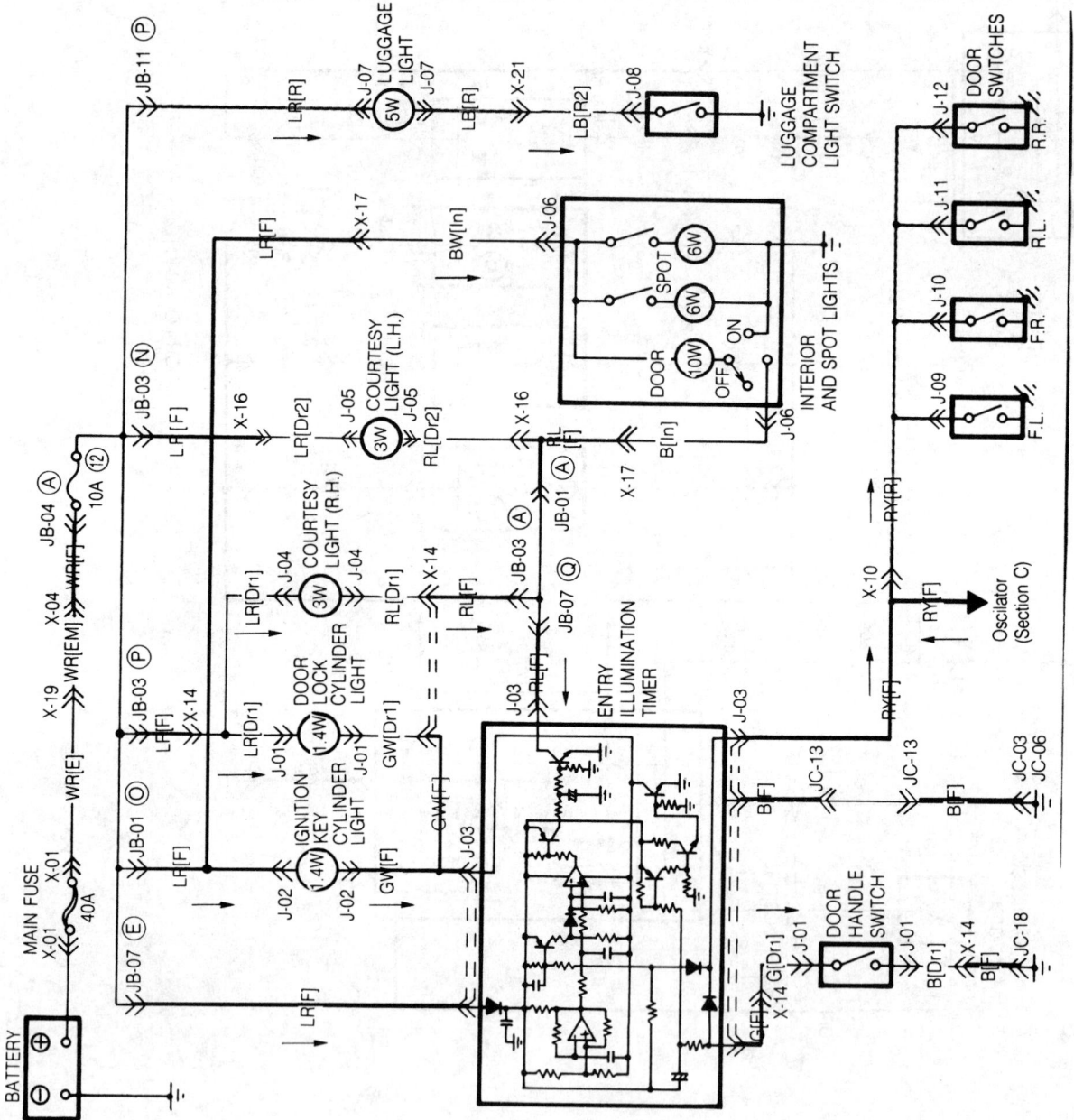

12.52 Typical wiring diagram – Courtesy, door lock, interior, ignition lock and luggage compartment lights (1987 on)

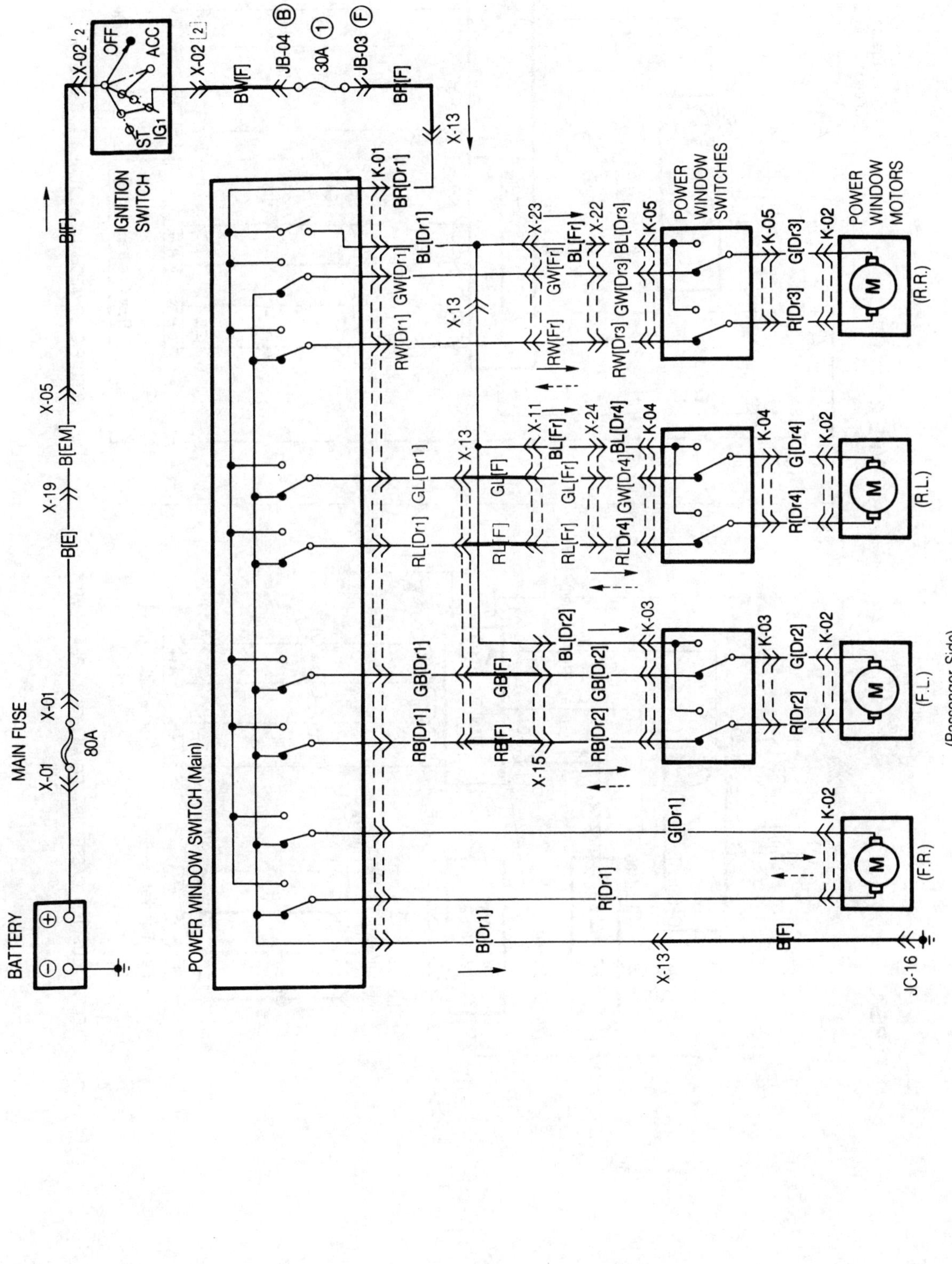

Chapter 12 Electrical system

12.53 Typical wiring diagram – Electric windows (1987 on)

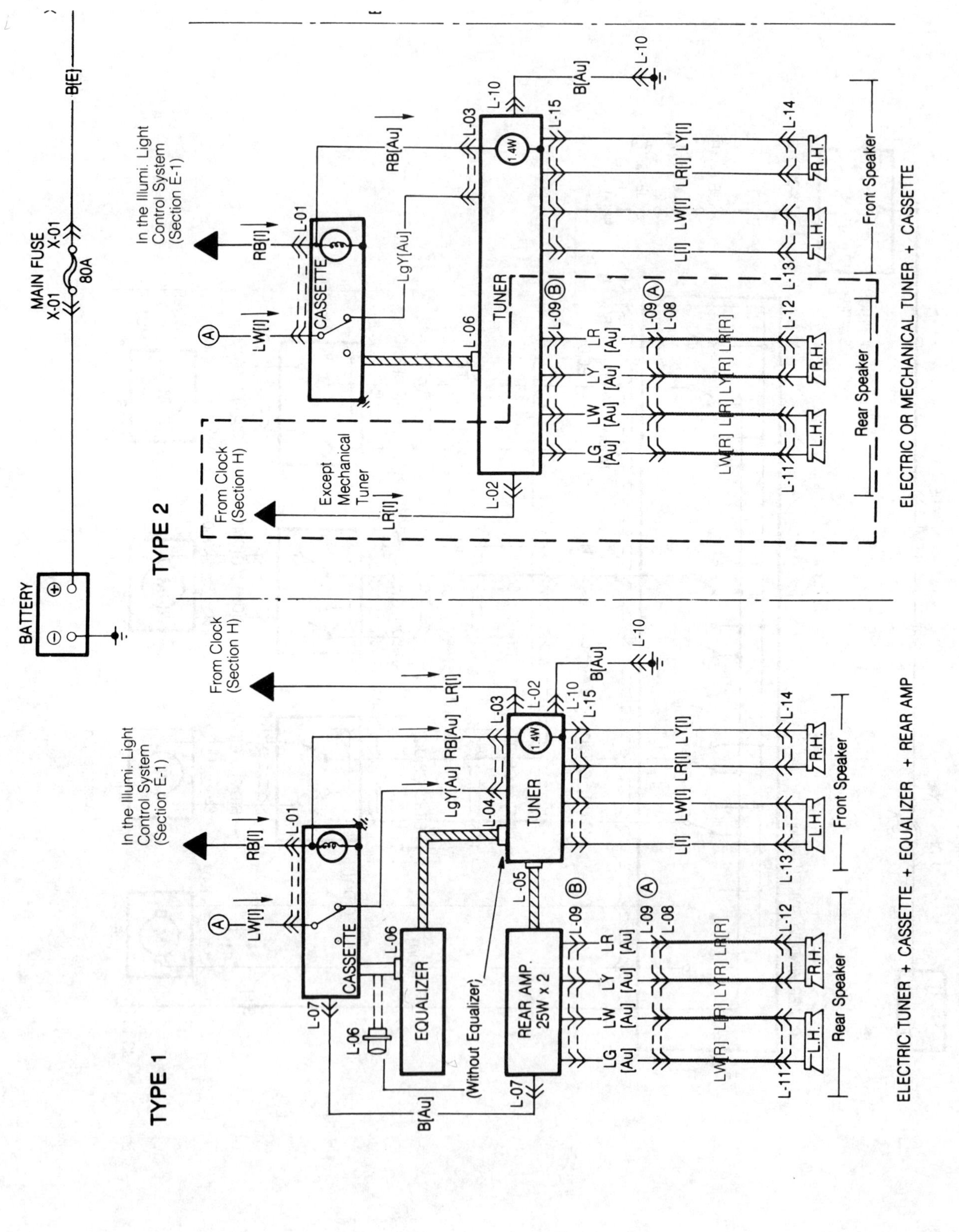

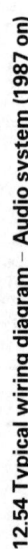

12.54 Typical wiring diagram – Audio system (1987 on)

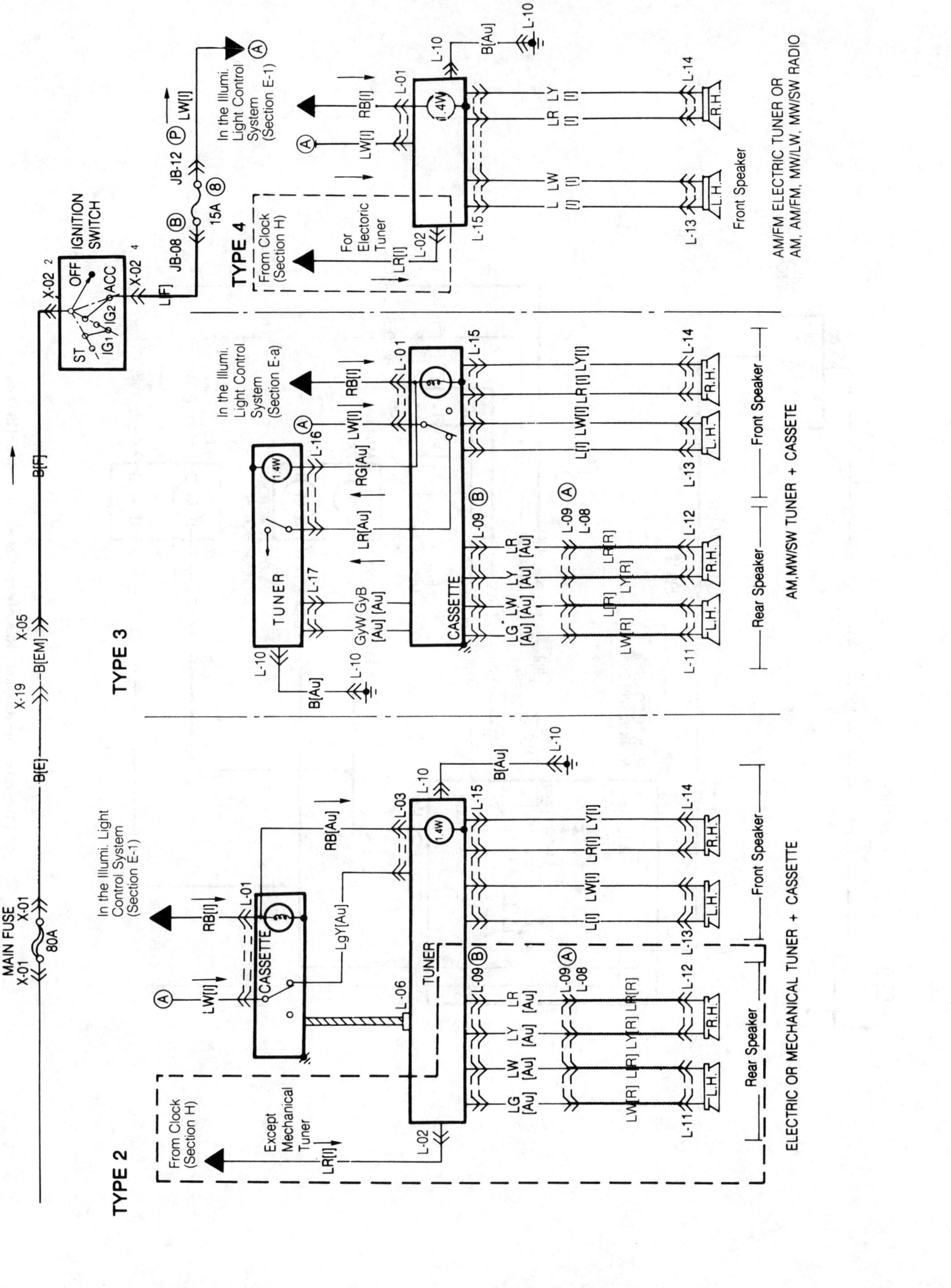

12.54 Typical wiring diagram – Audio system (1987 on) (continued)

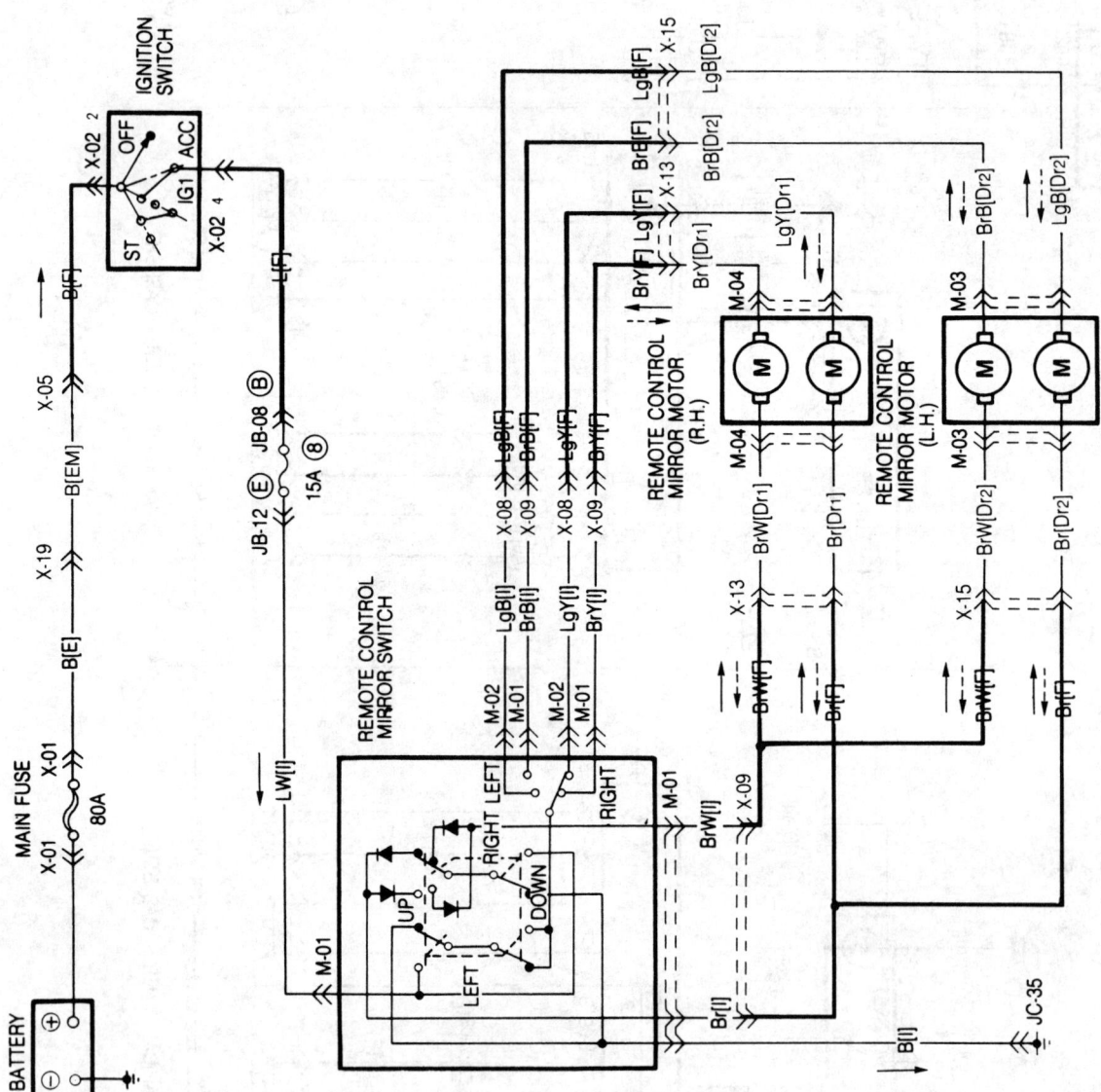

12.55 Typical wiring diagram – Remote control exterior mirror (1987 on)

Index

A

About this manual – 6
Accelerator cable
    carburettor engines – 132
    fuel injected engines – 150
Accelerator pedal
    carburettor engines – 133
    fuel injected engines – 150
Acknowledgements – 2
Aerial – 290
Air cleaner – 50
    carburettor engines – 131
    fuel injected engines – 150
Alternator – 64, 270, 271
Anti-roll bar
    front – 228
    rear – 233
Antifreeze – 50
Automatic transmission *see* **Manual gearbox and automatic transmission**

B

Balljoint
    front suspension – 230
    track rod – 241
Battery – 63, 270
Bearings
    big-end – 108
    clutch release – 177
    hub
        front – 63, 221
        rear – 63, 230
    main – 108, 109
Big-end bearings – 108
Bleeding
    brakes – 202
    clutch – 175
    power steering – 241
Bodywork and fittings – 63, 244 *et seq*
Bodywork repair – *see colour pages between pages 32 and 33*
Bonnet – 248, 250
Booster battery starting – 19
Boot lid – 256
Braking system – 25, 59, 198 *et seq*
Bulb renewal
    courtesy lamps – 280
    direction indicator – 278
    foglamp – 280
    headlamp – 277
    instrument panel lamp – 281
    luggage compartment lamp – 281
    map reading lamps – 281
    number plate lamp – 280
    rear lamp cluster – 278
    sidelamp – 278

Bumpers
    front – 246
    rear – 247

C

Cables
    accelerator
        carburettor engines – 132
        fuel injected engines – 150
    bonnet release – 250
    boot lid – 256
    choke (carburettor engines) – 133
    clutch – 56, 172
    fuel filler – 256
    handbrake – 216
    heater control – 124
    speedometer – 285
    tailgate – 256
Caliper
    front – 207
    rear – 215
Camshaft
    B series engine – 94, 95
    E series engine – 75, 78
Carburettor – 50, 52, 134, 136, 141
Cassette player – 289
Centre console – 262
Charging system – 270
Choke – 50
Choke cable (carburettor engines) – 133
Cigar lighter – 284
Clock – 285
Clutch – 24, 56, 171 *et seq*
CO adjustment – 50
Compression test
    B series engine – 87
    E series engine – 74
Condenser – 54
Connecting rods – 105, 107, 111
Console – 262, 277
Constant velocity joint – 58, 195
Contact breaker points – 54
Conversion factors – 20
Coolant level sensor – 119
Cooling fan – 117, 118
Cooling, heating and ventilation systems – 23, 48, 112 *et seq*
Courtesy lamp – 276, 280
Crankcase – 107
Crankshaft – 105, 108, 109
Crankshaft oil seals
    B series engine – 100
    E series engine – 84
CV joint – 58
Cylinder block – 107
Cylinder head – 103, 104
    B series engine – 87, 95
    E series engine – 75

D

Dimensions – 7
Direction indicators – 278, 281, 255
Disc
    front – 208
    rear – 216
Distributor – 55
    contact breaker ignition – 162
    electronic ignition – 165
Doors – 250, 254, 255, 276
Draining
    automatic transmission fluid – 58
    coolant – 49
    gearbox oil – 57
Drivebelt
    alternator – 64
    power steering pump – 62
    water pump – 50
Driveplate
    B series engine – 99
    E series engine – 83
Driveshaft oil seal
    automatic transmission – 190
    manual gearbox – 182
Driveshafts – 25, 58, 191 *et seq*
Dual proportioning valve (braking system) – 209

E

Economy drive indicator system
    carburettor engines – 145
    fuel injected engines – 158
Electrical system – 26, 63, 268 *et seq*
Emission control system – 159
Engine – 22, 47, 50, 66 *et seq*, 119
Exhaust manifold
    carburettor engines – 147
    fuel injected engines – 158
Exhaust system, 53
    carburettor engines – 147
    fuel injected engines – 158

F

Facia – 263, 275
Fault diagnosis – 21
    automatic transmission – 24
    braking system – 25
    clutch – 24
    cooling system – 23
    driveshafts – 25
    electrical system – 26, 269
    engine – 22
    fuel and exhaust system – 23
    manual gearbox – 24
    suspension and steering – 25
Filling
    automatic transmission – 58
    cooling system – 49
    gearbox – 57
Filter
    air – 50
    engine – 47
    fuel – 53
Fluid
    automatic transmission – 57
    power steering – 61
Flushing coolant system – 49
Flywheel
    B series engine – 99
    E series engine – 83

Foglamp – 280, 282
Fuel, exhaust and emission control systems – 23, 50, 127 *et seq*, 256
Fuses – 274

G

Gaiters
    constant velocity joint – 58, 195, 196
    steering gear – 238
Gearchange linkage/mechanism
    manual gearbox – 180
    automatic transmission – 186

H

Handbrake – 60, 216, 276, 277, 281, 288
Headlamp – 65, 277, 281
Heater – 120, 123, 124, 126
Horn – 285
HT leads – 55
Hub bearings – 63
    front – 221
    rear – 230
Hydraulic fluid
    braking system – 59
    clutch – 56
Hydraulic pipes and hoses – 202

I

Idle speed adjustment – 50
Idle up system (fuel injected engines) – 158
Ignition coil
    contact breaker ignition – 163
    electronic ignition – 169
Ignition system – 54, 160 *et seq*, 238, 275
Indicators – 278, 281, 282
Inlet manifold
    carburettor engines – 146
    fuel injected engines – 158
Inner constant velocity joint rubber gaiter – 58, 196
Instrument panel – 281, 282, 284
Intermediate shaft (steering column) – 238
Introduction to the Mazda 323 – 6

J

Jacking – 8
Jump starting – 19

K

Kickdown solenoid – 188
Kickdown switch – 188

L

Lateral links (rear suspension) – 234
Locks
    bonnet – 250
    boot lid – 256
    door – 254
    steering – 238, 275
    tailgate – 259
Lower arm (front suspension) – 229
Luggage compartment lamp – 281

Index

M

Main bearings – 108, 109
Manifolds
    exhaust
        carburettor engines – 147
        fuel injected engines – 158
    inlet
        carburettor engines – 146
        fuel injected engines – 158
Manual gearbox and automatic transmission – 24, 57, 179 *et seq*
Map reading lamps – 281
Master cylinder
    braking system – 203
    clutch – 174
Mirror – 255
Mixture adjustment – 50
MOT test checks – 28
Mountings
    B series engine – 99
    E series engine – 84
    transmission
        B series engine – 99
        E series engine – 84

N

Number plate lamp – 280, 282

O

Oil
    engine – 47
    filter – 47
    gearbox – 57
Oil pump
    B series engine – 97
    E series engine – 79
Oil seals
    camshaft (B series engine) – 95
    crankshaft
        B series engine – 100
        E series engine – 84
    driveshaft
        automatic transmission – 190
        manual gearbox – 182
Overhead console – 277

P

Pads, 59
    front – 204
    rear – 214
Pedals
    accelerator
        carburettor engines – 133
        fuel injected engines – 150
    brake – 60, 199
    clutch – 56, 173
Piston rings – 109
Pistons – 105, 107, 111
Points – 54
Power steering – 61, 62, 239
PTC heater system (Aisan carburettor) – 145

Q

Quarter window glass (Hatchback models) – 254

R

Radiator – 115
Radiator grille – 248
Radio – 289
Rear lamp cluster – 278, 282
Rear window – 65, 259, 287, 288
Relays – 274
Release bearing – 177
Repair procedures – 14
Reversing lamp switch (manual gearbox) – 183
Rocker gear
    B series engine – 88
    E series engine – 75, 78
Rotor arm – 55
Routine maintenance – 32 *et seq*
    bodywork – 63, 244, 245
    braking system – 59
    clutch – 56
    cooling, heating and ventilation systems – 48
    driveshafts – 58
    electrical system – 63
    engine – 47
    fuel and exhaust systems – 50
    ignition system – 54
    manual gearbox and automatic transmission – 57
    suspension and steering – 61

S

Safety first – 12
    electrical system – 269
    electronic ignition – 164
    fuel, exhaust and emission control systems (carburettor engines) – 130
Seat belts – 261
Seats – 260
Selector cable/mechanism (automatic transmission) – 186
Servo unit – 200, 201
Shoes – 59, 209
Shutter valve control system (carburettor engines) – 142
Shutter valve ignition advance mechanism (electronic ignition) – 169
Sidelamps – 278, 281
Slave cylinder (clutch) – 175
Spare parts – 11
Spark plugs – 55
Spark plug conditions – *see colour pages between pages 32 and 33*
Speakers – 290
Speedometer drive
    automatic transmission – 190
    cable – 285
    manual gearbox – 180
Starter inhibitor switch (automatic transmission) – 189
Starter motor – 272
Starting system – 271
Steering
    angles – 242
    column – 235, 275
    gear – 238
    lock – 238
    power – 61, 62, 239
    wheel – 235
Stop lamp switch – 218
Stub axle – 231
Sump
    B series engine – 96
    E series engine – 78
Sunroof – 262
Support strut (tailgate) – 259
Surge tank (fuel injected engines) – 152
Suspension and steering – 25, 61, 219 *et seq*
Suspension strut
    front – 226, 227
    rear – 232, 233
Switches
    courtesy – 276
    door – 276
    electric cooling fan – 118

facia – 275
handbrake – 276
ignition – 275
kickdown – 188
overhead console – 277
reversing lamp (manual gearbox) – 183
starter inhibitor (automatic transmission) – 189
steering column combination – 275
stop lamp – 218
throttle position (fuel injected models) – 52
Swivel hub assembly – 221

T

Tailgate – 65, 256, 257, 259, 287, 288
Temperature gauge sender unit – 118
Thermostat – 116
Throttle housing (fuel injected engines) – 152
Throttle positioner system – 52
Timing (ignition) – 54
Timing belt and sprockets (B series engine) – 90, 93
Timing chain and sprockets E series engine) – 82
Tools – 15
Top Dead Centre (TDC) for number one piston
    B series engine – 87
    E series engine – 74
Towing – 8
Track rod – 241
Trailing arms (rear suspension) – 234
Trim
    door – 250
    interior – 261
    exterior – 259
Tyres – 63

U

Unleaded petrol
    carburettor engines – 133
    fuel injected engines – 151

V

Vacuum diaphragm (automatic transmission) – 189
Vacuum servo unit – 200, 201
Valve – 47, 103
Vehicle identification numbers – 11
Vibration damper – 197

W

Washer system – 65, 288
Water pump – 50, 119
Weights – 7
Wheel alignment – 242
Wheel changing – 8
Wheel cylinder (drum brakes) – 212
Wheels – 63
Window glass and regulator – 254
Windscreen – 65, 259, 285, 288
Wiper blades and arms – 65
Wiper motor and linkage – 285, 287
Wiring diagrams – 290
Working facilities – 15